János Fodor, Ryszard Klempous, and Carmen Paz Suárez Araujo (Eds.)

Recent Advances in Intelligent Engineering Systems

Studies in Computational Intelligence, Volume 378

Further volumes of this series can be found on our homepage: springer.com

János Fodor, Ryszard Klempous,
and Carmen Paz Suárez Araujo (Eds.)

Recent Advances in Intelligent Engineering Systems

Editors

Prof. Dr. János Fodor
Óbuda University
Institute of Intelligent Engineering Systems
Bécsi út 96/b
H-1034 Budapest
Hungary
E-mail: fodor@uni-obuda.hu

Dr. Ryszard Klempous
Wrocław University of Technology
Institute of Computer Engineering,
Control and Robotics 27 Wybrzeze
Wyspianskiego st.
50-372 Wrocław
Poland
E-mail: ryszard.klempous@pwr.wroc.pl

Assoc. Prof. Dra. Carmen Paz
Suárez Araujo
Universidad de Las Palmas de Gran Canaria
Instituto Universitario de Ciencias y
Tecnologías Cibernéticas Departamento de
Informática y Sistemas Campus
Universitario de Tafira, s/n
35017 Las Palmas de Gran Canaria
Spain
E-mail: cpsuarez@dis.ulpgc.es

ISBN 978-3-642-27006-2
ISBN 978-3-642-23229-9 (eBook)

DOI 10.1007/978-3-642-23229-9

Studies in Computational Intelligence
ISSN 1860-949X

Typeset & Cover Design: Scientific Publishing Services Pvt. Ltd., Chennai, India.

Printed on acid-free paper

9 8 7 6 5 4 3 2 1

springer.com

Foreword

This book contains extended and updated versions of carefully selected contributions to INES 2010, the 14th IEEE International Conference on Intelligent Engineering Systems, held on May 5-7, 2010, in Las Palmas de Gran Canaria, Spain. INES 2010 was organized under the auspices of Óbuda University (Budapest, Hungary), University of Las Palmas of Gran Canaria (Spain), and Wroclaw University of Technology (Poland).

The series of INES conferences has a well-established tradition, started in Budapest in 1997. The next places are neighbors of Hungary: Austria (1998), Slovakia (1999) and Slovenia (2000). Then the geographical distribution of locations broadened with Finland (2001), Croatia (2002), Egypt (2003), Romania (2004), the Mediterranean Sea (2005), London (2006), again Budapest (2007), then Miami (2008) and Barbados (2009), and finally Las Palmas in 2010. This list together with the IEEE support is very impressive, giving prestige to the conference series.

The aim of the INES conference series is to provide researchers and practitioners from industry and academia with a platform to report on recent developments in the area of intelligent engineering systems. Intelligence covers the ability to adapt to varying situations by reasoning. A characteristic feature of intelligent systems is that they integrate competences from many different areas in analysis, design and implementation.

As technology and basic knowledge more closely merged, the demand for introducing intelligent techniques to various industrial problems became apparent and was encouraged. This is reflected in the contributions of the current volume made by various scientists, to whom I am much indebted.

I would like to thank the editors of this volume for their meritorious work in evaluating the proceedings papers and presented talks, and for the final product of this outstanding selection. It truly reflects the essential structure, spirit, and scientific quality of INES 2010.

Budapest, May 2011

Imre J. Rudas
Founding Honorary Chair of INES

Preface

It is our great pleasure and privilege to introduce this book entitled Recent Advances in Intelligent Engineering Systems to our collaborators, colleagues and the interested readers of computational intelligence. We do hope that everyone will find this book intellectually stimulating and professionally rewarding, and will benefit from the content.

The present edited volume is a collection of 19 invited chapters written by respectable experts of the fields. They contribute to diverse facets of intelligent engineering systems. The topics arc in full harmony with the general aims of the 14th IEEE International Conference on Intelligent Engineering Systems (INES 2010), held on May 5-7, 2010, in Las Palmas de Gran Canaria, Spain, where preliminary versions of these selected contributions were presented.

The book is divided into three parts: Foundation of Computational Intelligence, Intelligent Computation in Networks, and Applications of Computational Intelligence. Here we comment briefly on these parts and their chapters that make up this volume.

The first part is devoted to the foundational aspects of computational intelligence. It consists of 8 chapters that include studies in genetic algorithms, fuzzy logic connectives, enhanced intelligence in product models, nature-inspired optimization technologies, particle swarm optimization, evolution algorithms, model complexity of neural networks, and fitness landscape analysis.

The first chapter by Michael Affenzeller, Stefan Wagner, Stephan M. Winkler and Andreas Beham gives a look inside the internal behavior of several enhanced genetic algorithms (GAs). After a treatment of fundamental notions and results on evolutionary algorithms, the authors oppose the characteristic behavior of conventional GAs with the typical behavior of generic hybrids based upon self-adaptive selection pressure steering. The observations are discussed with respect to the ability of the algorithms to inherit essential genetic information in the sense of Hollands schema theory and the according building block hypothesis. The characteristics of an empirical building block analysis for a standard GA, for an offspring selection GA, as well as for a relevant allele preserving GA are also presented.

Chapter 2 is written by József Dombi on families of fuzzy logic connectives that he calls multiplicative pliant system. He studies special De Morgan classes of fuzzy conjunctions, disjunctions and negations represented by strict t-norms, strict t-conorms and strong negations, respectively. In such a class the product of the additive generator of the t-norm and the additive generator of the t-conorm is equal to 1. The general form of negations together with their representation is established, and De Morgan classes with infinitely many negations are determined. The author studies some relationships among aggregative operators, uninorms, strict t-norms and t-conorms.

In Chapter 3 the authors, László Horváth and Imre J. Rudas, introduce one of their contributions in knowledge assisted intelligent control of engineering object definition at product model based engineering activities. The proposed modeling is devoted as an extension to currently prevailing product modeling in leading product life cycle management systems. Transferring knowledge from the human product definition process to the product model, method for better communication at object definition, new content definition for engineering objects and multilevel structure in order to facilitate implementation are discussed.

Chapter 4 by Czesław Smutnicki presents a critical survey of methods, approaches and tendencies observed in modern optimization. It focuses on nature-inspired techniques recommended for particularly hard discrete problems arising in practice. Applicability of these methods, depending the class of stated optimization task and classes of goal function, are discussed. Quality of each particular method depends on space landscape, ruggedness, big valley, distribution of solutions in the space and the problem balance between intensification and diversification of the search. The best promising approaches are indicated with practical recommendation of using. Some numerical as well as theoretical properties of these algorithms are also shown.

Chapter 5 by Ján Zelenka studies a scheduling problem a jobs sequence and allocation to machines during a time period in a manufacturing company. The author highlights that the many of the existing approaches are often impractical in dynamic real-world environments where there are complex constraints and a variety of unexpected disruptions. Then cooperation of one meta-heuristic optimization algorithm with manufacturing model by the dynamical rescheduling is described. Particle Swarm Optimization algorithm solved scheduling problem of real manufacturing system. Model of the manufacturing system is represented as discrete event system created by SimEvents toolbox of MATLAB programming environment.

Chapter 6 by Andrzej Cichoń and Ewa Szlachcic presents an efficient strategy for self-adaptation mechanisms in a multi-objective differential evolution algorithm. The algorithm uses parameters adaptation and operates with two differential evolution schemes. Also, a novel DE mutation scheme combined with a transversal individual idea is introduced to support the convergence rate of the algorithm. The performance of the proposed algorithm, named DEMOSA, is tested on a set of benchmark problems. The numerical results confirm that the proposed algorithm performs considerably better than the one with simple DE scheme in terms of computational cost and quality of the identified non-dominated solutions sets.

Chapter 7 by Věra Kůrková investigates the role of dimensionality in approximation of functions by one-hidden layer neural networks. Methods from nonlinear approximation theory are used to describe sets of functions which can be approximated by neural networks with a polynomial dependence of model complexity on the input dimension. The results are illustrated by examples of Gaussian radial networks, where the author characterizes sets which can be tractably approximated in terms of suitable norms defined by constraints on magnitudes of derivatives.

In Chapter 8 the authors, Erik Pitzer and Michael Affenzeller, provide a comprehensive survey on fitness landscape analysis. They formally define fitness landscapes, provide an in-depth look at basic properties and give detailed explanations and examples of existing fitness landscape analysis techniques. Moreover, several common test problems or model fitness landscapes that are frequently used to benchmark algorithms or analysis methods are examined and explained and previous results are consolidated and summarized. Finally, the authors point out current limitations and open problems pertaining to the subject of fitness landscape analysis.

The second part of this book contains contributions to intelligent computation in networks, presented in 5 chapters. The covered subjects include the application of self-organizing maps for early detection of denial of service attacks, combating security threats via immunity and adaptability in cognitive radio networks, novel modifications in WSN network design for improved SNR and reliability, a conceptual framework for the design of audio based cognitive infocommunication channels, and a case study on the advantages of fuzzy and anytime signal- and image processing techniques.

In chapter 9 team leaded by Carmen Paz Suárez Araujo, is dealing with the Internet Cyber Attacks problems. Growing up Internet in an exponential way is associated, unfortunately, with the same rise of Denial of Service (DoS). The team presents a flexible method capable to overcome DoS attacks using Computer Intelligent System for DoS Attacks Detection (CISDAD). This is a hybrid intelligent system with a modular structure with a processing module based on the Kohonen Self-Organizing Artificial Neural Networks. It permits to detect several types of toxic traffics. Presented results prove the effectiveness of CISDAD managing traffic in highly distributed networking environment. It should be also underlined the integration of the CISDAD into a clinical workstation EDEVITALZH.

Chapter 10 written by Jan Nikodem et al. presents very important problems of security, immunity and adaptability of Cognitive Radio (CR) networks. An overview of available CR models, threats and their mitigation patter was provided. The proposed novel relational method permits reconciling two (often dichotomous) points of view: immunity and adaptability to neighborhood. Management of complex system in such environment yields in growing both adaptability and immunity. By modeling CR network activities using relational approach the authors have managed to precisely describe the complex characteristics of network interactions. Moreover, at the same time they have eliminated the CR networks parameters. This attempt guarantee to scale the complexity of interactions and model with much higher precision various aspects of CR networks.

Chapter 11 written by Kamil Staniec and Grzegorz Debita is devoted the construction of a Wireless Network Structure (WSN) of minimized intra-network interference and structure. Authors examined their simulation algorithm for WSN operating in 2,5GHz ISM band. They have also developed a simulator for determining the ZigBee performance as a function of the numbers of nodes. The simulation results confirm that creating a redundancy in a self-organizing WSN is not a simply task. Received from neighbors the knowledge of the electromagnetic power level is not sufficient information. The information about nodes location, as well as some support from the network designer could be essential support here.

Chapter 12 written by Ádám Csapó and Péter Baranyi develops engineering systems which are capable of using cognitive info-communication channels in order to convey feedback information in novel and meaningful ways. They describe the main challenge behind the development of cognitive info-communication channels as a two-part problem which consists of the design of a synthesis algorithm and the design of a parameter-generating function for the synthesis algorithm. They use formal concept algebra to describe the kinds of synthesis algorithms which are capable of reflecting realistic forms of interaction between the user and the information which is to be communicated. Through an experimental evaluation of the application, the authors demonstrate that their approach can be used successfully for the design of cognitive info-communication channels.

In Chapter 13 the author, Teréz A. Várkonyi, starts from the observation that in practical engineering problems the available knowledge about the information to be processed is usually incomplete, ambiguous, noisy, or totally missing, and the available time and resources for fulfilling the task are often not only limited, but can change during the operation of the system. This urge researchers and engineers to turn towards non-classical methods which are very advantageous. For this reason, the author gives an overview about various imprecise, fuzzy and anytime, signal- and image processing methods and their applicability is discussed in treating the insufficiency of knowledge of the information necessary for handling, analyzing, modeling, identifying, and controlling of complex engineering problems.

Computational intelligence represents a widely spread interdisciplinary research area with many applications in various disciplines including engineering, medicine, technology, environment, among others. The Part III of this book is devoted to this practical aspect of computational intelligence, the applications. This part is a very important section of the volume because in it the reader could find a wide range of fields where the computational intelligence plays a significant role.

The first chapter of the Part III, chapter 14, is devoted to clinical decision support systems based on computational intelligence, concretely on neural computation. Patricio García Báez, Carmen Paz Suárez Araujo, Carlos Fernández Viadero and Aleš Procházka focus their studies on a very hard problem in clinical medicine, the Differential Diagnosis of Dementias (DDD). They propose new automatic diagnostic tools based on a data fusion scheme and neural ensemble approach for facing the DDD. The authors present HUMANN-S ensemble systems with missing data processing capability, where the neural architecture HUMANN-S is the main module of these intelligent decision support systems. Their ability was explored using a novel information environment applied to DDD, different combinations of a battery of cognitive and functional/instrumental scales. In this chapter is also presented a comparative study between the proposed methods and a clinical expert, reaching these new intelligent systems a higher level of performance than the expert. Finally, in the chapter is shown that the proposal described is an alternative and effective complementary method to assist the diagnosis of dementia having important advantages referring to other computational solutions based on artificial neural networks.

Chapter 15, written by Martina Mudrová, Petra Slavíková and Aleš Procházka, presents an application in environmental engineering, the developments of new methods for air pollution detection. The authors deal with classification of microscope images of Picea Abies stomas. They base their proposal on the assumption that a stoma character strongly depends on the level of air pollution, so that stoma can stand for an important environmental bioindicator. The chapter is devoted to the development of an automatic algorithm which can recognize the level of stoma changes by means of methods of texture classification. In this study two basic principles are discussed, the application of gradient methods and the use of methods based on the wavelet transform. Several methods of image preprocessing as noise reduction, brightness correction and resampling are studied, as well. The authors present an algorithm validation study based upon the analysis of the set of about four hundred images collected from 6 localities in the Czech Republic and results achieved were compared with an experts sensual classification.

In Chapter 16, the authors Carlos M. Travieso, Juan C. Briceño and Jesús B. Alonso present a computational intelligence method in the Biometrics field. The main goal of this field is the identification of a person using her/his body features, in this concrete case the lips, which has a big interest in the security area. They propose a biometric identification approach based on lip shape with three main steps. The first step is devoted to detect the face and lips contour detection. In the second step, the lip features, based on angular coding, are extracted and afterwards transformed using Hidden Markov Model kernel (HMMK). Finally, a one-versus-all multiclass supervised approach based on Support Vector Machines (SVM) with RBF kernel is applied as a classifier.

The chapter 17 by Aleš Procházka, Martina Mudrová, Oldrich Vysata, Lucie Gráfová and Carmen Paz Suárez Araujo deals with an interdisciplinary research area based upon general digital signal processing methods and adaptive algorithms. It is the computational intelligence and signal analysis of multi-channel data working together. In this chapter is restricted to their use in biomedicine and particularly in electroencephalogram signal processing to find specific components of such multi-channel signals. The methods used by the authors included multi-channel signals de-noising, extraction of their components and the application of the double moving window for signal segmentation using its first principal component. During this preprocessing stage are used both the discrete Fourier transform and the discrete wavelet transform. Resulting pattern vectors will be classified using the artificial neural network approach, concretely self-organizing neural networks using a specific statistical criterion proposed to evaluate distances of individual feature vector values from corresponding cluster centers. Owing to the complexity of the multi-channel signal processing, distributed computing is mentioned in the chapter as well. Proposed methods are verified in the MATLAB environment using distributed data processing.

The chapter 18 introduces an ambient intelligence application, an Intelligent TeleCare System. The authors, Stoicu-Tivadar, L. Stoicu-Tivadar, S. Puşcoci, D. Berian and V. Topac, explore several existing telecare solutions mainly from the technological point of view, but also considering the degree in which seamless care is achieved. Having this study as a background, the chapter describes a teleassistance / telemonitoring system assisting elderly persons, an integral, holistic solution, with emphasis on the server component, TELEASIS. The TELEASIS platform architecture, which is based on a service-oriented architecture, the hardware platform and infrastructure, the software platform, the dispatcher component and the web services are briefly described. The main idea provides by this chapter is not only to obtain an intelligent telecare system, but to contribute to improvement of the management of care for a specific category of persons, as well as a future generation telecare networking applications.

Chapter 19 by Nicolaie Popescu-Bodorin and Valentina Emilia Balas, presents a new authentication system based on supervised learning of iris biometric identities. They use a neural-evolutionary approach to iris authentication, reaching an important power of discrimination between the intra- and inter-class comparisons. The authors show that when using digital identities evolved by a logical and intelligent artificial agent (Intelligent Iris Verifier/Identifier) the separation between inter- and intra-class scores is so good that it ensures absolute safety for a very large percent of accepts. They also make comparison to a result previously obtained by Daugman. The difference between both studies comes from a different understanding of what it means to recognize: Daugman sustained the idea of a statistical decision landscape of recognition and the authors sustaining the idea of a logically consistent approach to recognition. This chapter, also discusses the latest trends in the field of evolutionary approaches to iris recognition, and announces the technological advance from inconsistent iris verification to consistent iris identification. It finally shows that the future iris-based identification will be inevitably marked by multi-enrollment, and by the newly proposed concept of consistent, intelligent, adaptive, evolutionary biometric system.

We would like to thank the INES 2010 Technical Program Committee Chairs Michael Affenzeller, László Horváth and Aleš Procházka for their meritorious work in evaluating the selected papers, as well as all program committee members who assisted the TPC Chairs.

The editors are grateful to the authors for their superior work. Thanks are also due to Jan Nikodem for his excellent editorial assistance and sincere effort in bringing out the volume nicely in time. Last but not least we want to thank Springer-Verlag for smooth cooperation in publication of this volume.

Budapest, Las Palmas, Wrocław
June 2011

János Fodor
Ryszard Klempous
Carmen Paz Suárez Araujo

Contents

Part I
Foundation of Computational Intelligence

Chapter 1
Analysis of Allele Distribution Dynamics in Different Genetic Algorithms

Michael Affenzeller, Stefan Wagner, Stephan M. Winkler, and Andreas Beham

Abstract. This chapter exemplarily points out how essential genetic information evolves during the runs of certain selected variants of a genetic algorithm. The discussed algorithmic enhancements to a standard genetic algorithm are motivated by Holland's schema theory and the according building block hypothesis. The discussed offspring selection and the relevant alleles preserving genetic algorithm certify the survival of essential genetic information by supporting the survival of relevant alleles rather than the survival of above average chromosomes. This is achieved by defining the survival probability of a new child chromosome depending on the child's fitness in comparison to the fitness values of its own parents. By this means the survival and expansion of essential building block information information is supported also for problem representations and algorithmic settings which do not fulfill the theoretical requirements of the schema theory. The properties of these GA variants are analyzed empirically. The selected analysis method assumes the knowledge of the unique globally optimal solution and is therefore restricted to rather theoretical considerations. The main aim of this chapter is to motivate and discuss the most important properties of the discussed algorithm variants in a rather intuitive way. Aspects for meaningful and practically more relevant generalizations as well as more sophisticated experimental analyses are indicated.

1.1 Introduction

In this chapter we try to look inside the internal behavior of several enhanced genetic algorithms (GAs). The discussed generic hybrids couple aspects of genetic

Michael Affenzeller · Stefan Wagner · Stephan M. Winkler · Andreas Beham
Heuristic and Evolutionary Algorithms Laboratory
School of Informatics, Communications and Media
Upper Austria University of Applied Sciences, Campus Hagenberg
Softwarepark 11, 4232 Hagenberg, Austria
e-mail: {maffenze, swagner, swinkler, abeham}@heuristiclab.com

J. Fodor et al. (Eds.): Recent Advances in Intelligent Engineering Systems, SCI 378, pp. 3–29.
springerlink.com

algorithms with selection principles basically inspired by evolution strategies [8] and self adaptive measures comparable to the parameter-less GA [15]. For this purpose we use the information about globally optimal solutions which is only available for well studied benchmark problems of moderate dimension. Of course, the applied optimization strategies (in our case variants of GAs) are not allowed to use any information about the global optimum; we just use this information for analysis purposes in order to obtain a better understanding of the internal behavior and the dynamics of the discussed algorithmic concepts.

A basic requirement for this (to a certain extent idealized) kind of analysis is the existence of a *unique* globally optimal solution which has to be known. Concretely, we aim to observe the distribution of the alleles of the globally optimal solution (denoted as essential genetic information) over the generations in order to observe the ability of the proposed algorithms to preserve and possibly regain essential genetic material during the run of the algorithm.

The main aim of this contribution is not to give a comprehensive analysis of many different problem instances, but rather to highlight the main characteristics of the considered algorithms. For this kind of analysis we have chosen the traveling salesman problem (TSP), mainly because it is a well known and well analyzed combinatorial optimization problem and a lot of benchmark problem instances are available. We concentrate on the *ch130* TSP instance taken from the TSPLib [17], for which the unique globally optimal tour is known; the characteristics of the global optimum of this 130 city TSP instance are exactly the 130 edges of the optimal tour which denote the essential genetic information.

In a broader interpretation of the building block theory [12, 20] these alleles should on the one hand be available in the initial population of a GA run, and on the other hand be maintained during the run of the algorithm. If essential genetic information is lost during the run, then mutation is supposed to help regaining it in order to be able to eventually find the globally optimal solution (or at least a solution which comes very close to the global optimum). In order to observe the actual situation in the population, we display each of the 130 essential edges as a bar indicating the saturation of each allele in the population. The disappearance of a bar therefore indicates the loss of the corresponding allele in the entire population, whereas a full bar indicates that the certain allele occurs in each individual (which is the desired situation at the end of an algorithm run). As a consequence, the relative height of a bar stands for the actual penetration level of the corresponding allele in the individuals of the population. The observation of the dynamic behavior of these bars allows observing the distribution of essential genetic information during the run.

In the following, the distribution of essential genetic information and its impact on achievable solution quality will be discussed for the standard GA, a GA variant including offspring selection (OS) [3] as well as for the relevant alleles preserving GA (RAPGA) [4]. A definition of these algorithms as well as a discussion of their theoretical foundations and implications is given in an introductional theoretical section where the main results of Holland's schema theory [13] and the according building block hypothesis [12] are recapitulated. The results shown in the following sections are presented in a rather illustrative way and aim to provide a intuitive approach to the discussed algorithms. The results shown for the standard genetic algorithm and the offspring selection genetic alorithm are taken from [1]; additional results for the RAPGA are taken from [5]. For statistically more relevant experimental results the interested reader is referred to the homepage of our book[1] where numerous result tables are provided.

The rest of the chapter is organized as follows: In Section 2 an overview about those theoretical foundations of evolutionary algorithms is stated which are relevant for the further considerations in this chapter. The descriptions are mainly taken from [6]. Sections 3, 4, and 5 show the results and discuss the characteristics of an empirical building block analysis for as standard GA, for an offspring selection GA, as well as for a relevant allele preserving GA (taken from [5]). Finally, Section 6 summarizes the chapter and indicates some aspects for future research which are considered important by the authors.

1.2 Theoretical Foundations

1.2.1 Schema Theorem and Building Block Hypothesis

Researchers working in the field of GAs have put a lot of effort into the analysis of genetic operators (crossover, mutation, sclection). In order to achieve better analysis and understanding, Holland has introduced a construct called schema [13]:

Assuming the use of a canonical GA with binary string representation of individuals, the symbol alphabet {0,1,#} is considered where {#} (don't care) is a special wild card symbol that matches both, 0 and 1. A schema is a string with fixed and variable symbols. For example, the schema [1#11#00] is a template that matches the following four strings: [1011000], [1011100], [1111000], and [1111100]. The symbol # is never actually manipulated by the genetic algorithm; it is just a notational symbol that makes it easier to talk about families of strings.

Essentially, Holland's idea was that every evaluated string actually gives partial information about the fitness of the set of possible schemata of which the string is a member. Holland analyzed the influence of selection, crossover and mutation on the expected number of schemata, when going from one generation to the next. A detailed discussion of related analyses can be found in [12]; in the context of the present work we only outline the main results and their significance.

Assuming fitness proportional replication, the number m of individuals of the population belonging to a particular schema H at time $t+1$ (i.e., $m(H,t+1)$) can be calculated and depends on the average fitness value of the string representing schema H and the average fitness value over all strings within the population. Assuming that a particular schema remains above the average by a fixed amount for a certain number of generations, $m(H,t)$ can be calculated directly.

[1] `http://gagp2009.heuristiclab.com/material/statisticsTSP.html`

Considering the effect of crossover which breaks strings apart (at least in the case of canonical genetic algorithms) we see that short defining length schemata are less likely to be disrupted by a single point crossover operator. The main result is that above average schemata with short defining lengths will still be sampled at an exponential increasing rate; these schemata with above average fitness and short defining length are the so-called building blocks and play an important role in the theory of genetic algorithms.

Using several considerations and proofs given in [13], the effects of mutation, crossover, and reproduction can be described which results in Holland's well known schema theorem:

$$m(H,t+1) \geq m(H,t)\frac{f_H(t)}{\overline{f}(t)}[1 - p_c\frac{\delta(H)}{l-1} - o(H)p_m] \tag{1.1}$$

where $f_H(t)$ is the average fitness value of individuals represented by H, $\overline{f}(t)$ is the average fitness of all individuals in the population, $\delta(H)$ is the defining length of a schema H (i.e., the distance between the first and the last fixed string position), $o(H)$ represents the order of H (i.e., the number of non-wildcard positions in H), and p_c and p_m denote the probabilities of crossover and mutation, respectively.

This theorem essentially says that the number of short schemata with low order and above average quality grows exponentially in subsequent generations of a genetic algorithm.

The major drawback of the building block theory is given by the fact that the underlying GA (binary encoding, proportional selection, single-point crossover, strong mutation) is applicable only to very few problems, as it usually requires more sophisticated problem representations and corresponding operators to tackle challenging real-world problems.

Keeping in mind that the ultimate goal of any heuristic optimization technique is to approximately and efficiently solve highly complex real-world problems rather than stating a mathematically provable theory that holds only under very restricted conditions, our intention for an extended building block theory is a not so strict formulation that in return can be considered for arbitrary GA applications. At the same time, the enhanced variants of genetic algorithms proposed in this chapter aim to support the algorithms in their intention to operate in the sense of an extended building block interpretation discussed in the following sections.

1.2.2 Stagnation and Premature Convergence

Genetic algorithms encounter a problem which, at least in its effect, is quite similar to the problem of stagnating in a local, but not global, optimum. In the terminology of GAs this drawback is called premature convergence and occurs, if the population of a GA reaches such a suboptimal state that the genetic solution manipulation operators (crossover and mutation) are no longer able to produce offspring that outperform their parents (as discussed for example in [11] and [5]). In general, this happens mainly when the genetic information stored in the individuals of a population does not contain the genetic information which is sufficient to further improve the solution quality.

The term "population diversity" has been used in many papers to study premature convergence(e.g. [19, 12]) where the decrease of population diversity (i.e. a homogeneous population) is considered as the primary reason for premature convergence.

The basic approaches for avoiding premature convergence discussed in GA literature aim to maintain genetic diversity. The most common techniques for this purpose are based upon pre-selection [9], crowding [10], or fitness-sharing [12]. The main idea of these techniques is to maintain genetic diversity by the preferred replacement of similar individuals [9, 10] or by the fitness-sharing of individuals which are located in densely populated regions [12]. While methods based upon those discussed in [10] or [12] require some kind of neighborhood measure depending on the problem representation, the approach given in [12] is additionally quite restricted to proportional selection.

In natural evolution the maintenance of genetic diversity is of major importance as a rich gene pool enables a certain species to adapt to changing environmental conditions. In the case of artificial evolution, the environmental conditions, for which the chromosomes are optimized, are represented in the fitness function which usually remains unchanged during the run of an algorithm. Therefore, we do not identify the reasons for premature convergence in the loss of genetic variation in general, but more specifically in the loss of what we call essential genetic information, i.e. in the loss of alleles which are part of a globally optimal solution. If parts of this essential genetic information are missing or get lost, premature convergence is already predetermined in a certain way, as only mutation (or migration in the case of parallel GAs [7]) is able to regain this genetic information.

A very essential question about the general performance of a GA is whether or not good parents are able to produce children of comparable or even better fitness – after all, the building block hypothesis implicitly relies on this. Unfortunately, this property cannot be guaranteed easily for GA applications in general. It is up to the user to take care of an appropriate encoding in order to make this fundamental property hold.

Reconsidering the basic functionality of a GA, the algorithm selects two above average parents for recombination and sometimes (with usually rather low probability) mutates the crossover result. The resulting chromosome is then considered as a member of the next generation and its alleles are therefore part of the gene pool for the ongoing evolutionary process.

Reflecting the basic concepts of GAs, the following questions arise:

- Is crossover always able to fulfill the implicit assumption that two above-average parents can produce even better children?
- Which of the available crossover and mutation operators are best suited for a certain problem in a certain representation?
- Which of the resulting children are "good" recombinations of their parents chromosomes?

- What makes a child a "good" recombination?
- Which parts of the chromosomes of above-average parents are really worth being preserved?

In conventional GAs these issues are not always resolved in a satisfactory way. This observation constitutes the starting point for enhanced algorithms as stated in the following sections. The preservation of essential genetic information, widely independent of the actually applied representation and operators, plays a main role.

1.2.3 Offspring Selection (OS)

The goal of the extended algorithms described in this chapter is to support crossover-based evolutionary algorithms (i.e., evolutionary algorithms that are intended to work as building-block assembling machines) in their intention to combine these parts of the chromosomes that define high quality solutions. In this context we concentrate on selection and replacement which are the parts of the algorithm that are independent of the problem representation and the according operators. Thus, the application domain of the new algorithms is very wide; in fact, offspring selection can be applied to any task that can be treated by genetic algorithms (of course also including genetic programming). The unifying purpose of the enhanced selection and replacement strategies is to introduce selection after reproduction in a way that checks whether or not crossover and mutation were able to produce a new solution candidate that outperforms its own parents. Offspring selection realizes this by claiming that a certain ratio of the next generation (pre-defined by the user) has to consist of children that were able to outperform their own parents (with respect to their fitness values). OS implies a self-adaptive regulation of the actual selection pressure that depends on how easy or difficult it is at present to achieve evolutionary progress. An upper limit for the selection pressure provides a good termination criterion for single population GAs as well as a trigger for migration in parallel GAs.

The first selection step chooses the parents for crossover either randomly or in any other well-known way as for example roulette-wheel, linear-rank, or some kind of tournament selection strategy. After having performed crossover and mutation with the selected parents, we introduce a further selection mechanism that considers the success of the apparently applied reproduction. In order to assure that the progression of genetic search occurs mainly with successful offspring, this is done in such a way that the used crossover and mutation operators are able to create a sufficient number of children that surpass their parents' fitness. Therefore, a new parameter called success ratio ($SuccRatio \in [0,1]$) is introduced. The success ratio is defined as the quotient of the next population members that have to be generated by successful mating in relation to the total population size. Our adaptation of Rechenberg's success rule [16, 18] for genetic algorithms says that a child is successful, if its fitness is better than the fitness of its parents, whereby the meaning of "better" has to be explained in more detail: Is a child better than its parents, if it surpasses the fitness of the weaker parent, the better parent, or some kind of weighted average of both?

In order to answer this question, we have borrowed an aspect from simulated annealing: The threshold fitness value that has to be outperformed lies between the worse and the better parent and the user is able to adjust a lower starting value and a higher end value which are denoted as comparison factor bounds; a comparison factor (*CompFactor*) of 0.0 means that we consider the fitness of the worse parent, whereas a comparison factor of 1.0 means that we consider the better of the two parents. During the run of the algorithm, the comparison factor is scaled between the lower and the upper bound resulting in a broader search at the beginning and ending up with a more and more directed search at the end; this procedure in fact picks up a basic idea of simulated annealing.

In the original formulation of offspring selection (OS) we have defined that in the beginning of the evolutionary process an offspring only has to surpass the fitness value of the worse parent in order to be considered as "successful"; as evolution proceeds, the fitness of an offspring has to be better than a fitness value continuously increasing between the fitness values of the weaker and the better parent. As in the case of simulated annealing, this strategy results in a more exploratory search at the beginning, whereas at the end of the search process this operator acts in a more and more focused way. Having filled up the claimed ratio (*SuccRatio*) of the next generation with successful individuals using the success criterion defined above, the rest of the next generation ($(1 - SuccRatio) \cdot |POP|$) is simply filled up with individuals randomly chosen from the pool of individuals that were also created by crossover, but did not reach the success criterion. The actual selection pressure *ActSelPress* at the end of generation i is defined by the quotient of individuals that had to be considered until the success ratio was reached and the number of individuals in the population in the following way:

$$ActSelPress_{i+1} = \frac{|POP_{i+1}| + |POOL_i|}{|POP_i|} \tag{1.2}$$

An upper limit of selection pressure (*MaxSelPress*) defines the maximum number of offspring considered for the next generation (as a multiple of the actual population size) that may be produced in order to fulfill the success ratio. With a sufficiently high setting of *MaxSelPress*, this new model also functions as a detector for premature convergence:

If it is no longer possible to find a sufficient number ($SuccRatio \cdot |POP|$) of offspring outperforming their own parents, even if ($MaxSelPress \cdot |POP|$) candidates have been generated, premature convergence has occurred.

As a basic principle of this selection model, higher success ratios cause higher selection pressures. Nevertheless, higher settings of success ratio, and therefore also higher selection pressures, do not necessarily cause premature convergence. The reason for this is mainly that the new selection step does not accept clones that emanate from two identical parents per definition. In conventional GAs such clones represent a major reason for premature convergence of the whole population

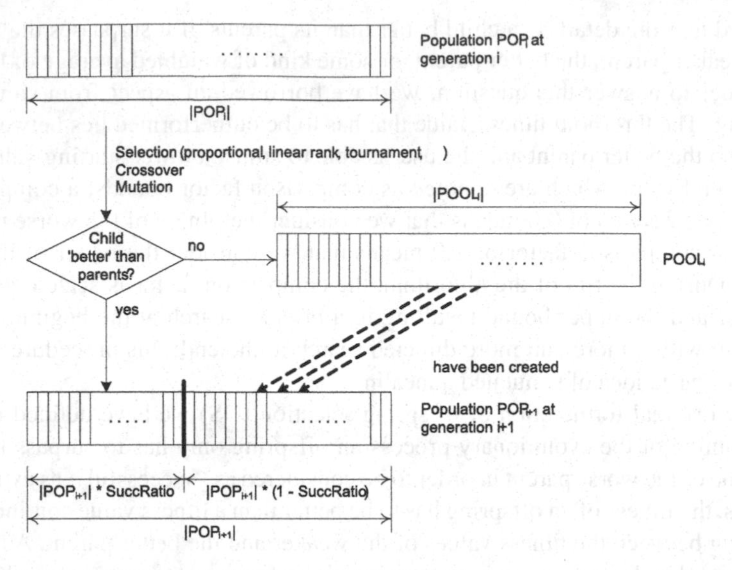

Fig. 1.1 Application flow of offspring selection (OS).

around a suboptimal value, whereas the new offspring selection works against this phenomenon [5].

1.2.4 The Relevant Alleles Preserving Genetic Algorithm (RAPGA)

Assuming generational replacement as the underlying replacement strategy, the most essential question at generation i is which parts of genetic information from generation i should be maintained in generation $i+1$ and how this could be done most effectively applying the available information (chromosomes and according fitness values) and the available genetic operators selection, crossover and mutation.

The presented enhanced algorithm based upon GA-solution manipulation operators aims to achieve this goal by trying to get as much progress out from the actual generation as possible and losing as little genetic diversity as possible at the same time.

This idea is implemented using ad-hoc population size adjustment: Potential offspring generated by the basic genetic operators are accepted as members of the next generation, if and only if they are able to outperform the fitness of their own parents and if they are new in that sense that their chromosome consists of a concrete allele alignment that is not represented in an individual of the next generation yet. As long as new and (with respect to the definition given previously) "successful" individuals can be created from the gene pool of the actual generation, the population size is allowed to grow up to a maximum size. A potential offspring which is not able

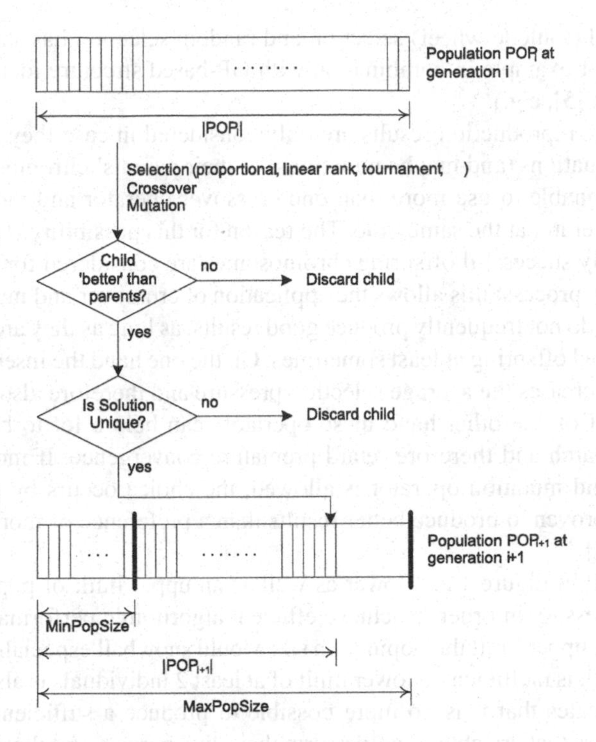

Fig. 1.2 Application flow of the relevant alleles preserving genetic algorithm (RAPGA).

to fulfill these requirements is simply not considered for the gene pool of the next generation.

Figure 1.2 illustrates the operating sequence of the relevant alleles preserving genetic algorithm (RAPGA). An offspring is included into the next generation if it is better than its own parents and as long as it is unique, i.e. not yet member of the next generation. If the next population becomes smaller or larger, depends on the success of the genetic operators crossover and mutation in the above stated claim to produce new and successful chromosomes. For reasons of effectiveness there is an upper limit for the adaptive population size. Also there is a user defined upper limit for the maximal effort spent per generation. The maximal effort is defined as the maximal number of generated solutions per iteration. The RAPGA terminates if even under maximal effort the algorithm is unable to generate at least a lower limit of new unique and better solution candidates out of the gene pool of the last generation.

For a generic, stable and robust realization of these RAPGA ideas some practical aspects have to be considered:

- The algorithm should offer the possibility to use different settings also for conventional parent selection, so that the selection mechanisms for the two parents do not necessarily have to be the same. In many examples a combination of proportional (roulette wheel) selection and random selection has shown a lot of potential (for example in combination with GP-based structure identification as discussed in [5], e.g.).

- The fact that reproduction results are only considered in case they are successful recombinations (and maybe mutations) of their parents' chromosomes, it becomes reasonable to use more than one crossover operator and more than one mutation operator at the same time. The reason for this possibility is given by the fact that only successful offspring chromosomes are considered for the ongoing evolutionary process; this allows the application of crossover and mutation operators which do not frequently produce good results, as long as they are still able to generate good offspring at least sometimes. On the one hand the insertion of such operators increases the average selection pressure and therefore also the average runtime, but on the other hand these operators can help a lot to broaden evolutionary search and therefore retard premature convergence. If more than one crossover and mutation operator is allowed, the choice occurs by pure chance which has proven to produce better results than a preference of more successful operators [5].
- As indicated in Figure 1.2, a lower as well as an upper limit of population size are still necessary in order to achieve efficient algorithmic performance. In case of a missing upper limit the population size would snowball especially in the first rounds which is inefficient; a lower limit of at least 2 individuals is also necessary as this indicates that it is no more possible to produce a sufficient amount of chromosomes that are able to outperform their own parents and therefore acts as a good detector for convergence.
- Depending on the problem at hand there may be several possibilities to fill up the next population with new individuals. If the problem representation allows an efficient check for genotypical identity, it is recommendable to do this and accept new chromosomes as members for the next generation, if there is no structurally identical individual included in the population yet. If a check for genotypical identity is not possible or too time-consuming, there is still the possibility to assume that two individuals are identical, if they have the same fitness values. However, the user has to be aware of the fact that this assumption may be too restrictive in case of fitness landscapes with identical fitness values for a lot of different individuals; in such cases it is of course advisable to check for genotypical identity.
- In order to terminate the run of a certain generation in case it is not possible to fill up the maximally allowed population size with new successful individuals, an upper limit of effort in terms of generated individuals is necessary. This maximum effort per generation is the maximum number of newly generated chromosomes per generation (no matter if these have been accepted or not).
- The question, whether or not an offspring is better than its parents, is answered in the same way as in the context of offspring selection.

Figure 1.3 shows the typical development of the actual population size during an exemplary run of RAPGA applied to the ch130 benchmark instance of the traveling

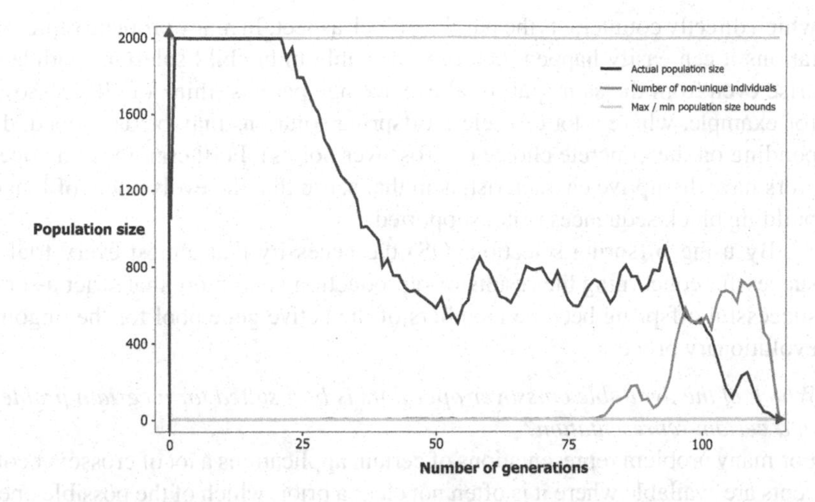

Fig. 1.3 Typical development of actual population size between the two borders (lower and upper limit of population size) displaying also the identical chromosomes that occur especially in the last iterations.

salesman problem taken from the TSPLib [17]. More sophisticated studies analyzing the characteristics of RAPGA are presented in [5].

1.2.5 Consequences Arising Out of Offspring Selection and RAPGA

Typically, GAs operate under the implicit assumption that parent individuals of above average fitness are able to produce better solutions as stated in Holland's schema theorem [13] and the related building block hypothesis. This general assumption, which ideally holds under the restrictive assumptions of a canonical GA using binary encoding, is often hard to fulfill for many practical GA applications. Some crucial question about the general behavior of GA-based methods shall be phrased and answered here in the context of offspring selection:

1. *Is crossover always able to fulfill the implicit assumption that two above-average parents can produce even better children?*
   Unfortunately, the implicit assumption of the schema theorem, namely that parents of above average fitness are able to produce even better children, is not accomplished for a lot of operators in many theoretical as well as practical applications. This disillusioning fact has several reasons: First, a lot of operators tend to produce offspring solution candidates that do not meet the implicit or explicit constraints of certain problem formulations. Commonly applied repair strategies included in the operators themselves or applied afterwards have the consequence that alleles of the resulting offspring are not present in the parents

which directly counteracts the building block aspect. In many problem representations it can easily happen that a lot of highly unfit child solution candidates arise even from the same pair of above average parents (think of GP crossover for example, where a lot of useless offspring solutions may be developed, depending on the concrete choice of crossover points). Furthermore, some operators have disruptive characteristics in that sense that the evolvement of longer building block sequences is not supported.

By using offspring selection (OS) the necessity that almost every trial is successful concerning the results of reproduction is no more that strict as only successful offspring become members of the active gene pool for the ongoing evolutionary process.

2. *Which of the available crossover operators is best suited for a certain problem in a certain representation?*
For many problem representations of certain applications a lot of crossover concepts are available where it is often not clear a priori which of the possible operators is suited best. Furthermore, it is often also not clear how the characteristics of operators change with the remaining parameter settings of the algorithm or how the characteristics of the certain operators change during the run of the algorithm. So it may easily happen that certain (maybe more disruptive) operators perform quite well at the beginning of evolution whereas other crossover strategies succeed rather in the final (convergence) phase of the algorithm.

In contrast to conventional GAs, for which the choice of usually one certain crossover strategy has to be done in the beginning, the ability to use more crossover and also mutation strategies in parallel is an important characteristic of OS-based GAs as only the successful reproduction results take part in the ongoing evolutionary process. It is also an implicit feature of the enhanced algorithms that when using more operators in parallel only the results of those will succeed which are currently able to produce successful offspring which changes over time. Even the usage of operator concepts that are considered evidentially weak for a certain application can be beneficial as long as these operators are able to produce successful offspring from time to time [5].

3. *Which of the resulting children are "good" recombinations of their parents chromosomes?*
Offspring selection has been basically designed to answer this question in a problem independent way. In order to retain generality, these algorithms have to base the decision if and to which extent a given reproduction result is able to outperform its own parents by comparing the offspring's fitness with the fitness values of its own parent chromosomes. By doing so, we claim that a resulting child is a good recombination (which is a beneficial building block mixture) worth being part of the active gene poo if the child chromosome has been able to surpass the fitness of its own parents in some way.

4. *What makes a child a "good" recombination?*
Whereas question 3 motivates the way how the decision may be carried out whether or not a child is a good recombination of its parent chromosomes,

question 4 intuitively asks why this makes sense. Generally speaking, OS directs the selection focus after reproduction rather than before reproduction. In our claim this is reasonable, as it is the result of reproduction that will be part of the gene pool and that has to keep the ongoing process alive. Even parts of chromosomes with below average fitness may play an important role for the ongoing evolutionary process, if they can be combined beneficially with another parent chromosome which motivates gender specific parents selection [21] as is for example applied in the experiments shown in [5]. In this gender specific selection step typically one parent is selected randomly and the other one corresponding to some established selection strategy (proportional, linear-rank, or tournament strategies) or even both parents are selected randomly. In that way selection pressure originating from parent selection is decreased and we balance this by increasing selection pressure after reproduction which is adjusted self-adaptively depending on how easy or difficult it is to achieve advancement.

5. *Which parts of the chromosomes of parents of above-average fitness are really worth being preserved?*
   Ideally speaking, exactly these parts of the chromosomes of above-average parents should be transferred to the next generation that make these individuals above average. What may sound like a tautology at the first view cannot be guaranteed for a lot of problem representations and corresponding operators. In these situations, OS is able to support the algorithm in this goal which is essential for GAs as building block assembling machines

1.3 Building Block Analysis for Standard Genetic Algorithms

For observing the distribution of essential alleles in a standard GA we have used the following test strategy: First, our aim was to observe the solution quality achievable with parameter settings that are quite typical for such kinds of GA applications (as given in Table1.1) using well known operators for the path representation, namely OX and ERX [14]; each algorithm has been analyzed applying no mutation as well as mutation rates of 5% and 10%.

Table 1.1 Parameters for test runs using a standard GA.

Parameter	Value
Generations	20'000
Population Size	100
Elitism Solutions	1
Mutation Rate	0.00 or 0.05 or 0.1
Selection Operator	Roulette
Crossover Operator	OX (Fig. 1.4) or ERX (Fig. 1.5)
Mutation Operator	Simple Inversion

The following Figures 1.4 and 1.5 show the fitness charts (showing best and average solution qualities of the GA's population as well as the best known quality) for a standard GA using order crossover OX (see Figure 1.4) and edge recombination crossover ERX (Figure 1.5); the parameter settings used for these experiments are given in Table 1.1.

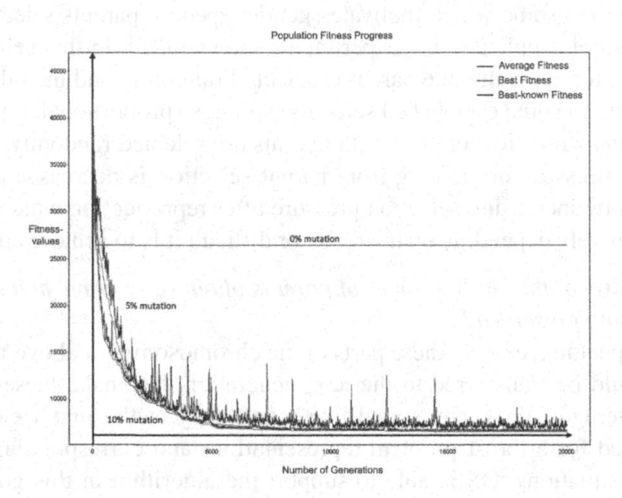

Fig. 1.4 Quality progress for a standard GA with OX crossover for mutation rates of 0%, 5%, and 10%.

For the OX crossover, which achieved the best results with these parameter settings, the results are shown in Figure 1.4; it is observable that the use of mutation rates of 5% and 10% leads to achieve quite good results (about 5% to 10% worse than the global optimum), whereas disabling mutation leads to a rapid loss of genetic diversity so that the solution quality stagnates at a very poor level.

The use of the more edge preserving crossover operator ERX (for which results are shown in Figures 1.5) shows different behavior in the sense that applying the same parameter settings as used for the OX the results are rather poor independent of the mutation rate. The reason for this is just that this operator requires more selection pressure (as for example tournament selection with tournament size 3); when applying higher selection pressure it is possible to achieve comparably good results also with ERX. Still, also when applying parameter settings which give good results with appropriate mutation rates, the standard GA fails dramatically when disabling mutation[2].

[2] In order to keep the discussion compact and on an explanatory level, detailed parameter settings and the corresponding statistically relevant result tables are not stated in this contribution; detailed results for the TSP tests are given at the website of the book [5] `http://gagp2009.heuristiclab.com/material/statisticsTSP.html`

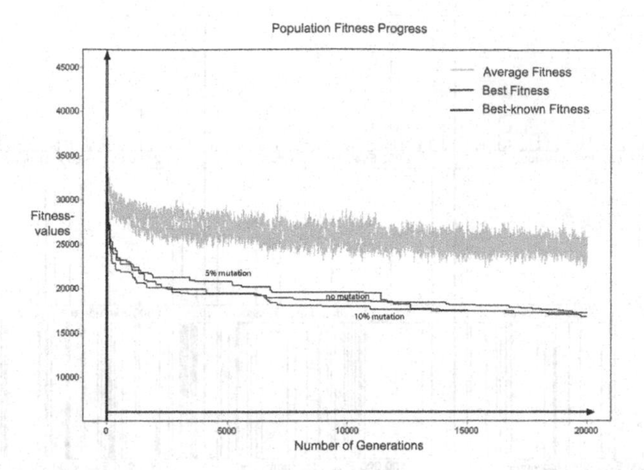

Fig. 1.5 Quality progress for a standard GA with ERX crossover for mutation rates of 0%, 5%, and 10%.

Summarizing these aspects we can state for the standard GA applied to the TSP that several well suited crossover operators[3] require totally different combinations of parameter settings in order to make the standard GA produce good results. Considering the results achieved with the parameter setting as stated in Table 1.1, the use of the OX yields good results (around 10% worse than the global optimum) whereas the use of ERX leads to unacceptable results (more than 100% worse than the global optimum). On the contrary, tuning the residual parameters (population size, selection operator) for ERX would lead to poor solution quality for OX.

Thus, an appropriate adjustment of selection pressure is of critical importance; as we will show in the following, self-adaptive steering of the selection pressure is able to make the algorithm more robust as selection pressure is adjusted automatically according to the actual requirements.

Figure 1.6 shows the distribution of the 130 essential alleles of the unique globally optimal solution over time for the overall best parameter constellation found in this section, i.e., the use of OX crossover with 5% mutation rate. In order to make the snapshots for the essential allele distribution within the standard GA's population comparable to those captured applying a GA with offspring selection or the RAPGA, the timestamps are not given in iterations but rather in the number of evaluations (which is in the case of the standard GA equal to the population size times the number of generations executed).

Until after about 10.000 evaluations, i.e., at generation 100, we can observe typical behavior, namely the rise of certain bars (representing the existence of edges of the global optimum). However, what happens between the 10.000th and 20.000th evaluation is that some of the essential alleles (about 15 in our test run) become fixed whereas the rest (here about $130 - 15 = 115$ in our test run) disappears in the

[3] OX and ERX are both edge preserving operators and therefore basically suited for the TSP.

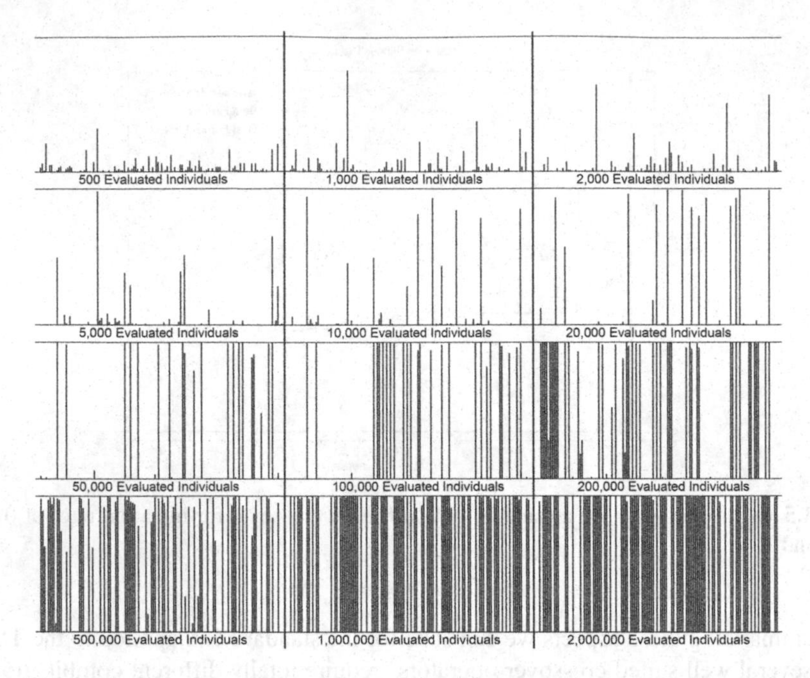

Fig. 1.6 Distribution of the alleles of the globally optimal solution over the run of a standard GA using OX crossover and a mutation rate of 5% (remaining parameters are set according to Table 1.1).

entire population. As we can see in Figure 1.6, without mutation the genetic search process would already be over at that moment due to the fixation of all alleles; from now on mutation is the driving force behind the search process of the standard GA.

The effects of mutation in this context are basically as follows: Sometimes high quality alleles are (by chance) injected into the population, and if they are beneficial (not even necessarily in the mutated individual), then a suited crossover operator to some degree is able to spread newly introduced essential alleles over the population and achieve a status of fixation quite rapidly. Thus, most of the essential alleles can be reintroduced and fixed approximately between the 20.000th and the 2.000.000th evaluation.

However, even if this procedure is able to perform reasonably good when applying adjusted parameters, it has not much in common with the desired behavior of a genetic algorithm as stated in the schema theorem and the according building block hypothesis. According to this theory, we expect a GA to systematically collect the essential pieces of genetic information which are initially spread over the chromosomes of the initial population as reported for the canonical GA[4]. As we will point out in the next sections, GAs with offspring selection as well as RAPGA are able to

[4] This statement is in fact restricted to binary encoding, single point crossover, bit-flip mutation, proportional selection and generational replacement.

considerably support a GA to operate in exactly that way even under not so idealized conditions as required in the context of the canonical GA.

1.4 Building Block Analysis for GAs Using Offspring Selection

When using offspring selection (OS), newly generated solution candidates are selected as individuals of the next generations's population if and only if they meet some given criterion. In its strict variant this means that children are discarded unless their fitness value is better than the fitness value of the better of the two parents and a new generation is filled up only with successful offspring. In each generation, selection pressure *selPres* is in this context calculated as the ratio of the number of produced offspring and the number of successful offspring that meet the given success criterion: $selPres = \frac{|generated\,solutions|}{|successful\,solutions|}$, with a maximum value of selection pressure acting as a quite intuitive and meaningful termination criterion.

The aim of this section is to highlight some characteristics of the effects of offspring selection. As the termination criterion of a GA with offspring selection is self-triggered, the effort of these test runs is not constant; however, the parameters given in Table 1.2 are adjusted in a way that the total effort is comparable to the effort of the test runs for the standard GA building block analyses discussed in Section 1.3.

Table 1.2 Parameters for test runs using a GA with OS. The results are graphically presented in Figures 1.7

Parameter	Value
Population Size	500
Elitism Solutions	1
Mutation Rate	0.00 or 0.05
Selection Operator	Random
Crossover Operator	OX , ERX, or comb. of OX, ERX, and MPX
Mutation Operator	Simple Inversion
Maximum Selection Pressure	500

Similarly as for the standard GA, we here take a look at the performance of some basically suited (edge preserving) operators. The results shown in Figure 1.7 highlight the benefits of self-adaptive selection pressure steering introduced by offspring selection: Independent of the other parameter settings, the use of all considered crossover operators yields results near the global optimum.

As documented in the previous section we have observed that the standard GA heavily relies on mutation, when the selection pressure is adjusted at a level that allows the standard GA to search in a goal oriented way. Therefore, we are now especially interested in how offspring selection can handle the situation when mutation is disabled. Figure 1.8 shows the quality curves for the use of the ERX crossover

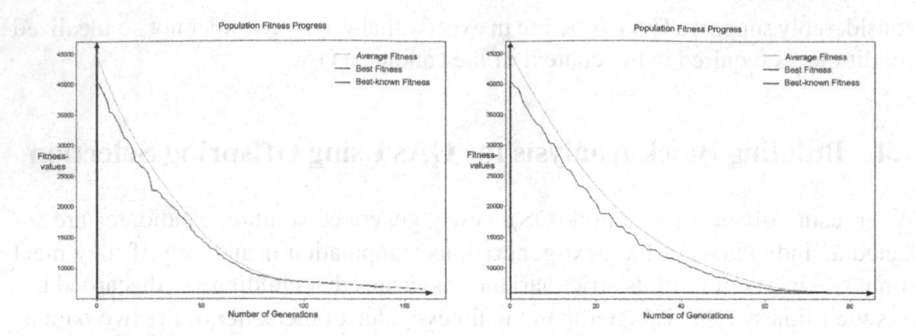

Fig. 1.7 Quality progress for a GA with offspring selection and a mutation rate of 5% for the OX (left curve) and ERX (right curve) crossover operators.

operator (which achieved the best results with 5% mutation) without mutation and the same settings for the remaining parameters. It is remarkable that the result is practically as good as with mutation, which at the same time means that offspring selection does not rely on the genetic diversity regaining aspect of mutation. Furthermore, this also means that offspring selection is able to keep the essential genetic information which in the concrete example is given by the alleles of the globally optimal solution. Offspring selection enhances the evolutionary process in such a way that only offspring of two parents take part on the ongoing evolutionary search that were successful recombinations. Therefore, in contrast to the standard GA the algorithm is not only able to keep the essential genetic information, but slowly merges the essential building blocks step by step which complies with the core statements of the building block theory and is not restricted to binary encoding or the use of certain operators.

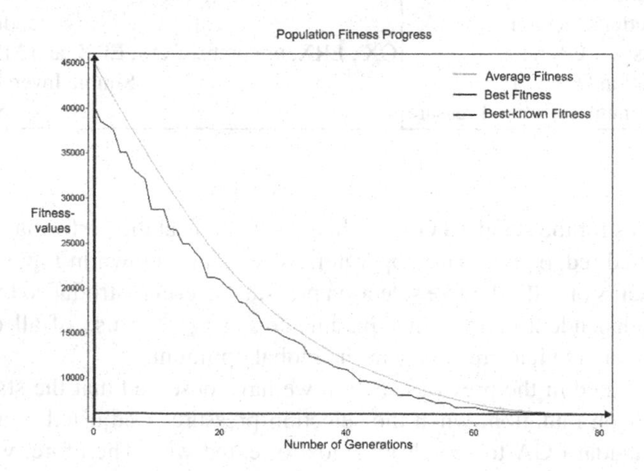

Fig. 1.8 Quality progress for a GA with offspring selection, ERX and no mutation.

In previous publications we have even gone one step further: We have been able to show in [2] that even with crossover operators basically considered unsuitable for the TSP (as they inherit the position information rather than the edge information like CX or PMX [14] it becomes possible to achieve high quality results in combination with offspring selection. The reason is that it is sufficient that a crossover operator is able to produce good recombinations from time to time (as only these are considered for the future gene pool); the price which has to be paid is that higher average selection pressure has to be applied, if the crossover operator is more unlikely to produce successful offspring.

As a proof of concept for applying more than one crossover at the same time, we have repeated the previous test runs with OX, MPX and ERX with the only difference that for these tests all crossover operators have been used. Figure 1.9 shows the quality curves (best and average results) for this test run and shows that the results are in the region of the globally optimal solution and therefore at least as good as in the test runs before. A further question that comes along using multiple operators at once is their performance over time: Is the performance of each of the certain operators relatively constant over the run of the algorithm?

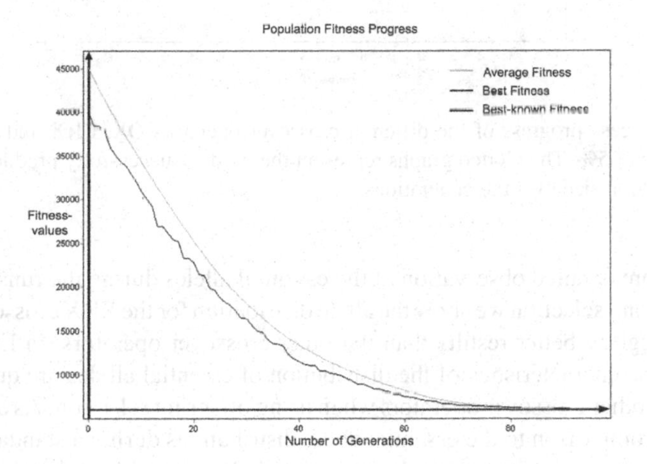

Fig. 1.9 Quality progress for a GA with offspring selection using a combination of OX, ERX and MPX and a mutation rate of 5%.

In order to answer this question, Figure 1.10 shows the ratio of successful offspring for each crossover operator used (in the sense of strict offspring selection which requires that successful children have to be better than both parents). Figure 1.10 shows that ERX performs very well at the beginning (approximately until generation 45) as well as in the last phase of the run (approx. from generation 75). In between (approximately from generation 45 to generation 75), when the contribution of ERX is rather low, MPX shows significantly better performance. The performance of OX in terms of its ability to generate successful offspring is rather mediocre during the whole run showing very little success in the last phase. The analysis of reasons of the behavior of the certain operators over time would be an interesting field of research; anyway, it is already very interesting to observe that the performance characteristics of the operators are changing over time to such an extent.

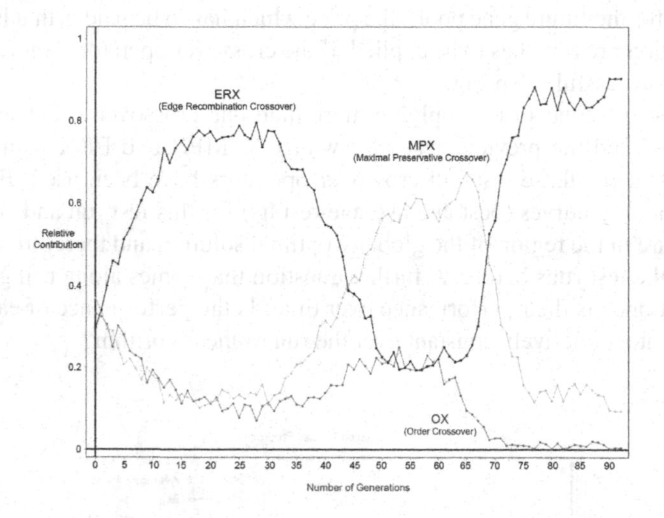

Fig. 1.10 Success progress of the different crossover operators OX, ERX and MPX and a mutation rate of 5%. The plotted graphs represent the ratio of successfully produced children to the population size over the generations.

For a more detailed observation of the essential alleles during the runs of the GA using offspring selection we show the allele distribution for the ERX crossover, which achieved slightly better results than the other crossover operators, in Figure 1.11. However, the characteristics of the distribution of essential alleles are quite similar also for the other crossover operators when using offspring selection. As a major difference in comparison to the essential allele distributions during a standard GA, we can observe that the diffusion of the essential alleles is established in a rather slow and smooth manner. The essential alleles are neither lost nor fixed in the earlier stages of the algorithm, so the bars indicating the occurrence of the certain essential allele (edges of the optimal TSP path) in the entire population are growing steadily until almost all of them are fixed by the end of the run. This behavior not only indicates a behavior in accordance with the building block hypothesis, but also implies that the algorithm performance no more relies on mutation to an extent as observed for the corresponding standard GA analyses. In order to confirm this assumption we have repeated the same test without mutation and indeed, as it can be seen when comparing Figure 1.11 and Figure 1.12, the saturation behavior of the essential building blocks is basically the same, no matter if mutation is used or not. This is a remarkable observation as it shows that offspring selection enables a GA to collect the essential building blocks represented in the initial population and compile high quality solutions very robustly in terms of parameters and operators like mutation, selection pressure, crossover operators etc. This property is especially important when exploring new fields of application where suitable parameters and operators are usually not known a priori.

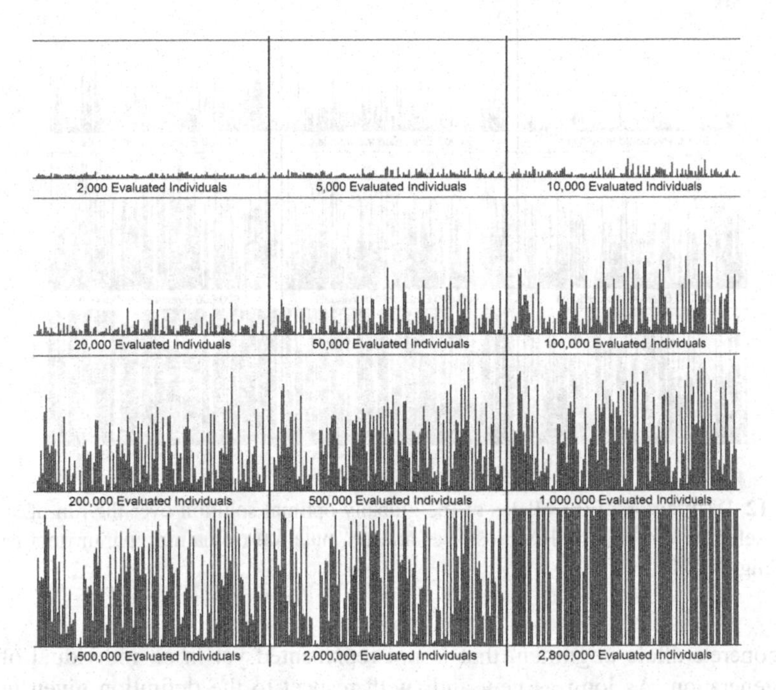

Fig. 1.11 Distribution of the alleles of the globally optimal solution over the run of an offspring selection GA using ERX crossover and a mutation rate of 5% (remaining parameters are set according to Table 1.2).

1.5 Building Block Analysis for the Relevant Alleles Preserving GA (RAPGA)

Similarly to the previous section we aim to highlight some of the most characteristic features of the relevant alleles preserving GA (RAPGA). The RAPGA works in such a way that new child solutions are added to the new population as long as it is possible to generate unique and successful offspring stemming from the gene pool of the last generation.

This idea is implemented using ad hoc population size adjustment in that sense that potential offspring generated by the basic genetic operators are accepted as members of the next generation if and only if they are able to outperform the fitness of their own parents and if they are new in that sense that their chromosome consists

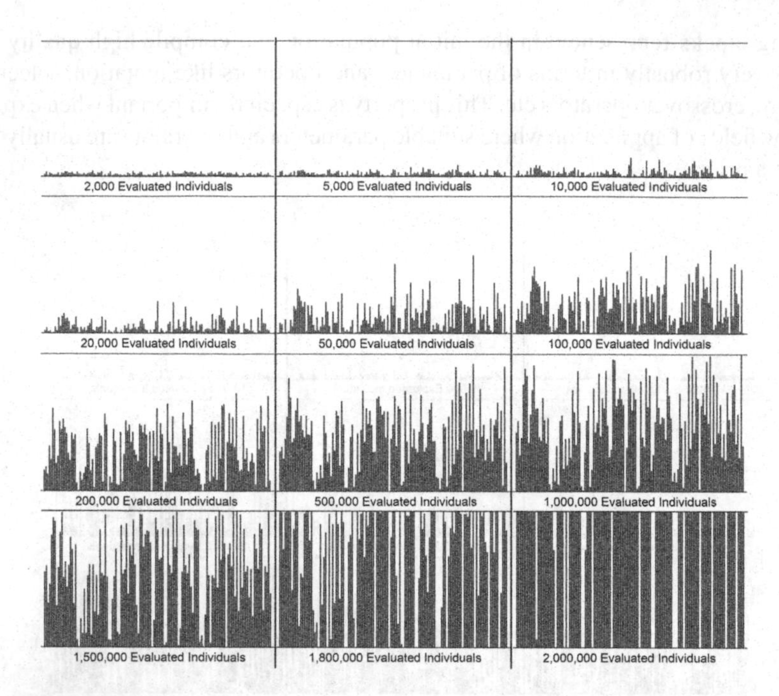

Fig. 1.12 Distribution of the alleles of the globally optimal solution over the run of an offspring selection GA using ERX crossover and no mutation (remaining parameters are set according to Table 1.2).

of a concrete allele alignment that is not represented yet in an individual of the next generation. As long as new and (with respect to the definition given in the context of OS) successful individuals can be created from the gene pool of the actual generation, the population size is allowed to grow up to a maximum size. Similar to OS, a potential offspring which is not able to fulfill these requirements is simply not considered for the gene pool of the next generation.

Still a lower as well as an upper limit of population size are necessary in order to achieve efficient algorithmic performance. In order to end a certain generation in case it is not possible to fill up the maximally allowed population size with new successful individuals, an upper limit of effort in terms of generated individuals is also necessary. This maximum effort per generation is the maximum number of newly generated chromosomes per generation (no matter if these have been accepted or not). In order to terminate the run of a certain generation in case it is not possible to fill up the maximally allowed population size with new successful individuals, an upper limit of effort in terms of generated individuals is necessary. This maximum effort per generation is the maximum number of newly generated chromosomes per generation (no matter if these have been accepted or not). The question, whether or not an offspring is better than its parents, is answered in the same way as in the context of offspring selection.

Table 1.3 Parmeters for test runs using the relevant alleles preserving genetic algorithm. The results are presented in Fig.1.13 and Fig.1.14.

Parameter	Value
Max. Generations	1'000
Initial Population Size	500
Mutation Rate	0.00 or 0.05
Elitism Rate	1
Male Selection	Roulette
Female Selection	Random
	OX
Crossover Operator	ERX
	combined (OX, ERX and MPX)
Mutation Operator	Simple Inversion
Minimum Population Size	5
Maximum Population Size	700
Effort	20'000
Attenuation	0

The following experiments are set up quite similar to the offspring selection experiments of the previous section. Firstly, the considered operators OX and ERX as well as the combination (OX, MXP, ERX) are applied to the *ch130* benchmark TSP problem taken from the TSPLib. Then the most successful operator or operator combination, respectively, is also exemplarily considered without mutation in order to show that the RAPGA (as offspring selection) does not rely on mutation to such an extent as conventional GAs.

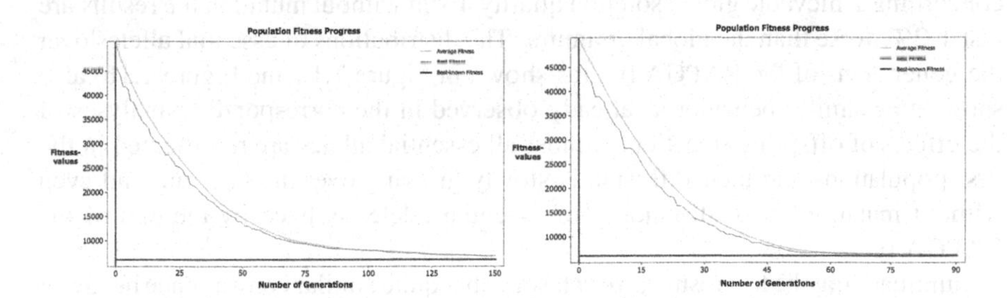

Fig. 1.13 Quality progress for a relevant alleles preserving GA with a mutation of 5% for the OX (left curve) and ERX (right curve) crossover operators.

Already the experiments using OX (see Figure 1.13) show good results (approximately 5% – 10% worse than the globally optimal solution) which are even slightly better than the corresponding offspring selection results. Even if only single test runs are shown in this chapter, it has to be pointed out that the authors have taken care that characteristical runs are shown. Besides, as described in the more systematical

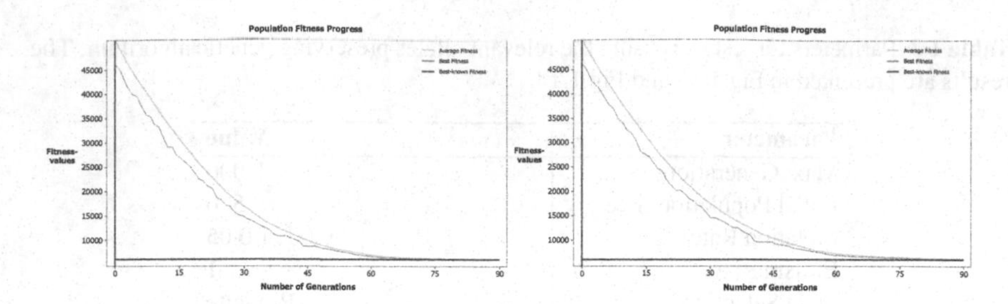

Fig. 1.14 Fitness curve for a relevant alleles preserving GA using a combination of OX, ERX and MPX and a mutation rate of 5% (left curve) and with no mutation (right curve).

experiments described in [5], especially due to the increased robustness caused by offspring selection and RAPGA the variance of the results' qualities is quite small.

Similarly to what we stated for the OS analyses, also for the RAPGA the best results could be achieved using ERX (as shown in Figure 1.13) as well as using the combination of OX, ERX and MXP (see Figures 1.14). The achieved results using these operators are about 1% or even less worse than the globally optimal solution. In the case of the RAPGA the operator combination turned out to be slightly better than ERX (in 18 of 20 test runs). Therefore, this is the operator combination we have also considered for a detailed building block analysis without mutation as well as applying 5% mutation.

Barely surprising, the results of RAPGA with the operator combination consisting of OX, ERX, and MPX turned out to be quite similar to those achieved using offspring selection and the ERX operator. Due to the name giving aspect of essential allele preservation, disabling mutation (see Figure 1.14) has almost no consequences concerning achievable global solution quality. Even without mutation the results are just 1-2% worse than the global optimum. The distributions of essential alleles over the generations of the RAPGA run (as shown in Figure 1.15 and Figure 1.16) also show quite similar behavior as already observed in the corresponding analyses of the effects of offspring selection. Almost all essential alleles are represented in the first populations and their diffusion is slowly growing over the GA run, and even without mutation the vast majority of essential alleles is fixed by the end of the RAPGA runs.

Summarizing the test results, we can state that quite similar convergence behavior is observed for a GA with offspring selection and the RAPGA, which is characterized by efficient maintenance of essential genetic information. As shown in Section 1.3, this behavior (which we would intuitively expect from any GA) cannot be guaranteed in general for GA applications where it was mainly mutation which helped to find acceptable solution qualities.

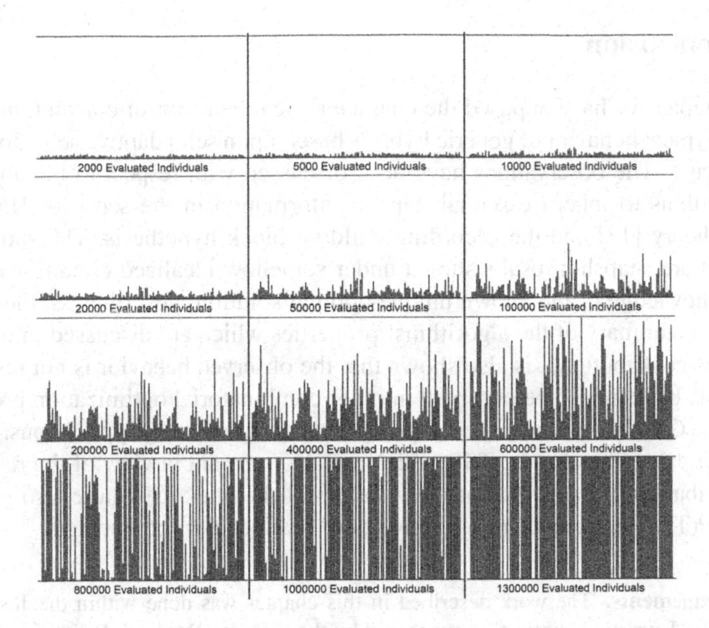

Fig. 1.15 Distribution of the alleles of the globally optimal solution over the run of a relevant alleles preserving GA using a combination of OX, ERX and MPX and a mutation rate of 5% (remaining parameters are set according to Table 1.3).

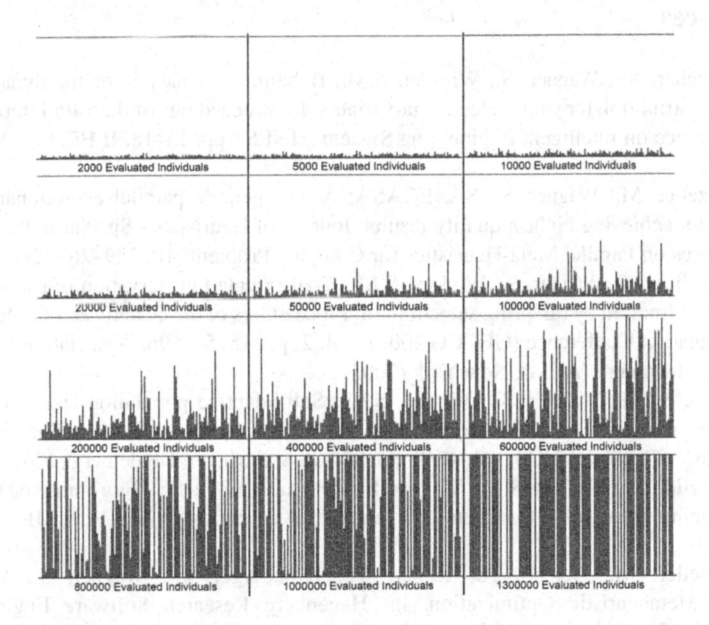

Fig. 1.16 Distribution of the alleles of the globally optimal solution over the run of a relevant alleles preserving GA using a combination of OX, ERX and MPX without mutation (remaining are set parameters according to Table 1.3).

1.6 Conclusion

In this chapter we have opposed the characteristical behavior of conventional GAs with the typical behavior of generic hybrids based upon self adaptive selection pressure steering. The observations have been discussed with respect to the ability of the algorithms to inherit essential genetic information in the sense of Holland's schema theory [13] and the according building block hypothesis. The shown experiments are snapshot results shown under somehow idealized conditions which use the knowledge of the known unique optimal solution; nevertheless, these tests represent a summary of the algorithms' properties which are discussed in depth in [5]. In this contribution it is also shown that the observed behavior is not restricted to the TSP, but can also be observed in other combinatorial optimization problems such as the CVRP(TW) and especially in genetic programming applications. Extensive empirical studies for the TSP, which are also relevant in view of the results of this contribution, are provided at the website[5] of the book [5] where also tests for the CVRP(TW) and genetic programming applications are documented.

Acknowledgements. The work described in this chapter was done within the Josef Ressel centre for heuristic optimization sponsored by the Austrian Research Promotion Agency (FFG).

References

1. Affenzeller, M., Wagner, S., Winkler, S.M., Beham, A.: Analysis of the dynamics of allele distribution for some selected ga-variants. In: Proceedings of the 14th International Conference on Intelligent Engineering Systems (INES), pp. 13–18. IEEE, Los Alamitos (2010)
2. Affenzeller, M., Wagner, S.: SASEGASA: A new generic parallel evolutionary algorithm for achieving highest quality results. Journal of Heuristics - Special Issue on New Advances on Parallel Meta-Heuristics for Complex Problems 10, 239–263 (2004)
3. Affenzeller, M., Wagner, S., Winkler, S.M.: Goal-oriented preservation of essential genetic information by offspring selection. In: Proceedings of the Genetic and Evolutionary Computation Conference (GECCO 2005), vol. 2, pp. 1595–1596. Association for Computing Machinery (ACM), New York (2005)
4. Affenzeller, M., Wagner, S., Winkler, S.M.: Self-adaptive population size adjustment for genetic algorithms. In: Moreno Díaz, R., Pichler, F., Quesada Arencibia, A. (eds.) EUROCAST 2007. LNCS, vol. 4739, pp. 820–828. Springer, Heidelberg (2007)
5. Affenzeller, M., Winkler, S.M., Wagner, S., Beham, A.: Genetic Algorithms and Genetic Programming: Modern Concepts and Practical Applications. CRC Press, Boca Raton (2009)
6. Affenzeller, M., Beham, A., Kofler, M., Kronberger, G., Wagner, S., Winkler, S.M.: Metaheuristic Optimization. In: Hagenberg Research Software Engineering, pp. 103–155. Springer, Heidelberg (2009)
7. Alba, E.: Parallel Metaheuristics: A New Class of Algorithms. Wiley Interscience, Hoboken (2005)

[5] http://gagp2009.heuristiclab.com/material.html

8. Beyer, H.G.: The Theory of Evolution Strategies. Springer, Heidelberg (2001)
9. Cavicchio, D.: Adaptive Search Using Simulated Evolution. PhD thesis, University of Michigan (1975)
10. DeJong, K.A.: An Analysis of the Behavior of a Class of Genetic Adaptive Systems. PhD thesis, University of Michigan (1975)
11. Fogel, D.B.: An introduction to simulated evolutionary optimization. IEEE Transactions on Neural Networks 5(1), 3–14 (1994)
12. Goldberg, D.E.: Genetic Algorithms in Search, Optimization and Machine Learning. Addison-Wesley Longman, Amsterdam (1989)
13. Holland, J.H.: Adaption in Natural and Artifical Systems. University of Michigan Press, Ann Arbor (1975)
14. Larranaga, P., Kuijpers, C.M.H., Murga, R.H., Inza, I., Dizdarevic, D.: Genetic algorithms for the travelling salesman problem: A review of representations and operators. Artificial Intelligence Review 13, 129–170 (1999)
15. Lobo, F.G., Goldberg, D.: The parameter-less genetic algorithm in practice. Information Sciences 167(1-4), 217–232 (2004)
16. Rechenberg, I.: Evolutionsstrategie. Friedrich Frommann Verlag (1973)
17. Reinelt, G.: TSPLIB - A traveling salesman problem library. ORSA Journal on Computing 3, 376–384 (1991)
18. Schwefel, H.-P.: Numerische Optimierung von Computer-Modellen mittels der Evolutionsstrategie. Birkhäuser Verlag, Switzerland (1994)
19. Smith, R.E., Forrest, S., Perelson, A.S.: Population diversity in an immune systems model: Implications for genetic search. In: Foundations of Genetic Algorithms, vol. 2, pp. 153–166. Morgan Kaufmann Publishers, San Francisco (1993)
20. Stephens, C.R., Waelbroeck, H.: Schemata evolution and building blocks. Evolutionary Computation 7(2), 109–124 (1999)
21. Wagner, S., Affenzeller, M.: SexualGA: Gender-specific selection for genetic algorithms. In: Callaos, N., Lesso, W., Hansen, E. (eds.) Proceedings of the 9th World Multi-Conference on Systemics, Cybernetics and Informatics (WMSCI 2005), vol. 4, pp. 76–81. International Institute of Informatics and Systemics (2005)

Chapter 2
Pliant Operator System

József Dombi

Abstract. We give a new representation theorem of negation based on the generator function of the strict operator. We study a certain class of strict monotone operators which build the DeMorgan class with infinitely many negations. We show that the necessary and sufficient condition for this operator class is $f_c(x)f_d(x) = 1$, where $f_c(x)$ and $f_d(x)$ are the generator functions of the strict t-norm and strict t-conorm. On the other hand our starting point is study of the relationship for Dombi aggregative operators, uninorms, strict t-norms and t-conorms. We present new representation theorem of strong negations where two explicitly contain the neutral value. Then relationships for aggregative operators and strong negations are verified as well as those for t-norm and t-conorm using the Pan operator concept. We will study a certain class of aggregative operators which build a self-DeMorgan class with infinitely many negation operators. We introduce the multiplicative pliant concept and characterize it by necessary and sufficient conditions.

2.1 Pliant Triangular Norms

2.1.1 Introduction and Some Elementary Considerations

In this article we will focus on the DeMorgan systems which correspond to infinitely many negations. These types of operators are important because the fix point of the negation (see Eq.(2.4) later) can be varied, this value can be interpreted as a decision level and this kind of logic is very flexible. Such logic is very important. Cintula, Klement, Mesiar and Pap focus on fuzzy logic with an additional involutive negation operator [11], but in our case we have infinitely many.

József Dombi
Department of Informatics, University of Szeged, 6720 Szeged, Árpád tér 2., Hungary
e-mail: dombi@inf.u-szeged.hu

J. Fodor et al. (Eds.): Recent Advances in Intelligent Engineering Systems, SCI 378, pp. 31–58.
springerlink.com

This general characterisation makes it possible for us to construct a new type of operator system.

In the introductory part we give an elementary discussion for readers not familiar with this topic. In the Section 2 we extend the operators with weights and then we describe the relation between strict t-norm, strict t-conorm generator function and negation. This result is a reformulation of the known results. We show that the involutive properties of the negation (given $f_c(x)$ and $f_d(x)$) ensure that $k(x)$ is a function (see Fig.1). We give the general form of the negation by using $k(x)$. We show that all involutive negation operators can be represented in this form and we will give some examples. The main result of the article can be found in Section 3. We show that a DeMorgan triplet is valid with an infinitely many negations if and only if $f_c(x)f_d(x) = 1$ (i.e. $k(x) = \frac{1}{x}$) . We call such a system a pliant system. In Section 4 we characterize the pliant operators.

From an application point of view, the strict monotonously increasing operators are useful. They have many applications. This is the reason why in this article we will focus on strictly monotonously increasing operators.

Triangular Norms and Conorms

Here we summarize the necessary notations and some previous results which will be used in the sequel.

For the basic properties of triangular norms (t-norms for short) and triangular conorms (t-conorms) Klement, Mesiar and Pap [29, 30, 31] refer to [28, 47]. By definition, a t-norm $c(x,y)$ and a t-conorm $d(x,y)$ turn $[0,1]$ into an abelian, fully ordered semigroup with neutral element 1 and 0, respectively. Here Klement, Mesiar and Pap [29, 30, 31] will restrict themselves to continuous t-norms and t-conorms. Let us only recall that a continuous t-norm $c(x,y)$ is Archimedean if it satisfies $c(x,x) < x$ for all $x \in]0,1[$. A continuous Archimedean t-norm is called strict if $0 \leq x < y \leq 1$ and $0 < z \leq 1$ implies $c(x,y) < c(y,z)$. Non-strict continuous Archimedean t-norms are called nilpotent. The basic result can be found in the book of Aczl [1]. From [33, 42] Klement, Mesiar and Pap [29, 30, 31] state that a t-norm $c(x,y)$ is continuous Archimedean if and only if it has a continuous additive generator, i.e.,there is a continuous, strictly decreasing function $t : [0,1] \to [0,\infty]$ satisfying $t(1) = 0$ such that for all $(x,y) \in [0,1]^2$

$$c(x,y) = t^{(-1)}\,(t(x)+t(y)) \tag{2.1}$$

where the pseudo-inverse $t^{(-1)} : [0,\infty] \to [0,1]$ of t in this special context is given by $t^{(-1)}(x) = t^{-1}\,(\min(t(x),t(0)))$. Observe that the additive generator of a continuous Archimedean t-norm is unique up to a positive multiplicative constant. The case $t(0) = \infty$ occurs if and only if $c(x,y)$ is strict (in which case the pseudo-inverse in Eq.(2.1) is an ordinary inverse).

In this section, besides the min/max and the drastic operators, we will be concerned with strict t-norms, that is

$$c(x,y) < c(x',y) \quad \text{if} \quad x < x' \quad x,y \in (0,1]$$

and t-conorms, that is

$$d(x,y) < d(x',y) \quad \text{if} \quad x < x' \quad x,y \in [0,1)$$

Later on we will look for the general form of $c(x,y)$ and $d(x,y)$. We assume that the following conditions are satisfied:

1. Continuity:
   $c\colon [0,1] \times [0,1] \to [0,1]$ $\qquad$ $d\colon [0,1] \times [0,1] \to [0,1]$
2. Strict monotonous increasing:
   $c(x,y) < c(x,y')$ if $y < y'$ $x \neq 0$ $\qquad$ $d(x,y) < d(x,y')$ if $y < y'$ $x \neq 0$
3. Compatibility with two-valued logic:
   $c(0,0) = 0$ $\quad c(1,1) = 1$ $\qquad$ $d(0,0) = 0$ $\quad d(1,1) = 1$
   $c(0,1) = 0$ $\quad c(1,0) = 0$ $\qquad$ $d(0,1) = 1$ $\quad d(1,0) = 1$
4. Associativity:
   $c(x,c(y,z)) = c(c(x,y),z)$ $\qquad$ $d(x,d(y,z)) = d(d(x,y),z)$
5. Archimedean:
   $c(x,x) < x, \quad x \in (0,1)$ $\qquad$ $d(x,x) > x, \quad x \in (0,1)$

So

$$c(x,y) = f_c^{-1}(f_c(x) + f_c(y)). \tag{2.2}$$

Similarly, the strict t-conorm on $(0,1] \times (0,\infty]$ has the form:

$$d(x,y) = f_d^{-1}(f_d(x) + f_d(y)). \tag{2.3}$$

Here $f_c(x) : [0,1] \to [0,\infty]$ $(f_d(x) : [0,1] \to [0,\infty])$ are continuous and strictly increasing (decreasing) monotone functions and they are the generator functions of the strict t-norms and strict t-conorms.

Those familiar with fuzzy logic theory will find that the terminology used here is slightly different from that used in standard texts [28, 9, 3, 6, 40, 22]. This is because I would like to distinguish between fuzzy logic and pliant logic.

Consistent many-valued (fuzzy) operators have to satisfy of certain Boole identities. The most important one is the DeMorgan law. Esteva [18] and Dombi [12] were the first two researchers who carefully analysed the DeMorgan identity. It corresponds to the conjunction, disjunction and negation operators.

In order to analyse the DeMorgan identity we first need a good definition of negation. Strong negations are order reversing automorphisms of the unit interval. Because here we deal only with strong negations we will refer to them as *negation*. The usual requirements for such a negation (η) are the following.

Definition 2.1. We say that $\eta(x)$ is a negation if $\eta : [0,1] \to [0,1]$ satisfies the following conditions:

C1: $\eta : [0,1] \to [0,1]$ is continuous (Continuity)
C2: $\eta(0) = 1$, $\eta(1) = 0$ (Boundary conditions)
C3: $\eta(x) < \eta(y)$ for $x > y$ (Monotonicity)
C4: $\eta(\eta(x)) = x$ (Involution)

From C1 and C3 it follows that there exists a fix point $\nu_* \in [0,1]$ of the negation where

$$\eta(\nu_*) = \nu_* \tag{2.4}$$

So another possible characterization of negation, is when we assign a so-called decision value ν for a given ν_0, i.e. one can specify a point (ν, ν_0) that the curve must intersect. This tells us something about how strong the negation operator is.

$$\eta(\nu) = \nu_0 \tag{2.5}$$

If $\eta(x)$ has a fix point ν_*, we use the notation $\eta_{\nu_*}(x)$ and if the decision value is ν, then we use the notation $\eta_\nu(x)$. If $\eta(x)$ is used without a suffix for then the parameter has no importance in the proof. Later on we will characterize the negation by the ν_*, ν_0 and ν parameters.

In the following, we will examine the relations between f_c, f_d and η, to see whether c, d and η satisfy the DeMorgan law.

The ν_* value has several terminologies. It is called the fixpoint eigenvalue, or equilibrium point. In the article of De Baets and Fodor [5] the negations are induced by the uninorm. Here ν_* is called the neutral element. In multicriteria decision making 'expectation value' has a meaning and the pliant logic can be applied in this area.

2.1.2 DeMorgan Law and General Form of Negation

We will use the generalized operator based on strict t-norms and strict t-conorms introduced by the authors. Calvo [10] and Yager [55].

Definition 2.2. Generalized operators based on strict t-norms and t-conorms which are

$$c(\mathbf{w},\mathbf{x}) = c(w_1,x_1;w_2,x_2;\ldots;w_n,x_n) = f_c^{-1}\left(\sum_{i=1}^{n} w_i f_c(x_i)\right), \tag{2.6}$$

$$d(\mathbf{w},\mathbf{x}) = d(w_1,x_1;w_2,x_2;\dots w_n,x_n) = f_d^{-1}\left(\sum_{i=1}^{n} w_i f_d(x_i)\right), \qquad (2.7)$$

where $w_i \geq 0$.

If $w_i = 1$ we get the t-norm and t-conorm. If $w_i = \frac{1}{n}$, then we get mean operators. If $\sum_{i=1}^{n} w_i = 1$, then we get weighted operators.

Definition 2.3. The DeMorgan law holds for the generalized operator based strict t-norms and strict t-conorms and for negation if and only if the following equation holds.

$$c(w_1,\eta(x_1);w_2,\eta(x_2);\dots;w_n,\eta(x_n)) = \eta(d(w_1,x_1;w_2,x_2;\dots;w_n,x_n)), \qquad (2.8)$$

We call Eq.(2.8) later on generalized DeMorgan law.

Theorem 2.1 (DeMorgan Law). *If $\eta(x)$ and $f_d(x)$ are given, then $c(x,y),d(x,y)$ and $\eta(x)$ form a DeMorgan triplet iff*

$$f_c^{-1}(x) = \eta(f_d^{-1}(ax)), \qquad (2.9)$$

where $a \neq 0$.

Eq.(2.9) can be written in other form, see later (Eq.(2.10)).

Remark 2.1. It is well-known that based on Eq.(2.9) from a generator function of a strict t-norm (or strict t-conorm) operator using negation we can get a generator function os the strict t-conorm (or strict t-norm), see [6].

Proof. See [15].

Remark 2.2. On the basis of Theorem 2.1 and given $f_c(x)$ and $\eta(x)$, $f_d(x)$ can be determined, so that c, d and η is a DeMorgan triple. Similar to the above-mentioned consideration, with a given $f_d(x)$ and $\eta(x)$, $f_c(x)$ can be determined.

2.1.2.1 Form of Negations

Here the following question naturally arises. If f_c and f_d are given, what kind of condition ensures that η is a negation (i.e. fulfills C1-C4)? From Theorem 2.1 we know that the necessary and sufficient condition of the DeMorgan Law is Eq. (2.9). Substituting the $x := f_d^{-1}(ax)$, we have

$$\eta(x) = f_c^{-1}\left(\frac{1}{a} f_d(x)\right), \qquad a \neq 0. \qquad (2.10)$$

Let us give a parametric form of negation.

Theorem 2.2. *If $f_c(x)$ and $f_d(x)$ are given, then $c(x,y), d(x,y)$ and $\eta(x)$ form a DeMorgan triplet iff*

$$\eta_{v_*}(x) = f_d^{-1}\left(\frac{f_d(v_*)}{f_c(v_*)} f_c(x)\right), \tag{2.11}$$

$$\eta_{v_*}(x) = f_c^{-1}\left(\frac{f_c(v_*)}{f_d(v_*)} f_d(x)\right). \tag{2.12}$$

Proof. See [15].

Negation is not always necessary involutive if it has the form (2.11) or (2.12), i.e.

if

$$f_c(x) = \ln(x), \quad f_d(x) = \ln(1-x), \quad i.e.: \quad f_c^{-1}(x) = e^x, \quad f_d^{-1}(x) = 1 - e^x$$

then

$$\eta(x) = f_d^{-1}(K f_c(x)) = 1 - e^{K \ln x} = 1 - x^K \tag{2.13}$$

$$\eta(x) = f_c^{-1}(K f_d(x)) = e^{K \ln(1-x)} = (1-x)^K \tag{2.14}$$

If $K \neq 1$, then Eq.(2.13) and Eq.(2.14) are not involutive.

This negation satisfies (C1-C3). The next important question is whether they obey the involution condition C4: $\eta(x) = \eta^{-1}(x)$.

Theorem 2.3. *Let η be given by (2.10). Then $\eta(x)$ is involutive if and only if*

$$f_c(x) = \frac{1}{a} k(f_d(x)) \quad or \quad k(x) = a f_c\left(f_d^{-1}(x)\right), \qquad a \neq 0, \tag{2.15}$$

where $k\colon [0,\infty] \to [0,\infty]$ is a strictly decreasing, continuous function and

$$k^{-1}(x) = k(x). \tag{2.16}$$

Proof. See [15].

We can obtain a new representation theorem for the negation using Theorem (2.3).

Theorem 2.4 (General form of the negation). *We have that $c(\boldsymbol{w},\boldsymbol{x})$, $d(\boldsymbol{w},\boldsymbol{x})$ and $\eta(x)$ is a DeMorgan triple if and only if*

$$\eta(x) = f^{-1}(k(f(x))), \tag{2.17}$$

where $f(x) = f_c(x)$ or $f(x) = f_d(x)$ and $k(x)$ is a strictly decreasing function with the property

$$k(x) = k^{-1}(x). \tag{2.18}$$

where $k\colon [0,\infty] \to [0,\infty]$.

Proof. See [15].

Corollary 2.1. *From Eq. (2.17) it is easy to see that*

$$k(x) = f(\eta(f^{-1}(x))), \tag{2.19}$$

i.e. if $f(x)$ and $\eta(x)$ is given, then $k(x)$ is determined by Eq. (2.19).

2.1.2.2 Representation Theorem of Negation

Another interesting question is whether Eq. (2.17) is a general representation form of the negation? The following theorem ensures that all negations can be written in this form.

While Trillas' theorem [50] represents negations (from our point of view) for the nilpotent class of t-norms and t-conorms, our next result provides a representation theorem for the strict t-norms and t-conorms.

Let $k(x) = \frac{A}{x}$. In the next theorem we show that all negation can be expressed by using this $k(x)$.

Theorem 2.5 (Representation theorem of negation). *For any given $\eta(x)$ there exists an $f(x)$ such that*

$$\eta(x) = f^{-1}\left(\frac{A}{f(x)}\right), \tag{2.20}$$

where f is the generator function of some strict t-norm, or strict t-conorm and $A > 0$.

Remark 2.3. This theorem is similar to what Trillas' theorem states, i.e. for any given $\eta(x)$ there exists a $f(x)$ such that $\eta(x) = f^{-1}(1 - f(x))$, where $f(x)$ is the generator function of some non-strict operator. In Theorem 2.5 the generator function is the generator function a strict monotonously increasing operator.

Proof. See [15].

Remark 2.4. A DeMorgan triple can be built by using one the generator function of just one operator and choosing a $k(x)$. That is,

$$\eta(x) = f_c^{-1}\left(k\left(f_c(x)\right)\right) \tag{2.21}$$

$$c(\mathbf{w},\mathbf{x}) = f_c^{-1} \sum_{i=1}^{n} (w_i f(x_i)) \tag{2.22}$$

$$d(\mathbf{w},\mathbf{x}) = f_c^{-1} \left(k \left(\sum_{i=1}^{n} w_i k(f_c(x_i)) \right) \right) \tag{2.23}$$

form a DeMorgan triple, and

$$f_c(x) = k(f_d(x)), \tag{2.24}$$

so $k(x)$ can be understood as a kind of negation.

2.1.2.3 Examples of DeMorgan Systems

Using the above results, we can construct classical systems and also new operator systems.

- If $f_c(x) = -\ln(x)$ and $\eta(x) = 1 - x$, then
  $$k(x) = f_c(\eta(f_c^{-1}(x))) = -\ln(1 - e^{-x}) \quad c(x,y) = xy, \quad d(x,y) = x + y - xy.$$
- If $f_d(x) = -\ln(1 - x)$ and $\eta(x) = 1 - x$, then
  $$k(x) = f_d(\eta(f_d^{-1}(x))) = -\ln(1 - e^{-x}) \quad c(x,y) = xy, \quad d(x,y) = x + y - xy.$$

We give a new example of where $\eta(x)$ can be varied.

If $f_c(x) = -\ln(x)$ and $f_d(x) = -\dfrac{1}{\ln(x)}$, then $k(x) = \dfrac{1}{x}$.

$$c(x,y) = xy \qquad d(x,y) = e^{\frac{1}{\frac{1}{\ln x} + \frac{1}{\ln y}}} \qquad \eta(x) = e^{\frac{a}{\ln(x)}}$$

where $a > 0$.
Here $d(1,x) = d(x,1) = \lim_{y \to 1} d(x,y) = 1$ and $\eta(1) = \lim_{x \to 1} \eta(x) = 0$ and $\eta(0) = \lim_{x \to 0} y(x) = 1$.

2.1.2.4 Parametric Form of the Negations

Lemma 2.1. *The parametric form of the negation is*

$$\eta(x) = f^{-1} \left(f(\nu_*) \frac{k(f(x))}{k(f(\nu_*))} \right) \tag{2.25}$$

$$\eta(x) = f^{-1}\left(f(\nu_0)\frac{k(f(x))}{k(f(\nu))}\right). \tag{2.26}$$

Proof. See [15].

2.1.3 Operators with Infinitely Many Negations

Now we will characterize the operator class (strict t-norm and strict t-conorm) for which various negations exist and build a DeMorgan class. The fixpoint ν_* or the neutral value ν can be regarded as decision threshold. Operators with various negations are useful because the threshold can be varied.

It is straightforward to see that the min and max operators belong to this class, as does the drastic operator. The next theorem characterizes those strict operator systems that have infinitely many negations and build a DeMorgan system. It is easy to see that $c(x,y) = xy$, $d(x,y) = x+y-xy$ and $\eta(x) = 1-x$ build a DeMorgan system. There are no other negations for building a DeMorgan system, as we will see below.

Theorem 2.6. *$c(x,y)$ and $d(x,y)$ build a DeMorgan system for $\eta_{\nu_*}(x)$ where $\eta_{\nu_*}(\nu_*) = \nu_*$ for all $\nu_* \varepsilon (0,1)$ if and only if*

$$f_c(x) f_d(x) = 1. \tag{2.27}$$

Proof. See [15].

2.1.4 Multiplicative Pliant Systems

From Dombi's result [13] we know that if $f(x)$ is a generator function, then $f^{\alpha}(x)$ is a generator function. As we saw earlier $k(x)$ plays an important role in DeMorgan systems. Let us define the multiplicative pliant system by one of the simplest $k(x)$ functions.

Definition 2.4. If $k(x) = 1/x$, that is

$$f_c(x) f_d(x) = 1, \tag{2.28}$$

then we call the generated connectives a multiplicative pliant system.

If we have a generator function, then its power is also a generator function. Therefore in the pliant system we can use the power function of the generator function and define $f_c(x)$ by

$$f_c(x) = f^{\alpha}(x).$$

Remark 2.5. A similar operator system was in fact presented by Roychowdhury [45]. Theorem 2.6 above gives the necessary and sufficient conditions for a such system.

Theorem 2.7. *The general form of the multiplicative pliant system is*

$$o_\alpha(x,y) = f^{-1}\left(\left(f^\alpha(x)+f^\alpha(y)\right)^{1/\alpha}\right) \tag{2.29}$$

$$\eta_\nu(x) = f^{-1}\left(f(\nu_0)\frac{f(\nu)}{f(x)}\right) \quad \text{or} \tag{2.30}$$

$$\eta_{\nu_*}(x) = f^{-1}\left(\frac{f^2(\nu_*)}{f(x)}\right), \tag{2.31}$$

where $f(x)$ is the generator function of the strict t-norm operator and $f : [0,1] \to [0,\infty]$ continuous and strictly decreasing function. Depending on the value of α, the operator is

This operator called the drastic operator.

Proof. See [15].

Remark 2.6. In the multiplicative pliant system it is vital that negation be independent of the value and the sign of α. (In other words, it does not depend on whether the generator function belongs to the strict t-norm or strict t-conorm.)

Remark 2.7. The limit values of the pliant operators (min, max and drastic) also have the property that the DeMorgan triplet is valid for infinitely many negations.

Theorem 2.8. *If $g(x) = f^\alpha(x)$ is the generator function, negation does not change in the pliant system.*

Proof. See [15].

Theorem 2.9. *Let $c(x,y) = f^{-1}(f(x)+f(y))$, $d(x,y) = f^{-1}\left(\dfrac{1}{\dfrac{1}{f(x)}+\dfrac{1}{f(y)}}\right)$ and $\eta(x) = f^{-1}\left(f(\nu_0)\dfrac{f(\nu)}{f(x)}\right)$, then $c(x,y)$, $d(x,y)$ and $\eta(x)$ form a DeMorgan triplet.*

Proof. See [15].

2.2 Uninorms, Aggregative Operators

2.2.1 Introduction

The term *uninorm* was first introduced by Yager and Rybalov [53] in 1996. Uninorms are generalization of t-norms and t-conorms by relaxing the constraint on the identity element from $\{0,1\}$ to the unit interval. Since then many articles have focused on uninorms, both from a theoretical [34, 35, 26, 37, 8, 41] and a practical point of view [54]. The paper of Fodor, Yager and Rybalov [19] is important since it defined a new subclass of uninorms called representable uninorms. This characterization is similar to the representation theorem of strict t-norms and t-conorms, in the sense that both originate from the solution of the associativity functional equation given by Aczél [1].

The aggregative operators were introduced in the paper [13] by selecting a set of minimal concepts which must be fulfilled by an evaluation like operator.

As mentioned in [19], there is a close relationship between Dombi's aggregative operators and uninorms.

We will distinguish between logical operators (strict, continuous t-norms and t-conorms) and aggregative operators, where the former means strict, continuous operators.

The first goal is to show the close correspondence between strong negations, aggregative and logical operators. The second goal is to introduce and characterize multiplicative pliant operator systems.

The reader may recall that the field of uninorm, t-norm, t-conorm and its application were discussed in recent Information Science issues. For example see articles [2, 4, 7, 23, 24, 25, 32, 36, 38, 44, 52, 56].

This chapter is organized as follows. First we give some basic definitions. We emphasis the role of the neutral value in Section 2. Section 3 describes the correspondence between strong negations, aggregative and logical operators. Lastly, we present and give a characterization the so-called multiplicative pliant systems using examples in Section 4.

2.2.1.1 Basic Definitions

In 1982 Dombi [13] defined the aggregative operator in the following way:

Definition 2.5. An aggregative operator is a function $a : [0,1]^2 \to [0,1]$ with the properties:

1. Continuous on $[0,1]^2 \backslash \{(0,1),(1,0)\}$
2. $a(x,y) < a(x,y')$ if $y < y', x \neq 0, x \neq 1$
   $a(x,y) < a(x',y)$ if $x < x', y \neq 0, y \neq 1$
3. $a(0,0) = 0$ and $a(1,1) = 1$ (boundary conditions)
4. $a(x,a(y,z)) = a(a(x,y),z)$ (associativity)
5. There exists a strong negation η such that $a(x,y) = \eta(a(\eta(x),\eta(y)))$ (self DeMorgan identity) if $\{x,y\} \neq \{0,1\}$ or $\{x,y\} \neq \{1,0\}$
6. $a(1,0) = a(0,1) = 0$ or $a(1,0) = a(0,1) = 1$

We note that the original definition of aggregative operators has the condition of correct cluster formation instead of the self DeMorgan identity (see [13]), which later proved to be equivalent.

For the sake of completeness, strong negation will be defined by the following:

Definition 2.6. $\eta(x)$ is strong negation iff $\eta : [0,1] \to [0,1]$ satisfies the following conditions:

1. $\eta(x)$ is continuous
2. $\eta(0) = 1$, $\eta(1) = 0$ (boundary conditions)
3. $\eta(x) < \eta(y)$ for $x > y$ (monotonicity)
4. $\eta(\eta(x)) = x$ (involution)

The definition of uninorms, originally given by Yager and Rybalov [53] in 1996, is the following:

Definition 2.7. A uninorm U is a mapping $U : [0,1]^2 \to [0,1]$ having the following properties:

- $U(x,y) = U(y,x)$ (commutativity)
- $U(x_1,y_1) \geq U(x_2,y_2)$ if $x_1 \geq x_2$ and $y_1 \geq y_2$ (monotonicity)
- $U(x,U(y,z)) = U(U(x,y),z)$ (associativity)
- $\exists \nu_* \in [0,1]\ \forall x \in [0,1]\ U(x,\nu_*) = x$ (neutral element)

A uninorm is a generalization of t-norms and t-conorms. By adjusting its neutral element, a uninorm is a t-norm if $\nu_* = 1$ and a t-conorm if $\nu_* = 0$. The following representation theorem of strict, continuous on $[0,1] \times [0,1] \setminus (\{0,1\},\{1,0\})$ uninorms (or *representable uninorms*) was given by Fodor et al. [19].

Theorem 2.10. *Let $U : [0,1] \to [0,1]$ be a function and $\nu_* \in]0,1[$. The following are equivalent:*

1. *U is a uninorm with neutral element ν_* which is strictly monotone on $]0,1[^2$ and continuous on $[0,1]^2 \setminus \{(0,1),(1,0)\}$.*
2. *There exists a strictly increasing bijection $g_u : [0,1] \to [-\infty,\infty]$ with $g_u(\nu_*) = 0$ such that for all $(x,y) \in [0,1]^2$ we have*

$$U(x,y) = g_u^{-1}\left(g_u(x) + g_u(y)\right), \tag{2.32}$$

where, in the case of a conjunctive uninorm U, we use the convention $\infty + (-\infty) = -\infty$, while, in the disjunctive case, we use $\infty + (-\infty) = \infty$.

If Eq.(2.32) holds, the function g_u is uniquely determined by U up to a positive multiplicative constant, and it is called an additive generator of the uninorm U.

Proof. See [19].

Strong negation plays an important role. Besides Trillas' representation theorem, we introduced another form of negation [13]. In the following theorem which is well-known, we show that this representation is universal.

Theorem 2.11. *Let $\eta : [0,1] \to [0,1]$ be a continuous function, then the following are equivalent:*

1. *η is a strong negation.*
2. *There exists a continuous and strictly monotone function $g : [0,1] \to [-\infty,\infty]$ with $g(\nu_*) = 0$, $\nu_* \in]0,1[$ such that for all $x \in [0,1]$*

$$\eta(x) = g^{-1}(-g(x)). \tag{2.33}$$

If Eq.(2.33) holds, then ν_* is called the neutral value of the strong negation, i.e. for which $\eta(\nu_*) = \nu_*$.

Sketch of the proof. See [17] under refree process.

2.2.2 *The Neutral Value*

Theorem 2.11 tells us that all strong negations have the form $g^{-1}(-g(x))$ for a suitable g generator function. In this formula the neutral value of the strong negation is implicitly present the generator function. The following representation theorem of strong negations explicitly contains the neutral value.

Theorem 2.12 (Additive form of strong negations). *Let $\eta : [0,1] \to [0,1]$ be a continuous function, then the following are equivalent:*

1. *η is a strong negation with neutral value ν_*.*
2. *There exists a continuous and strictly monotone function $g : [0,1] \to [-\infty,\infty]$ and $\nu_* \in]0,1[$ such that for all $x \in [0,1]$*

$$\eta(x) = g^{-1}(2g(\nu_*) - g(x)). \tag{2.34}$$

Proof. See Dombi [13].

Similar to strong negations, the representation theorem of aggregative operators (Theorem 2.10) does not explicitly contain the neutral value of the aggregative operator. With the help of the following lemma, a new representation can be given.

Definition 2.8. (On the neutral element of the aggregative operator) The neutral element ν_* of the aggregative operator has the property

$$a(x, \nu_*) = x. \tag{2.35}$$

With the aggregative operator we have

$$g(\nu_*) = 0, \quad g^{-1}(0) = \nu_*. \tag{2.36}$$

From $g_a^{-1}(g_a(x) + g_a(\nu_*)) = x$, we get the above result.

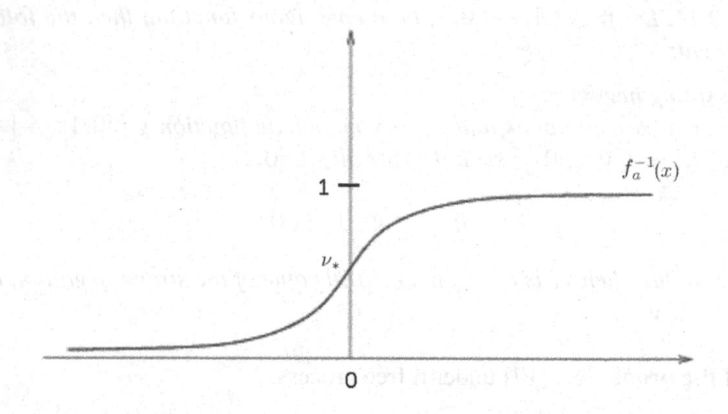

Fig. 2.1 v_* is the neutral element of the aggregative operator

Lemma 2.2 (Dombi [13]). *If g is the additive generator function of an aggregative operator a, then the function displaced by* $d \in \mathbb{R}$, $g_*(x) = g(x) + d$ *is also a generator function of an aggregative operator with neutral value* $v_* = g^{-1}(-d)$.

Proof. See Dombi [13].

Theorem 2.13 (Dombi [13]). *Let* $a : [0,1]^n \to [0,1]$ *be a function and let a be an aggregative n-valued operator with additive generator g. The neutral value of the aggregative operator is* v_* *if and only if* $\mathbf{x} \in [0,1]^n, \forall x$. *It has the following form:*

$$a(\mathbf{x}) = g^{-1}\left(\sum_{i=1}^{n} g(x_i) - (n-1)g(v_*)\right). \tag{2.37}$$

Proof. See Dombi [13].

In the next we will study the aggregative operator when v_* neutral element is given. Let

$$g_1(x) = g_a(x) - g_a(v_*), \quad \text{then} \quad g_1^{-1}(x) = g_a(x + g(v_*)), \tag{2.38}$$

where $v_* \in (0,1)$.

$g_1(x)$ is also a generator function on a certain aggregative operator. Se we get

$$a_1(x,y) = g_1^{-1}(g_1(x) + g_1(y)) = g_1^{-1}(g(x) + g(y) - g(v_*)) =$$
$$a_1(x,y) = g^{-1}(g(v_*) + (g(x) - g(v_*)) + (g(y) - g(v_*))) = g^{-1}(g(x) + g(y) - g(v_*))$$

For $a_1(x,y)$, the following is valid:

$$a_1(\nu_*,x) = \nu_*.$$

If an aggregative operator has a ν_* neutral value we denote it by $a_{\nu_*}(\mathbf{x})$.

The general form of the n-valued aggregative operator is:

$$a_{\nu_*}(\mathbf{x}) = g^{-1}\left(\sum_{i=1}^{n} g(x_i) - (n-1)g_a(\nu_*)\right) \tag{2.39}$$

$$a_{\nu_*}(\mathbf{x}) = g^{-1}\left(g(\nu_*) + \sum_{i=1}^{n}(g_a(x_i) - g_a(\nu_*))\right)$$

According to this, one can construct an aggregative operator from any given generator function that has the desired neutral value.

There are infinitely many possible neutral values, and with each different neutral value, a different aggregative operator can be given.

Definition 2.9. Let a_f and a_g be aggregative operators, with the additive generator functions f and g, respectively. The functions a_f and a_g belong to the same family if $f(x) = g(x)+d$, for all $x \in [0,1]$ and a suitable $d \subset \mathbb{R}$. Note that a_f and a_g do not necessarily have the same neutral value.

The following theorem shows that there is a one-to-one correspondence between aggregative operators and strong negations. (see [13] as well)

Theorem 2.14. *Let a be an aggregative operator with generator function g and neutral value ν_*. The only strong negation satisfies the self DeMorgan identity is*

$$\eta(x) = g^{-1}(2g(\nu_*) - g(x)). \tag{2.40}$$

Proof. See [17].

Definition 2.10. Let a be an aggregative operator with the additive generator function g and neutral value ν_*. Let us call the strong negation $\eta(x) = g^{-1}(2g(\nu_*) - g(x))$ the corresponding strong negation of the aggregative operator.

By Theorem 2.14 every aggregative operator has exactly one corresponding strong negation, which is a strong negation that fulfils the self DeMorgan identity. Conversely, every strong negation has exactly one corresponding aggregative operator.

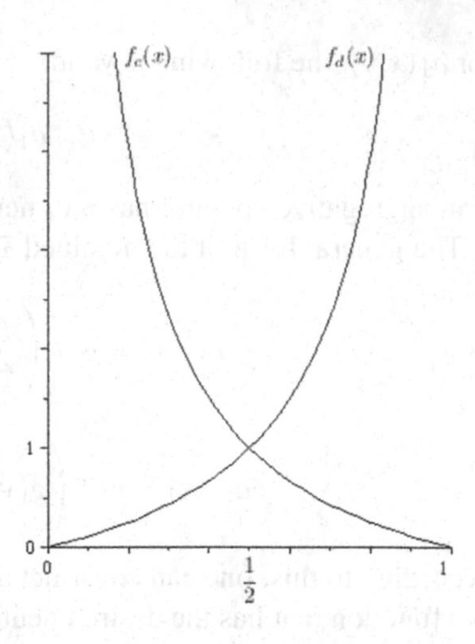

Fig. 2.2 The generator function of the conjunctive and disjunctive operators (additive representation)

2.2.3 On Additive and Multiplicative Representations of Strict Conjunctive and Disjunctive Operators

Let

$$c(x,y) = f_c^{-1}\left(f_c(x) + f_c(y)\right) \qquad d(x,y) = f_d^{-1}\left(f_d(x) + f_d(y)\right),$$

where f_c and f_d are the generator functions of the operators. The shape of these function can be seen in Figure 2.2.

Let

$$g_c(x) = e^{-f_c(x)} \qquad g_d(x) = e^{-f_d(x)} \tag{2.41}$$

Then

$$f_c(x) = -\ln(g_c(x)) \qquad f_d(x) = -\ln(g_d(x)). \tag{2.42}$$

So

$$c(x,y) = f_c^{-1}\left(-\ln(g_c(x)) - \ln(g_c(y))\right) = g_c^{-1}\left(e^{-(-\ln(g_c(x)) - \ln(g_c(y)))}\right)$$

In the next we will study the multiplicative form of the aggregative operator. We will use the transformation defined in (2.41) and (2.42) to get the multiplicative operator

$$a_{\nu_*}(x) = f^{-1}\left(f(\nu_*)\prod \frac{f(x_i)}{f(\nu_*)}\right) = f^{-1}\left(f^{1-n}(\nu_*)\prod f(x_i)\right) \tag{2.43}$$

In the Dombi operator case, we get

$$a_{\nu_*}(\mathbf{x}) = \frac{1}{1 + \frac{1-\nu_*}{\nu_*}\prod\limits_{i=1}^{n}\left(\frac{1-x_i}{x_i}\frac{\nu_*}{1-\nu_*}\right)} \tag{2.44}$$

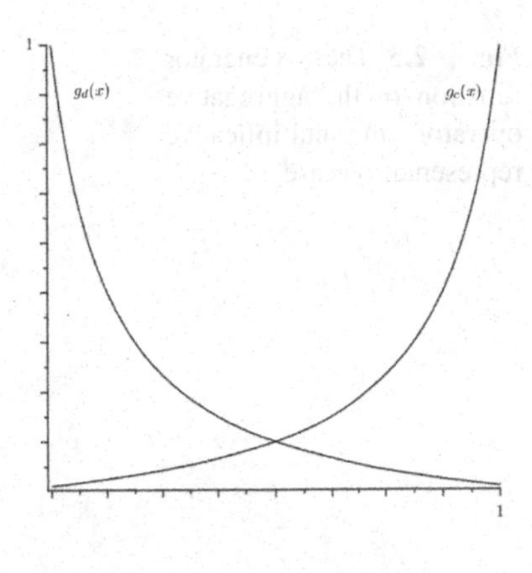

Fig. 2.3 The generator function of the conjunctive and disjunctive operators (multiplicative representation)

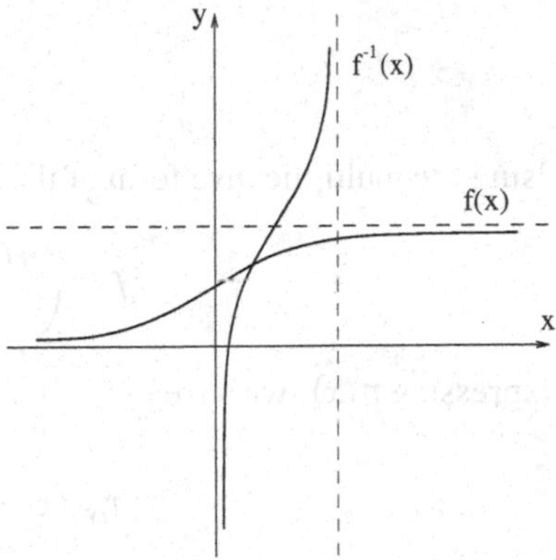

Fig. 2.4 The generator function of the aggregative operator in additive representation case

If $\nu_* = \frac{1}{2}$, then we get

$$a_{\frac{1}{2}}(\mathbf{x}) = \frac{\prod_{i=1}^{n} x_i}{\prod_{i=1}^{n} x_i + \prod_{i=1}^{n} (1 - x_i)}. \tag{2.45}$$

In the next we will make an issue on the negation of the aggregative operator. The basic identity of the aggregative operator is the self-De Morgan identity:

$$\eta(a_{\nu_*}(x,y)) = a_{\nu_*}(\eta(x),\eta(y)). \tag{2.46}$$

Let $y = \eta(x)$. Then

$$\eta(a_{\nu_*}(x,\eta(x))) = a_{\nu_*}(\eta(x),x).$$

Because $a_{\nu_*}(x,y) = a_{\nu_*}(y,x)$, we get

$$a_{\nu_*}(x,\eta(x)) = \nu_*.$$

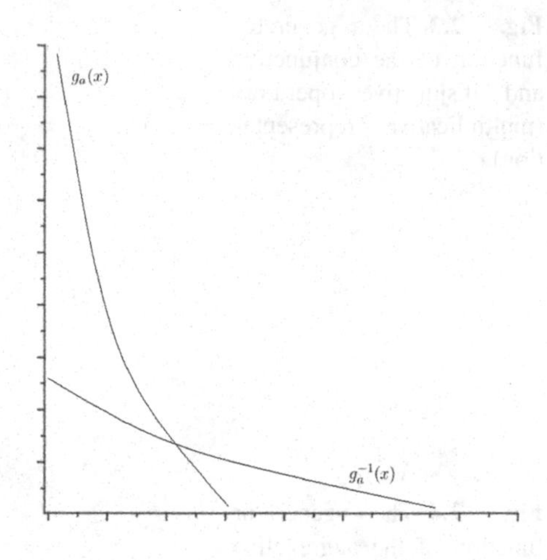

Fig. 2.5 The generator function of the aggregative operator in multiplicative representation case

Using the multiplicative form of the aggregative operator

$$f^{-1}\left(\frac{f(x)f(\eta(x))}{f(\nu_*)}\right) = \nu_*. \tag{2.47}$$

Expressing $\eta(x)$, we have

$$\eta_{\nu_*}(x) = f^{-1}\left(\frac{f^2(\nu_*)}{f(x)}\right). \tag{2.48}$$

In the next we will determine the ν_* value in the aggregative operator case.

Given $x_1, x_2, \ldots, x_n$ and also z, its aggregative value, we can express z in terms of the x_i variables like so:

$$z = f^{-1}\left(f^{1-n}(\nu_*)\prod_{i=1}^{n} f(x_i)\right). \tag{2.49}$$

For ν_*, we have

$$\nu_* = f^{-1}\left(\left(\frac{f(z)}{\prod_{i=1}^{n} f(x_i)}\right)^{\frac{1}{1-n}}\right) = f^{-1}\left(\left(\frac{\prod_{i=1}^{n} f(x_i)}{f(z)}\right)^{\frac{1}{n-1}}\right) \tag{2.50}$$

In the Dombi operator case

$$\nu_* = \frac{1}{1 + \left(\frac{z}{1-z}\prod_{i=1}^{n}\frac{1-x_i}{x_i}\right)^{\frac{1}{n-1}}} \tag{2.51}$$

It is interesting when the self-DeMorgan identity is valid with two different negation operators.

Let us suppose that the following identity holds:

$$a_\nu\left(\eta_{\nu_1}(x), \eta_{\nu_1}(y)\right) = \eta_{\nu_2}(a_\nu(x,y)), \tag{2.52}$$

in analogy to (2.46), where

$$\eta_\nu(x) = f^{-1}\left(\frac{f^2(\nu)}{f(x)}\right). \tag{2.53}$$

Theorem 2.15. *(2.52) is valid if and only if*

$$\nu = \eta_{\nu_1}(\nu_2). \tag{2.54}$$

Proof. See [17].

2.2.4 *The Weighted Aggregative Operator*

The general form of the weighted operator in the additive representation case is

$$a(\mathbf{w},\mathbf{x}) = f^{-1}\left(\sum_{i=1}^{n} w_i f(x_i)\right). \tag{2.55}$$

We will derive the weighted aggregative operators when ν_* is given.

First, we use the construction described by (2.38), i.e.

$$f_1(x) = f_a(x) - f_a(\nu_*) \qquad f_1^{-1}(x) = f_a(x + f(\nu_*)),$$

where $\nu_* \in (0,1)$.

$$a_{\nu_*}(\mathbf{w},\mathbf{x}) = f_a^{-1}\left(\sum_{i=1}^{n} w_i f_a(x_i) + f(\nu_*)\left(1 - \sum_{i=1}^{n} w_i\right)\right) \tag{2.56}$$

To get the multiplicative form of the aggregative operator, we will use (2.42)

$$a_{\nu_*}(\mathbf{w},\mathbf{x}) = f^{-1}\left(f^{1-\sum_{i=1}^{n} w_i}(\nu_*)\prod_{i=1}^{n} f^{w_i}(x_i)\right) \tag{2.57}$$

From (2.2.4) if $\sum_{i=1}^{n} w_i = 1$, then $a_{\nu_*}(\mathbf{w},\mathbf{x})$ is independent of ν_* and

$$a(\mathbf{w},\mathbf{x}) = f^{-1}\left(\prod_{i=1}^{n} f^{w_i}(x_i)\right). \tag{2.58}$$

In the Dombi Operator Case

$$a_{\nu_*}(\mathbf{w},\mathbf{x}) = \frac{\nu_*(1-\nu_*)^{\sum_{i=1}^{n} w_i} \prod_{i=1}^{n} x_i^{w_i}}{\nu_*(1-\nu_*)^{\sum_{i=1}^{n} w_i} \prod_{i=1}^{n} x_i^{w_i} + (1-\nu_*)\nu_*^{\prod_{i=1}^{n} w_i} \prod_{i=1}^{n}(1-x_i)^{w_i}} \tag{2.59}$$

If $\sum_{i=1}^{n} w_i = 1$, then

$$a(\mathbf{w},\mathbf{x}) = \frac{\prod_{i=1}^{n} x_i^{w_i}}{\prod_{i=1}^{n} x_i^{w_i} + \prod_{i=1}^{n}(1-x_i)_i^{w}}. \tag{2.60}$$

Theorem 2.16. *If $\sum_{i=1}^{n} w_i = 1$, then for $a(\boldsymbol{w},\boldsymbol{x})$ the self-De Morgan identity hold for all ν_*, i.e. $a(x,y)$ is independent of the ν_* value.*

Proof. See [17].

Remark 2.8. If $\sum_{i=1}^{n} w_i = 1$, then infinite many negation operators exist for the aggregative operator, which has the self-De Morgan property.

2.3 Infinitely Many Negation When the Self-De Morgan Identity Holds

The self-De Morgan identity is

$$a(w_1,\eta(x_1);\ldots;w_n,\eta(x_n)) = \eta(a(\mathbf{w},\mathbf{x})).$$

Theorem 2.17. *The self identity is valid if and only if*

a) $\sum w_i = 1$, *then* $\eta(x) \neq f^{-1}\left(\frac{1}{f(x)}\right)$

b) $\sum w_i = 1$, *then* $\eta(x) = f^{-1}\left(\frac{c}{f(x)}\right) = f^{-1}\left(f(\nu_0)\frac{f(\nu)}{f(x)}\right)$

Remark 2.9. In case a) infinitely many negations fulfills the self-DeMorgan identity.

Proof. See [17].

2.3.1 Strict t-Norms, t-Conorms and Aggregative Operators

From an application point of view, the strict monotonously increasing operators are useful. They have many applications. This is the reason why in this article we will focus on strictly monotonously increasing operators.

T-norms are commutative, associative and monotone operations on the real unit interval with 1 as the unit element. t-conorms are in some sense dual to t-norms. A t-conorm is a commutative, associative and monotone operation with 0 as the unit element [28]. In this section, besides the min/max and the drastic operators, we will be concerned with strict t-norms and t-conorms, that is,

$$\begin{aligned} c(x,y) &< c(x',y) \quad \text{if} \quad x < x' \quad x,y \in (0,1] \\ d(x,y) &< d(x',y) \quad \text{if} \quad x < x' \quad x,y \in [0,1) \end{aligned}$$

We will call the elements of pliant logic conjunctive, disjunctive and negation operators denote them by $c(x,y)$ and $d(x,y)$ respectively. Those familier with fuzzy logic theory will find that the terminology used here is slightly different from that used in standard texts [28, 9, 3, 6, 40, 22].

The so-called pan operator concept were introduced by Mesiar and Rybárik [39]. The next theorem is based on this result.

In the following we will show that from any strict continuous t-norm or t-conorm we can derive an aggregative operator by changing addition to multiplication in their additive generator functional forms.

The following theorem can be found in Klement, Mesiar and Pap's paper [27].

Theorem 2.18. *The following are equivalent:*

1. $o(x,y) = f^{-1}(f(x)+f(y))$ is a strict t-norm or t-conorm.
2. $a(x,y) = f^{-1}(f(x)f(y))$ is an aggregative operator.

where f is the generator function of the strict t-norm or t-conorm.

Sketch of the Proof. Suppose $o : [0,1]^2 \to [0,1]$ is a strict t-norm or t-conorm with additive generator function $f(x)$. Let $g(x) = \log_s f(x)$, $g^{-1}(x) = f^{-1}(s^x)$. Then $g(x)$ fulfils the conditions of Theorem 2.10 i.e.

$$a(x,y) = g^{-1}(g(x)+g(y)) = f^{-1}(f(x)f(y)) \tag{2.61}$$

is an aggregative operator.

Similarly, let $a : [0,1]^2 \to [0,1]$ be an aggregative operator with additive generator function g, and let $f(x) = s^{g(x)}$. It is easy to see that $f : [0,1] \to [0,\infty]$ is strictly monotone, continuous and either $f(0) = 0$ or $f(1) = 0$. ■

Definition 2.11. Let f be the additive generator of a strict t-norm or t-conorm. Then the corresponding aggregative operator is $a(x,y) = f^{-1}(f(x)f(y))$.

We note that pan-operators, introduced by Wang and Klir [51], with a non-idempotent unit element (see [39] and [49]) have properties not unlike to a corresponding pair of strict t-norm or t-conorm and aggregative operators.

Corollary 2.2. *A strict t-norm or t-conorm is distributive with its corresponding aggregative operator, i.e.*

$$\begin{aligned} a(x,c(y,z)) &= c(a(x,y),a(x,z)), \\ a(x,d(y,z)) &= d(a(x,y),a(x,z)). \end{aligned} \tag{2.62}$$

We can find this and some other similar results in the paper by Ruiz and Torrens [46]. The following statement shows the relationship between the set of strict t-norm or t-conorm and the set of aggregative operators.

Corollary 2.3. *Let $f(x)$ be the additive generator of a strict t-norm or t-conorm. The aggregative operators gene-rated by $f(x)$ and $f_*(x) = cf(x)$ $(c > 0)$ are the same.*

Corollary 2.4. *Let $f(x)$ be a generator function of a strict t-norm or t-conorm and let $f_\alpha(x) = (f(x))^\alpha$ $(\alpha > 0)$. Then its corresponding aggregative operator is independent of α.*

Proof. See [17].

By Theorem 2.18 and Corollaries 2.3 and 2.4, every strict t-norm or t-conorm has infinitely many corresponding aggregative operators because its generator function is determined up to a multiplicative constant. Conversely, every aggregative operator has infinitely many corresponding strict t-norm or t-conorm because a generator function on different powers generates different strict t-norm or t-conorm and identical aggregative operators.

A direct consequence of Theorem 2.12 and Theorem 2.18 is the following representation theorem of strong negations.

Corollary 2.5 (Multiplicative form of strong negations). *The function $\eta : [0,1] \to [0,1]$ is a strong negation with neutral value ν_* if and only if*

$$\eta(x) = f^{-1}\left(\frac{f^2(\nu_*)}{f(x)}\right), \tag{2.63}$$

where f is a generator function of a strict t-norm or t-conorm.

Summarizing the above statements, there is a well-defined correspondence between strict t-norm or t-conorm, aggregative operators and strong negations. Every strict t-norm or t-conorm has corresponding aggregative operators, and corresponding strong negations as well.

The next theorem gives a necessary and sufficient condition for a pair of strict t-norm and strict t-conorm to have identical corresponding aggregative operators.

Theorem 2.19. *Let f_c be an additive generator function of a strict t-norm, and f_d be an additive generator function of a strict t-conorm. Their corresponding aggregative operators a_c and a_d are equivalent on $[0,1]^2 \backslash \{(0,1),(1,0)\}$ if and only if $f_d(x) = f_c^k(x)$, where $k \in \mathbb{R}^-$.*

Proof. See [17].

Corollary 2.6. *Let c and d be a strict t-norm and a strict t-conorm with additive generator functions f_c and f_d. Let a_c and a_d be their corresponding aggregative operators, and let η_c and η_d be their corresponding strong negations. The strong negations η_c and η_d are equivalent if and only if $f_d(x) = f_c^k(x)$, $k \in \mathbb{R}^-$.*

2.3.2 Pliant Operators

If the condition $f_d(x) = f_c^k(x)$ with $k < 0$ is fulfilled then the strict t-norm or t-conorm have a common aggregative operator and strong negation. This set of strict t-norm or t-conorm is still general. DeMorgan's law is a condition which must be fulfilled by a "good" triplet of connectives. Persanding that they satisfy of DeMorgan's law further restricts the given set of strict t-norm or t-conorm.

Theorem 2.20. *Let c and d be a strict t-norm and a strict t-conorm with additive generator functions f_c and f_d. Suppose their corresponding strong negations are equivalent (i.e. $f_d(x) = f_c^k(x)$, $k < 0$), denoted by η $(\eta(\nu_*) = \nu_*)$. The three connectives c,d and n form a DeMorgan triplet if and only if $k = -1$.*

Proof. See [17].

Definition 2.12. A system of strict t-norm or t-conorm which have the property $f_c(x)f_d(x) = 1$ is called a multiplicative pliant system.

In multiplicative pliant systems the corresponding aggregative operators of the strict t-norm and strict t-conorm are equivalent, and DeMorgan's law is obeyed with the (common) corresponding strong negation of the strict t-norm or t-conorm.

It was shown in [16] that the multiplicative pliant system fulfils the DeMorgan identity and the correct strong negation is defined by Eq.(2.68).

For example, let $f_c(x) = -\ln x$, the additive generator of the product operator. Assuming we have a pliant system, $f_d(x) = (-\ln x)^{-1}$ is a valid generator of a strict t-conorm. Their corresponding strong negations are the same as $\eta_c(x) = \eta_d(x) = \eta(x) = \exp[\frac{(\ln(v_*))^2}{\ln x}]$, so that $\eta(1) = \lim_{x\to 1}\eta(x)$, for which $c(x,y) = xy$ and

$$d(x,y) = \exp\left[\frac{\ln x \ln y}{\ln xy}\right] \tag{2.64}$$

form a DeMorgan triplet.

2.3.3 *Summary*

We can summarize the elements of the multiplicative pliant system (operators and their weighted form) like so:

$$c(\mathbf{x}) = f^{-1}\left(\sum_{i=1}^{n} f(x_i)\right) \qquad c(\mathbf{w},\mathbf{w}) = f^{-1}\left(\sum_{i=1}^{n} w_i f(x_i)\right) \tag{2.65}$$

$$d(\mathbf{x}) = f^{-1}\left(\frac{1}{\sum_{i=1}^{n}\frac{1}{f(x_i)}}\right) \qquad d(\mathbf{w},\mathbf{x}) = f^{-1}\left(\frac{1}{\sum_{i=1}^{n}\frac{w_i}{f(x_i)}}\right) \tag{2.66}$$

$$a(\mathbf{x}) = f^{-1}\left(\prod_{i=1}^{n} f(x_i)\right) \qquad a(\mathbf{w},\mathbf{x}) = f^{-1}\left(\prod_{i=1}^{n} f^{w_i}(x_i)\right) \tag{2.67}$$

$$\eta(x) = f^{-1}\left(\frac{f^2(v_*)}{f(x)}\right), \tag{2.68}$$

where $f(x)$ is the generator function of the strict t-norm operator and in (2.67) $a(\mathbf{x})$ is the aggregative operator (i.e. representable uninorms, see [13, 19, 26].)

The Operator System of Dombi. The Dombi operators form a pliant system and the operators are:

$$c(\mathbf{x}) = \frac{1}{1+\left(\sum_{i=1}^{n}\left(\frac{1-x_i}{x_i}\right)^{\alpha}\right)^{1/\alpha}} \qquad c(\mathbf{w},\mathbf{x}) = \frac{1}{1+\left(\sum_{i=1}^{n} w_i\left(\frac{1-x_i}{x_i}\right)^{\alpha}\right)^{1/\alpha}} \tag{2.69}$$

$$d(\mathbf{x}) = \frac{1}{1+\left(\sum_{i=1}^{n}\left(\frac{1-x_i}{x_i}\right)^{-\alpha}\right)^{-1/\alpha}} \qquad d(\mathbf{w},\mathbf{x}) = \frac{1}{1+\left(\sum_{i=1}^{n} w_i\left(\frac{1-x_i}{x_i}\right)^{-\alpha}\right)^{-1/\alpha}} \tag{2.70}$$

$$a(\mathbf{x}) = \frac{1}{1+\prod_{i=1}^{n}\frac{1-x_i}{x_i}} \qquad a(\mathbf{w},\mathbf{x}) = \frac{1}{1+\prod_{i=1}^{n}\left(\frac{1-x_i}{x_i}\right)^{w_i}} \tag{2.71}$$

$$\eta(x) = \frac{1}{1+\left(\frac{1-\nu_*}{\nu_*}\right)^2 \frac{x}{1-x}} \tag{2.72}$$

where $\nu_* \in]0,1[$, with generator functions

$$f_c(x) = \left(\frac{1-x}{x}\right)^{\alpha} \qquad f_d(x) = \left(\frac{1-x}{x}\right)^{-\alpha} \tag{2.73}$$

where $\alpha > 0$. The operators c, d and η fulfil the De Morgan identity for all ν, a and η fulfill the self De Morgan identity for all ν and the aggregative operator is distributive with the logical operators.

Eqs.(2.69), (2.70), (2.71), (2.72) can be found in different articles by Dombi. Eqs.(2.69) and Eq.(2.70) can be found in [12], (2.71) in [13] and Eq.(2.72) can be found in [14]. These are all previous results by Dombi.

Eq.(2.71) is called 3π operator because it can be written in the following form:

$$a(\mathbf{x}) = \frac{\prod_{i=1}^{n} x_i}{\prod_{i=1}^{n} x_i + \prod_{i=1}^{n}(1-x_i)} \tag{2.74}$$

The main results of this article can be summarized in the following way.

Given a strict t-norm c and a strict t-conorm d with generators f_c, f_d the paper determines the conditions for which a strong negation η exists such that c, d, η form a DeMorgan triple. To this end, a helper negation function $k : [0,\infty] \to [0,\infty]$ is required. For one particular $k(x) = \frac{1}{x}$ the conditions on f_c, f_d are given such that c, d, η form a DeMorgan triple, where η was obtained using k.

- We employ weighted operators. See eqs.(2.6), (2.7).
- We provide an involutive negation operator given $f_c(x)$ and $f_d(x)$. See in Eq.(2.15).
- We give the general form of the DeMorgan triplet using the $k(x)$ function. See eqs.(2.21), (2.22), (2.23).
- We give the parametric form of the negation operator. See eqs. (2.25), (2.26).
- We show that the DeMorgan triplet has infinitely many negation operators if and only if $f_c(x) f_d(x) = 1$ (the main result) and such a system is called a pliant system. This condition is the same if the representable uninorm (aggregative operator) corresponds to the strict t-norms and strict t-conorms.

- We give the general form of the pliant operators. See eqs.(2.65), (2.66), (2.67), (2.68).
- We show that consistent aggregation can be achieved.
- The special case of the pliant system is the Dombi operator class.
- In the second part of the chapter we demonstrated the equivalence of the class of representable uninorms and the class of uninorms that are also aggregative operators.
- Three new representation theorems of strong negations were given.
- The relationships for strict, continuous operators, aggregative operators and strong negations were clarified.
- We show the correspondence between the elements of the three classes.
- The concept of multiplicative pliant system was introduced in the aggregative case.
- The multiplicative pliant system was characterized by necessary and sufficient conditions.

Acknowledgements. This study was partially supported by the TÁMOP-4.2.1/B-09/1/KONV-2010-0005 program of the Hungarian National Development Agency.

References

1. Aczél, J.: Lectures on Functional Equations and Applications. Academic Press, New York (1966)
2. Aguil, I., Suner, J., Torrens, J.: A characterization of residual implications derived from left-continuous uninorms. Information Sciences 180(20), 3992–4005 (2010)
3. Alsina, C., Schweizer, B., Frank, M.J.: Associative functions: triangular norms and copulas. Word Scientific Publishing, Singapore (2006)
4. De Baets, B., De Meyer, H., Mesiar, R.: Piecewise linear aggregation functions based on triangulation. Information Sciences (2010) (in Press)
5. De Baets, B., Fodor, J.: Residual operators of uninorms. Soft Computing 3, 89–100 (1999)
6. Beliakov, G., Pradera, A., Calvo, T.: Aggregation Functions: A Guide for Practitioners. Studies in Fuzziness and Soft Computing, vol. 221. Springer, Heidelberg (2007)
7. Boulkroune, A., M'Saad, M., Chekireb, H.: Design of a fuzzy adaptive controller for MIMO nonlinear time-delay systems with unknown actuator nonlinearities and unknown control direction. Information Sciences 180(24), 5041–5059 (2010)
8. Calvo, T., Baets, B.D., Fodor, J.: The functional equations of frank and alsina for uninorms and nullnorms. Fuzzy Sets and Systems 120, 385–394 (2001)
9. Calvo, T., Mayor, G., Mesiar, R.: Aggregation Operators. New Trends and Applications Studies in Fuzziness and Soft Computing 97 (2002)
10. Calvo, T., Mesiar, R.: Weighted triangular norms-based aggregation operators. Fuzzy Sets and Systems 137(1), 3–10 (2003)
11. Cintula, P., Klement, E.P., Mesiar, R., Navara, M.: Fuzzy logics with an additional involutive negation. Fuzzy Sets and Systems 161, 390–411 (2010)
12. Dombi, J.: General class of fuzzy operators, the demorgan class of fuzzy operators and fuzziness included by fuzzy operators. Fuzzy Sets and Systems 8, 149–168 (1982)

13. Dombi, J.: Basic concepts for a theory of evaluation: The aggregation operator. European Journal of Operation Research 10, 282–293 (1982)
14. Dombi, J.: Towards a General Class of Operators for Fuzzy Systems. IEEE Transaction on Fuzzy Systems 16, 477–484 (2008)
15. Dombi, J.: DeMorgan systems with an infinitely many negations in the strict monotone operator case. Information Sciences (under print, 2011)
16. Dombi, J.: DeMorgan systems with infinite number of negation. Information Science (2010) (under Review Process)
17. Dombi, J.: On a certain class of aggregative operators. Information Sciences (2011) (under Refree Process)
18. Esteva, F.: Some Representable DeMorgan Algebras. Journal of Mathematical Analysis and Applications 100, 463–469 (1984)
19. Fodor, J., Yager, R.R., Rybalov, A.: Structure of uninorms. International Journal of Uncertainty, Fuzziness and Knowledge-Based Systems 5(4), 411–427 (1997)
20. Garcia, P., Valverde, L.: Isomorphisms between DeMorgan triplets. Fuzzy Sets and Systems 30, 27–36 (1989)
21. Gehrke, M., Walker, C., Walker, E.: DeMorgan Systems on the Unit Interval. International Journal of Intelligent Systems 11, 733–750 (1990)
22. Grabisch, M., Marichal, J.-L., Mesiar, R., Pap, E.: Aggregation Functions. Encyclopedia of Mathematics and Its Applications, vol. 127. Cambridge University Press, Cambridge (2009)
23. Grabisch, M., Marichal, J.-L., Mesiar, R., Pap, E.: Aggregation functions: Means. Information Sciences (2010) (in Press)
24. Grabisch, M., Marichal, J.-L., Mesiar, R., Pap, E.: Aggregation functions: Construction methods, conjunctive, disjunctive and mixed classes. Information Sciences (2010) (in Press)
25. Guadarrama, S., Ruiz-Mayor, A.: Approximate robotic mapping from sonar data by modeling perceptions with antonyms. Information Sciences 180(21), 4164–4188 (2010)
26. Hu, S.-K., Li, Z.-F.: The structure of continuous uni-norms. Fuzzy Sets and Systems 124, 43–52 (2001)
27. Klement, E.P., Mesiar, R., Pap, E.: On the relationship of associative compensatory operators to triangular norms and conorms. Uncertainty, Fuzziness and Knowledge-Based Systems 4, 129–144 (1996)
28. Klement, E.P., Mesiar, R., Pap, E.: Triangular norms. Kluwer, Dordrecht (2000)
29. Klement, E.P., Mesiar, R., Pap, E.: Triangular norms. Position paper I: basic analytical and algebraic properties 143(1), 5–26 (2004)
30. Klement, E.P., Mesiar, R., Pap, E.: Triangular norms. Position paper II: general constructions and parameterized families 145(3), 411–438 (2004)
31. Klement, E.P., Mesiar, R., Pap, E.: Triangular norms. Position paper III: continuous t-norms 145(3), 439–454 (2004)
32. Kolesrov, A., Mesiar, R.: Lipschitzian De Morgan triplets of fuzzy connectives. Information Sciences 180(18), 3488–3496 (2010)
33. Ling, C.M.: Representation of associative functions. Publ.Math. Debrecen 12, 189–212 (1965)
34. Li, Y.-M., Shi, Z.-K.: Weak uninorm aggregation operators. Information Sciences 124, 317–323 (2000)
35. Li, Y.M., Shi, Z.K.: Remarks on uninorms aggregation operators. Fuzzy Sets and Systems 114, 377–380 (2000)
36. Li, J., Mesiar, R., Struk, P.: Pseudo-optimal measures. Information Sciences 180(20), 4015–4021 (2010)

37. Mas, M., Mayor, G., Torrens, J.: The distributivity condition for uninorms and t-operators. Fuzzy Sets and Systems 128, 209–225 (2002)
38. Mas, M., Monserrat, M., Torrens, J.: The law of importation for discrete implications. Information Sciences 179(24), 4208–4218 (2009)
39. Mesiar, R., Rybárik, J.: Pan-operations structure. Fuzzy Sets and Systems 74, 365–369 (1995)
40. Mesiar, R., Kolesrov, A., Calvo, T., Komornkov, M.: A Review of Aggregation Functions. In: Fuzzy Sets and Their Extension: Representation, Aggregation and Models. Studies in Fuzziness and Soft Computing, vol. 220 (2008)
41. Monserrat, M., Torrens, J.: On the reversibility of uninorms and t-operators. Fuzzy Sets and Systems 131, 303–314 (2002)
42. Mostert, P.S., Shields, A.L.: On the structure of semi-groups on a compact manifold with boundary. Ann. Math. II. Ser. 65, 117–143 (1957)
43. Nguyen, H.T., Walker, E.: A First Course in Fuzzy Logic. CRC Press, Boca Raton (1997); 2nd. edn. (1999), 3rd edn. (2006)
44. Ouyang, Y., Fang, J.: Some results of weighted quasi-arithmetic mean of continuous triangular norms. Information Sciences 178(22), 4396–4402 (2008)
45. Roychowdhury, S.: New triangular operator generators for fuzzy systems. IEEE Trans. of Fuzzy Systems 5, 189–198 (1997)
46. Ruiz, D., Torrens, J.: Distributivity and Conditional Distributivity of a Uninorm and a Continous t-Conorm. IEEE Transactions on Fuzzy Systems 14(2), 180–190 (2006)
47. Schweizer, B., Sklar, A.: Probabilistic Metric Spaces. North-Holland, New York (1983)
48. Silvert, W.: Symmetric summation: a class of operations on fuzzy sets. IEEE Transactions on Systems, Man and Cybernetics 9, 659–667 (1979)
49. Tong, X., Chen, M., Li, H.: Pan-operations structure with non-idempotent pan-addition. Fuzzy Sets and Systems 145, 463–470 (2004)
50. Trillas, E.: Sobre funciones de negación en la teoría de conjuntos difusos. Stochastica III, 47–60 (1979)
51. Wang, Z., Klir, G.J.: Fuzzy Measure Theory. Plenum Press, New York (1992)
52. Waegeman, W., De Baets, B.: A transitivity analysis of bipartite rankings in pairwise multi-class classification. Information Sciences 180(21), 4099–4117 (2010)
53. Yager, R.R., Rybalov, A.: Uninorm aggregation operators. Fuzzy Sets and Systems 80(1), 111–120 (1996)
54. Yager, R.R.: Uninorms in fuzzy systems modeling. Fuzzy Sets and Systems 122, 167–175 (2001)
55. Yager, R.R.: Weighted triangular norms using generating functions. International Journal of Intelligent Systems 19(3), 217–231 (2004)
56. Zhang, K.-L., Li, D.-H., Song, L.-X.: Solution of an open problem on pseudo-Archimedean t-norms. Information Sciences 178(23), 4542–4549 (2008)
57. Zimmermann, H.J., Zysno, P.: Latent connectives in human decision making. Fuzzy Sets and Systems 4, 37–51 (1980)

Chapter 3
Coordinated and Recorded Human Interactions for Enhanced Intelligence in Product Model

László Horváth and Imre J. Rudas

Abstract. Product definition on the basis of STEP IPIM, ISO 10303 product model standard grounded a new technology in engineering. One of the main trends in product modeling is improving communication between human and modeling procedure by including intelligent computing methods both in modeling procedures and model descriptions. This is a great challenge because definition of product objects must be transparent for engineers at any time in an industrial engineering process. In this chapter, the authors introduce one of their contributions in knowledge assisted intelligent control of engineering object definition at product model based engineering activities. The proposed modeling is devoted as an extension to currently prevailing product modeling in leading product life cycle management systems. Chapter starts with an introduction to the proposed modeling by its preliminaries in work of the authors and others. As the main contribution, transferring knowledge from the human product definition process to the product model, method for better communication at object definition, new content definition for engineering objects, and multilevel structure in order to facilitate implementation are discussed. In the main methodology of the proposed modeling contextual human intent, engineering object, and decision entities are applied in the form of spaces extending the conventional model space in product models.

3.1 Introduction

Engineering has been changed by the introduction of powerful and efficient product modeling that produced highly integrated product descriptions utilizing recent

László Horváth · Imre J. Rudas
University of Óbuda, John von Neumann Faculty of Informatics, Institute of Intelligent Engineering Systems, H-034 Budapest, Bécsi u. 96/b, Hungary
e-mail: horvath.laszlo@nik.uni-obuda.hu, rudas@uni-obuda.hu

J. Fodor et al. (Eds.): Recent Advances in Intelligent Engineering Systems, SCI 378, pp. 59–78.
springerlink.com

achievements in informatics and its technology. On the road to recent product life-cycle management (PLM) systems milestones were among others partial solutions in CAD/CAM, CAE and other systems, the STEP IPIM (Standard for the Exchange of Product Model Data, Integrated Product Information Model) ISO 10303 product model standard, and the ISO STEP FFIM (Form Feature Information Model) for form feature definition.

By now, including intelligence in the product definition practice is a reality. One of the main trends in product modeling is including intelligent computing at object definition methods both in modeling procedures and model descriptions. The main objective is improving communication between human and modeling procedure. Despite numerous approaches in the literature, it is hard to define that what the purpose of intelligence is in engineering systems. Intelligence in product model is urged by virtual prototyping for product evaluation at the extending virtual stage of product development. Other challenges such as flexible configuration of product, well-engineered construction, handling frequent product changes, and engineering for short innovation cycles stimulate need for intelligent product definition.

Achievements in product modeling established a new technology in engineering where highly integrated object model serves engineering activities for the life cycle of a product. The authors of this chapter analyzed the latest product modeling technology and recognized several new characteristics. Starting from these recognitions, they developed a methodology for knowledge transfer from human to product model. Concept of this methodology is producing a contribution to currently prevailing product modeling and industrial product PLM systems.

Integration of all embedded mechanical, electrical, electronic, hardware, and software subsystems of products in a single model caused a demand for the application of different engineering disciplines at the same modeling process. Advanced product modeling, life cycle management of product information, connection of this virtual world with a related physical world, and advanced simulations constitute a new environment that is ready to accept intelligent engineering methods.

Communication between two engineers is outlined in Fig. 3.1. Engineer A is an authorized and responsible human interacting product object handling processes in order to define product objects according to own intent. Object handling processes together with knowledge support processes create and modify objects in product model for life cycle engineering. These processes apply approved knowledge sources. Authorized and responsible human Engineer B understands product object definitions by Engineer A. Work of Engineer B includes relating new objects to objects defined by Engineer A and sometimes modification of those objects according to changed circumstances. For this purpose, Engineer B applies knowledge about intent and thinking process on the objects defined by Engineer A. Consequently, product model must serve as an advanced medium for the communication between engineers.

Knowledge representation and knowledge based problem solving are constantly developing since eighties in the past century. However, most of the achievements are realized in processes those are closed for human interaction during generation of a result. In an industrial engineering system, definition of product entities must be transparent for responsible engineers at any time. Knowledge must be locally verified and it is always legal property of a project, a product, a company, or an engineer. The above characteristics make application of knowledge based problem solving at product modeling to a very specific task. This area still needs development of new and robust product modeling methods. In this chapter, the authors introduce one of their contributions to these efforts.

3.2 Preliminaries

The authors of this chapter analyzed product modeling procedures, product model structures and entities as well as application of product model at problem solving in industrial PLM systems. They published the results in [6] and applied them at an

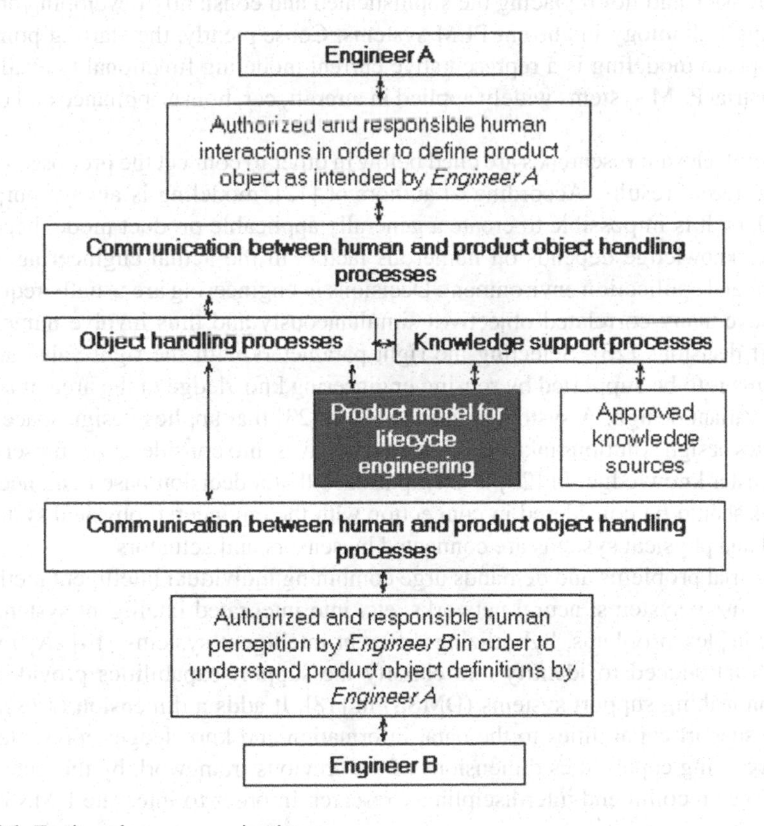

Fig. 3.1 Engineering communication

analysis for possibilities to include problem solving centered knowledge in product models in order to establish better knowledge assistance for development and application of product models. They proposed a knowledge based extension to currently prevailing object based product models in the form of new content as mapped to engineering objects [7]. As a step towards intelligent definition of product objects, processes based on the new content take the direct control of engineering objects from engineers. The authors developed method for the definition of engineering objectives in the form of representation of situation dependent behaviors. They applied an extended behavior definition [9] and prepared their earlier elaborated human intent modeling for human intent based behavior definition [24]. In paper [4], the authors of this chapter characterized intelligent engineering features in current industrial product modeling systems, discussed possibilities for adding intelligence to modeling by the definition of human intent driven engineering objectives and contextual connections of engineering objects.

In the area of product modeling new concepts and methods are mostly realized as application developments of professional PLM products. The method introduced in this chapter is also planned as an extension to industrial product modeling systems. Purpose of the criticism of current product modeling serves the development of this extension and not replacing the sophisticated and constantly developing object modeling technology in current PLM systems. Consequently, the starting point of the proposed modeling is a representative current modeling functionality available in industrial PLM systems widely applied in aircraft, car, house appliance and other industries.

Several relevant researches are cited below in order to connect the proposed modeling to recent results. According to authors of [12], modeling is always purpose dependent. It is impossible to create a generally applicable product model because building knowledge depends on numerous factors in the actual engineering, production, and application environment. Decisions in engineering are usually required to achieve many correlated objectives simultaneously and thus involve numerous tradeoff decisions [23]. Selecting the right parameters with the right value at the right time is to be supported by reusing engineering knowledge in the area of adaptive or variant design. A system is introduced in [23] that applies design space and generates design solutions taking all design objectives into consideration, preserving and transfer knowledge. In [22] it is emphasized that a decision based engineering systems should be considered in connection with the engineered physical systems. Virtual and physical systems are connected by sensors and actuators.

Industrial problems and demands urge combining individual intelligent methods such as fuzzy systems, neural networks, etc. into integrated intelligent systems to solve complex problems, hybridizing different intelligent systems [14]. A framework is introduced to identify and classify the support capabilities provided by decision making support systems (DMSS) in [18]. It adds a dimension of user interface support capabilities to the data, information and knowledge representation and processing capabilities dimensions of the previous framework by the same authors. They recommend interdisciplinary research in order to integrate DMSS and knowledge based concepts to allow for more complex representations for data, information, an knowledge models in intelligent processing.

Authors of [21] show the personal and organizational characteristics of knowledge and propose method for its handling in engineering processes. In [8], personalization and codification are emphasized in the development of a multidisciplinary framework for engineers. In [20], multiple expert sources are proposed in engineering as a more feasible alternative.

In [9] the ontological aspect of intelligent CAD systems is analyzed. This is in close connection with the proposed modeling, especially in case of communication demand with connected systems for different purposes. Ontological aspect is in close connection with change management in product modeling. The authors of this chapter grounded their method for human intent based management of changes in [19]. Representation of engineering objectives as behaviors is based on modeling of situations for the handling of product functions. Product related situational modeling and control are introduced in [1]. Product modeling is also in close connection with enterprise modeling. As an example, authors of [2] analyze enterprise modeling and introduce intelligent methods.

3.3 Knowledge in Product Model

One of the basic concepts of the proposed modeling is bringing knowledge from the human product definition process to the model. In this way, the knowledge in the model can be applied at modification of the related product objects and at relating new objects to these objects. At product definition Engineer A (Fig. 3.2) applies knowledge from different sources such as documents, handbooks, experience, and mind (KLA). Engineer A includes knowledge in the product model to define object parameters and their relationships. Currently, most of the object parameters and relationships are defined directly because knowledge is not available in a form that can be included in the product model or product model is not in the possession of suitable capabilities for knowledge representation. Typically, most of the knowledge (KLrA) remains at Engineer A. Other knowledge is communicated with knowledge support processes by Engineer A (KLcA.). Object handling processes work together with knowledge support processes and utilize the communicated knowledge. Capabilities of product modeling for knowledge representation do not allow including all communicated knowledge in the product model. Some of this knowledge remains at knowledge support (KLrs). As a result of this knowledge utilization process, knowledge is not communicated with Engineer B among others about way to decisions, allowed modifications, strength of decisions, and relationship definitions. Knowledge must be originated from locally approved sources.

Despite high level automation by using of product definition and application of intelligent engineering methods, authorized human must have absolute control above product model generation processes. A product object definition is normally affected by several humans, directly or indirectly. Human affects and their main connections are summarized in Fig. 3.3. Direct affecting responsible engineers, deciding supervisors, experts, and scientists control product definition, accept works, and makes decisions. Indirect affects are realized through authorized human defined company strategies, law, and standards. Compliance with them is mandatory for engineers who are responsible for the relevant product definition. Earlier proved and accepted results also represent indirect affects such as stored product models, model units, features, generic object models, and experience in knowledge entities.

The next question is how to include knowledge in product model (Fig. 3.4). In order to keep absolute control over model object definition process, application of any knowledge must be decided by authorized engineer who is responsible for the included knowledge. No general, book, journal or other unproved and unaccepted knowledge is allowed. At the same time, engineer who is responsible for the relevant

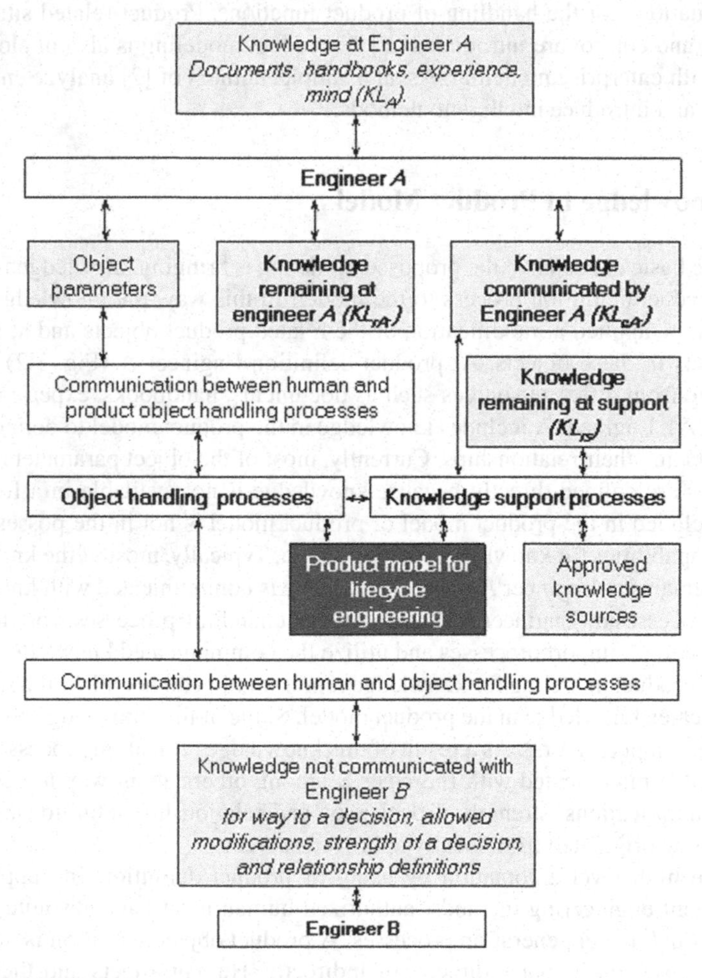

Fig. 3.2 Utilization of knowledge in product model

product definition may be authorized to accept and prove knowledge from these and similar sources. Consequently, any knowledge entity is emerged at an authorized human who directly applies or links it to the product model. Knowledge is considered as a medium of communication between human and product object handling processes and acts as advanced form of relating object parameters and their groups. Knowledge support processes coordinate and process knowledge for embedding in the product model. In case of knowledge for embedding in product model, knowledge source is identified and characterized by its origination, owner, purpose, and weight. Following this, condition information is placed in the product model for the application of knowledge such as context, parameter range, product view, authorization, and responsibility. Finally, knowledge entity is recorded in the product model in one of the representations available in the actual PLM system.

Formulas, rules, checks, reactions are easy to understand for engineers. A rule or rule set is applied to calculate parameters depending on situation. A check or check set is applied to recognize situation. Reaction is applied to react defined event by defined activity. Fuzzy aggregates, neural networks, Petri nets, genetic algorithms,

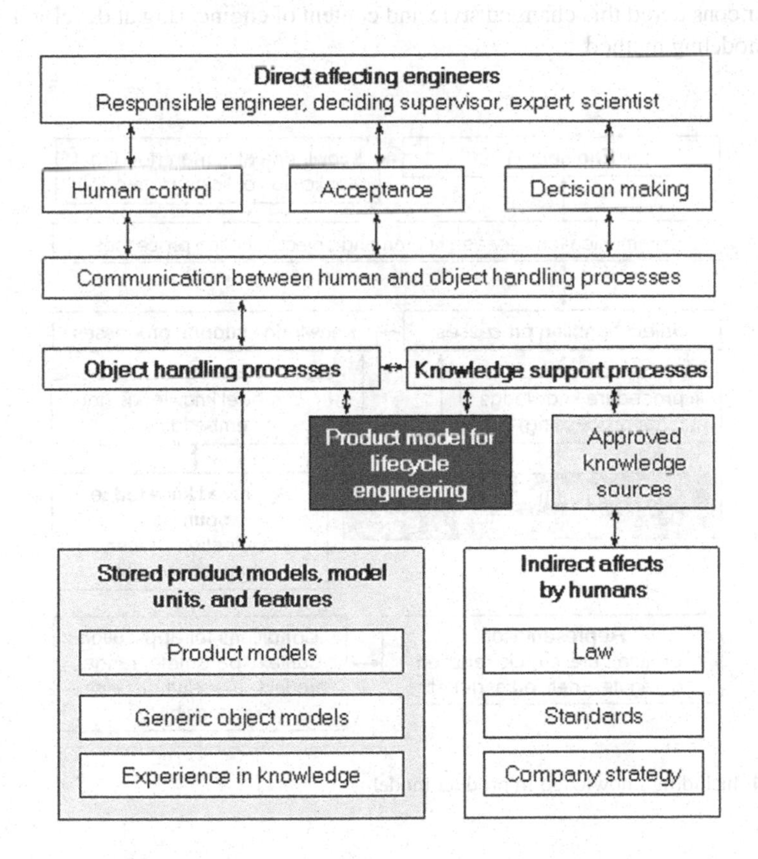

Fig. 3.3 Scenario of human affects

optimizing procedures, and their combinations represent advanced methods from the intelligent computing.

Quality of the engineering activities is protected by a minimum level of knowledge, known as threshold knowledge that is mandatory to consider. Generally applicable part of this knowledgev is embedded in object handling procedures. The product depended part is embedded in product model. By using these knowledge records, object handling processes enforce threshold knowledge. Sometimes a knowledge entity is not allowed to include in procedures or models because its owner keeps it at own environment. In this case expert contribution request is needed to start at any application.

The above mechanism of knowledge utilization may be considered as an overcomplicated procedure. However, property of knowledge and application of the best practice and science at any work are critical for success of products and avoiding legal consequences in the industry. Leading companies collect proven practice in knowledge ware of PLM systems. Because model construction was substantially simplified in industrial PLM systems during the past decade, task of engineer is to compose existing results and add some new contribution to them. The authors of this chapter considered this changed style and content of engineering at development of their modeling method.

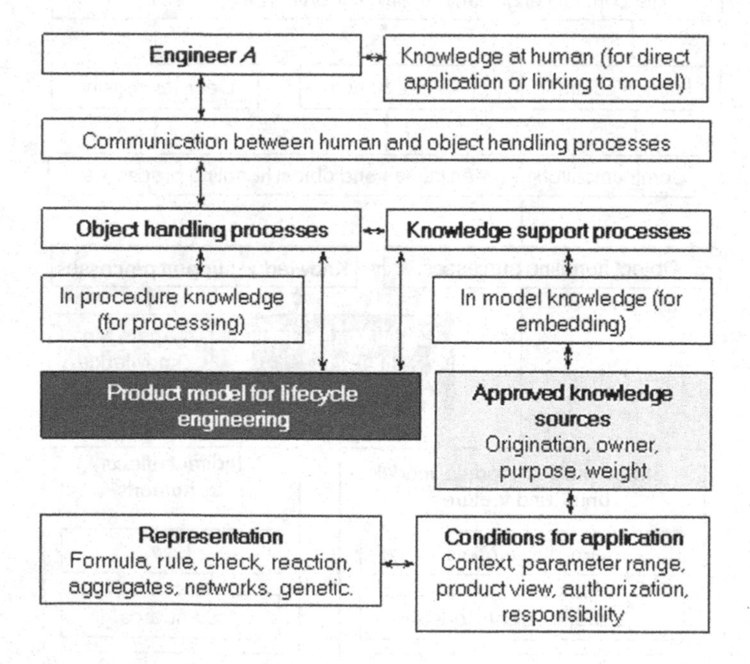

Fig. 3.4 Including knowledge in product model

3.4 Definition of Product Objects

Product model serves engineering for life cycle and it is traditionally shape centered. Shape is represented in a three dimensional Cartesian space called as model space. Environment of product modeling is constituted by human influence environment and physical environment (Fig. 3.5). Human influence environment is connected with object and knowledge handling processes for object definition, simulation definition, and interactive model space visualization for humans. Simulations are organized for virtual prototyping. Interactive space visualization presently means interactive computer graphics that projects three dimensional model space onto the two dimensional viewport. Recent efforts are aimed to develop three dimensional interactive visualization spaces. However, the results are still not available for the every day engineering applications. Physical environment connection serves two way information exchange between the capturing and controlled equipment and the model space. Application of individual and organized sensors and shape capturing devices is relative new achievement. The method introduced in this chapter serves improved human-computer communication at engineering object definition.

Figs. 3.6 and 3.7 explain the currently prevailing and the proposed methods for object definition in product modeling, respectively. In the product model of current PLM systems (Fig. 3.6) product objects are divided into engineering (EO1EOn) and

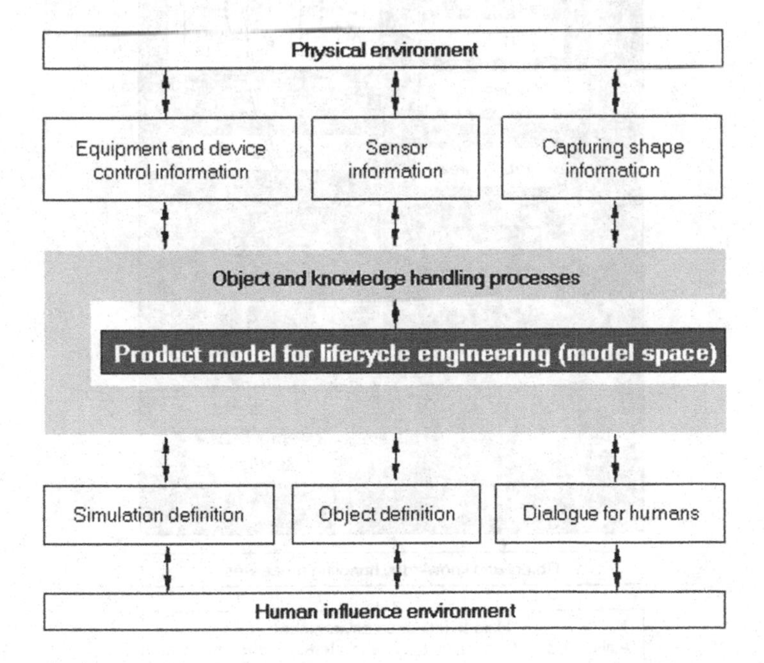

Fig. 3.5 Environment of product modeling

knowledge (Ke1Kem) objects. Knowledge objects relate engineering objects and are mapped to them.

All of the objects are arranged in a common product structure [6]. On the other hand, engineering objects are organized in taxonomy. While engineering objects are applied for product definition and represent units, parts, analysis results, control programs, etc., knowledge objects are applied at the generation of engineering objects.

In a knowledge object, relation sets organize knowledge representation entities. These relation sets are applied for given parameters of given engineering objects in case of a situation defined by engineering object parameters. In this schema, environment affect objects are also defined as engineering object. In Fig. 3.6, rules, checks, and reactions are exampled as knowledge representation entities.

In the object definition proposed by the authors of this chapter (Fig. 3.7) the direct human control of the object and knowledge handling processes is replaced by

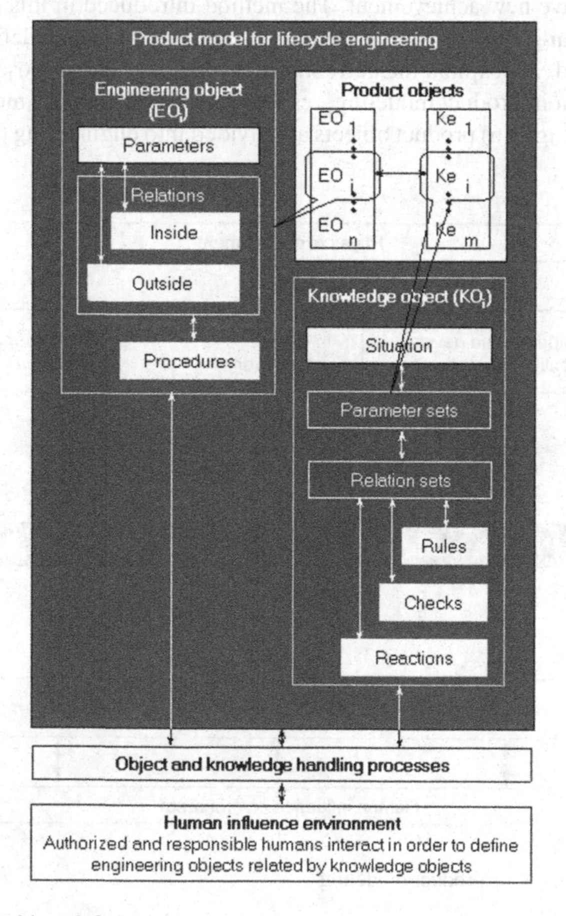

Fig. 3.6 Current object definition practice

new content based adaptive object definition. For this reason, new content definition for engineering objects and new content definition processes are to be included in the modeling. The proposed new content definition for engineering objects includes organized contextual structure of human intent, engineering object, and decision entities. The new elements of the modeling make it possible to record important information from the human thinking process for the definition of engineering objects. Knowledgev is communicated by humans with the new content definition processes. It is embedded in human thinking process in order to give the capability for product model to be information source for reconstruction of human intent at application of product model.

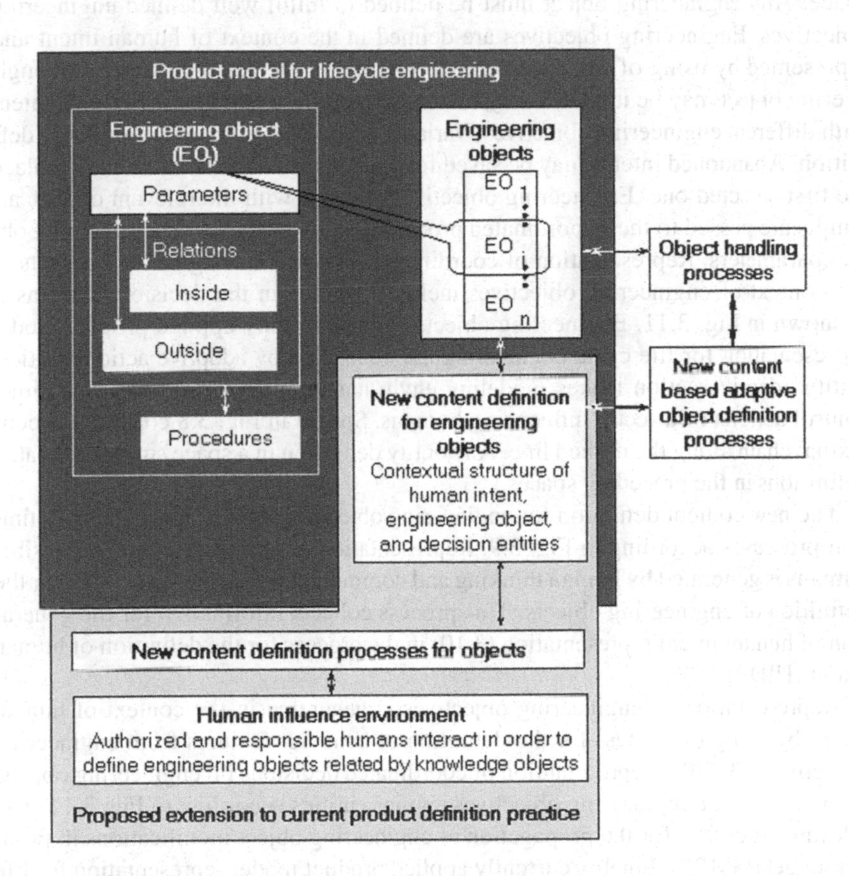

Fig. 3.7 The proposed object definition

3.5 Representation of New Content Definition

As it can be seen from the comparison of Fig. 3.6 and Fig. 3.7, content definition is integrated with the currently applied product modeling. Representation of the new content entities, their mapping to engineering objects, and connection of the extension to the currently applied modeling are explained in Fig. 3.8. Representation of authorized and responsible humans is mapped to engineering objects in the influence space. The purpose is identification of owners of human intent entities. Human contributes to model construction by change attempt. When this attempt is well defined, it serves one or more of the engineering objectives , and the consequences of the attempted change on indirectly affected engineering objects are allowable, the attempt is executable. The way to this status of an attempt leads through the representation chain in the Fig. 3.8. Representation of human intent carries knowledge for engineering object definition. Human intent entities are placed in the human intent space. Any engineering object must be defined to fulfill well defined engineering objectives. Engineering objectives are defined in the context of human intent and represented by using of extended behavior entities in the behavior space. An engineering object may be influenced by several, sometimes contradicting human intent with different engineering objectives. Variants are processed by behavior entity definition. Abandoned intents may be saved for changed situation where it may replace the first selected one. Engineering objectives together with the relevant change attempts are passed to their coordinated processing into decisions on engineering object parameters. Representation of coordinated decisions on engineering objects in the context of engineering objectives includes entities in the decision space as it is shown in Fig. 3.11. Engineering objects of the currently applied product model representation for life cycle engineering are controlled by adaptive action entities. Entity adaptive action makes modeling environment adaptive and replaces direct control actions of directly influencing humans. Spaces in Fig. 3.8 constitute a contextual chain along the dashed line. Any entity definition in a space supposes related definitions in the preceding spaces.

The new content definition for engineering objectsv requires new content definition processes according to Fig. 3.9. Representation of authorized and responsible humans is generated by human thinking and communication process (HTCP) for the definition of engineering objects. This process collects information for the generation of human intent representation (3.10) in the process for the definition of human intent (HIDP).

Representation of engineering objectives is generated in the context of human intent by using of process for the definition of behavior features from engineering objectives (EODP). Representation of coordinated decisions on engineering objects in the context of engineering objectives applies entities according to Fig. 3.11. The relevant process is for the propagation of engineering object modifications in product model (OMPP). Finally, currently applied product model representation for life cycle engineering is controlled by adaptive action with executable status. This control is done by using of the process for adaptive content based control of engineering objects (EOCP).

The human intent related entities are placed in the influence and human intent spaces as it is shown in Fig. 3.10. Authorization and access is a developed strategy for the definition of allowed and agreed human activity areas in PLM environments. It can be connected with the related product data management (PDM) information. PDM system constitutes environment for product modeling at handling of product data, definition of engineering processes, and management of group work and engineering project tasks. Implementation of the proposed modeling must be extended to this area in PLM systems.

Status of human carries important information for the evaluation of change attempt including knowledgev definition proposals. It inform about strength of human through role and position of humans. Characteristics of intent definition include its status and purpose. These entities were proposed in a previous work of the authors [19]. They represent related concepts from long experience in industry. Choice must be elaborated and verified in the application environment as part of the corporate knowledge.

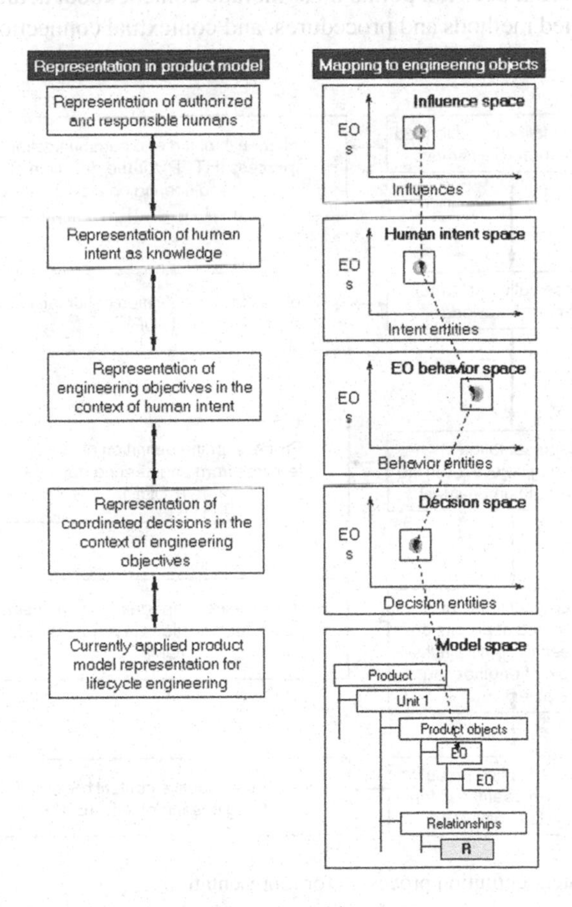

Fig. 3.8 New content definition for engineering objects

Considering the represented content, an opinion for example may have essential affect on product quality despite its weakness in comparison with a decision. Allowable range of a parameter and similar range type purposes make decision easier for the engineer who applies the product model.

The main part of human intent definition is extraction of content from the human thinking process. Because any engineering objectb definition in the product model is in context of human intent, this part of the model essential and critical. Moreover, this is the point of human intervention in an increasingly automated intelligent object definition in the future.

Relevant elements of human thinking process must include any content that is necessary to evaluate, change, or connect the related intent on engineering object definition. It is an answer for the question that what engineer did during engineering object definition. Most important content entities are considered circumstances, emerged and selected methods, as well as applied knowledge and experience. Recorded elements of human thinking process are in the background of partial or interim decision point records. Partial decision points must include content about actual engineering objectives, applied methods and procedures, and contextual connections.

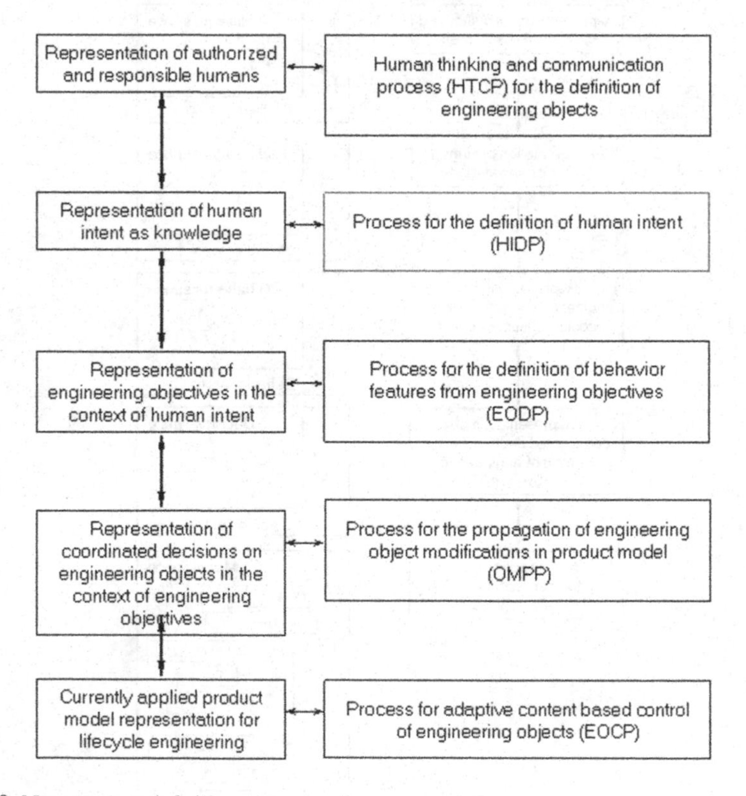

Fig. 3.9 New content definition processes for representations

In order to better understand the motivation of the human intent based engineering object definition, attention of reader should draw for two important characteristics of the current industrial product modeling systems. One is collecting knowledge in knowledge ware both for application at construction of the actual product model and for future application as experience, proven practice, and scientific content. Second is application of the feature principle for all modeled objects. Any engineering and knowledge object acts as a feature to modify an existing product model. The proposed method applies these well-proven modeling methods.

Decision space entities serve decision on change attempts for engineering objects including modification of object structure by modeling functions for insert, delete, replace, activate, and inactivate engineering objects. Change attempt is recorded as initialized change (IC). Before execution of the attempted change in the product model, it must be coordinated with existing model entities and with other attempts

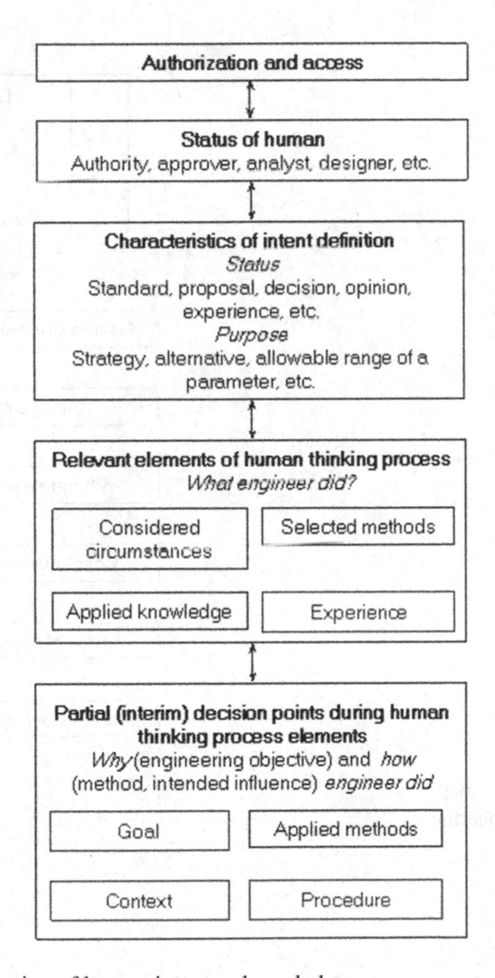

Fig. 3.10 Representation of human intent as knowledge

under processing (Fig. 3.11). For this purpose, change chains (CHC) are defined along contextual chains in the product model. Change chains for the same attempt are integrated in a change affect zone (CAZ). Change chains also include consequence changes (CC). Content of a change attempt from the relevant human intent and actual status of object change (OS) are recorded in adaptive action entity (AA). Status of object change for an adaptive action may be under processing such as under discussion, under revision, and argued and concluded such as accepted, decided, and executed. Product model entities are controlled by adaptive actions with decided status. Executed status later can be modified, mainly potential alternatives in the related human intent entity descriptions.

Transparency and feasibility of the above process should be supported by appropriate grouping of engineering objects in change attempt definitions and selection of leading parameters with relationship driven automatic correction of other parameters those are not important for the actual decision.

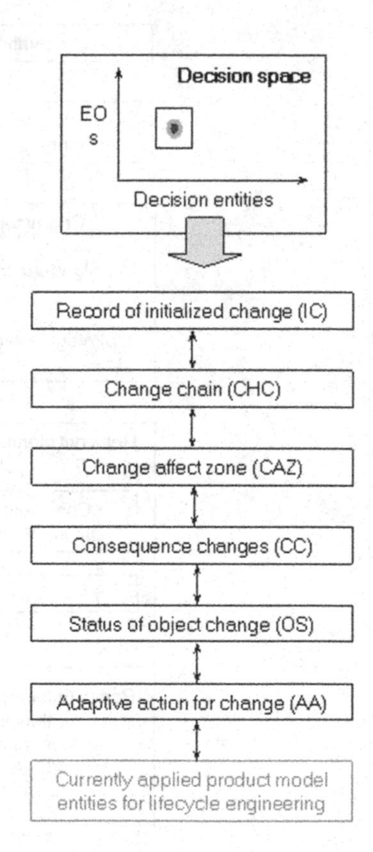

Fig. 3.11 Entities for adaptive control of engineering object definition

3.6 Implementation

The proposed product model extension can be considered as a multilevel structure where solution proceeds level by level (Fig. 3.12). This approach assists implementation of this modeling. In order to better connection between the content based modeling and the modeling in current PLM systems, an additional new multilevel structure was proposed by the authors of this chapter. These levels organize information about engineering objects. Consequently, content levels constitute the controlling, while information levels constitute the controlled section of the extended product model. In the controlling sector intent of human is applied to define new or modified objects for the product model. Processing human intent and engineering objective entities at decisions on adaptive actions requires organized contextual definitions across the product model. In the controlled section, engineering object

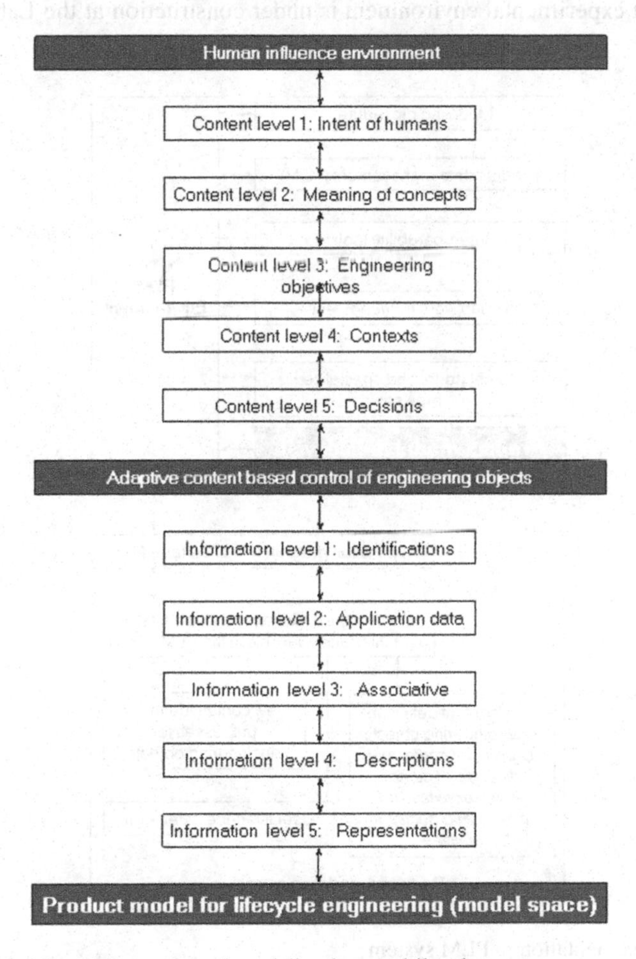

Fig. 3.12 Definition of the proposed modeling in a multilevel structure

information is organized on identification, application, associative, description, and representation levels.

Implementation of the proposed modeling is planned in open industrial PLM system environments where user development tools are available to extend the object model and the related modeling procedures to the proposed modeling (Fig. 3.13). This PLM environment includes modeling procedures, product data management (PDM) functionality, interoperability tools to make connections with other modeling and scientific systems, group work management functionality, and Internet browser functionality in order to develop and handle product model for lifecycle product information management. Procedures, model entities, and user interface available in the actual PLM environment can be accessed by user developed system elements through application programming interface (API).

In order to establish an up-to-date laboratory environment to make experiments with the proposed modeling in a suitably configured industrial PLM environment, a pilot experimental environment is under construction at the Laboratory of

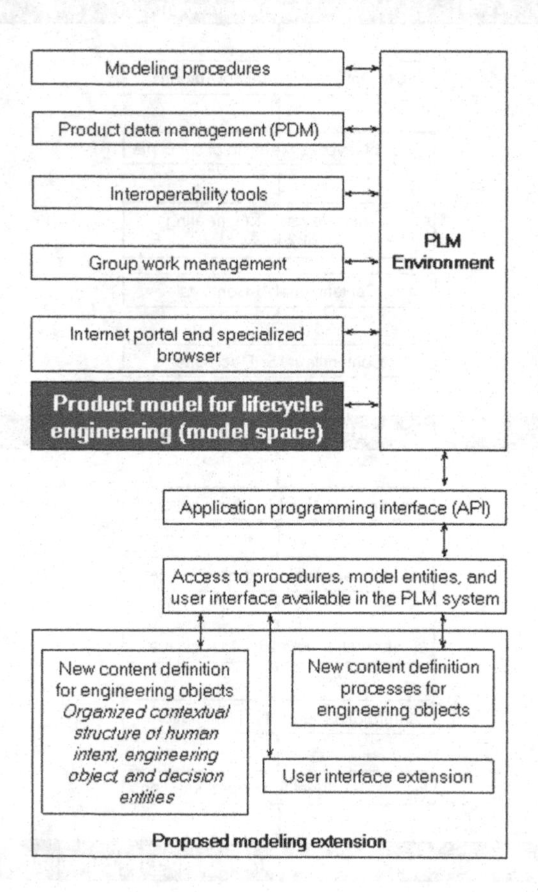

Fig. 3.13 Implementation in PLM system

Intelligent Engineering Systems (LIES) of the Institute of Intelligent Engineering Systems, John von Neumann Faculty of Informatics, Óbuda University. Within this program, LIES will be equipped with one of the leading industrial PLM and intelligent problem solving software.

3.7 Conclusions

In this chapter, a new content based product modeling method was introduced. The authors defined model entities and modeling processes in order to establish a modeling in which product specific knowledge is included in the product model as the background of decision making during computer modeling based definition of products. In current PLM systems, most of this knowledge is lost because product modeling is not in possession of appropriate representation capabilities.

The proposed modeling method serves improved human-computer communication at engineering object definition by using of changing the direct human control of the object and knowledge handling processes for new content based adaptive object definition. The new content definition for engineering objects includes organized contextual structure of human intent, engineering object, and decision entities. Object definition is controlled by elements and partial decisions from human thinking process. In this way, knowledge is communicated with the new content definition processes in order to give the capability for product model to be information source for reconstruction of human intent at application of product model.

Human defines change attempt serving one or more engineering objectives. Before execution of an attempted change in the product model, it must be coordinated with existing model entities and with other attempts under processing. Entity adaptive action carries change information to engineering object generation processes and takes control action from directly influencing humans.

The proposed modeling does not require new approach in representation of engineering objects and it can be implemented as user defined extension to current advanced and well-proved industrial PLM systems.

Acknowledgements. The authors gratefully acknowledge the grant provided by the OTKA Fund for Research of the Hungarian Government. Project number is NKTH OTKA K68029.

References

1. Andoga, R., Madarász, L., Főző, L.: Situational modeling and control of a small turbojet engine MPM 20. In: Proc. of the IEEE International Conference on Computational Cybernetics, Tallinn, Estonia, pp. 81–85 (2006)
2. Avgoustinov, N.: Modelling in mechanical engineering and mechatronics: towards autonomous intelligent software. Springer, Heidelberg (2007)
3. Colomboa, G., Moscaa, A., Sartori, F.: Towards the design of intelligent CAD systems: An ontological approach. Advanced Engineering Informatics 2(21), 153–168 (2004)

4. Horváth, L., Rudas, I.J.: Modeling and Problem Solving Methods for Engineers. Elsevier, New York (2004)
5. Horváth, L.: A New Method for Enhanced Information Content in Product Model. WSEAS Transactions on Information Science and Applications 3(5), 277–285 (2007)
6. Horváth, L.: Supporting Lifecycle Management of Product Data by Organized Descriptions and Behavior Definitions of Engineering Objects. Journal of Advanced Computational Intelligence and Intelligent Informatics 11(9), 277–285 (2008)
7. Horváth, L., Rudas, I.J.: Human Intent Description in Environment Adaptive Product Model Objects. Journal of Advanced Computational Intelligence and Intelligent Informatics 4(9), 415–422 (2005)
8. Horváth, L., Rudas, I.J., Hancke, G.: New Content behind the Concept Intelligent Engineering. In: Proc. of the IEEE 14th International Conference on Intelligent Engineering Systems, Las Palmas of Gran Canaria, pp. 73–78 (2010)
9. Horváth, L.: New Design Objective and Human Intent Based Management of Changes for Product Modeling. Acta Polytechnica Hungarica 1(4), 17–30 (2007)
10. Kitamura, Y., Kashiwase, M., Fuse, M., Mizoguchi, R.: Deployment of an ontological framework of functional design knowledge. Advanced Engineering Informatics 2(18), 115–127 (2004)
11. Liu, S.C., Tomizuka, M., Ulsoy, G.: Challenges and opportunities in the engineering of intelligent systems. In: Proc. of the 4th International Workshop on Structural Control, New York, pp. 295–300 (2004)
12. Madarász, L., Racek, M., Kovač, J., Timko, M.: Tools and Intelligent Methods in Enterprise Modeling. In: Proc. of the IEEE 9th International Conference on Intelligent Engineering Systems, Mediterranean Sea, pp. 187–192 (2005)
13. McMahon, C., Lowe, A., Culley: Knowledge management in engineering design: personalization and codification. Journal of Engineering Design 4(15), 307–325 (2004)
14. Mora, M., Forgionne, G., Gupta, F.: A Framework to Assess Intelligent Decision-Making Support Systems. In: Book Knowledge-Based Intelligent Information and Engineering Systems, pp. 59–65. Springer, Heidelberg (2003)
15. Richardson, M., Domingos, P.: Learning with Knowledge from Multiple Experts. In: Proc. of the Twentieth International Conference on Machine Learning, Washington DC, pp. 624–631 (2003)
16. Yang, Q., Reidsema, C.: Supporting Lifecycle Management of Product Data by Organized Descriptions and Behavior Definitions of Engineering Objects. Journal of Advanced Computational Intelligence and Intelligent Informatics 6(37), 609–634 (2006)
17. Zja, X.F., Howlett, R.J. (eds.): Integrated Intelligent Systems for Engineering Design. IOS Press, Amsterdam (2006)

Chapter 4
Optimization Technologies for Hard Problems

Czesław Smutnicki

Abstract. This chapter presents critical survey of methods, approaches and tendencies observed in modern optimization, focusing on nature-inspired techniques recommended for particularly hard discrete problems generated by practice. Applicability of these methods, depending the class of stated optimization task and classes of goal function, have been discussed. The best promising approaches have been indicated with practical recommendation of using. Some numerical as well as theoretical properties of these algorithms are also shown.

4.1 Introduction

The philosophy of approaches employed to solve optimization tasks completely changed during recent 10-20 years. Problems with unimodal, convex, differentiable scalar goal functions disappeared from research labs, because a lot of satisfactory efficient methods were already developed for them. Now researchers focus on particularly hard cases: multimodal, multi-criteria, non-differentiable, NP-hard, discrete, with huge dimensionality, derived from practice (control, planning, timetabling, scheduling, transporting, designing, management, etc.). These tasks, generated by industry and market, have caused serious troubles in process of seeking global optimum. In recent years, great effort has been done by scientist to reinforce power of solution methods and to fulfill expectations of users from practice. The moderate success in algorithms development strike practitioners imagination, so there are still needs for further research in this area. In the sequel of this chapter we refer to the following form of optimization task: find $x^* \in \mathscr{X}$ so that

$$K^* \stackrel{\text{def}}{=} K(x^*) = \min_{x \in \mathscr{X}} K(x) \tag{4.1}$$

Czesław Smutnicki
Institute of Computer Engineering, Control and Robotics, Wrocław University of Technology, Wrocław, Poland
e-mail: czeslaw.smutnicki@pwr.wroc.pl

J. Fodor et al. (Eds.): Recent Advances in Intelligent Engineering Systems, SCI 378, pp. 79–104.
springerlink.com

where x, x^*, $\mathscr{X}$ and $K(x)$ are respectively solution, optimal solution, set of feasible solutions and scalar goal function. The form of x, $\mathscr{X}$ and $K(x)$ depends on the type of optimization task. We focus mainly on practical cases, where $\mathscr{X}$ is discrete, $K(x)$ is nonlinear, non-differentiable while the problem is strongly NP-hard.

4.2 Troubles of Optimization

There have been recognized a few reasons responsible for optimization troubles causing high calculation cost and/or low quality of obtained solutions. Although one can easily pinpoint a lot of problems from the practice (industry, management, etc.) possessing such features, nonetheless we discuss and exemplify crucial problem properties referring to plain common benchmarks in 2D, being the best perceiving case for the reader (those instances should be treated as illustrative). As to applications, we will refer to real problems from the practice of job scheduling.

Huge number of extremes. This problem is illustrated well by Griewank's benchmark function (4.2) which exhibits huge number of local extremes quite regularly distributed, see Fig. 4.1,

$$K(x) = \frac{1}{4000}\sum_{i=1}^{n} x_i^2 - \prod_{i=1}^{n}\cos(\frac{x_i}{\sqrt{i}}) + 1. \tag{4.2}$$

Its global minimum $f(x^*) = 0$ is obtainable for $x_i^* = 0$, $i = 1, \ldots, n$. Function interpretation changes with the scale of the view. The general overview shows classical convex function, medium-scale view suggests existence of a few local extremes, whereas only precise zoom indicates complex structure of numerous local extremes. Theoretically, due to regularity of the surface, one can easily define strategic search directions, which lead quickly the search process to the most promising part of the solution space. Unfortunately, by (4.2) the number of local extremes grows exponentially with the dimension of the space n. This fact practically eliminates methods, which completely examine even a small fraction of local extremes, and completely disqualifies exhaustive search methods.

For the benchmark ta51 of the flow-shop scheduling problem, the number of solutions x with $(K(x) - K^*)/K^* \leq 0.5\%$ ((local extremes) is $L \approx 0.2 \cdot 10^{57}$ and constitutes infinitisimal fraction $L/50! \approx 0.8 \cdot 10^{-8}$ of the whole space, [26].

Distribution of extremes. This problem is illustrated well by Langermann's benchmark function (4.3), which exhibits numerous local minima unevenly distributed, see Fig. 4.2, depending on parameters m, a_{ij}, c_i unknown a priori

$$K(x) = \sum_{i=1}^{m} c_i \exp[-\frac{1}{\pi}\sum_{j=1}^{n}(x_j - a_{ij})^2]\cos[\pi\sum_{j=1}^{n}(x_j - a_{ij})^2]. \tag{4.3}$$

It means that strategic search directions in the solution space have no regular character and have to be set in an adaptive way, depending on the space landscape.

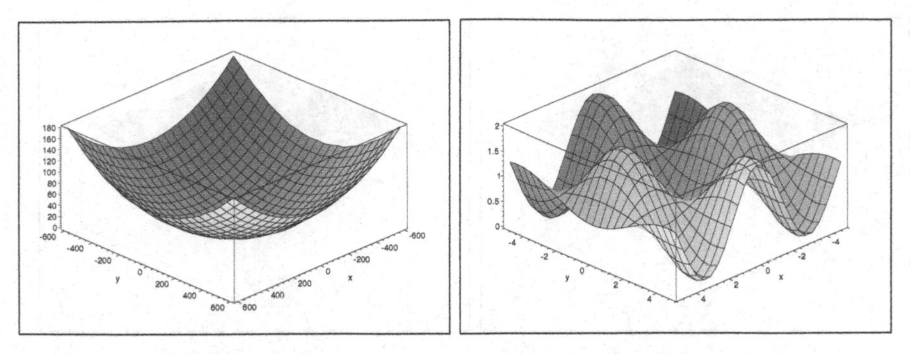

Fig. 4.1 Griewank's test function in 2D: overview (left), zoom (right)

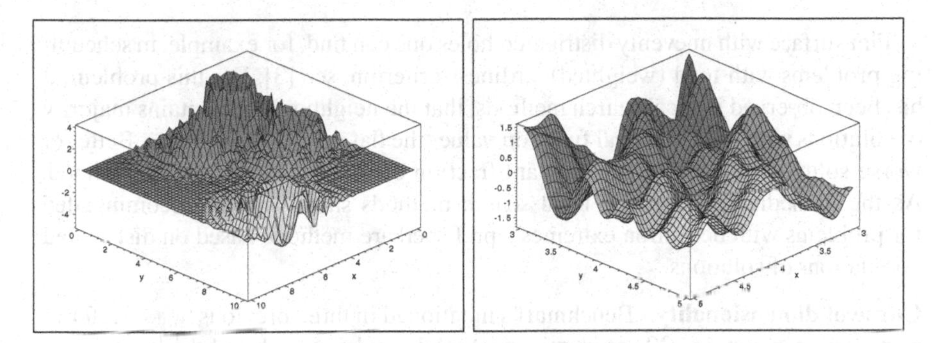

Fig. 4.2 Langerman's test function in 2D: overview (left), zoom (right)

The hybrid flow shop scheduling problem, [33], which models e.g. automatic line for producing printed packages, exhibits huge number of local extremes unevenly distributed, see Fig. 4.4. It is clear, that the skillful solution method have to examine the landscape of space $\mathscr{X}$ to adjust properly promising search directions.

Deception extremes. This problem is illustrated well by Shakel's benchmark function (4.4), which exhibits quite deep local minima (holes) unevenly distributed on an almost flat surface, see Fig. 4.3,

$$K(x) = -\sum_{i=1}^{m}(\sum_{j=1}^{n}[(x_j - a_{ij})^2 + c_j])^{-1}, \tag{4.4}$$

where (c_i, $i = 1,\dots,m$), (a_{ij}, $j = 1,\dots,n$, $i = 1,\dots,m$) are constant numbers fixed in advance; it is recommended to set $m = 30$. The behavior of the function between holes (significant part of the surface) provides no information about minima expected in the vicinity. Moreover, iterative search methods (walking step-by-step) are often unable to get out from so deep minima, which results in premature convergence and/or search stagnation.

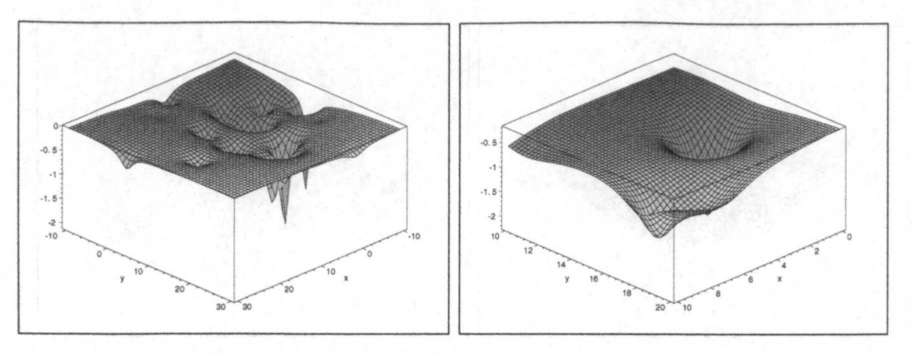

Fig. 4.3 Shekel's fox holes test function in 2D: overview (left), zoom (right)

Flat surface with unevenly distributed holes one can find, for example, in scheduling problems with total (weighted) tardiness criterion, see [3]. For this problem, it has been observed in local search methods, that the neighborhood contains majority of solutions with the same goal function value (the flat part of the surface). Better or worse solutions constitute insignificant fraction of solutions in this neighborhood. As the immediate results, the local search methods should not be recommended for problems with deception extremes - preferred are methods based on distributed populations of solutions.

Curse of dimensionality. Benchmarks mentioned in three previous heads, refer to a space dimension $n \leq 20$ (academic size). Nobody have analyzed the behavior of the suitable methods for greater space dimension (real-case size). In the discrete optimization the difference between *academic* instance and *real* case is more evident.

Let us consider the oldest historically real-case benchmark FT10 (10 jobs, 10 machines, 100 operations) from the scheduling theory, one of the smallest considered now. The solution space $\mathscr{X}$ is discrete, finite, has dimension 90, which comparing to academic 20, one can consider already as a significant size (the biggest solvable now benchmarks of this type have dimension 1980). If we use so called permutation-and-graph representation for x leading to the smallest space cardinality, [24, 25, 27], one can find approximately $4 \cdot 10^{65}$ different solutions, however some of them can be unfeasible. The fraction of feasible solutions to all of them, depends on data and varies from 1 to 10^{-17} (for the data in FT10). We would like to make a 2D projection of the $4 \cdot 10^{48}$ feasible solutions from $\mathscr{X}$ and then plot it as a colorful map using 2400dpi printer in which single dot is of size 0.01×0.01 mm. We hope that this map help us to detect unevenly distributed local extremes or at least to define strategic directions during the search guided by the landscape. To this end we'll use a transformation (unspecified here) preserving the distance, i.e. so that distant solutions in the space correspond to distant points on 2D plane and close solution to close points. As the result of printing we'll obtain completely filled area of order $4 \cdot 10^{32}$ km^2. For comparison the biggest known planet in our solar system Jupiter has only $6.4 \cdot 10^{10}$ km^2 of the surface. Typical searching procedure is able to check in a reasonable time at most billion solutions, which corresponds to area $0.1\mathrm{m}^2$ or can be represented by spiderweb 0.01 mm width and 10 km length, [27].

Fraction of feasible solution to all solutions is infinitesimal in FT10 and drastically decreases with increasing instance size. On the other hand, for this class of problems feasible solutions in the space provide huge number of local optima unevenly distributed. The designed solution algorithm should not only preselect feasible solutions, but should identify the most promising space regions for exploration.

NP hardness. Most of discrete optimization problems derived from practice (scheduling, transport, time-tabling, etc.) are NP-hard, which immediately implies exponential-time computational complexity of solution algorithm. Since the power of processors increases linearly in recent years, while the cost of calculations as well as the number of local extremes increases exponentially with the size of problem, there is no hope to solve real instances in the acceptable amount of time in practice.

Calculation cost. NP-hardness implies unacceptable large calculation cost measured by the processor running time. Moreover, discrete problems are considered as superfluously rigid, in that sense that small perturbation of data destroys optimality of solution found in an expensive way, which force the user to make expensive calculation once again. That's why seeking optimal solution is not popular in the society of practitioners.

4.3 Space Landscape

It is commonly known that the behavior of the solution algorithm have to be adjusted to the landscape of the solution space. Detection of several recognized properties of the landscape (mentioned in the previous section) allow us to design efficient algorithms fitted well to the problem and landscape.

Distributions. Distribution of solutions as well as local minima in the discrete space is usually uneven. One can verify this fact by observing *random*, *local search* or *goal oriented search* trajectories walking through the space by neighboring solutions (distant by one unit), see Fig. 4.4. Then, random sampling is usually used for testing the space and landscape properties.

Let us consider the instance ta45 of the job-shop scheduling problem with makespan criterion. Random sampling has at least two goals: (1) identifying regions containing feasible solutions, (2) identifying the promising search regions in terms of $K(x)$. Distance of any solution to the optimal one is distributed normally with the average value of 50% of the space diameter, see Fig. 4.5. Feasible solutions are located significantly closer distant, approximately 20% of the space diameter. The distribution of feasible solutions in terms of $K(x)$ is shown in Fig. 4.6. Relative error RE is defined by the equation (4.10). The distribution is also close to normal. The approximation by the least square method, provides average 115.8 and standard deviation 17.8, which means that random feasible solution is statistically more than

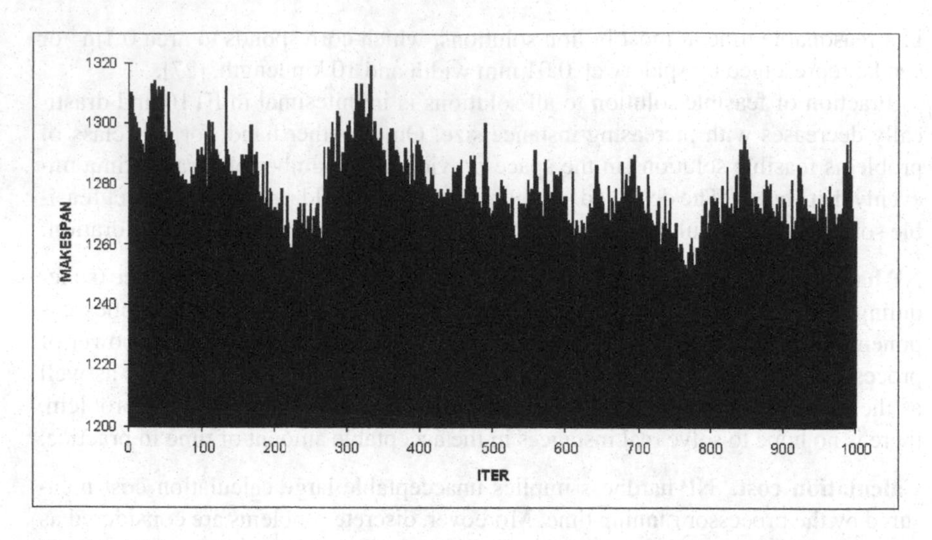

Fig. 4.4 $K(x)$ values on local search trajectory for the hybrid scheduling problem; distance between successive solutions equals one unit, [33]

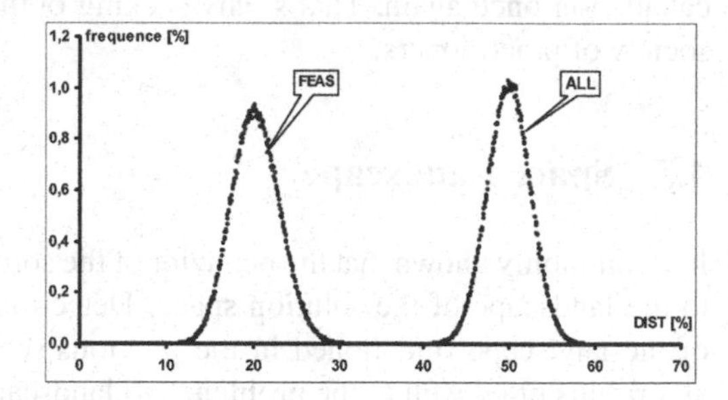

Fig. 4.5 Distribution of Hamming distance DIST of all (ALL) and feasible (FEAS) solutions to x^*; sample of 500,000 random solutions for ta45 instance of the job-shop scheduling problem

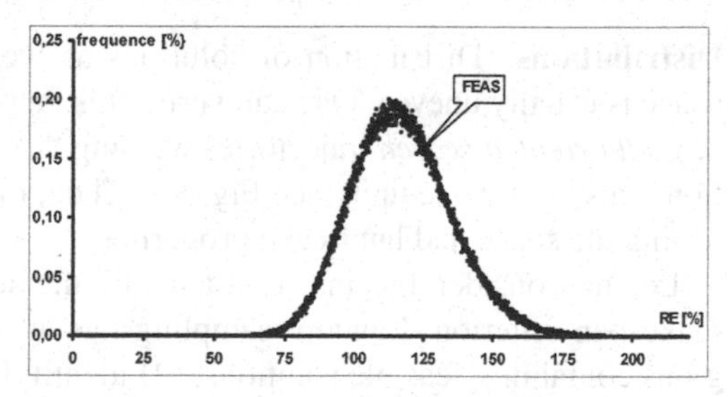

Fig. 4.6 Distribution of relative error RE defined by equation (4.10) for feasible solutions; sample of 500,000 random solutions for ta45 instance of the job-shop scheduling problem

twice worse than the optimal one. Interestingly, the probability of finding solution with $RE \leq 1\%$ by random sampling of the space $\mathscr{X}$ is approximately $5 \cdot 10^{-11}$, so practically infinitesimal, even considering 10^6 solutions. On the other hand in $\mathscr{X}$, there exist approximately $2 \cdot 10^{39}$ solutions with $RE \leq 1\%$. There is no way to enumerate them, even partially. This is a serious drawback of random search methods.

Big valley. Problem is suspected to own *big valley* phenomenon if there exists positive correlation between goal function value and the distance to optimal solution (the best found solution), see e. g. Fig. 4.7 for its geometric interpretation. In the big valley a densification of local extremes appears, so the search processes should be direct there and continue around this area. Big valley phenomenon has been detected in many discrete optimization problems, to mention at least TSP, scheduling, see [30]. Recent papers, [25], suggest that the size of the valley is usually relatively very small with respect to the size of the whole solution space.

Ruggedness. This is a measure characterizing diversity of goal function values of related (usually neighboring) solutions, [8]. Greater ruggedness means sharper and unpredicted changes of $K(x)$ for neighboring solutions, in comparison with Fig. 4.4. Smaller ruggedness means flat or slow-changeable landscape. For distance $d(x,y)$, $x,y \in \mathscr{X}$ (there are many possible definition of distance measure) the proposed ruggedness measure is the autocorrelation coefficient, [35],

$$\rho(d) = 1 - \frac{AVE((K(x)-K(y))^2)_{d(x,y)=d}}{AVE((K(x)-K(y))^2)} \tag{4.5}$$

where $AVE((K(x)-K(y))^2)$ is an average value of $(K(x)-K(y))^2$ on the set of pair of solutions (x,y), $x,y \in \mathscr{X}$, and $AVE((K(x)-K(y))^2)_{d(x,y)=d}$ is an average value of $(K(x)-K(y))^2$ found on the set of pairs of solutions (x,y), $x,y \in \mathscr{X}$, so that distance from x to y equals exactly d. Value $\rho(d)$ defines correlation between solutions $x,y \in \mathscr{X}$ distant by d each other. Taking into account special properties of local search algorithms, the most interesting is the value of $\rho(1)$, being the correlation between solutions located in the single neighborhood. Value $\rho(1)$ close to zero means no correlation, which implies strong differentiation of $K(x)$ in the neighborhood (rough landscape); value $\rho(1)$ close to one means strong correlation and flat

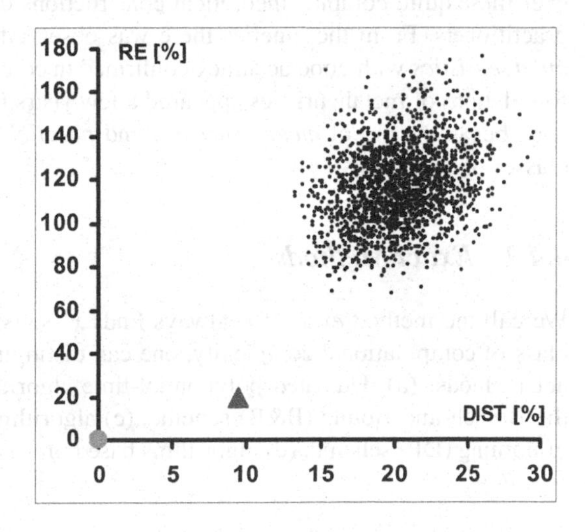

Fig. 4.7 Big valley. Distribution of local extremes for ta45 instance of the job-shop scheduling problem. RE is the measure defined by equation (4.10), DIST is the distance to the optimal solution (in % of the space diameter).

or slow-changeable landscape. Because of the cardinality of $\mathscr{X}$, finding the proper average values is troublesome or even impossible. Therefore in order to evaluate, in practice, the ruggedness of the landscape, one proposes to use random search and autocorrelation function defined as follows

$$r(s) = 1 - \frac{AVE((K(x_i) - K(x_{i-s}))^2)}{2(AVE(K^2) - (AVE(K))^2)} \tag{4.6}$$

where $(x_1, x_2, ..., x_k)$ is the trajectory (sequence of solutions) generated by the random search algorithm. On the base of $r(1)$ we define autocorrelation coefficient ξ as $\xi = 1/(1 - r(1))$. In accordance to previous comments, greater value of ξ means flat landscape.

4.4 Solution Methods

The evolution of solution approaches for discrete problems has long and rich history. It started from the commonly used *heuristics* based chiefly on various *priority rules* in the fifties and sixties, and went through the *theory of NP-completeness* (the seventies) which classified problems and algorithms into *polynomial-time* and *exponential-time*, [14]. Significant development of *exact algorithms* in the seventies and eighties (as an example B&B scheme, DP scheme, Integer Linear Programming, Binary Programming, etc.) moved slightly the border of instance sizes which can be solved by these methods but ultimately restricted their applicability. Pessimistic experience with exact methods stimulated, among others, the development of *approximation algorithms* (the eighties and nineties) and *approximation theory*. Besides the theoretical results, a lot of *approximation schemes* (AS), *polynomial-time AS* (PTAS) and *fully polynomial time AS* (FPTAS) were proposed. For the class of on-line algorithms the similar role plays so called *competitive analysis*. However these quite complex theoretical constructions did not gain acceptance among practitioners. From the nineties there was observed the intensive development of *metaheuristics* with good accuracy confirmed in computer benchmarks. Theoretical foundations of metaheuristics appeared a few years later. From 2000, in the natural way, began the era of *meta*^{2}*heuristics* and *parallel metaheuristics*, being the new classes of algorithms.

4.4.1 Exact Methods

We call the method *exact* if it always finds x^* satisfying (4.1). Depending on the class of computational complexity, one can distinguish the following types of exact methods: (a) dedicated polynomial-time algorithms, (b) algorithms based on the Branch-and-Bound (B&B) scheme, (c) algorithms based on the Dynamic Programming (DP) scheme, (d) algorithms based on Integer Linear Programing (ILP),

(e) algorithms based on Binary Linear Programing (BLP), (f) subgradient methods. Methods (a) are considered as computationally cheap specialized methods for problems from P-class or NP-hard numeral. Methods (b) – (f) are computationally expensive, dedicated for strongly NP-hard problems. Up to the end of the eighties one considered them as "sole right" approaches for strongly NP-hard problems, after that time there appeared barrier of dimension. Although significant was done in their development, practitioners still consider them as unattractive, or limit their applications to a narrow scope. These methods are time- and memory- consuming, whereas size of instances which can be solved in a reasonable time is still too small for practical use. Moreover, implementation of more complex algorithms of this type needs skillful programmers. The validity of the instance data is also a serious problem; frequently they have been perturbed just after the expensive finding of optimal solution and so called superfluous rigidness of the problem. One can say that the cost of finding optimal solution is still too high in comparison with the profits obtained from its implementation. Nevertheless, there exist several problems where application of exact methods are justified and recommended. For example, seeking for the optimal cycle time in repetitive manufacturing system with small number of various products, provides profits proportional to the number of performed cycles.

4.4.2 Approximate Methods

Approximate algorithm A provides solution x^A, so that

$$K(x^A) = \min_{x \in \mathscr{X}^A} K(x) \geq K(x^*) \tag{4.7}$$

where $\mathscr{X}^A \subset \mathscr{X}$ is the subset of solutions checked by A. In the extreme case $\mathscr{X}^A$ may have single element. The overall aim is to find x^A so that $K(x^A)$ is *close* to $K(x^*)$ by examining the smallest possible $\mathscr{X}^A$, [14]. The closeness to $K(x^*)$ (accuracy) can be either guaranteed a priori or evaluated a posteriori. It is clear that accuracy has opposing tendency to running time, i.e. finding better approximate solution needs longer running time (greater $\mathscr{X}^A$), and this dependence owns strongly nonlinear character. Therefore, discrete optimization manifests a variety of models and solution methods, usually dedicated for narrow classes of problems or even separate problems. Reduction of the generality of models allow us to find special features for some problem. Those features allow us to improve numerical properties of the algorithm such that running time, speed of convergence. Quite often, a strongly NP-hard problem has several various algorithms in the literature with different numerical characteristics. Knowledge about models and algorithms allow us to fit satisfactory algorithm for each newly stated problem. Bear in mind, that in the considered research area the goal *is not* to formulate whatsoever model and method, but to provide *simple* model and solution method *reasonable* from the implementation point of view.

Approximation accuracy. The set of data specifies the *instance* Z of the problem. Denote by $\mathscr{X}(Z)$ the set of feasible solutions for the problem, and by $K(x;Z)$ value of criteria K for solution x in the instance Z. Solution $x^* \in \mathscr{X}(Z)$ so that $K(x^*;Z) = \min\{K(x;Z) : x \in \mathscr{X}(Z)\}$ is the optimal solution for the instance Z. Let $x^A \in \mathscr{X}(Z)$ denote the approximate solution generated by algorithm A for the instance Z. One can evaluate the approximation accuracy of algorithm A referring to suitable optimal value, e.g.

$$B^A(Z) = \left|K(x^A;Z) - K(x^*;Z)\right|, \tag{4.8}$$

$$S^A(Z) = \frac{K(x^A;Z)}{K(x^*;Z)}, \tag{4.9}$$

$$T^A(Z) = \frac{K(x^A;Z) - K(x^*;Z)}{K(x^*;Z)}. \tag{4.10}$$

Dependence of the accuracy measure on Z can be examined either experimentally or analytically. Experimental analysis is a subjective method, since it depends on the chosen sample of instances. However, only this analysis is able, to justify, in the context of *no free lunch theorem*, [38], observed superiority of the selected algorithm over the remains on a subclass of instances Z. More sophisticated worst-case and probabilistic analysis provided rating independently on Z. These two analyses provide different, complementary, sometimes more adequate characteristics of the behavior of the algorithm. Results of three mentioned analyzes, completed by computational complexity, create the full numerical characteristics of the algorithm.

Experimental analysis. This is the most popular method, despite its imperfections. Evaluates behavior of the algorithm a posteriori (chosen accuracy measure, running time) basing on the results obtained for limited *representative sample* of instances Z. Since various researchers diversely understand notion "representative sample", obtained results are quite often incomparable. Instances in the sample can be fixed (common benchmarks) or randomly generated. For strongly NP-hard problems value $K(x^*;Z)$ can be replaced by certain reference value $K(x^{Ref};Z)$.

Worst case analysis. It evaluates a priori behavior of the chosen measure of accuracy on *entire* infinite population of instances Z. Usually, it is applied for accuracy $S^A(Z)$, for which is also defined the *worst-case ratio*

$$\eta^A = \min\{y : S^A(Z) \leq y, \forall Z\} \tag{4.11}$$

and *asymptotic worst-case ratio*

$$\eta^A_\infty = \min\{y : S^A(Z) \leq y, \forall Z \in \{W : K(x^*;W) \geq L\}\}, \tag{4.12}$$

where L is a positive number.

Probabilistic analysis. It assumes that each instance Z was obtained as the result of realization of n independent random variables with known distributions (usually uniform) of probability; this fact will be denoted by writing Z_n instead of Z. Then, values $K(x^*;Z_n)$, $K(x^A;Z_n)$ and measures (4.8) – (4.10) are random variables as functions defined on Z_n. Let denote by $M^A(Z_n)$ the accuracy, in sense of a chosen measure (4.8) – (4.10), of algorithm A for instance Z_n. Probabilistic analysis provides basic information about behavior of random variable $M^A(Z_n)$, namely its distribution, moments, etc. Further interesting characteristics refer to *type* and *speed of convergence* of $M^A(Z_n)$ to constant m (or to zero) with the increasing n. There are fundamentally considered the following types of convergence:
(a) *almost sure*

$$P(\lim_{n\to\infty} M^A(Z_n) = m) = 1 \tag{4.13}$$

(b) *in probability*

$$\lim_{n\to\infty} P(|M^A(Z_n) - m| > \varepsilon) = 0, \text{ for any } \varepsilon > 0 \tag{4.14}$$

(c) *in average*

$$\lim_{n\to\infty} |E(M^A(Z_n)) - m| = 0 \tag{4.15}$$

where $P()$, $E()$ denote the probability and the average value. Convergence (a) implicates (b), but (b) implicates (a) if only for any $\varepsilon > 0$ holds

$$\sum_{n=1}^{\infty} P(|M^A(Z_n) - m| > \varepsilon) < \infty \tag{4.16}$$

Similarly, (c) implies (b) but (b) implies (c) if only condition (4.16) holds. The best algorithm is the one having accuracy convergent almost sure to zero. Probabilistic analysis provides average evaluations over the whole population of instances. Hence, results observed in experiments can differ from theoretical instances. In practice, most of instances show the decrease of accuracy to zero with the increasing size of instances. Since the probabilistic analysis is usually rather complicated, only few algorithms have developed sufficient results.

Approximation schemes. Approximation scheme (AS) is the family of algorithms A_ε, such that A_ε provides for the given $\varepsilon > 0$ solution x^A satisfying $S^{A_\varepsilon}(Z) \leq 1+\varepsilon$, $\forall Z$. AS is the *polynomial-time approximation scheme* (PTAS), if for any fixed ε it owns polynomial computational complexity. If additionally this complexity is a polynomial of $1/\varepsilon$, then scheme is *fully polynomial-time approximation scheme* (FPTAS). Beside positive results (AS has been constructed) theory may provide negative results (AS does not exist for ε less than certain threshold value). The latter case is usually analyzed with the help of small-size instance, therefore one expect contrary behavior of the algorithm on large instances. In practice, ASes turned out to be rather complex algorithmic constructions. Some of them have the computational complexity provided in the asymptotic notation which pass over the issue of its practical applicability. Moreover, the degree of polynomial increases very quickly with $\varepsilon \to 0$. That's why these schemes still remain not competitive to other approaches.

4.5 Modern Approaches

In recent 10 – 20 years, simultaneously with the development of mathematically perfect theories, there has been observed rapid development of *metaheuristics*, i.e. approximate methods without excessive theory but with good or even excellent numerical properties, confirmed in numerous computer tests. These methods are more interesting for practitioners, since they quickly provide solutions with quality 1 – 5% in terms of (4.10), than mathematically perfect approximation schemes, offering perfect mathematical guarantee with rather poor practical quality. These methods are classified as either *constructive* or *improvement*. The former are fast, easily implementable, but provide solutions of poor quality. The later are slower, need starting solution improved next iteratively, but provide solutions with good or excellent quality. They also allow to form in a flexible way the compromise between the solution quality and the algorithm's running time. Theoretical guarantee of quality were found, up till now, for numerous constructive methods but only for few improvement methods. For some improvement methods there have been proved convergence to the optimal solution, but because of fact that sufficient conditions do not hold in practice, thus these results have rather theoretical than practical significance. Finally, the practical usefulness of approaches and/or algorithms follows from various theoretical as well as experimental analysis.

Constructive algorithm generates $\mathscr{X}^A$ for (4.7) containing either the single solution or a set with few predefined a priori solutions. These algorithms are based on the following approaches: (a) priority rules, (b) adaptation of the solution obtained for the relaxed problem, (c) approximation of the solution by solution obtained for a relative problem, (d) others. Algorithms from this class can be used alone or in the group of competitive methods. There are computationally cheap. Commonly are used as generators of starting solutions for improvement algorithms. By combining, in the parametric way, certain number of this type algorithms into so called space of heuristics, we can synthesize new algorithms for newly formulated problems

Priority rules (PR) are the oldest and commonly used technique for quick finding solutions of moderate quality, see e.g. applications in scheduling [18, 28]. There are simple, computationally cheap, dimly sensitive to data perturbation, and are recommended also for nondeterministic systems. They can be formed as *static* rules (priority index does not change the value) and *dynamic* (priority index changes value during the task waiting and service). Depending on the problem type, form of the goal function and constraints, priority rules provide solutions distant from 20% (in average) to 500% (extremely) in terms of measure (4.10).

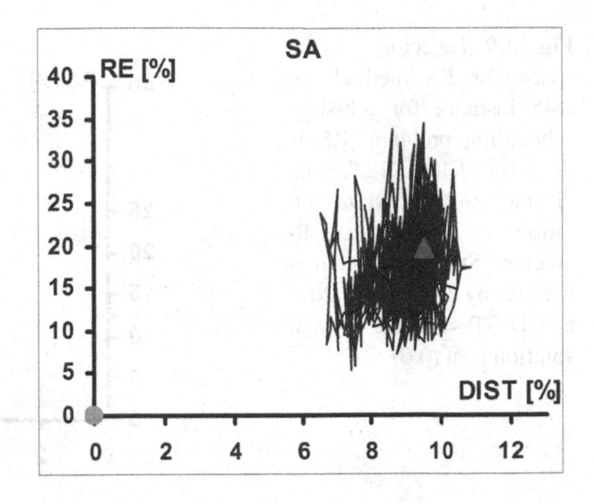

Fig. 4.8 Trajectory in the space for SA method for ta45 instance for job-shop scheduling problem. RE is the error (4.10), DIST is the distance to the optimal solution (in % od space diameter). Starting solution is marked by $\triangle$ at RE $\approx$ 20% and DIST $\approx$ 9.5%; optimal solution is at (0,0)

Local search (LS) defines the general group of methods which check in (4.7) certain subset of solutions $\mathscr{X}^A \stackrel{\text{def}}{=} \mathscr{N}^*(x) \subseteq \mathscr{N}(x) \subset \mathscr{X}$ selected from the *local neighborhood* $\mathscr{N}(x)$ of the given *starting solution* x. The process is usually repeated iteratively, providing the *search trajectory* $x^0, x^1, \ldots, x^s$ being successive starting solutions. Usually, the next starting solution is selected among those from the neighborhood, i.e. $x^{k+1} \in \mathscr{N}(x^k)$, $k = 0, 1, \ldots, s-1$. It is clear that in the iterative case we check solutions from $\mathscr{X}^A - \cup_{k=0}^{s} \mathscr{N}^*(x^k)$. By using various techniques of creating $\mathscr{N}^*(x^k)$ and selecting x^{k+1} we obtain several particular approaches known in the literature, see excellent monograph [2]. LS methods offer many advantageous properties: simple implementation, high speed of convergence, good quality of generated solutions. Hence, they are considered as the very promising for hard discrete optimization task.

Descending search (DS) is the oldest, simplest and the most intuitive method of improving the given solution $x \in \mathscr{X}$. This is the LS method, which generates search trajectory $x^0, x^1, \ldots x^s$ such, that $K(x^0) > K(x^1) > K(x^2) > \ldots > K(x^s) < K(x^{s+1})$, [2]. Successive solution on the trajectory $x^{k+1} \in \mathscr{N}^*(x^k)$ is the best one (in the sense of the criterion value) among solutions in the neighborhood $\mathscr{N}(x^k)$. There are at least two variants of DS, both greedy: (a) to the first improvement, (b) to the greatest improvement. The former variant looks through solutions from the neighborhood $\mathscr{N}(x^k)$ in certain order $y^1, y^2, \ldots, y^t$, to find the first solution better than $K(x^k)$, i.e. solution y^l so that $K(y^i) \geq K(x^k)$, $i = 1, 2, \ldots, l-1$, $K(y^l) < K(x^k)$. If only such solution exists, we set $x^{k+1} = y^l$, $\mathscr{N}^*(x^k) = \{y_1, \ldots, y_l\}$ and continue this process; otherwise DS ends its activity in the local extreme x^s. The latter variant choses x^{k+1} so that $K(x^{k+1}) = \min\{K(y) : y \in \mathscr{N}(x^k)\}$, thus we have $\mathscr{N}^*(x^k) = \mathscr{N}(x^k)$. By starting DS several times from various x^0 we can eliminate, in some scope, its sensitivity to local extremes (observed as premature convergence). Certain mutation of this method is the *descending search with drift* (DSD), which accepts on the search trajectory also unvisited solutions with the same criterion value, i.e $K(x^{k+1}) = K(x^k)$. DS is relative to *hill climbing* technique known for maximization problems.

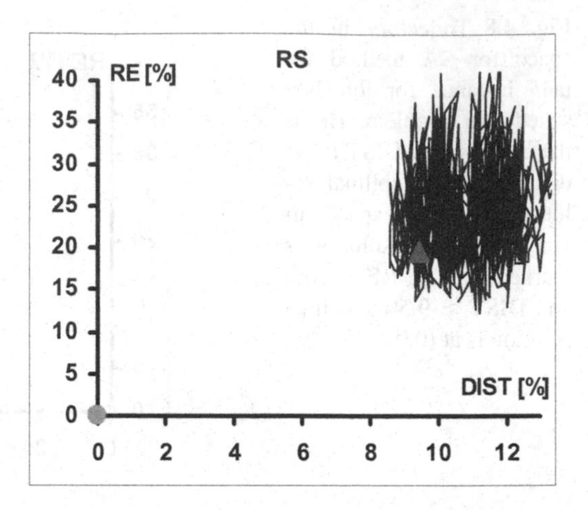

Fig. 4.9 Trajectory in the space for RS method for ta45 instance for job-shop scheduling problem. RE is the error (4.10), DIST is the distance to the optimal solution (in % od space diameter). Starting solution is marked by $\triangle$ at RE $\approx$ 20% and DIST $\approx$ 9.5%; optimal solution is at (0,0)

Random search (RS), called sometimes Monte Carlo method, defines the general class of algorithms which generate for (4.7) the subset $\mathscr{X}^A$ by random sampling of $\mathscr{X}$, [31]. The most common variant relies on the LS method generating random search trajectories $x^0, x^1, \ldots, x^s$, where x^{k+1} is selected randomly in $\mathscr{N}(x^k)$ with uniform distribution, and thus $\mathscr{N}^*(x^k) = \{x^{k+1}\}$. Some theoretical properties referring to the conditions of convergence and the speed of convergence were shown. Although RS is not sensitive to local extremes, its convergence to good local optimum is generally poor in practice, see e.g. Fig. 4.9 where RS ends its activity very far from optimal solution. Generalizations consider: special distribution of probability estimated on the base of the search history, larger (than single-element) samples.

Simulated annealing (SA) simulates process of annealing, by slow cooling, ferromagnetic or anti-ferromagnetic solid in order to eliminate internal stretches, [22]. There are built analogy between energy of the solid and criterion function $K(x)$, and between internal configuration of particles and the solution x of the optimization problem (4.1). SA is the LS, which generates search trajectory by using the following scenario. At first, a *perturbed solution* $x' \in \mathscr{N}(x^k)$ is chosen randomly with uniform distribution of probability. If $K(x') \leq K(x^k)$ then x' is accepted immediately as the new starting solution, i.e. $x^{k+1} := x'$. Otherwise, x' is accepted as the new starting solution with probability $\min\{1, e^{-\Delta_k/T_k}\}$, where $\Delta_k = K(x') - K(x^k)$ and T_k is the parameter called *temperature* at step k. If x' has not been accepted (neither by former nor by latter condition) we keep the old solution, i.e. $x^{k+1} := x^k$. Temperature T_k is being changed in accordance to *cooling scheme*. Usually, at each fixed temperature, a number of m steps is being performed. There are commonly used several cooling schemes: geometric $T_{k+1} = \lambda_k T_k$, logarithmic $T_{k+1} = T_k/(1 + \lambda_k T_k)$, harmonic $T_k = \Delta_{\max}/\log(k+2)$, $k = 0, \ldots, N-1$, where N is the total iteration number, λ_k is a parameter so that $T_N \to 0$, T_0 is the starting temperature. Clearly $\mathscr{N}^*(x^k) = \{x'\}$.

SA has a few parameters usually tuned experimentally to the problem type and size. Parameter λ_k is often chosen as constant $\lambda_k = \lambda = const$, whereas m is set one to avoid time consuming calculations. Assuming known (estimated) values of T_0, T_N and N, one can simply calculate fixed parameter λ. For the logarithmic scheme it is equal $\lambda = (T_0 - T_N)/(NT_0T_N)$, whereas for the geometric scheme we have $\lambda = (T_N/T_0)^{1/N}$. Interesting is the *automatic self-tuning and adaptive* SA, provided in the literature, [1]. The proposal is to set x^0 randomly. Starting temperature is set in accordance to changes of the goal function value observed on the sample of M trial solutions. To this order, starting from x^0 we generate random sample trajectory $x^0, x^1, \ldots, x^{M-1}$, where $x^{k+1} \in \mathcal{N}(x^k)$ is chosen randomly, and then calculate $\Delta_{\max} = \max_{0<k<M}(K(x^k) - K(x^{k-1}))$. Te initial temperature is set as $T_0 = -\Delta_{\max}/\ln(p)$, where $p \approx 0.9$. At each fixed temperature we perform $m > 1$ steps and then we set $\lambda_k = \ln(1+\delta)/(3\sigma_k)$ for the logarithmic cooling scheme, where δ is the parameter of reaching equilibrium (0.1 – 10.0) and σ_k is the standard deviation of goal function values for m solutions x generated at temperature T_k. The stop criteria is based on T_k close to zero. Values m and M are $\Theta(|\mathcal{N}(x)|)$.

Cooling scheme has significant influence on the behavior of the method. If cooling is to fast, SA is similar to DS and premature converges to local minimum of poor quality. If cooling is too slow, then running time becomes unacceptable long. Under some conditions (unfortunately not realistic in practice), SA converges with probability one to optimal solution. The necessary condition is *connectivity property*, see TS method for its meaning. There are several modifications of the basic scheme of SA, by allowing: (1) several perturbed solution, i.e. $|\mathcal{N}^*(x^k)| > 1$, (2) changed distribution of acceptance, (3) multiple restarts from the initial temperature.

SA generally is better than RS, however still owns disadvantageous features of its ancestor RS, see e.g. Fig. 4.8, where SA ends its activity closer to optimum than RS (compare with Fig. 4.9), however still too far.

Simulated jumping (SJ) is the LS which simulates the process of annealing the mixture of ferromagnetic and anti-ferromagnetic materials, called the spin-glass system, having numerous local states without long-range periodic order, [7]. The system owns many metastable states with large energy barriers in between the domain, called domain walls. SJ tries to break these energy barriers by a process of repetitive, alternative heating and cooling. It is recommended for problems with very rough landscape. One important difference of SJ from SA is that the system is not allowed to to reach an equilibrium state at any temperature. SJ starts generating perturbed solution in low temperature (completely reverse to SA) and realizes heating followed by cooling several times; the recommended starting temperature is $T_0 = 0.001$. Perturbed solutions are generated in the same way as in SA and acceptance criteria are almost the same; the only difference is in expression on probability $\exp(-\Delta_k/(\Omega T_k))$, where Ω is a parameter analogous to conventional Boltzman constant however adapted during the run. Heating phase contains N ($N \approx 300$) successive iterations (generating perturbed solutions) and results in increasing the temperature $T_{k+1} = T_k + R/\xi$ if only x' has not been accepted, where $r \in [0, \omega]$, $\xi \in \{1, 2, \ldots, N\}$, $\omega = 0.15$. After heating phase, the system is once cooled $T_{k+1} = \lambda T_k$, where $\lambda \in (0, \varepsilon]$, $\varepsilon \in [0.001, 0.2]$. The recommended technique of adjusting $\Omega = \overline{E}/\log(A\eta_2/\eta_1)$, where $A \in [16, 20]$; $\eta_1 = 1$ at the initial temperature and increased by one if the perturbed solution has been accepted; $\eta_2 = 1$ at the initial temperature and increased by one if $\Delta_k > 0$; average energy of the system at low temperature considering only positive jumps, is $\overline{E} = \sum_{k=1;\, \Delta_k>0}^{\eta_2} \Delta_k$.

SJ method owns more than SA tuning parameters and their proper setting is the key of SJ success (tuning is more troublesome, parameters are more sensitive). Although SJ offers, stronger than SA, space penetration, its popularity is rather small.

Stochastic tunneling (ST) tries to overcome poor behavior of SA observed on strongly rough optimization surfaces. Because of notorious freezing, the escape rate of SA from local minima diverges with decreasing temperature. The general idea of ST is to allow the particle to pass through tunnel under high energy regions between two low energy regions. This can be accomplished by applying of non-linear transformation of the criterion $K(x)$ into $K'(x)$

$$K'(x) = 1 - \exp(-\gamma(K(x) - K^*)), \tag{4.17}$$

where $\gamma > 0$ is a tunneling parameter and K^* is the goal function value, for the best solution found so far, [34]. This transformation preserves the location of minimum, but reduces the range of criteria value changes. At a given finite temperature, the search process can pass through energy barriers of arbitrary height, while the low energy-region is still penetrate well.

Generally, ST overcomes SA both in efficiency and speed of convergence, thus can be recommended instead of SA. No comparison to SJ has been made till now.

Tabu search (TS) is the LS method which mimic imitate of seeking solution by a man. It generates the search trajectory $x^0, x^1, \ldots, x^s$, by using the he greatest improvement variant of LS with slight modification, [16]. It choses x^{k+1} so that $K(x^{k+1}) = \min\{K(y) :\ y \in \mathcal{N}(x^k),\ y \text{ is not tabu}\}$, thus we have $\mathcal{N}^*(x^k) = \mathcal{N}(x^k) \setminus \{y :\ y \text{ is tabu}\}$. To avoid cyclic repetition of solutions already generated, becoming trapped in the local minimum, TS guides the search trajectory into most promising regions of the solution space by introducing the *memory of the search history*. Among many memory classes used in TS, the commonly used is short-term memory called *tabu list*. The list stored in limited time interval, the most recent solutions from the search trajectory, selected attributes of these solutions, transitions (moves) leading to these solutions or attributes of these moves, treating them as prohibition for solutions visited in the future. Performing the move having tabu attributes is prohibited (tabu), so suitable solution have to be skipped in the searching process. Tabu status can be canceled if the *aspiration function* evaluates the solution as sufficiently profitable. The stop criteria is conventional: limit of iterations has been reached, running time has been exhausted, no improvements has been observed recently, there has been found solution accepted already by the user. It has been proved that TS is theoretically convergent to optimal solution. The necessary condition is known as *connectivity property* which ensures, for any x^0, the existence of trajectory $x^0, x^1, \ldots, x^r$ so that $x^{k+1} \in \mathcal{N}(x^k)$ and $x^r = x^*$.

Currently, TS is the most promising for many practical applications, see e.g. Fig. 4.10 where TS goes quickly to optimum.

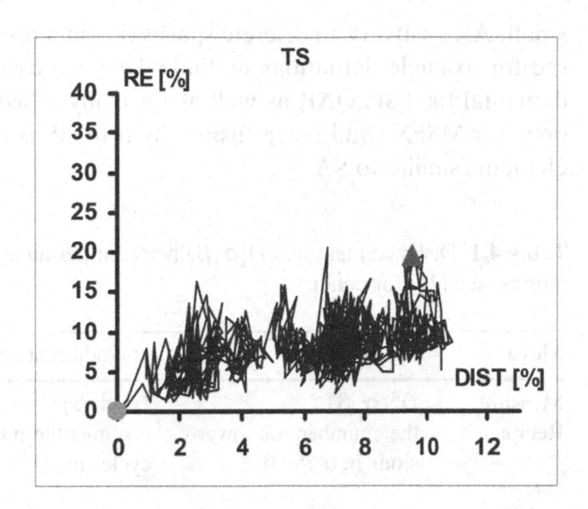

Fig. 4.10 Trajectory in the space for TS method for ta45 instance for job-shop scheduling problem. RE is the error (4.10), DIST is the distance to the optimal solution (in % od space diameter). Starting solution is marked by $\triangle$ at RE $\approx$ 20% and DIST $\approx$ 9.5%; optimal solution is at (0,0)

Adaptive memory search (AMS) is used for determining more advanced schemes of TS, going behind set of components described in the previous section, [15]. It has been confirmed in many computational tests that the skillful search process should ensure the balance between *intensification* (detailed checking small areas of the solution space) and *diversification* (ability to penetrate distant areas of the solution space). Thus AMS introduced more sophisticated memory techniques (e.g. tactic medium and strategic long-term memory, frequency/recency based memory, hash list), new strategic elements (e.g. candidate list, elite solutions, back jump tracking, strategic oscillations, paths relinking) and adaptive behavior with learning (e.g. reactive tabu).

Path search (PS) is the LS method which generates from the single solution x^0 a set of trajectories called *searching paths*, [36]. Each trajectory $x^0,\ldots,x^s$ contains solutions so that $H(x^{k+1}) = \min\{H(y);\ y \in \mathscr{N}(x^k)\}$, where $H(x)$ is an *evaluation function*, e. g. $H(x) = K(x)$. Trajectory is ended (closed) if a number of recent solutions do not improve the best solution found so far. Immediately after closing the non-perspective search trajectory, the new trajectory is generated starting from some (unvisited) solution located on already generated trajectories. Since the image of space penetration by disjoin paths imitate to star paths or decay of a particle, thus the name of the method.

Goal oriented tracing paths (GOTP) defines the LS method where search trajectory ends exactly in or around the given goal solution $x^g \in \mathscr{X}$, [27, 30]. Trajectories can be performed in either deterministic or stochastic way. In the deterministic trajectory, in the neighborhood $\mathscr{N}(x^k)$ of the current solution x^k there has been selected x^{k+1} so the distance $D(x^{k+1},x^g)$ to x^g is the smallest one among $D(y,x^g)$ for all $y \in \mathscr{N}(x^k) \setminus \{x^k\}$. In the probabilistic case, distance influences on the distribution of probability of choosing the successive solution on the GOTP. Distance to the goal is evaluated by the legal distance measure between solutions dedicated for the space $\mathscr{X}$. In the Euclidian spaces $\mathscr{X}$ the variety of recommended measures is rather small. Alternatively, in discrete spaces the final result depends on the used measure, see for example definitions in Table 4.1 for measures between permutations, fundamental for TSP, QAP, as well as for many scheduling problems. Notice, genetic operator MSFX (multi-step fusion) used in GS is in fact GOTP with some random elements similar to SA.

Table 4.1 Distance measures $D(\alpha,\beta)$ between permutations α and β in the space of permutations, see [27] for detail

Move	adjacent swap	non-adjacent swap	insert
Measure	$D^A(\alpha,\beta)$	$D^S(\alpha,\beta)$	$D^I(\alpha,\beta)$
Recipt	the number of inversions in $\alpha^{-1}\circ\beta$	n minus the number of cycles in $\alpha^{-1}\circ\beta$	n minus the length of the maximal increasing subsequence in $\alpha^{-1}\circ\beta$
Mean	$n(n-1)/4$	$n-H_n$	$n-2\sqrt{n}$ *)
Variance	$n(n-1)(2n+5)/72$	$H_n-H_n^{(2)}$	$\theta(n^{1/3})$
Complexity	$O(n^2)$	$O(n^2)$	$O(n\log n)$

*) – asymptotically $H_n=\sum_{i=1}^{n}\frac{1}{i}$ $H_n^{(2)}=\sum_{i=1}^{n}\frac{1}{i^2}$

Curtailed branch-and-bound (CB) we obtain by restricting arbitrary resources of computational process performed by classical B&B scheme, which results in the algorithm having certain compromise between running time and quality. Limitations can refer to, among others, the number of nodes in the solution tree, depth of penetration, running time, elimination threshold. CB algorithms preserve features of ancestors – exhibit explosion of calculations for instances with large size.

Randomized method (RM) does not determine particular algorithm but a general scheme which replaces in known approximate algorithms some deterministic parameters by their random counterparts, [6]. For such randomized algorithms one can find theoretical, asymptotic evaluation of quality. Randomized algorithms quite often appears in the context of so called on-line algorithms with applications in queening networks, computer networks, manufacturing systems.

Beam search (BS) can be considered as the special case of CB. It light up by beam solution space. Each illuminated point (solution) become the source of directed, limited beam containing several descending solutions, usually the most promising from the search point of view. Evaluations of the usefulness of a solution can be provided either by the goal function or by the another utility function defined by the user. Frequently, BS appears in the context of CB, where the limit refers to the limited number of immediate successors of the node in the solution tree (so called limited width of the beam) and selected successor nodes in the solution tree (so called filtered beam search), [29].

Guided local search (GLS) defines the method which is problem-oriented and directs the search toward the *most promising part of the solution space*. The desired search region is defined for particular problem using specific problem properties, [9]. Known techniques of GLS link BS with LS, which implies the side effect of calculation explosion.

Genetic search (GS) refers to the Nature and Darwin theory of evolution, assuming that non-evident goal of evolution is to find an individual best fitted to environment, [19, 17]. GS uses dispersed set of solutions $\mathscr{W} \subset \mathscr{X}$ called *population* in order to carry on the search simultaneously in many areas of the solution space $\mathscr{X}$. Each solution $x \in \mathscr{W}$ called *individual* is codded by set of its attributes, written in genetic material (chromosomes, genes). Population is controlled fundamentally by natural successive cyclic processes: *reproduction*, *crossing over*, *mutation* and *selection*. In the reproduction phase all individuals are copied proportionally to their measure of *fitness* (usually the goal function value). This process ensures that individuals better fitted will have more descendants in the next generation. Selected individuals from this extended population creates *matting pool*. From this pool we link pairs of parents, which by *crossing over* provides the *offspring*. Each individual from offspring is the solution x' obtained from parent solutions x and y through application of *crossing over operator*. Mutation is the insurance for losing important attributes of solutions and slow mechanism of introducing innovation attributes. It introduces occasional, random (with small probability), changes in genetic material. In the selection phase, the set of individuals for the next population (among individuals from the old population and offspring) has been selected.

GS owns may points of freedom. Their proper composition is a key for the success of the method. Despite many research, some shortcomings of GS algorithms are still observed in practice. These are premature convergence to local minimum of poor quality, or slow convergence to the best solution. While GS behaves well for small instances, errors of approximation for large-size instances can be significant. It has been verified experimentally that premature convergence to local extrema results from wrong control of population development dynamics. In order to improve the population fitness, the best individuals are preferred for crossing over in each generation, which results in genetic diversity reduction. This, in order, decreases the possibility of finding essentially different better solution and stops the progress till the moment when just mutation (after huge number of generations) introduces necessary change in genetic material. So the overall aim of GS is to keep fitness the best without losing diversity of population.

There are proposed technologies to control convergence of GS through *strategies of matching parents*, *structured populations* and *patterns of social behavior*. There has been introduced *sharing function* in parents fitness to prevent so close genotype similarities. Another direct approaches of *incest prevention* uses Hamming distance to evaluate genotype affinity. Structures in population can be obtained by partitioning the population onto islands (migration model) or by *overlapping neighborhoods* covering the population (diffusion model). Both models assume limited direct exchange of data between individuals from different areas, which implies *niches* ensuring desired diversity of population. Social approach assigns for each individual suitable social attitude, e.g. satisfied, glad, disappointed. The pattern of attitude follows from the criteria value and has influence on th attitude in the future, which means that individuals may react differently in the same situation. Social attitude influences on procreation model: crossing-over, mutation, cloning.

Memetic search (MS) refers to the Lamarck early theory of evolution, which allows transfer inherent as well as acquired features of an individual by using *memes*, being the informative analogy to genes in GS, [23]. MS extends GS by introducing *learning*, before crossing over, applied for each individual (usually by using DS or LS) to change permanently acquired features having also influence on the fitness evaluation. The side effect of MS is constant reduction of population diversity. To avoid this phenomenon one can apply Baldwin theory of evolution, which is slight modification of that of Lamarck. After learning, memes of an individual are not changed, only fitness value of the best one is changed.

Obtained results are better than for pure GS (the speed of convergence is better), however the running time is significantly longer. Because learning provides local extremes in place of individuals in the population, then special attention need to be paid for the preserving suitable diversity of the population.

Differential evolution (DE) works with the subset of solution called population $\{x^1,\ldots,x^s\} = \mathscr{P} \subset \mathscr{X}$. Starting population is set randomly. Instead of democracy characteristic for GS, DE fathoms space $\mathscr{X}$ by generating trial solutions built as the specific combination of selected solutions from $\mathscr{P}$, [11]. To this order, for each $i = 1,2,\ldots,s$ there is generated a trial solution

$$y^i = x^c + k(x^b - x^a), \tag{4.18}$$

where a,b,c are random numbers from the set $\{1,2,\ldots,s\}$ mutually different and also different than i, whereas k is a random parameter. Solutions x^a and x^b plays role of parents in GS. If the trial solution y^i is better than x^i, it replaces x^i; otherwise is released. Iterations are repeated until a fixed a priori number of repetitions has been reached, or stagnation has been detected.

Artificial immunological system (AIS) mimics, in construction and activity, a biological defense system, [37]. In Nature, immunological system is the complex, distributed, self-controlled, self-adaptive system which uses learning, memory, evolution and data processing in order to defense host organism against pathogens and toxins. Its fundamental aim is to recognize and classify cells into own (host) and alien (invasive). Intruder is recognized by the presence of specific alien proteins (antigens). This activates the reaction of immunological system, beginning from creation of specific antibodies which blocks receptors of antigen till activating the killers. Lymphocyte B and T are the main defense cells; thay are able to recognize host, have memory and diversity.

AIS implements only few components from the real system. *Antigen* (invasive protein) represents temporary constraints for the solution (e.g. range of some or all components of vector x) or data for some instance. The number of various possible antigens is huge (sometimes infinite) and sequence of presented antigens unknown a priori. *Antibody* (protein blocking antigen) is the procedure or algorithm how to create solution satisfying constraints owned by antigen. Variety of antibodies is small, however there exist evolutionary mechanism allowing their aggregation and recombination in order to create new antibodies with new unlike features. Patterns of antibodies are collected in the *library of antibodies* being the *immunological memory* of the system. *Matching* is the selection of antibody to antigen. *Score of matching* is ideal if antibody allows to generate solution satisfying antigen constraints. Otherwise, score is a measure of deviation of obtained solution from the ideal case. Bad score forces the system for seeking new antibodies.

Path relinking (PR) is the technique used for fathoming the solution space in order to examine its topology and landscape, [30]. Detection if the discrete problem owns big valley, location of the valey, as well as examining the topology of the space is possible by the use of trajectories going between selected pairs of distant solutions. Depending on the problem type, trajectories can be generated by simple linear combination or as the GOTP with specific measure between solutions.

Biochemical random search (BRS) starts with coding data of the instance by segments of DNA (sequences of length 10-30 nucleotides) with clammy ends, which fit each other in unique way, [4]. Solution is the sequence of segments, the optimal (feasible) solution is the sequence having the length less than a threshold value. Searching by random matching of segments is performed in the laboratory (test tube) in medium containing sufficiently large number of copies of segments. This reaction corresponds to huge number (e.g. 10^{10}) of parallel random search processes. Detection if test tube contains optimal solution is performed by specific biochemical test seeking for the DNA sequence of defined size.

Ant search (AS, ant colony optimization ACO) refers to intelligent behavior of non-intelligent but cooperating colony of individuals, having the analogy to ant flock. Searching is distributed among individuals with simple ability, which mimic real ants. The method is inspired by detecting the significance of *pheromone trail* in seeking by an ant the shortest way between anthill and food, see the source of the idea [12]. Whereas the isolated ant moves on the ground randomly, detecting the the ground pheromone path moves along this trail with probability proportional to the intensity of pheromone track. Each ant during the move signifies path by his own pheromone, increasing its intensity. Simultaneously, pheromone evaporated form ground to prevent process explosion. Assuming certain feasible paths for ants (depending on the problem type), certain model of ant behavior (ant has short term memory, limited vision, pheromone detector), starting distribution of ants on limited area and starting intensity of pheromone in the ground, there is carried out simulation of ant activity.

AS builds the analogy between solution x and the path from ant-hill to to food, and analogy between path length and the goal function value $K(x)$. The ant path is created in the connected graph $G = (N^*, E)$, with set of nodes $N^* = N \cup \{o, *\}$ and set of arcs $E \subseteq N^* \times N^*$, where o is the source (ant-hill) and $*$ is the target (food). We have $|N| = n = |x|$. Paths from o to $*$ in G represent all solutions in the space $\mathscr{X}$. Each arc $(i, j) \in E$ has assigned pheromone intensity $\tau_{ij}(t)$ which changes in time t. The path is built step-by-step from o to $*$ taking into account current intensity of the pheromone, so the probability of choosing the arc (i, j), $j \in U$, beginning from node i is equal $p_{ij} = \tau_{ij}(t) / \sum_{s \in U} \tau_{is}(t)$, for some reachable by the ant points $U \subset N^*$ and $p_{ij} = 0$ for $j \notin U$. If the arc (i, j) has been chosen by an ant, the amount of the pheromone is increased $\tau_{ij}(t) \leftarrow \tau_{ij}(t) + \Delta\tau$. The evaporation proces is given by equation $\tau_{ij}(t) \leftarrow (1 - \rho)\tau_{ij}(t)$, where $\rho \in (0, 1]$. Each ant chooses his own path, so AS operates on th set of solutions called *colony*. Ants going by shorter paths reaches goal faster. This implies that due to more intensive circulation of ants, these path becomes more intensively saturated by pheromone. There are at least two schemes of simulations: (1) ant pass from ant-hill to food is performed in one cycle and and then paths are saturated inversely proportional to path length, (2) ant pass is a dynamic process, ant velocity and amount of left pheromone are constant in time.

Up till now, ACO has been applied to many different problems with various level of the success, see the monograph [13] and the review [11].

Scatter search (SS) operates on the set of solutions called *reference solutions* containing elite solutions collected during the search, see e.g. [29]. SS generates new solutions as the *linear* or *nonlinear combination* of selected reference solutions. Instead of democracy characteristic for GS, SS introduces new technology of creating descendant solutions without losing possibility of preserving sufficient diversity of solutions. In the basic version, SS checks only reference solutions. In the advanced search, SS is supported by PR technique, [30], taking into account also solutions located on trajectories (paths) in the space between elite solutions. Nonetheless, the paper [25] showed that SS+PR is still to week to solve the problem efficiently. SS+PR offers a good combination for diversification process, and should be supported also by suitable intensification process e.g. DS or TS.

Constraint satisfaction (CS) considers the optimization problem (4.1) as the problem of satisfiability *does there exist solution* $x \in \mathscr{X}$ *so that* $K(x) \leq L$ *for the given* L, [10]. L may follow from the user expectation. It is clear that such formulation does not reduce hardness of the problem, however in some cases may provide answer *no* significantly quicker. The method operates on *decision variables* V (e.g. $(x, K(x))$ in our case), *domain* $\mathscr{D}$ for these variables and *constraints* $\mathscr{C}$ imposed on two or more variables from V. The basic step of the method constructs, in certain ordered way, the solution by *depth first* extension of current partial solution. Each such extension defines new CS problem having modified $\mathscr{D}$ i $\mathscr{C}$. New form of $\mathscr{D}$ is obtained by *constraint propagation* from $\mathscr{C}$. If the newly obtained problem is inconsistent) the search process backtracks from this stage, and the process is continued from another consistent state. CS differs by type and level of forcing consistency, mechanisms of backtracing from inconsistent stages (e.g. *back jumps*, *dynamic bactrack*, *direction-dependent backtrack*, heuristics used for generation solution extensions. CS is similar to B&B scheme and CB, therefore it shows advantageous as well disadvantageous features of these approaches.

Geometric approach (GES) refers to geometric interpretation of solution and some general theorems from probability and algebra, [32]. There is commonly applied the following scheme; (1) find obvious solution x not necessary feasible, (2) disturb randomly solution x to obtain x', (3) transform x' into feasible solution $x'' \in \mathscr{X}$ by using special approach. Although GES provides algorithms with poor experimental quality, it usually allow us to make theoretical quality analysis.

Particle swarm optimization (PSO) is inspired by movement of *swarm* (birds flock, fish school) in Nature, see source of the idea [21] or the review in [11]. It uses subset of solutions $\mathscr{W} \subset \mathscr{X}$ called *swarm of particles* for diffuse searching some areas of the solution space. The solution $x \in \mathscr{W}$ corresponds to the *location* of a particle. *Particle* goes through the space by trajectory $x_1, x_2, \ldots$ being the sequence of successive positions of this particle

$$x_{k+1} = x_k + v_{k+1} \tag{4.19}$$

where v_{k+1} is the *velocity* of the particle, defined by equation

$$v_{k+1} = w \cdot v_k + c_1 r_1 (pb_k - x_k) + c_2 r_2 (gb - x_k). \tag{4.20}$$

Pre-defined coefficients w (inertia), c_1, c_2 are set experimentally, whereas r_1, r_2 are set randomly in each step. Value gb is location of current *global best* solution (swarm leader, usually the best in the swarm), whereas pb_k the best in local neighborhood surrounding x_k (local best, usually the best in some small neighborhood of x_k, e.g. $\{x_{k-1}, x_k, x_{k+1}\}$). Although the number of PSO applications strongly increases in recent years, no major efficiency of this approach has been discovered.

Artificial bee swarm (ABS) is the population neighborhood search combined with random search and supported by cooperation (learning), mimics to the behavior of society of bees, [20]. Bee swarm collects honey in hive. Each bee performs the random path (solution) to the region of nectar penetration. After the come back, quality of bee is evaluated on the base of collected nectar. Champions are selected to perform in hive *waggle dance* in order to inform other bees about promising search regions (direction, distance, quality). The next day bees will be going into directions modified by obtained informations. The analogy to optimization problem is clear. The criterion value is inversely proportional to the collected nectar. Location of flowers being the goal of the bee is the solution. Waggle dance is the statistical characteristics of the particular region of the solution space.

Neural nets (NN) in discrete application is usually problem-oriented and so common. Most applications refers to analog model of Hopfield and Tank. Components of the solution are treated as analogous to potentials in selected points in the net, constraints are represented by neurons interactions, and goal function value $K(x)$ is analogous to net energy. For optimization process there is used net self-convergence to the minimal energy state, starting from any initial stage, [39]. In fact, NN realizes very fast DS in the single run. Starting NN from various solutions, one can obtain multi-starting DS with all their advantages and disadvantages.

Parallel methods In recent years the increase of computer power evolves via parallel architectures. Since the increase of number of processors or cores in single computer is still to slow comparing it with the increase of solutions in the space, there is no hope to vanquish NP-hardness barrier. That's why parallel metaheuristics becomes the most desired class of algorithms, [5]. In parallel programing there has been performed already several fundamental steps: (1) theoretical models of parallel calculation (SISD, SIMD, MISD, MIMD), (2) theoretical models of memory access (EREW, CREW, CRCW), (3) practical parallel calculation environments (hardware, software, GPGPU), (4) shared memory programming (Pthreads in C, Java threads, Open MP in FORTRAN, C, C++), (5) distributed memory programming, message-passing, object-based, (6) Internet computing (PVM, MPI, Sockets, Java RMI, CORBA, Globus, Condor), (6) measures of quality of parallel algorithms (runtime, speedup, efficiency, cost), (7) single/multiple searching threads, (8) granularity evaluation, (9) independent/cooperative search threads, (10) distributed (reliable) calculations in the net.

Notice, each of the mentioned in the previous paragraph methods can be implemented in parallel environment in the non-trivial way in different manner. Let us consider SA approach. We can adapt this method as follows: (a) single thread, conventional SA, parallel calculation of the goal function value, fine grain, theory of convergence, (b) single thread, pSA, parallel moves, subset of random trial solutions selected in the neighborhood, parallel evaluation of trial solutions, theory of convergence, (c) exploration of equilibrium state at fixed temperature in parallel, (d) multiple independent threads, coarse grain, (e) multiple cooperative threads, coarse grain. For GS we have: (a) single thread, conventional GA, parallel calculation of the goal function value, small grain, theory of convergence, (b) single thread, parallel evaluation of population, (c) multiple independent threads, coarse grain, (d) multiple cooperative threads, (e) distributed subpopulations, migration, diffusion, island models. These means that from several sequential methods we can create many parallel methods, so the final number of possible technologies of seeking solution is quite large. Beside the natural shortening of the running time, also phenomenon of *superlinear speedup* has been observed.

4.6 Conclusion

The given survey does not cover whole list of algorithms proposed in the literature (there has been skipped e.g. harmony, electromagnetic, intelligent water drops, hybrid search). Quality of each particular method depends on space landscape, ruggedness, big valley, distribution of solutions in the space and the problem balance between intensification and diversification of the search. Currently the promising are SA, SJ, GS, MS – for problems without any particular properties (SA and SJ for problems having high cost of evaluating single solution) and TS, AMS – for problems having special properties. For flat landscapes with small number of local extremes simply multi-start diversification mechanism is enough. Considering rough landscapes we need to ensure sufficient diversification as well as intensification mechanisms. For instances of grater size there are recommended parallel method, possible to implement already on a PC.

References

1. Aarts, E.H.L., van Laarhoven, P.J.M.: Simulated Annealing: a Pedestrian Review of the Theory and Some Applications. In: Deviijver, P.A., Kittler, J. (eds.) Pattern Recognition and Applications. Springer, Heidelberg (1987)
2. Aarts, E.H.L., Lenstra, J.K.: Local Search in Combinatorial Optimization. Princeton University Press, Princeton (2003)
3. Abdul-Razaq, T.S., Potts, C.N., Van Wassenhove, L.N.: A Survey of Algorithms for the Single Machine Total Weighted Tardiness Scheduling Problem. Discrete Applied Mathematics 26, 235–253 (1990)
4. Adleman, L.M.: Molecular Computation of Solutions to Combinatorial Problems. Science 266, 1021–1024 (1994)
5. Alba, E.: Parallel Metaheuristics: a New Class of Algorithms. John Wiley & Sons, Chichester (2005)
6. Albers, S.: On-Line Algorithms: a Survey. Mathematical Programming 97, 3–24 (2003)
7. Amin, S.: Simulated Jumping. Annals of Operations Research 86, 23–38 (1999)
8. Angel, E., Zissimopoulos, V.: On the Landscape Ruggedness of the Quadratic Assignment Problem. Theoretical Computer Science 263, 159–172 (2001)
9. Balas, E., Vazacopoulos, A.: Guided Local Search with Shifting Bottleneck for Job-Shop Scheduling. Management Science 44, 262–275 (1998)
10. Bartak, R.: On-line guide to Constraint programming (2010), http://ktiml.mff.cuni.cz/bartak/constraints/
11. Corne, D., Dorigo, M., Glover, F.: New Ideas in Optimization. McGraw Hill, Cambridge (1999)
12. Dorigo, M., Maniezzo, V., Colorni, A.: The Ant System: Optimization by a Colony of Cooperating Agents. IEEE Tansactions on Systems, Man, and Cybernetics: Part B 26, 29–41 (1996)
13. Dorigo, M., Stützle, T.: Ant Colony Optimization. Bradford Books (2004)
14. Garey, M.R., Johnson, D.S.: Computers and Intractability: A Guide to the Theory of NP-Completeness. W.H.Freeman and Co., New York (1979)
15. Glover, F.: Tabu Search and Adaptive Memory Programing - Advances, Application and Challenges. In: Barr, R.S., Helgason, R.V., Kennington, J.L. (eds.) Interfaces in Computer Science and Operations Research, Kluwer, Dordrecht (1996)
16. Glover, F., Laguna, M.: Tabu Search. Kluwer Academic Publishers, Boston (1997)
17. Goldberg, D.E.: Genetic Algorithms in Search, Optimization and Machine Learning. Addison-Wesley, Reading (1989)
18. Haupt, R.: A Survey of Priority Rule-Based Scheduling. OR Spectrum 11, 3–16 (1989)
19. Holland, J.H.: Adaptation in Natural and Artificial Systems. University of Michigan Press, MI (1975)
20. Karaboga, D., Basturk, B.: A Powerful and Efficient Algorithm for Numerical Function Optimization: Artificial Bee Colony (ABC) Algorithm. Journal of Global Optimization 39, 459–471 (2007)

21. Kennedy, J., Eberhart, R.C.: Particle Swarm Optimization. In: Proc. IEEE International Conference on Neural Networks (Perth, Australia), vol. IV, pp. 1942–1948. IEEE Service Center, Piscataway (1942)
22. Kirkpatrick, S., Gelatt, C.D., Vecchi, M.P.: Optimization by Simulated Annealing, Science. Science 220, 671–680 (1983)
23. Merz, P., Freisleben, B.: Fitness Landscapes and Memetic Algorithms Design. In: Corne, D., Dorigo, M., Glover, F. (eds.) New Ideas in Optimization. McGraw-Hill, New York (1999)
24. Nowicki, E., Smutnicki, C.: A Fast Taboo Search Algorithm for the Job Shop Problem. Management Science 42, 797–813 (1996)
25. Nowicki, E., Smutnicki, C.: An Advanced Tabu Search Algorithm for the Job Shop Problem. Journal of Scheduling 8, 145–159 (2005)
26. Nowicki, E., Smutnicki, C.: Some Aspects of Scatter Search in the Flow-Shop Problem. European Journal of Operational Research 169, 654–666 (2006)
27. Nowicki, E., Smutnicki, C.: Some New Ideas in TS for Job Shop Scheduling. In: Rego, C., Alidaee, B. (eds.) Metaheuristic Optimization via Memory and Evolution. Tabu Search and Scatter Search, pp. 165–190. Kluwer, Dordrecht (2005)
28. Panwalker, S.S., Iskander, W.: A Survey of Scheduling Rules. Operations Research 25, 45–61 (1977)
29. Pinedo, M.: Scheduling: Theory, Algorithms, and Systems. Springer, Heidelberg (2008)
30. Reeves, C., Yamada, T.: Genetic Algorithms, Path Relinking, and the Flowshop Sequencing Problem. Evolutionary Computation 6, 45–60 (1998)
31. Schumer, M., Steiglitz, K.: Adaptive Step Size Random Search. IEEE Transactions on Automatic Control 13, 270–276 (1968)
32. Sevast'janov, S.V.: On some geometric methods in scheduling theory: a survey. Discrete Applied Mathematics 55, 59–82 (1994)
33. Smutnicki, C.: Optimization and Control in JIT Manufacturing Systems. Oficyna Wydawnicza PWr, Wroclaw (1997)
34. Wenzel, W., Hamacher, K.: A Stochastic Tunneling Approach for Global Minimization of Complex Potential Energy Landscapes. Physical Review Letters 82, 3003 (1999)
35. Weinberger, E.D.: Correlated and Uncorrelated Fitness Landscapes and How to Tell the Difference. Biological Cybernetics 63, 325–336 (1990)
36. Werner, F., Winkler, A.: Insertion Techniques for the Heuristic Solution of the Job Shop Problem. Discrete Applied Mathematics 58, 191–211 (1995)
37. Wierzchon, S.T.: Artificial Immune Systems. Theory and application. EXIT, Warsaw (2001) (Polish)
38. Wolpert, D.H., Macready, W.G.: No Free Lunch Theorems for Optimization. IEEE Transactions on Evolutionary Computation 1, 67–82 (1997)
39. Zhou, D., Cherkassky, V., Baldwin, T.R., Olson, D.E.: A Neural Network Approach to Job-shop Scheduling. IEEE Transactions on Neural Networks 2, 175–179 (1991)

Chapter 5
Cooperation Particle Swarm Optimization with Discrete Events Systems by Dynamical Rescheduling

Ján Zelenka

Abstract. Currently, materials flow optimization and creating of optimal schedule are one of the main tasks of all companies for increase of competitiveness. A schedule problem in a manufacturing company is characterized as jobs sequence and allocation to machines during a time period. A variety of approaches have been developed to solve the problem of scheduling. However, many of these approaches are often impractical in dynamic real-world environments where there are complex constraints and a variety of unexpected disruptions. In this chapter cooperation of one meta-heuristic optimization algorithm with manufacturing model by the dynamical rescheduling is described. Particle Swarm Optimization algorithm solved scheduling problem of real manufacturing system. Model of the manufacturing system is represented as discrete event system created by SimEvents toolbox of MATLAB programing enviroment.

5.1 Scheduling Problem

The problem of scheduling is concerned with searching for optimal (or near-optimal) schedules subject to a number of constraints. A variety of approaches have been developed to solve the problem of scheduling. The principles of several dynamic scheduling techniques, their application and comparisons, namely dispatching rules, heuristics, meta-heuristics, artificial intelligence techniques and multi-agent systems are described in many publications [12]. A multi machine job-shop scheduling problem is to assign each operation to a machine and to find a sequence of jobs (operations) on machines that the maximal production time is minimized [5]. Scheduling is defined as the allocation of resources to jobs over time. It is a decision-making with the goal of optimizing one or more objectives [13]. The objectives can

Ján Zelenka
Institute of Informatics, Slovak Academy of Sciences
Dúbravská cesta 9, Bratislava, SK-84507, Slovak Republic
e-mail: jan.zelenka@savba.sk

J. Fodor et al. (Eds.): Recent Advances in Intelligent Engineering Systems, SCI 378, pp. 105–130.
springerlink.com

be the minimization of the completion time of jobs (makespan), mean flow time, lateness of jobs, processing cost, etc.

Scheduling plays an important role in manufacturing and production systems. The typical mathematical formulation of the schedule task is defined by the following sets:

- set of operations $O = \{1, 2, ..., j\}$;
- set of jobs $J = \{1, 2, ..., i\}$;
- set of machines $M = \{1, 2, ..., k\}$.

Finite sequence of operations which must be made within one order represents one job. One operation is a basic unit of technological process, and it is characterized by type and processing time. In case that the same operation can be made on several machines, processing time for all machines must be determined. Creation of job-shop schedule must meet some constraints, which we divide into hard and soft constraints. Hard constraints represent the technological process. They describe constraints of individual operation time dependence, manufacturing machine capacity and other manufacturing machine characteristics. Soft constraints represent the precedences which need not be fulfilled, but from the scheduling point of view it is convenient to fulfil them. The correct definition of soft constraints can reduce the search space and find better solution of the schedule. Precedences are an operation sequence which can minimize downtime or waste due to changing of manufacturing parameters. The time continuity of several operations is expressed by the precedence (the time continuity of several operations is expressed by the precedence $o_{xy} \prec o_{xy+1}$ ($x = 1, 2, ..., i$; $y = 1, 2, ..., j$), where operation o_{xy+1} cannot start sooner than operation o_{xy} finishes. Other technological constraints of operations processing on individual machines can be described as follows:

- an operation can be made on one machine only;
- an operation can be atomic, it means the producing process of operation cannot be interrupted by an arrival of other operation;
- it is specified in several processes, that two operations cannot be made on one machine at one time unit.

Every job can be specified by next input information [11]:

- operations O_j processed on machines;
- processing time p_{ijk} is the time to process j^{th} operation, i^{th} job on k^{th} machine;
- r_i is the time when it is possible to start processing job J_i;
- d_i is the required end time of job J_i;
- a_i is maximal allowed time of a job in the system $a_i = d_i - r_i$;
- w_i is the weight of job importance J_i.

Output information of a scheduling problem represents the data, which can be calculated for every job J_i of the schedule. The information consists of the following data:

- C_i is the finishing time of job J_i;
- F_i is the processing time of job J_i;

- W_i is the waiting time of job J_i in the system;
- L_i is the delay calculated by $L_i = C_i - d_i$;
- T_i is the maximal 0, L_i delay of job J_i ;
- E_i is the maximal 0, $-L_i$ advance of job J_i;
- $U_i = 0$, if $C_i <= d_i$, else $U_i = 1$ is penalty function of job J_i.

The task defined in this way belongs to the optimization problems. Optimal (near-optimal) solution of the schedule is found if several criteria and constraints are valid. Traditional approach of static schedule assumes static environment and does not assume failure of machine. Real manufacturing system assumes several types of unpredictable events, this result in the creation of a new schedule. According to [16] in manufacturing systems there are two types of events in real time:

- events related to source (failure of source, material ageing and his failure, human operator, defective materials, etc.);
- events related to job (new jobs arrival, changing job priorities, changing job deadline, etc.);

According to [2, 9, 10] dynamic scheduling is divided into four basic types:

- *reactive scheduling*: schedules are easily generated using dispatching rules. However, the solution quality is poor due to the myopic nature of these rules, which fail to provide any plan for other activities, and it is hard to predict system performance as decisions are made locally in real-time and they typically do not use global information. No firm schedule is generated in advance and decisions are made locally in real-time. Priority dispatching rules are frequently used. A dispatching rule is used to select the next job with highest priority to be processed from a set of jobs awaiting service at a machine that becomes free. The priority of a job is determined based on job and machine attributes. Dispatching rules are quick, and are usually intuitive and easy to implement. However, global scheduling has the potential to significantly improve shop performance compared to localised or myopic dispatching rules, where it is hard to predict system performance as decisions are made locally in real-time [12];
- *predictive-reactive scheduling*: this is the common approach in dynamic scheduling. It is a scheduling/rescheduling process in which schedules are revised in response to real-time events. Predictive-reactive scheduling is a two-step process. First, a predictive schedule is generated in advance with the objective of optimizing shop performance without considering possible disruptions on the shop floor. This schedule is then modified during execution in response to real-time events [12]. Predictive-reactive approaches search in a larger solution space, generate high quality schedules, and generate better system performance to increase productivity and to minimize operating costs compared with on-line scheduling and predictive scheduling. Simple schedule adjustments require little effort and are easy to implement. However, they may lead to poor system performance;
- *predictive-reactive (robust) scheduling*: most of the predictive-reactive scheduling strategies are based on simple schedule adjustments which consider shop

efficiency only. The new schedule may deviate significantly from the original schedule, which can seriously affect other planning activities that are based on the original schedule and may lead to poor performance of the schedule. It is therefore desirable to generate predictive-reactive schedules that are robust. Robust predictive-reactive scheduling focuses on building predictive-reactive schedules to minimize the effects of disruption on the performance measured value of the realized schedule [8, 17, 18]. Numerous publications or works have been published on the robust predictive-reactive scheduling topics. For example, Abumaizar and Svestka (1997) used efficiency (makespan) and stability measures to define a robust schedule. The scheduling objective is to maximize shop efficiency, and at the same time minimize the system impact caused by schedule changes. Jensen (2001) investigated different robustness measures to improve tardiness and total flow-time for machine breakdowns. In publication [8] robustness measures and robust scheduling are developed to deal with machine breakdowns and processing time variability in the case where a rightshift repair strategy is used;
- *pro-active (robust) scheduling*: robust pro-active scheduling approaches focus on building predictive schedules which satisfy performance requirements predictably in a dynamic environment. The main difficulty of this approach is the determination of the predictability measures [12].

Solution of dynamic scheduling problem can be broken into a series of static problems to be dealt by help of static scheduling methods. Depending on when we need to create a new job schedule, we use scheduling at regular intervals (rescheduling), upon arrival of new event (event rescheduling), or the hybrid way, when new job schedule is created periodically but in case urgent event arrives for job scheduling, a new job schedule is created. Dynamic scheduling uses five basic approaches for job scheduling:

- dispatching rules;
- heuristics;
- meta-heuristics (Tabu search, simulated annealing, genetic algorithm);
- artificial intelligence (neural networks, case-based reasoning, fuzzy logic, Petri nets);
- multi-agents systems.

Dispatching rules - the literature describes several types of rules, from simple to very complex rules. No set of rules can capture the complexity of scheduling requirements in dynamic environment. Therefore, to verify the efficiency and effectiveness of the rules simulation techniques are used. Experimental results show that the correct choice of rules depends not only on the characteristics of the manufacturing systems, but also on other factors, such as material flow, etc.

Heuristics - this is a frequently used approach in dealing with scheduling tasks. In combination with the set of dispatching rules, it may contribute to finding appropriate solution to scheduling tasks very significantly.

Meta-heuristics - this technique includes methods such as Tabu search, simulated annealing or genetic algorithms. All methods have been used successfully to solve different types of scheduling tasks.

Artificial intelligence - this field includes methods such as knowledge-based systems, neural networks, case reasoning, fuzzy logic, Petri nets, etc. The use of the technology is very successful in the field of machine learning and adaptive learning.

Multi-agent systems - MAS technologies are among the most progressive evolving technologies, from which we expected major benefits in addressing the job scheduling. Initial expectations have caused frustration and certain scepticism in the application of this theory into practice. Nevertheless, this approach is still a current topic, especially in research toward the development of comprehensive, robust and costeffective solutions for new generation businesses.

The following part of the chapter describes an approach to solving job scheduling based on a real manufacturing company problem.

5.2 Manufacturing System Description and Scheduling Problem Classification

The manufacturing system which is the focus of the chapter itself consists of several different machines recycling plastic materials. As a result of the process low-density polyethylene (LDPE) film is obtained, which serves as base material for waste bags production. Within one week it is necessary to establish a schedule of about 20 orders, which may change every day. This manufacturing system consideres the next real time events, when it starts to create a new schedule, namely arrival of a new order, planed order finishing, changing the order parameters, unplanned longmachine outage, unplanned buying of a primary granulate. The recycling capacity is 180 tons per month. The manufacturing system itself consists of several different machines recycling plastic materials and producing waste bags from the LDPE film. The system consists of the following machines:

- mill: allows to mince thicker pieces of plastic materials (e.g., the waste produced by the blowing machine during filter exchange);
- regranulation machine: polymerization of waste plastics leads to production of different color granulates; granulation process is marked as wet way production, i. e. waste film is cutt up into small pieces, which are washed, dried and then granulates are produced by the polymerization process;
- mixing machine: higher granulate quantity and other input additives are mixed to obtain a constant mixture for entire order (if it's possible with respect the mixing machine capacity);
- blowing machine (extruder): the polymerization process using granulates and other input additives produces LDPE film of desired shape, thickness, width and color (four different blowing machine types - extruders are available); the film is scrolled into rolls (maximum roll weight depends on blowing machine type);
- scroll machine: LDPE film is welded, punched and scrolled to desired size (two different scroll machine types are available);

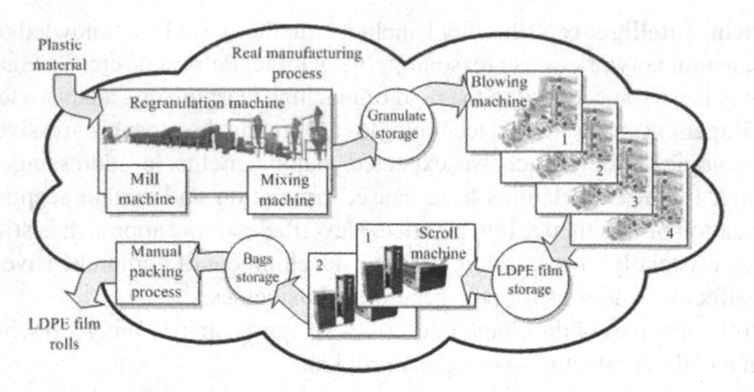

Fig. 5.1 Real manufacturing process

- manual packing process: desired quantity of rolls made on the scroll machine are marked by an identification label, packaged into bags or boxes, and stored on palettes.

The mill machine will not be considered in the schedule creation, because it is the additional component of the regranulation machine. If the thickness of plastic waste is greater than regranulation machine can process, the material is cut by the mill and then processed on the regranulation machine. The schedule problem of the manufacturing system can be defined by the next sets:

- set of operations $o = \{o_{gran}, o_{blow}, o_{scroll}, o_{pac}\}$;
- set of jobs $j = \{j_{gran}, j_{foil}, j_{bags}\}$;
- set of machines $m = \{m_{gran}, m_{ex1}, m_{ex2}, m_{ex3}, m_{ex4}, m_{scroll1}, m_{scroll2}, m_{pac}\}$.

Job (order) j_i consists of a sequence of l operations $o_{x1}, o_{x2}, \ldots, o_{xl}$ which must be processed in this job. For every operation o_{xl} and machine m_{xyl} a processing time T_{xyl} ($x = 1,2,\ldots,i$, $y = 1,2,\ldots,k$) is defined. In this case, the processing time of the blowing machine is not strictly determined, but it depends on the state of the machine after previous process operation, blown film width and thickness, on blowing machine type and granulates quality. In this case, use of conventional scheduling methods is impossible, because accurate determination of producing and processing time of the machine can be determined based on simulation only. The calculation of each processing or producing time can be simplified by introducing and using traditional scheduling method, but the scheduling result is poor. Therefore, in the following section the scheduling methods will move in heuristics or meta-heuristics field. The manufacturing system is defined by the following set of jobs:

$$j = \{j_{gran}, j_{foil}, j_{bags}\}; \tag{5.1}$$

where j_{gran} represents granulate production, j_{foil} represents the production of LDPE film roll and j_{bags} represents the production of waste bags from LDPE film. Every job is defined by the following operations:

$$j_{gran} \rightarrow \{o_{gran}\}; \tag{5.2}$$

$$j_{foil} \rightarrow \{o_{gran}, o_{blow}\}; \tag{5.3}$$

$$j_{bags} \rightarrow \{o_{gran}, o_{blow}, o_{scroll}, o_{pac}\}. \tag{5.4}$$

Creation of job-shop schedule must meet some constraints, which we divide into hard and soft constraints. Hard constraints represent the technological process. Soft constraints represent the precedences which need not be fulfilled, but from the scheduling point of view it is convenient to fulfil them. Correct definition of soft constraints can reduce the search space and quickly find better solution of the job-shop schedule. For j_{foil} or j_{bags} job we can define the following precedence set (5.5) or (5.6):

$$\{o_{granj} \prec o_{blowj}\}; \tag{5.5}$$

$$\left\{ \begin{array}{c} o_{granj} \prec o_{blowj}, o_{blowj} \prec o_{scrollj}, o_{scrollj} \prec o_{pacj} \\ o_{granj} \prec o_{pacj}, o_{blowj} \prec o_{pacj}, o_{granj} \prec o_{scrollj} \end{array} \right\}. \tag{5.6}$$

A possibility to find the optimal job-shop schedule is to search the entire space of possible combinations of loaded jobs on the operation level. If the number of loaded jobs to produce a granulate j_{gran} is n, the size of possible combinations space is n factorial combination, but in case of n jobs to produce waste bags j_{bags}, we are talking about variations with repetition, and the number of variations is expressed by the following equation:

$$spc_{jl} = (n!)^2 \times \prod_{i=1}^{n} (m_{ex_i} \times m_{scroll_i}); \tag{5.7}$$

where spc_{jl} is search space size generated on the jobs level, m_{ex} is the number of blowing machines allowing to process o_{blow} operation of the ith job, m_{scroll} is the number of scroll machines allowing to process o_{scroll} operation of the ith job. The number of possible combinations for n jobs to produce LDPE film j_{foil} is expressed by the equation (5.8).

$$spc_{jl} = (n!)^2 \times \prod_{i=1}^{n} m_{ex_i}; \tag{5.8}$$

Equations (5.7) and (5.8) are valid if each operation can be made on each machine defined for this operation. The search space size will multiply increase if jobs distributed to each operation. Because, the time generation of the search space increases exponentially with the number of operations (jobs), in the next simulation the search space is generated on the jobs level only. Fig.5.2 shows the search space size in dependence on the number of jobs (in this case only j_{bags} jobs were considered). The spc curve represents the number of possible combinations calculated by

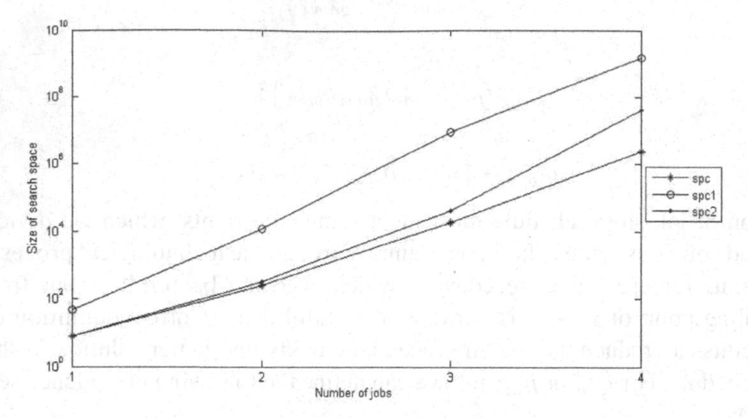

Fig. 5.2 Size of search space generated for j_{bags} jobs

equation (5.7). The increasing number of jobs in the system exponentially increases the area of the search space and also the requirements on computational complexity. Optimal schedule solution can be found by scheduling on the operation level, but at the cost of calculation time of the job-shop schedule. If individual jobs are divided to operations, then the search space can be calculated by the following equation (only j_{bags} jobs were considered):

$$spc_{ol} = \prod_{i=1}^{n} \left(m_{gran_i}, m_{ex_i}, m_{scroll_i}, m_{pac_i}\right) \times (n \times x)!; \tag{5.9}$$

where first part of the equation represents the product of all machines on which the ith job can be processed, n is a number of jobs and the x value depends on the ways of operations processing (more described in the fifth part of the chapter) and it may have value 3 or 4. The *spc1* curve represents the search space size generated on the operation level calculated by equation (5.9) where x=3. This search space contains many prohibited combination which is not possible to made, because soft constraints (precedence set) are not met. The application of the precedence set (5.6) can reduce the search space. This situation is represented by *spc2* curve in Fig.5.2. On the basis of simulation results [19, 21] the schedule result will move only in a near-optimal solution. In case, when the search space is generated on the operation level (with or without applying a set of precedence) the search space has many combinations and the simulation of all combinations takes a long time. This is why we want to use the stochastic search techniques such as evolutionary algorithms, an artificial immune system or particle swarm optimization algorithm (PSO). Application of an artificial immune system or PSO algorithm to the scheduling process reduces calculation time, but at the cost of finding the optimal solution.

The scheduling process applied in real manufacturing systems and transmitted information between real manufacturing process and scheduling process is shown

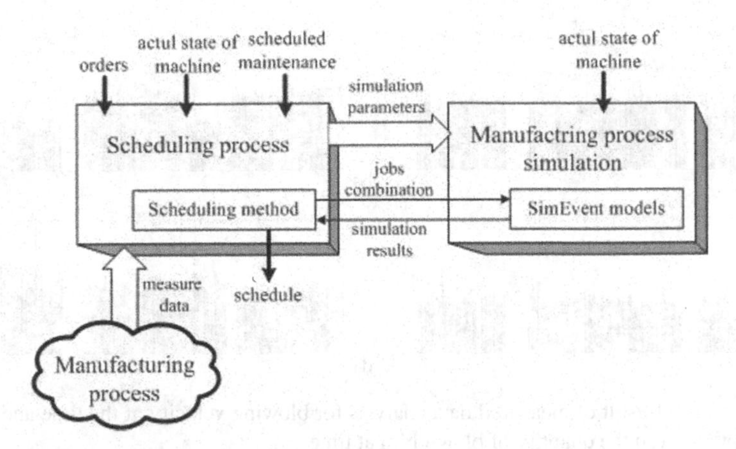

Fig. 5.3 Block diagram of the scheduling process

in Fig.5.3. Real manufacturing process is dealing with producing of recycled low-density polyethylene (LDPE) film and waste bags of desired parameters (i.e. color, shape, thickness, width,...). Detailed system description is shown in Fig.5.1. Required data are measured in the manufacturing process (i.e. power consumtion, blowing and scrolling velocity,...). Measured data are analyzed and detected the next dependences, which can influence the scheduling process. Detected parameters from the measured data are transmitted to a model as simulation parameters. A change in the manufacturing process is automatically reflected in the simulation model (i.e. machine innovation can achieve a higher blowing or scrolling velocity or the power consumption is reduced). This property is an advantage of this approach.

The scheduling process block searches for optimal (near-optimal) schedule using the implemented methods. Information about orders (jobs), actual state of machines and scheduled maintenance are input information of the scheduling process. Simulation of individual jobs is to determine which machine is suitable for the simulated order (job). A blowing machine model is created by the SimEvents toolbox, which allows to model a discrete event system . In this chapter a particle swarm optimization algorithm is considered.

Measured data contains all input information about operation processed on the machine, operation processing time, power consumption, waste and weight of the blow film (for blow machine only). The measured data are analyzed and the next dependences are detected (only dependences that directly enter into the model will be mentioned):

- blowing velocity [kg/min] versus time;
- power consumption versus quantity of blown film at time;
- filter change versus processing time or blown film quantity;
- blowing velocity [m^2/min] versus LDPE film thickness;
- blowing velocity [m^2/min] (desired thickness) versus filter purity.

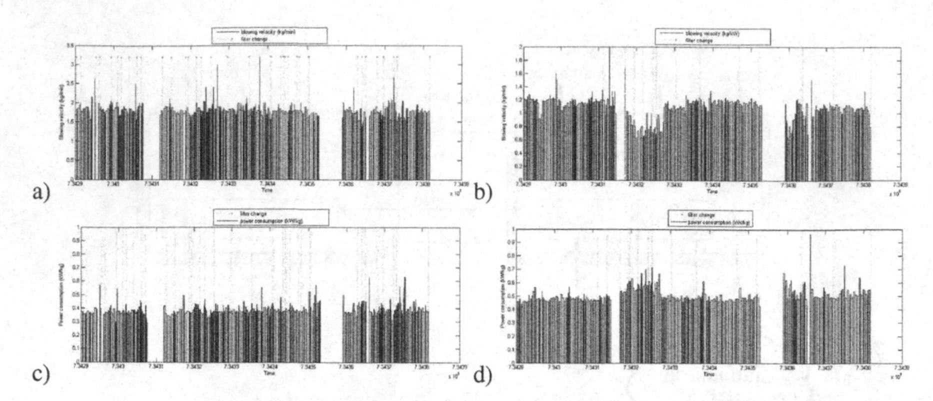

Fig. 5.4 a) - d) Result of measured data analysis for blowing velocity at the time and power consumption given the quantity of blown film at time

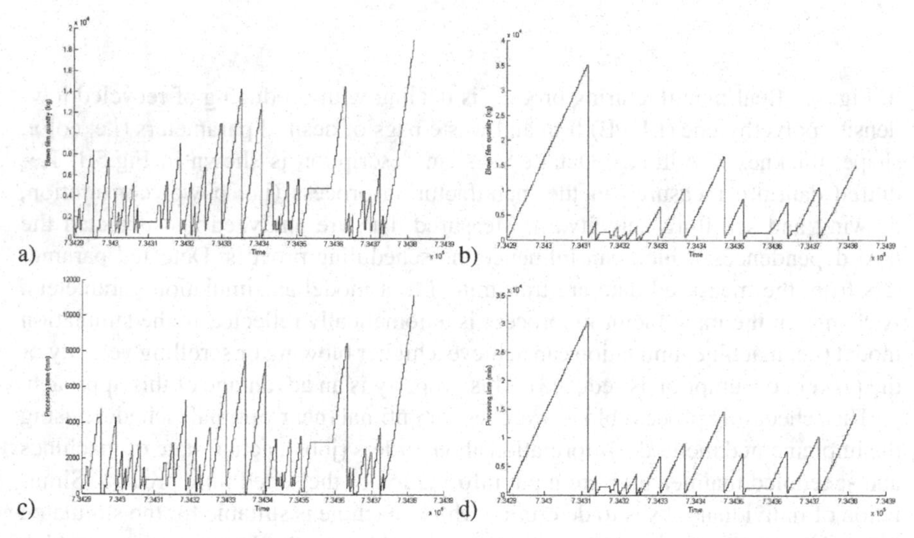

Fig. 5.5 (a - d) Result of measured data analysis for filter change given the time or blown film quantity

The results shown in Fig.5.4 -5.7 illustrate the data measured for three months. It is evident from the first two dependencies result that the blowing velocity (Fig.5.4a, b) is not a constant value and values of other parameters (i.e. ambient temperature, granulate purity) affect the velocity. Power consumption is approximately constant value (Fig.5.4c, d). In some cases minor changes may be due the granulate quality as shown in Fig.5.4d). It is evident from the result analysis of the third dependence (Fig.5.5) that the filter change is not dependent only on the blown film quantity (Fig.5.5a, b) or processing time length (Fig.5.5c, d) but it depends on the granulate quality too. Fig.5.6 shows that the blowing velocity is decreasing exponentially with

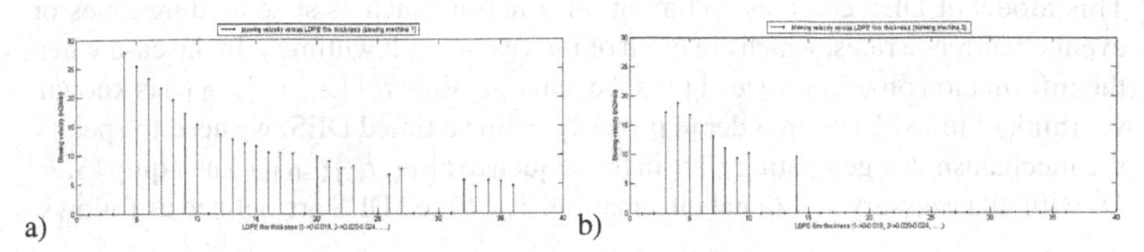

Fig. 5.6 a) - b) Result of measured data analysis for blowing velocity given the LDPE film thickness

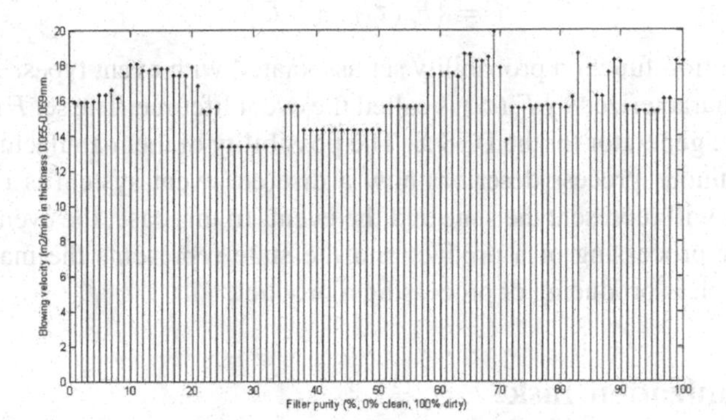

Fig. 5.7 Result of measured data analysis for blowing velocity (desired thickness) given the filter purity (0% clean - 100% dirty)

increasing LDPE film thickness. As shown in Fig.5.6a) or Fig.5.6b), the dependencies are different for each machine. The model described below simulates a LDPE film production where the blowing velocity decreases in dependence on blown film quantity. The filter is changed at velocity boundary value. The change blowing velocity given the filter purity is shown in Fig.5.7.

5.3 Manufacturing System as Discrete Event System

Discrete systems can be controlled by time (time-driven) or event (event-driven), i.e. the system state changes depend on time or event. This chapter deals with discrete event system (DES; the discrete systems controlled by events). Cassandras and Strickland [3] characterized discrete event system by the state space S (a countable set), set of asynchronous event E (a countable set), which correspond to the transition between states and a transition function D (D: $S \times E \rightarrow S \cup \Lambda$), where Λ denotes a null element used to indicate that the transition is undefined).Then the discrete event system can be defined as follows:

$$\Sigma = (S,E,D); \tag{5.10}$$

This model of DES contains no timing information, such as state holding times or event occurrence rates, which are often of interest (e.g. downtime). In the case when the information on occurrence of the kth event e^k at time t_k ($\{ e_k, t_k \}_{k=0,1,}$) is known, we think of timed DES. In order to properly define a timed DES, we need to specify the mechanism for generating all t_k in the sequence ($\{ e_k, t_k \}_{k=0,1,}$), and equip (S, E, D) with all necessary additional information. The timed DES are defined as follows:

$$\Sigma = (S, E, D, F); \qquad (5.11)$$

where

$$F = \{F_e(\tau_e) : e \in E\}; \qquad (5.12)$$

is the distribution function probability set associated with event types. The random variable τ_e characterized by $F_e(\tau_e)$ is called the event lifetime. The set F refers to the event lifetime generator for the DES Σ. The possibility of the manufacturing system with a continuous process describes how a discrete event system is necessary to specify what will represent the state and the event. In this case, the event represents producing or processing of a product and the state represents the manufacturing machine which is producing or processing a product.

5.4 Optimization Task

A multi machine scheduling problem is to assign each operation to a machine and to find a sequence of jobs (operation) on machines that the maximal production time is minimized [5]. By creating the convenient criterion we can consider the following:

- material loss minimization by color change (color change from dark to light by granulate or LDPE film production is used);
- material loss minimization by parameters change (width, thickness or color modification during LDPE film production, but the material loss is increasing);
- downtime minimization;
- production time minimization (we can divide one order into two or more lines and thus to reduce the production time);
- power consumption minimization.

In this case, we target production time minimization on individual machines, downtime and power consumption minimization and material loss minimization by parameters or color change.

5.5 Scheduling Algorithm and Algorithm Flow

Particle swarm optimization was developed by Kennedy and Eberhart (1995) as a stochastic optimization algorithm based on social simulation models. The algorithm employs a population of search points that moves stochastically in the search space. Concurrently, the best position ever attained by each individual, also called its experience, is retained in memory. This experience is then communicated to a part of or to whole population, biasing its movement towards the most promising regions detected so far. The communication scheme is determined by a fixed or adaptive social network that plays a crucial role as to the convergence properties of the algorithm [4]. The PSO algorithm searches for the best solution over the complex space through cooperation and competition. First of all, the PSO algorithm creates the initial particle swarm, namely, it initializes a swarm of particle randomly in the available solution space, making each particle an available solution of the optimization problem. Furthermore, the target function determines the fitness value through the target function. Each particle will move in the space of the solution, with its direction and distance determined by speed. The general particle will move following the best current particle, obtaining the best solution by searching generation by generation. In each generation, the particle will trace two limited values, one of which is the best solution, p_{best}, which is found so far by the particle itself. The other is the best solution, g_{best}, which has been found so far by general group swarm.

PSO is a population-based algorithm, i.e., it exploits a population of potential solutions to probe the search space concurrently. The population is called the swarm and its individuals are called the particles; a notation retained by nomenclature used form similar models in social sciences and particle physics. Suppose the searching space is D-dimensional. The swarm set is composed by l particles, where the ith particle represents a D-dimensional vector $x_i = (x_{i1}, x_{i2}, ..., x_{iD})^T$, $i = 1, 2, ..., l$. The ith particle's "flying" velocity is also a D-dimensional vector, denoted as $V_i = (v_{i1}, v_{i2}, ..., v_{iD})^T$. Denote the best position of the ith particle as $P_i = (p_{i1}, p_{i2}, ..., p_{iD})^T$, and the best position of the colony as $P_g = (p_{g1}, p_{g2}, ..., p_{gD})^T$. According to the principle of current best particle, the velocity and the position of the particle x_i will change according to (5.13) and (5.14),

$$v_{id}^{(t+1)} = v_{id}^{(t)} + c_1 r_1 \left(p_{id}^t - x_{id}^t\right) + c_2 r_2 \left(p_{gd}^t - x_{id}^t\right); \tag{5.13}$$

$$x_{id}^{(t+1)} = x_{id}^{(t)} + v_{id}^{(t+1)}; \tag{5.14}$$

where t represents the iterative number, r_1 and r_2 are random variables uniformly distributed within [0,1], c_1 and c_2 are weighting factors (the cognitive and social acceleration coefficient), usually c_1=c_2=2. Particles velocities on each dimension are limited to a maximum velocity v_{max}. If the sum of acceleration would cause the velocity on that dimension to exceed v_{max}, which is a user specified parameter, then the velocity on that dimension is limited to v_{max}. The v_{max} is therefore an important parameter. It determines the resolution, or fineness, with which regions between the present position and the target (best so far) position are searched. If v_{max} is too high, particles might fly past good solutions. If v_{max} is too small, on the other hand, particles may not explore sufficiently beyond locally good regions. In fact, they could become trapped in local optima, unable to move far enough to reach a better position in the problem space [4]. We often set it at about 10%-20% of the maximal number of possible combination.

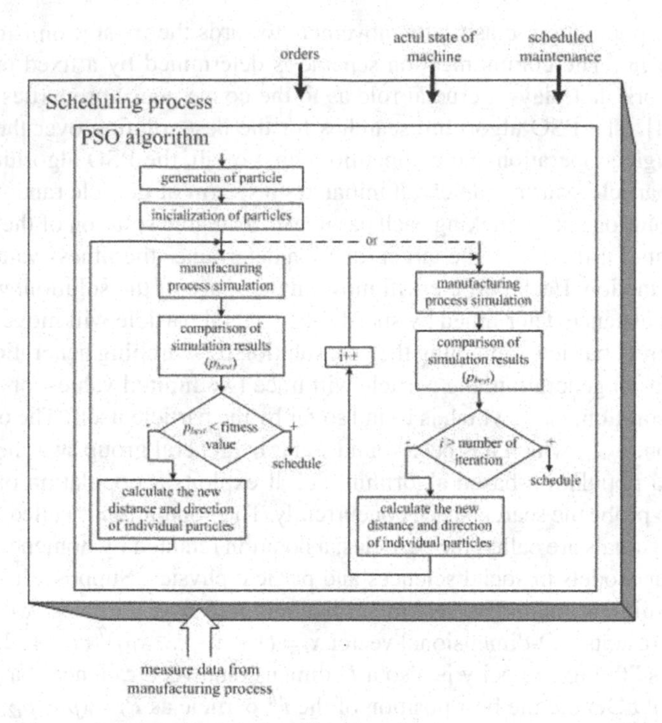

Fig. 5.8 Particle swarm optimization algorithm flowchart

For manufacturing system simulation and calculating job-shop schedule a software application in MATLAB programming environment has been created. This application allows to simulate the manufacturing process and to find near-optimal (optimal) solution of the schedule. Algorithm flowchart is shown in Fig.5.8. At the start of simulation, required quantity of particles are generated, each particle is initialized by an operations or jobs sequence, which is subsequently simulated. These sequences can be selected from the search space reduced by the set of precedence (5.6), but generation of the search space on the operation level for 4 jobs j_{bags} takes 98,5 hours, this has negative influence on the calculation time [19]. On the other hand, the search space contains many prohibited operation sequences if it is not reduced by the set of precedence. This solution positively influences the calculation time but negatively influences the particle swarm optimization algorithm parameters. All combinations are generated by the factorial number system, also called factoradic. This is done because we need to generate only one combination of operation or job sequence and not all combinations at once or a random sequence, as implemented in MATLAB programming environment. Fig.5.2 shows the search space size in dependence of the number of jobs (in this case only j_{bags} jobs were considered).

Each job (operation) sequence is simulated by the SimEvents model. By the simulation of j_{bags} jobs the sequence of operations can consider the following ways of operations processing:

- i^{th} operation o_{scroll} of the ith j_{bags} job is processed on the scroll machine only after production of the ith operation o_{blow} on the blowing machine; in this case the simulation time of one sequence is less but the size of searching space is greater than the following way. The number of possible combination can be calculated by equation (5.9) where parameter x=4 (one j_{bags} job contains 4 operation (5.4));
- in this case an optimal solution of operations sequence o_{blow}, o_{scroll} of the ith job processed on the blowing and scroll machines is searched, i. e. operation o_{scroll} may be processed on the scroll machine during o_{blow} operation processing already; each o_{blow} operation includes the information how many LDPE film rolls must be produced; in case if the number of rolls which must be produced during o_{blow} operation is greater than one and the processing time on the scroll machine is shorter than production time on the blowing machine, it is preferable then when the same job operations follow in order, in this case it is necessary to determine when rolls processing on the scroll machine starts. If the found number is too small then downtime on the scroll machine occurs and the scroll machine will be unnecessarily reserved for this operation. On the other side, rolls processing on the scroll machine finishes later if the found number will be large. In some cases it is a compromise choice between the downtime and subsequent finish time. Using this way can save the production time of the job, approximately about 20-30 percents for one job [19]. The number of possible combination can be calculated by equation (5.9) where parameter x=3 (operations o_{blow} and o_{scroll} are merged together) what is represented by *spc1* curve in Fig.5.2 (*spc2* curve in Fig.5.2 has represented reduced search space by the precedence set (5.6));

Each operation is assigned to a machine on the basis of simulation results. If one operation can be processed on several machines (i.e. o_{blow} operation can be processed on four blowing machines or o_{scroll} operation can be processed on two scroll machines) then a penalty value for each machine is calculated and the machine with the smallest/largest penalty value is assigned to an operation. Penalty value contains deadline date delay, power consumption, downtime and waste produced. Machine selection may be made by mutual comparison of individual variables (it is necessary to set the priority of individual variables) or individual variables may be converted for a single value (i.e. price). After the simulation of the entire sequence of operations (one particle) a penalty value of the schedule is calculated. Individual particle swarm schedules are compared with each other. The particle with the best penalty value is marked as g_{best} value. The best swarm value g_{best} is compared with the p_{best} value, which represents the best operations sequence found during the whole algorithm. If a penalty value of operation sequence schedule saved in p_{best} value is less than the required target function or if a number of iterations is greater than the required number of iterations, then searching for the best operation sequence in the search space finishes. The p_{best} value contains an operation sequence with near-optimal schedule. If the previous conditions are not met, then new velocity

Table 5.1 Job parameters

Num.	LDPE film width	LDPE film length	LDPE film thickness	Number of rolls	Number of bags on the roll	Color	Type	Start date	Delivery date	Quantity
1	450	1200	0.035	1000	100	Black	H	2.9.2010	13.9.2010	400
2	550	1100	0.045	1200	85	Blue	H	2.9.2010	13.9.2010	400
3	500	900	0.050	900	80	Red	H	2.9.2010	13.9.2010	400

value and new position of the particle according to (5.13) and (5.14) are calculated. Each particle initializes a new operation sequence, which is simulated as follows. These steps are repeated until the required target function or the maximum number of iterations is met.

In case of violation, the corresponding velocity value is set directly to the closest velocity bound, i.e.,

$$v_{id}^{(t+1)} = \begin{cases} v_{max}, & if\, v_{id}^{(t+1)} > v_{max} \\ -v_{max}, & if\, v_{id}^{(t+1)} < -v_{max} \end{cases} \tag{5.15}$$

For illustration only three jobs for waste bags producing were simulated. The parameters of individual jobs are shown in Table5.1. The first part of the simulation result confirms that using the PSO algorithm it is possible to find near-optimal solution of job-shop schedule. The search space size was 40320 possible combinations, the velocity range was [-10,10], the swarm contained five particles and the simulation finished after fifty iterations. The calculation time of the PSO algorithm

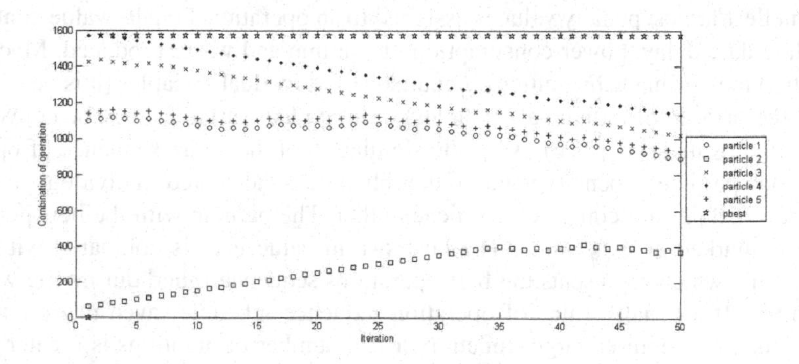

Fig. 5.9 Particle transition through searching space, swarm of five particles, velocity range [-10,10], fifty iterations and the search space consists of 40320 combinations

Table 5.2 Summary of results

	T_{PST}	T_{PFT}	BM	SM	PC [kW]	WF [kg]	SA	T_{CAL} [s]	PV
PSO_1	1-2.9.2010 16:55:59	1-9.9.2010 12:50:59	BM_3	SM_1	3658	154	1680	7560	+9
	2-3.9.2010 5:41:59	2-13.9.2010 6:55:00	BM_1	SM_2	4628	247			
	3-2.9.2010 6:00:00	3-7.9.2010 12:09:59	BM_1	SM_1	2683	190			
PSO_2	1-3.9.2010 11:40:59	1-8.9.2010 12:50:59	BM_3	SM_1	3658	154	1680	7565	+8
	2-2.9.2010 16:55:59	2-10.9.2010 6:55:00	BM_1	SM_2	4638	273			
	3-2.9.2010 6:00:00	3-10.9.2010 13:04:59	BM_2	SM_1	3077	129			

took 2 hours. The simulation result is shown in Fig.5.9. In the second simulation the velocity range was changed to [-35,35] (search space size and simulated jobs remained unchanged). The simulation result is shown in Fig.5.10. It is obvious from this simulation that the velocity value influences the result considerable. If velocity is high, then the simulation finds near-optimal solution during several iterations (see Fig.5.10), but at the cost of solution. Table5.2 shows the simulation results of three waste bags production jobs, where T_{PST} is the processing start time, T_{PFT} is the processing finish time, *BM* is the blowing machine processing a job, *SM* is the scroll machine processing a job, *PC* is power consumption, *WF* is the quantity of LDPE film waste due to rolls changing, *SA* is the search space size (the number of possible combinations), T_{CAL} is the calculation time and *PV* is the penalty value calculated for every job-shop schedule. In this case better solution can be omitted (see Fig. 5.9 or Table 5.2.). If the velocity is low, then the simulation needs more iterations to find a solution (see Fig. 5.9). It is evident from the comparison of the first (PSO_1 in Table 5.2) with the second (PSO_2 in Table 5.2) simulation result that the velocity value influences the solution found. In this case the PSO_1 schedule has lower power consumption (by about 400kW), higher waste of LDPE film, and the penalty value is higher too. The penalty value expresses the number of days of each job before the deadline. Fig. 5.11 shows the simulation result with 150 iterations. In this case the penalty value was the same as in PSO_1.

The disadvantage of the PSO algorithm (in this case) is that, in some cases, the same combination is simulated repeatedly by other particle, which influences the calculation time negatively throughout the simulation cycle. The solution would be to save the individual particles simulation result but a longer simulation could have problem with memory addressing. The next disadvantage is initialization of particle at simulation start. If individual particles are close together, the result may end up in a local minimum. Therefore, each particle will contain an operation of various jobs

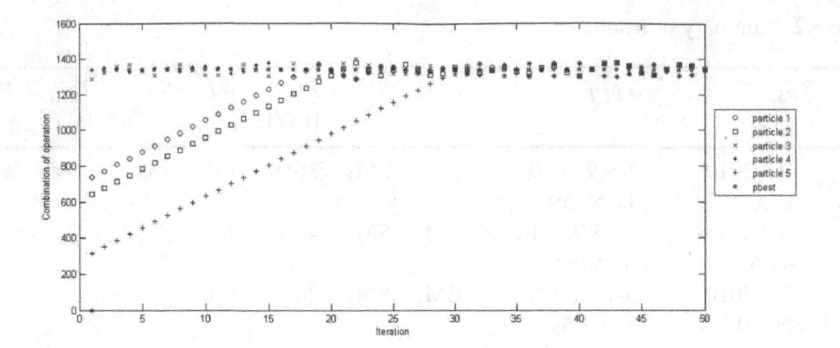

Fig. 5.10 Particle transition through searching space, swarm of five particles, velocity range [-35,35], fifty iterations and the search space consists of 40320 combinations.

on the first place of a combination at initializing of particles. Simulation result of PSO algorithm is represented by operations sequence with assigned machine in text or graphical form. In the text form there is information about operation's processing start and finish time, power consumption, the quantity of LDPE film waste due to rolls changing or downtime.

Finding optimal (near-optimal) schedule solution in search space with many possible combinations is a calculation time task demanding. Scheduling method finishes if the required target function is met. Wrong selection of the target function may cause an endless algorithm loop. Fixed setting of the number of iteration steps achieves secure algorithm ending. The PSO algorithm parameter, namely v_{max} value has to assimilate to fix a number of iteration steps. Particular solution of this problem can be the parallel computing.

Matlab programming environment supports parallel computing on the local machine or the cluster. Parallel Computing Toolbox software allows to run as many as eight MATLAB workers on local machine. MATLAB Distributed Computing Server software allows to run as many MATLAB workers on a remote cluster of computers as the licence allows. Parallel Computing Toolbox software allows to use the following ways for parallel computing:

- parallel for-loops (parfor): useful in situations where many loop iterations of a simple calculation are needed. The parfor command can be uniformly distributed between individual cores;
- single program multiple data (spmd): allows seamless interleaving of serial and parallel programming. The spmd that identical code runs on multiple labs. One program is run by the MATLAB client, and those parts of it labeled as spmd blocks run on the labs. When the spmd block is complete, a program continues running in the client;
- parallel jobs in a cluster: a parallel job consists of only a single task that is run simultaneously by several MATLAB workers, usually with different data.

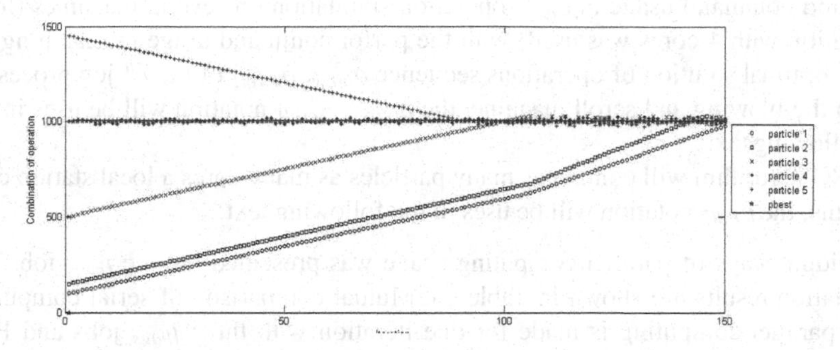

Fig. 5.11 Particle transition through searching space, swarm of five particles, velocity range [-10,10], 150 iterations and the search space consists of 40320 combinations

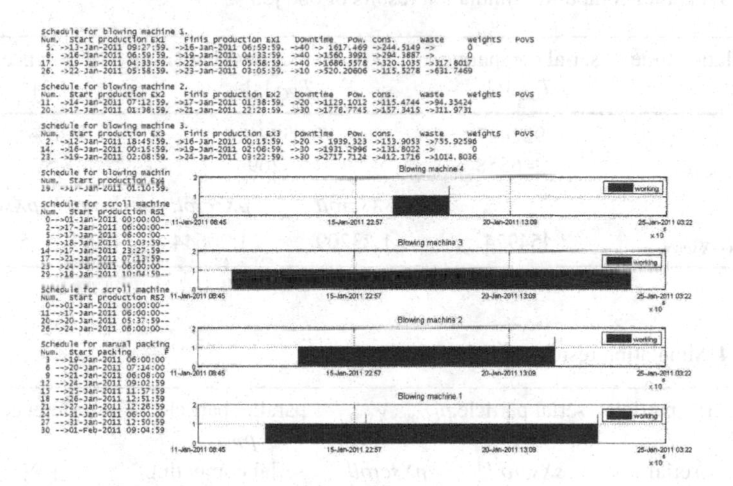

Fig. 5.12 Schedule in text or graphical form

The first two ways have an advantage, they can run on local station, but their application has to fulfil the conditions with regard to variables. The third way does not need to fulfil the conditions with regard to variables as the first two ways, but the communication between the local station and cluster can be longer than at the local station. For parallel computing application in scheduling process the following locations were identified:

- spmd command usage in o_{blow} operation simulation on several machines (local station with 4 cores was used), the p_{blow} notation will be uses in the following text;
- the parfor command usage in searching for an optimal solution of operations sequence o_{blow}, o_{scroll} of the ith job processed on the blowing and scroll machines, the $p_{Xscroll}$ notation will be uses in the following text;

- spmd command usage in o_{blow} operation simulation on several machines (local station with 4 cores was used) with the parfor command usage in searching for an optimal solution of operations sequence o_{blow}, o_{scroll} of the ith job processed on the blowing and scroll machine, the $p_{blow+Xscroll}$ notation will be uses in the following text;
- PSO algorithm will contain as many particles as many cores a local station contains, the p_{PSO} notation will be uses in the following text.

Individual ways of parallel computing usage was presented on one j_{bags} job. The simulation results are shown in Table5.3. Mutual comparison of serial computing with parallel computing is made for one iteration with three j_{bags} jobs and PSO algorithm contains four particles. The simulation results are shown in Table 5.4.

Table 5.3 Parallel computing simulation results of one job

calculation time	serial computing T_{SC} [s]	parallel computing T_{PC} [s]		difference [%]	
p_{blow}	0.922027	0.658911		28%	
$p_{Xscroll}$	1.246558	1.202699		3.5%	
		sXscroll	*pXscroll*	*sXscroll*	*pXscroll*
$p_{blow-Xscroll}$	2.454974	1.23209	1.20644	49.8%	50.8%

Table 5.4 Simulation results of one iteration

	serial particle	serial particle,$p_{blow-Xscroll}$		parallel particle p_{PSO}	difference		
	serial computing	*sXscroll*	*pXscroll*	serial computing	[%]		
T_{cal}	37.432529s	31.36009s	30.53902s	14.617437s	16.2	18.4	60.6

From the simulation result two positions in the parallel computing application have been identified. The use of the SPMD function for o_{blow} operation simulations and of parfor loop for the use in searching for an optimal solution of o_{blow}, o_{scroll} operations sequence saves approximately 51 percent of the calculation time compared to serial computing. A better result was achieved in parallel computation of individual particles, and jobs simulation was computed in series. The calculation time of one iteration took 14.6 seconds. Compared with serial calculation the time is reduced by 61 percent.

5.6 The SimEvents Toolbox and Model Description

The SimEvent toolbox extends utilization of Simulink which is a part of Matlab software tools for modeling and simulating a discrete-event system. With SimEvents we can create a model of discrete event systems to simulate passing of entities through a model created by queues, servers and other event based blocks.

A discrete-event simulation (event-based simulation) permits the system's state transition to be ruled by asynchronous discrete incidents called events. By contrast, a simulation based solely on differential equations in which time is an independent variable is a time-based simulation because state transitions depend on time. Simulink is designed for time-based simulation, while SimEvents is designed for discrete-event simulation. Discrete-event simulations involve discrete items of interest. These items are called entities in SimEvents. As noted previously, entities can pass through a network of queues, servers, gates, switches, etc.

Event occurrences cannot be represented graphically, but their occurrence can be assessed by observation of the consequences of using the Scope blocks (e.g. Instantaneous Event Counting Scope). In SimEvent, models are created by the "drag and drop" method. SimEvents uses two connection line types, an entity and a signal connection line. The difference is ensured by the different type of port at block input and output. Interconnection of entity and signal connections is excluded. Before we start to create a model, the setting of Simulink must be changed to simulation of discrete-event system by the command "*simeventsstartup('des')*" in Matlab command line. If we need to combine discrete-event simulation with a continuous system we have to use the hybrid setting through the command "*simeventsstartup('hybrid')*". In the following part the manufacturing models will be described.

Fig.5.13 shows model of blowing and scroll machine, which allows to find a compromise between the downtime on the scroll machine and subsequent finish time of job. In other words, the model allows to adjust a number LDPE film rolls produced by the blowing machine when the LDPE film rolls processing on the scroll machine can start. The input information of this model is the information about film thickness, LDPE film width, total weight of produced film, blowing velocity, blowing and scroll machine power consumption and dependence obtained by analyzing measured data from the manufacturing process (blowing film velocity versus blown film quantity, blowing film velocity versus blown film thickness, power consumption versus blow film thickness). The first block of the model represents the blowing machine. On the basis of input information, this block generates a command to generated an entity. One generated entity represents one LDPE film roll. The third block sets attributes to generated entity (length and weight of LDPE film roll). Then the entity is transmitted to a server marked "Storage of LDPE film roll". If the number of entities in the server equals to X parameter (the number of LDPE film rolls produced by the blowing machine when the LDPE film rolls processing on the scroll machine can start) then one entity from the LDPE film roll storage is transmitted trough Gate and Get attribute block to the Delay block. This block is used to hold entity as long as it will be moved to the server marked as the scroll machine. During this delay the subsystem block calculates the processing time of the roll by the scroll

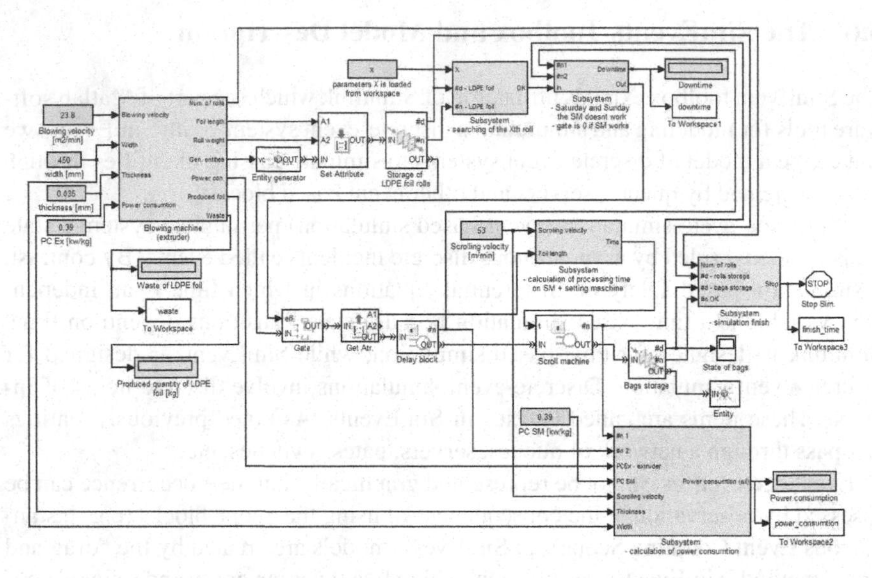

Fig. 5.13 Blowing machine and scroll machine model created by Simulink

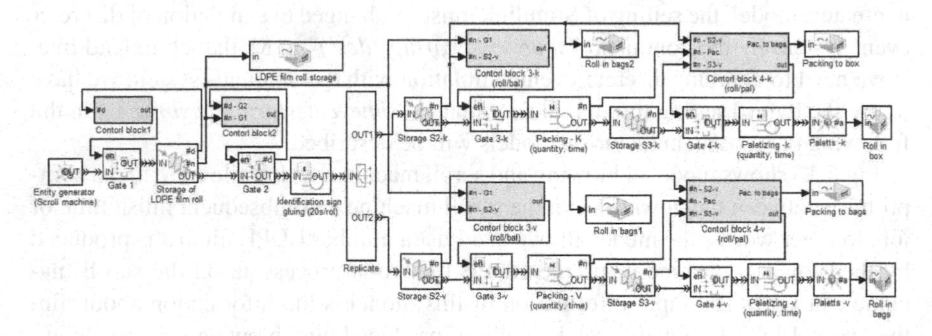

Fig. 5.14 Pack model of LDPE film rolls

machine. The simulation process ends when all entities are processed by the scroll machine. Information about power consumption (separately for each machine or together), downtime on the scroll machine and waste produced by blowing machine represent the model simulation results.

The pack model is shown in Fig.5.14. The first two blocks (Entity generator and Gate 1) are used to produce the required number of entities (the number of entities is equal to the number of rolls produced by the scroll machine). If the block marked "Storage of LDPE film roll" includes a required number of entities, the entity generation stops and the simulation rolls packing starts. In this model one entity represents one LDPE film roll produced by the scroll machine. Each roll has to be marked by identification label. This process is conducted in server block "Identification label glueing". Then entities are transmitted through replicate block to the storage block "Storage S2". The upper (lower) part of the model represents roll packing into boxes (bags) and subsequent palletizing. If there are more rolls in the storage than the required quantity of rolls per a box (bag), then the required number of rolls is transferred to the next storage block (box or bag with the required number of rolls). Entities from this storage are transmitted to the "Pallets" block by the N server "Palletizing" block. Individual processing times and required roll quantities per a box, bag or pallets are set up before the simulation starts. Transfer of entities through the pack model is shown in Fig.5.15 a) - d). Actual state the transfer of entities to storage boxes with LDPE film rolls, which is represented by the "Storage S3" block on the pack model, is given in Fig.5.15a). The time needed to fill individual pallets can be deduced from Fig.5.15b). In this case, four pallets were filled by 400 rolls each, except the last one which contained 300 rolls only, because it was necessary to produce 1500 rolls only. Actual state of storage with marked rolls is shown in Fig.5.17a). Here we can see, how the storage is being filled up with marked rolls and how the person packing rolls cannot manage to pack rolls into bags.

By the SimEvents model it is easy to determine the behavior of the manufacturing process to change the manufacturing system structure (i.e. to add another person packing rolls into bags). Fig.5.16 shows a part of modified pack model, where two persons pack rolls to bags. Operation processing time is shorter by about a third compared to the previous case Fig.5.15c), d). Actual state of storage with marked rolls is shown in Fig.5.17b). It is evident from the graph that in this case the storage with marked rolls won't be very full as in the previous case (Fig.5.17a)). Downtime of the person glueing labels on the roll is reduced to minimal value. The label gluing in the first case took approximately 500 minutes and packaging process finished in 790 minutes. The difference of these times is the downtime, in first case

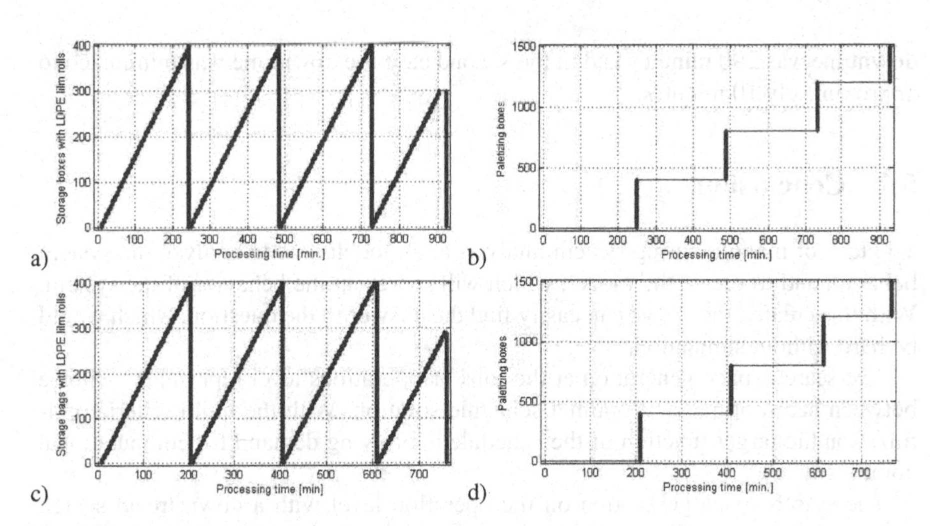

Fig. 5.15 a) - d) Simulation results

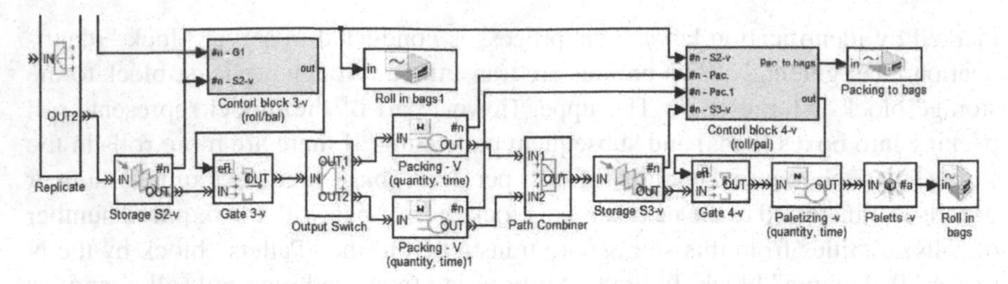

Fig. 5.16 Pack model with two workers packing rolls into bags

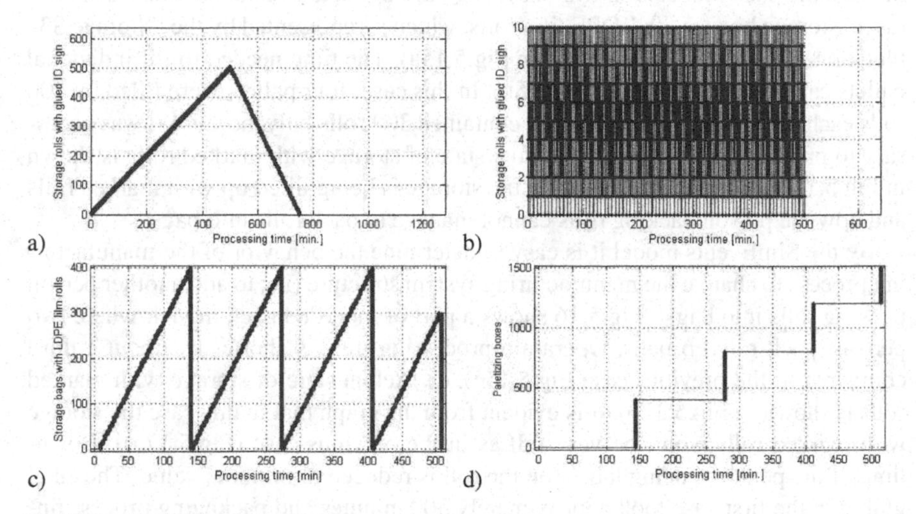

Fig. 5.17 a) - d) Simulation results

downtime was 290 minutes and in the second case the downtime was minimized to approximately 10 minutes.

5.7 Conclusion

The task of manufacturing system analysis and modeling is to analyze the system behavior and to create the model, which will represent the behavior of the system. With this created model we can easily find the answer to the question, which would be hard without simulation.

The search space generation at the jobs or operations level represents a choice between near-optimal and optimal schedule solution. With the multi-criteria optimization the target function of the schedule is growing demand for computational complexity.

The search space generation on the operation level with a downstream search space reduction using a precedence set and searching of entire area allows to find optimal schedule solution, but at the cost of computational complexity. This task cannot be solved for a larger number of orders in real time. The search space generation on the job level with followed manufacturing system simulation allows to find only near-optimal solution of the schedule. The speed of finding appropriate solutions largely depends on the chosen approach. The approach presented in the chapter was used for scheduling machine performance in time. The overall production cost (time) must include cost and time of optimization. Cost and time of optimization must be directly proportional to a found schedule, otherwise the optimization loses its meaning.

Acknowledgements. This work has been supported by the APVV scientific grant agency, grant No. 0168-09 "Non-conventional methods for recycling production lines optimization" and by VEGA scientific grant agency, grant No. 2/0197/10 "Smart methods and algorithms for intelligent integrated control of production systems.".

References

1. Abumaizar, R.J., Svestka, J.: Rescheduling job shops under random disruptions. International Journal of Production Research 35(7), 2056–2082 (1997)
2. Aytug, H., Lawley, M.A., McKay, K., Mohan, S., Uzsoy, R.: Execting production schedulesin the face of uncertainties: A review and some future directions. European Journal of Operational Research 161(1), 86–110 (2005)
3. Cassandras, G.C., Strickland, G.S.: Sample Path Properties of Timed Discrete Event Systems. Discrete Event Dynamic Systems Analyzing Complexity And Performance in The Modern World, 21–33 (1989); ISBN 0-87942-281-5
4. Ebenhart, R.C., Shi, Y.: Particle Swarm optimization: Developments, Applications and Resources. Evolutionary Computation 1, 81–86 (2001)
5. Frankovic, B., Budinska, I.: Single and multi-machine scheduling of jobs in production systems. In: In Advances in Manufacturing: Decision, Control and Information Technology. ch. 3, pp. 25–36. Springer, Heidelberg (1998)
6. Jensen, M.T.: Improving robustness and flexibility of tardiness and total flow-time job shops using robustness measures. Applied Soft Computing 1(1), 35–52 (2001)
7. Lian, Z., Jiao, B., Gu, X.: A similar particle swarm optimization algorithm for job-shop scheduling to minimize makespam. Applied Mathematics and Computation 183, 1008–1017 (2006)
8. Leon, V.J., Wu, S.D., Storer, R.H.: Robustness measures and robust scheduling for job shops. IIE Transactions 25(6), 32–41 (1994)
9. Leus, R., Herroelen, W.: The complexity of machine scheduling for stability with a single disrupted job. Operations Research Letters 33(2), 151–156 (2005)
10. Mehta, S.V., Uzsoy, R.: Predictable scheduling of a single machine subject to breakdown. International Journal of Computer Integrated manufacturing 12(1), 15–38 (1999)
11. Mičunek, P.: Metastratégie pri riešení problému JOB-Shop (2002), Available via DIALOG,
`http://frcatel.fri.utc.sk/pesko/Studenti/micunekP2.ps.gz`
(cited January 15, 2011)
12. Ouelhadj, D., Petrovic, S.: Survey of dynamic scheduling in manufacturing systems. Journale of Scheduling 12(4), 417–431 (2008)

13. Pinedo, M.: Schedulling theory, algorithms and systems, 1st edn. Prentice-Hall, Englewood Cliffs (1995)
14. Settles, M.: An introduction to Particle Swarm Optimization, Department of Computer Science, University of Idaho (2005)
15. SimEvents - Getting started, The MathWorks, Inc.
16. Viera, G.E., Hermann, J.W., Lin, E.: Rescheduling manufacturing systems: a frame-work of strategies, policies and methods. Journale of Scheduling 6(1), 36–92 (2003)
17. Wu, S.D., Storer, R.H., Chang, P.C.: A rescheduling procedure for manufacturing systems under random disruptions. In: Proceedings Joint Usa/German Conference On New Directions For Operations Research in Manufacturing, pp. 292–306 (1991)
18. Wu, S.D., Storer, R.H., Chang, P.C.: One machine rescheduling heuristics with efficiency and stability as criteria. Computers Operations Research 20(1), 1–14 (1993)
19. Zelenka, J.: Discrete event Dynamic Systems Framework for Analysis and Modeling of Real Manufacturing System. In: CD Intelligent Engineering Systems Conference, INES 2010, Las Palmas of Gran Canaria, pp. 287–291 (2010); ISBN 978-1-4244-7651-0, IEEE Catalog Number: CFP10IES-CDR
20. Zelenka, J.: Application of Particle Swarm Optimization in Job-Shop Scheduling problem in the Recycling Process. In: CINTI 2010, Budape 2010, pp. 137–140 (2010)
21. Zelenka, J., Kasanicky, T.: Comparison of Artificial Immune Systems with the Particle Swarm Optimization in Job-Shop Scheduling Problem (in press 2011)

Chapter 6
Multiobjective Differential Evolution Algorithm with Self-Adaptive Learning Process

Andrzej Cichoń and Ewa Szlachcic

Abstract. This chapter presents an efficient strategy for self-adaptation mechanisms in a multiobjective differential evolution algorithm. The algorithm uses parameters adaptation and operates with two differential evolution schemes. Also, a novel DE mutation scheme combined with a transversal individual idea is introduced to support the convergence rate of the algorithm. The performance of the proposed algorithm, named DEMOSA, is tested on a set of benchmark problems. The numerical results confirm that the proposed algorithm performs considerably better than the one with simple DE scheme in terms of computational cost and quality of the identified nondominated solutions sets.

6.1 Introduction

Most of the traditional optimization techniques are centered on evaluating the first derivatives to locate the optima. In recent times, several derivative-free optimization algorithms have emerged to locate the optima for many rough and discontinuous optimization surfaces, because of the difficulties in evaluating the first derivative [6]. Fortunately, an alternative exists. A family of stochastic optimizers called evolutionary algorithms (EAs) proved in last few decades to be an excellent means of solving multiobjective optimization problems. They have a unique ability to find a good approximation of optimal solutions sets in a single run. The theory of evolutionary algorithms demonstrated how biological crossovers and mutations of chromosomes can be used in computer algorithms to improve the quality of the solutions over successive iterations.

At the same times, some alternative operators replacing the classical probabilistic crossover and mutation operators were proposed, based on the suitable differential

Andrzej Cichoń · Ewa Szlachcic
Institute of Computer Engineering, Control and Robotics, Wroclaw University of Technology, Wroclaw, Poland
e-mail: {andrzej.cichon, ewa.szlachcic}@pwr.wroc.pl

J. Fodor et al. (Eds.): Recent Advances in Intelligent Engineering Systems, SCI 378, pp. 131–150.
springerlink.com

mutation operator. Ideas based on differential evolution have been of great interest, as in [6, 7, 10, 15, 19, 21].

In recent years, many multiobjective evolutionary algorithms (MOEAs) have been proposed [5, 8, 11, 12, 25, 27, 28]. NSGA-II [8], PAES [11], and SPEA2 [27, 28] are some of the most popular among them. Recently, differential evolution algorithms have also been proposed for constrained multiobjective optimization problems [13, 15, 23, 24].

With Differential Evolution (DE) being so popular and efficient algorithm, self-adaptation of key parameters (*Cr* responsible for crossover operation, and *F* influencing mutation) has been investigated by many researchers to make it even better and easier to use on various single- and multiobjective optimization problems.

Abbass in [2] simply uses *Cr* and *F* that change randomly according to Gaussian normal distribution $N(0,1)$, repairing those which fall out of range $[0,1]$. In [3], an idea of self-adaptation mechanism from evolution strategies is used. Each individual has its own pair of *Cr* and *F* control parameters, whose values are adjusted during the evolving procedure—the parameters are first mutated and then used to produce a new individual. A unique version of DE is described in [14]—it employs a wavelet-based mutation operation to dynamically control the scaling factor *F*. Its values decrease in subsequent generations, promoting explorative behavior of the algorithm at the beginning, and switching to local search at the latter stages of optimization procedure. In [18], a 'semi-adaptive' DE is proposed. A 'sensitivity analysis' is done for *Cr* and it finally takes a fixed value of 0.2 as the most suitable for test problems investigated in the paper. On the other hand, the parameter *F* takes values according to the Laplace distribution. Differential Evolution using Cauchy distribution may also be a good choice, as in [22]. Papers [16, 17] propose a very interesting modification of DE algorithm consisting of applying local search to the scaling factor *F*. Two simple local search approaches are successfully tested: the first one uses the golden section search, and the second one: a hill-climber idea.

It has been stated in literature that fixed values of DE control parameters is a poor idea in general case. All of the above mentioned improvements show competitive results in terms of convergence speed and solutions quality. Therefore, in this chapter, an extension of the differential evolution multi-objective algorithm is proposed. The differential evolution idea is used with some significant improvements concerning the DE strategies and parameters adaptation.

6.2 Multiobjective Optimization Problem

A multiobjective optimization problem (MOP) is considered when a simultaneous optimization of two or more objective functions is required. Typically, a whole *set* of compromising solutions is the solution of a MOP, as opposed to a one objective problem, in which a single solution is obtained. Thus, the goal is to find *a set* of optimal vectors $\mathbf{x}^* \in \mathbf{X} \subset \mathbb{R}^n$, which minimize (or maximize) a k-dimensional vector of objective functions:

$$F(\mathbf{x}^*) = [f_1(\mathbf{x}^*), f_2(\mathbf{x}^*), \ldots, f_k(\mathbf{x}^*)], \tag{6.1}$$

subject to inequality constraints $g_i(\mathbf{x}) \leq 0$, $i = \{1,\ldots,m\}$ and equality constraints $h_j(\mathbf{x}) = 0$, $j = \{1,\ldots,p\}$.

A vector $\mathbf{x} = \left[x^1, x^2, \ldots, x^n\right]^T$ denotes an n-dimensional decision variable vector, and $f_i(\mathbf{x})$ is one of the k objectives functions. $\mathbf{X}$ is called a parameter space (or variable space), while $\mathbf{y} = F(\mathbf{x})$ belongs to the objective space $\mathbf{Y}$.

The objective functions are usually conflicting. In order to compare solution vectors, the idea of Pareto-dominance is used, i.e., given two vectors $\mathbf{u} = F(\mathbf{x}) = [f_1(\mathbf{x}), f_2(\mathbf{x}), \ldots, f_k(\mathbf{x})]$ and $\mathbf{v} = F(\mathbf{x}') = [f_1(\mathbf{x}'), f_2(\mathbf{x}'), \ldots, f_k(\mathbf{x}')]$ we say that $\mathbf{u}$ dominates $\mathbf{v}$ if and only if $\mathbf{u}$ is partially less than $\mathbf{v}$:

$$\mathbf{u} \prec \mathbf{v} \iff \forall i \in \{1,\ldots,k\}\ u_i \leq v_i \wedge \exists i \in \{1,\ldots,k\} : u_i < v_i. \tag{6.2}$$

This means that $\mathbf{u} \prec \mathbf{v}$ holds if $\mathbf{u}$ has a better performance than $\mathbf{v}$ in at least one objective, and is not worse with respect to the remaining functions.

A solution vector $\mathbf{x}^* \in X$ is said to be *nondominated* (or Pareto optimal) if there is no $\mathbf{x}' \in \mathbf{X}$ for which $\mathbf{v} = F(\mathbf{x}') = [f_1(\mathbf{x}'), f_2(\mathbf{x}'), \ldots, f_k(\mathbf{x}')]$ dominates $\mathbf{u} = F(\mathbf{x}) = [f_1(\mathbf{x}), f_2(\mathbf{x}), \ldots, f_k(\mathbf{x})]$.

The P^* set is defined as Pareto Optimal Set, where:

$$P^* := \left[\mathbf{x}^* \in X | \neg\exists \mathbf{x}' \in X : F(\mathbf{x}') \prec F(\mathbf{x}^*)\right]. \tag{6.3}$$

Pareto optimal solutions are those solutions in the variable space $\mathbf{X}$, for which no other vectors can be found that dominate them in the objective space $\mathbf{Y}$. The image of the Pareto Optimal Set in the objective space constitutes a Pareto Front containing all nondominated solutions:

$$PF^* := \{\mathbf{u} = F(\mathbf{x}^*) | \mathbf{x}^* \in P^*\}. \tag{6.4}$$

Pareto Front determines the space in $\mathbb{R}^k$ formed by the objective solutions of the Pareto Optimal Set. A good approximation of the Pareto Front has to achieve two main requirements: precision and diversity. At each step of the optimization process in a multiobjective evolution approach, a set of solutions is constructed, and nondominated points are chosen from all the solutions.

6.3 Differential Evolution and Its Multiobjective Variant

Since its introduction in 1997, Differential Evolution has proved many times to be a very effective and robust algorithm for continuous optimization problems. The following paragraphs present in detail the idea of the basic Differential Evolution algorithm and then give an overview of one of DE multiobjective versions, which was used to develop DEMOSA algorithm.

6.3.1 Differential Evolution

Differential Evolution was first proposed in [19] and quickly became a very popular and broadly used algorithm. What is unique about DE is the reproduction procedure. New candidate solutions are created by combining the parent individual and several other individuals of the same population. At each generation and for each individual DE employs special mutation and crossover operations to produce a donor vector in the current population.

Differential Evolution algorithm aims at evolving a population of NP D-dimensional individuals represented by vectors $\mathbf{x}^i(t) = \left[x_1^i(t), x_2^i(t), \ldots, x_D^i(t)\right]$, $i = 1, \ldots, NP$. The initial population is generated randomly with uniform distribution within the search space constrained by predefined lower and upper variable bounds: $\mathbf{L} = [L_1, L_2, \ldots, L_D]$ and $\mathbf{U} = [U_1, U_2, \ldots, U_D]$. This means that the j-th variable of the i-th individual in the initial population is calculated as:

$$x_j^i(0) = L_j + rand(0,1) \cdot (U_j - L_j) \tag{6.5}$$

where $rand(0, 1)$ is a uniformly distributed random variable in the range $[0, 1]$.

For a user-defined number of generations GEN, the initial population evolves with the help of operations of mutation (also called *differentiation*, or *differential mutation* [9]), crossover, and selection.

6.3.1.1 Mutation

For each individual $\mathbf{x}^i(t)$, a special *mutant vector* is created as:

$$\mathbf{v}^i(t) = \mathbf{x}^{r1}(t) + F \cdot \left(\mathbf{x}^{r2}(t) - \mathbf{x}^{r3}(t)\right), \tag{6.6}$$

where $r1$, $r2$, $r3 \in [1, \ldots, NP]$ define indices of three randomly chosen individuals that are mutually different and also different form the individual i ($r1 \neq r2 \neq r3 \neq i$). F is a real number scaling the difference vector $\left(\mathbf{x}^{r2}(t) - \mathbf{x}^{r3}(t)\right)$ and thus controlling the mutation step. It typically takes values in the range $(0, 1)$, although some researchers suggest $F \in (0, 2+)$, or even negative values [9].

Equation (6.6) is the very basic mutation scheme (or strategy) of Differential Evolution called *DE/rand/*1, but several more strategies have been proposed, e.g.:

- *DE/rand/*2:

$$\mathbf{v}^i(t) = \mathbf{x}^{r1}(t) + F \cdot \left(\mathbf{x}^{r2}(t) - \mathbf{x}^{r3}(t)\right) + F \cdot \left(\mathbf{x}^{r4}(t) - \mathbf{x}^{r5}(t)\right), \tag{6.7}$$

- *DE/best/*1:

$$\mathbf{v}^i(t) = \mathbf{x}^{best}(t) + F \cdot \left(\mathbf{x}^{r1}(t) - \mathbf{x}^{r2}(t)\right), \tag{6.8}$$

- *DE/rand-to-best/*1:

$$\mathbf{v}^i(t) = \mathbf{x}^i(t) + F \cdot \left(\mathbf{x}^{best}(t) - \mathbf{x}^{r1}(t)\right) + F \cdot \left(\mathbf{x}^{r2}(t) - \mathbf{x}^{r3}(t)\right), \tag{6.9}$$

- *DE/current-to-best/2*:

$$\mathbf{v}^i(t) = \mathbf{x}^i(t) + F \cdot \left(\mathbf{x}^{best}(t) - \mathbf{x}^i(t)\right) + F \cdot \left(\mathbf{x}^{r1}(t) - \mathbf{x}^{r2}(t)\right), \quad (6.10)$$

- *DE/rand/1/either-or*:

$$\mathbf{v}^i(t) = \begin{cases} \mathbf{x}^{r1}(t) + F \cdot \left(\mathbf{x}^{r2}(t) - \mathbf{x}^{r3}(t)\right), & \text{if } rand(0,1) < P_F \\ \mathbf{x}^{r1}(t) + 0.5(F+1) \cdot \left(\mathbf{x}^{r2}(t) - \mathbf{x}^{r3}(t) - 2\mathbf{x}^{r1}(t)\right), & \text{otherwise} \end{cases}, \quad (6.11)$$

where $r1 \neq r2 \neq r3 \neq r4 \neq r5 \neq i$, $\mathbf{x}^{best}(t)$ is the best individual found thus far, and P_F is the strategy's parameter.

6.3.1.2 Crossover

Once a mutant vector is created, it is 'mixed' with $\mathbf{x}^i(t)$ to yield a *trial vector* $\mathbf{u}^i(t)$:

$$u^i_j(t) = \begin{cases} v^i_j(t), & \text{if } (rand(0,1) \leq Cr) \text{ or } (j = j_{rand}) \\ x^i_j(t), & \text{otherwise} \end{cases}, \quad (6.12)$$

for $j = 1, \ldots, D$. Typically, DE uses two kinds of crossover schemes, namely 'binomial' (as in (6.12)) and 'exponential'. Based on Cr value, which is a crossover rate $\in [0, 1]$, binomial crossover operator determines which parameter values are copied from $\mathbf{v}^i(t)$ or $\mathbf{x}^i(t)$ respectively to the corresponding element in the trial vector $\mathbf{u}^i(t)$. The condition $j = j_{rand}$ ensures that at least one element is copied from the mutant vector, so that $\mathbf{x}^i(t)$ and $\mathbf{u}^i(t)$ always differ by at least one parameter. Exponential crossover is scarcely used; its description can be found in DE literature, e.g. [20].

6.3.1.3 Selection

After obtaining a new trail vector, one must check if its variables do not exceed the lower and upper bounds. If some of them do, they must be fixed, usually by generating a new feasible value, as in (6.5). Then a greedy selection scheme is used, and for a minimization problem it is performed as:

$$\mathbf{x}^i(t+1) = \begin{cases} \mathbf{u}^i(t), & \text{if } f\left(\mathbf{u}^i(t)\right) \leq f\left(\mathbf{x}^i(t)\right) \\ \mathbf{x}^i(t), & \text{otherwise} \end{cases}. \quad (6.13)$$

A trial vector $\mathbf{u}^i(t)$ replaces the parent individual $\mathbf{x}^i(t)$ if its fitness value is not worse than the parent objective function value. Otherwise, the individual remains the same.

6.3.2 Differential Evolution for Multiobjective Optimization

There are many multiobjective applications of DE, such as Pareto-frontier Differential Evolution (PDE), Multiobjective Differential Evolution (MODE), Vector Evaluated Differential Evolution for Multiobjective Optimization (VEDE), or ε-MyDE. For the basic multiobjective engine yet another algorithm was chosen. It is called Differential Evolution for Multiobjective Optimization (DEMO), and was proposed by Robič and Filipič in [23]. DEMO showed to be a simple and competitive algorithm, outperforming several other MOEAs. DEMO combines NSGA-II selection scheme [8] with the strength and simplicity of Differential Evolution. It uses DE mechanisms of mutation and crossover but for selection the following rule is proposed:

1. The offspring replaces the parent if it dominates it.
2. If the parent dominates the offspring, such candidate solution is rejected.
3. If the parent and the offspring are incomparable, i.e. they neither dominate nor are dominated by the other, the offspring is added to the population.

With such a principle, after one generation the number of individuals varies between the initial size NP and $2 \cdot NP$. Thus, a truncation operation must be applied. To do that, DEMO employs NSGA-II mechanisms of *nondominated sorting*, *crowding-distance*, and *crowded-comparison*. These mechanisms keep the best individuals in the population with respect both to the *nondominated rank* ("closeness" to the true Pareto front) and diversity rank (uniform spread of solutions along the Pareto front), and thereby allow the algorithm to find a good approximation of the optimal solutions set. NSGA-II mechanisms can be roughly described in the following way. First, nondominated sorting assigns each individual with a nondominated rank equal to the number of other solutions dominating it. All individuals with this domination count as zero are nondominated. They are the best members of the entire population and constitute the first *front* of solutions. Later on, the other fronts of individuals with higher domination counts are identified.

Nondominated sorting uses elitism and thereby supports the convergence to the Pareto-optimal set. But it is also expected that the algorithm maintains a diverse set of solutions. To achieve that, a diversity measure called crowding distance is employed. It estimates the density of solutions surrounding a particular individual in the population with respect to all k objective functions. These mechanisms are applied to the augmented population of solutions, so at the end some of them must be discarded. To select the best NP individuals, the population is sorted with help of the crowded-comparison operator. It always prefers solutions with lower (better) nondominated rank. But if two solutions belong to the same front (i.e., have the same nondominated rank), the one located in a less crowded region is selected.

6.4 Differential Evolution for Multiobjective Optimization with Self Adaptation

All evolutionary algorithms need several parameters to be fine-tuned, so that satisfactory results can be achieved. It might be a difficult and time-consuming task because many factors need to be taken into consideration—potential interactions between the control parameters, dimensionality of the problem at hand, shape of the Pareto front, etc.

Furthermore, given the fact that evolution of solutions is a dynamic process, different parameters values might be preferred at different stages of evolution. Also, as it was said before, Differential Evolution as the basic algorithm provides various strategies to create new solution vectors, so allowing the algorithm to choose the most appropriate one makes it more flexible and suitable to apply on different optimization problems.

These issues have been addressed in the algorithm proposed in this chapter, which was named Differential Evolution for Multiobjective Optimization with Self Adaptation (DEMOSA) [4].

The main steps of the DEMOSA algorithm are summarized in Algorithm 6.1.

Algorithm 6.1. DEMOSA

1: Evaluate the initial population of NP randomly generated individuals
2: **while** stopping criterion not met **do** ▷ *
3: **for** each individual $\mathbf{x}^i(t)$ $(i = 1, \ldots, NP)$ **do**
4: Generate values of Cr and F parameters
5: Probabilistically determine the scheme that will be used to produce the offspring
6: Create and evaluate the offspring individual $\mathbf{u}^i(t)$
7: **if** $\mathbf{u}^i(t)$ dominates $\mathbf{x}^i(t)$ **then** ▷ **
8: replace $\mathbf{x}^i(t)$ with $\mathbf{u}^i(t)$
9: **else if** $\mathbf{x}^i(t)$ dominates $\mathbf{u}^i(t)$ **then** ▷ ***
10: discard the offspring individual $\mathbf{u}^i(t)$
11: **else** $\mathbf{x}^i(t)$ and $\mathbf{u}^i(t)$ are incomparable, so add $\mathbf{u}^i(t)$ to the population
12: **end if**
13: **end for**
14: **if** the number of individuals in the current population is greater than NP **then**
15: choose the best NP individuals using the mechanisms of nondominated sorting,
16: crowding-distance, and crowded-comparison operator
17: **end if**
18: After each learning period update the mean value of Cr and probability values assigned to
19: DE schemes
20: **end while**

* the number of generations GEN
** Cr value and chosen DE scheme were successful
*** Cr value and the scheme were not successful

6.4.1 Strategy Adaptation

The basic idea of self adaptation follows the mechanisms proposed by Qin and Suganthan for one objective DE algorithm [21]. After numerous tests, two schemes were integrated into DEMOSA algorithm. The first scheme is the basic strategy of Differential Evolution, called *DE/rand/1*, but the second of them is a novel strategy, proposed by the authors and named *DE/none/1*. The schemes are assigned with probability values (p_1 for *DE/rand/1* and p_2 for *DE/none/1*) of being applied in the next generations. Initially, $p_1 = p_2 = 0.5$ and both strategies have equal chances to be used. Before a new candidate individual is created, a uniformly distributed random number is generated in the range [0, 1]. If it is smaller than or equal to p_1, the strategy *DE/rand/1* will be used. Otherwise, *DE/none/1* will be applied. The trial vector is then evaluated and compared to its parent. If the candidate dominates the parent, a special counter is incremented, depending on the strategy used. These counters are denoted as ns_1 for *DE/rand/1* and ns_2 for *DE/none/1*. Such a procedure is repeated for all members of the population.

After a predefined number of generations called *learning period* (*LP*), the probabilities are recalculated as:

$$p_1 = \frac{ns_1}{ns_1 + ns_2}, \tag{6.14}$$

$$p_2 = \frac{ns_2}{ns_1 + ns_2} = 1 - p_1. \tag{6.15}$$

These probabilities can be interpreted as the percentage values of creating successful offspring. Hence, the better performance a strategy has during a particular learning period, the more likely it will be used in the next one. It is possible to use a greater number of DE strategies, however scheme selection and probabilities calculation must be changed. In this context, the Roulette Wheel selection may be the simplest choice.

As mentioned before, a new DE strategy, named *DE/none/1* is introduced here. During preliminary research, the performance of several existing DE strategies was analyzed, but also several new ideas were tested. Finally, one of the new strategies that showed to increase the convergence rate for some test problems was added to the algorithm. The chosen scheme is defined as:

$$\mathbf{v}^i(t) = F \cdot \left(\mathbf{x}^{r1}(t) - \mathbf{x}^{r2}(t)\right). \tag{6.16}$$

This strategy lacks the base vector present in all other DE schemes, which may diminish explorative abilities of the algorithm. To minimize such an effect, the idea of *transversal individual* is used [9], which means that a candidate vector evolves several times before it can replace its parent. In DEMOSA algorithm this procedure requires two steps:

1. The first mutant vector is created with *DE/none/1* (6.16). Then it is mixed with crossover operation with the parent vector to produce the transversal individual,

2. Another mutant vector is created but this time using *DE/rand/1* (6.6). The crossover operation is performed between the second mutant vector and transversal individual, and by that means a trial vector is obtained and later compared to the parent vector.

(Note that five random individuals are needed—two for the first step and three for the second one).

6.4.2 Parameters Adaptation

To adapt the crossover parameter Cr, an additional value Crm is introduced. Initially, during the first learning period, $Crm = 0.4$—in accordance with many observations that lower Cr values are often preferable for multiobjective optimization problems. Before the crossover operation is performed, Cr is generated as a Gaussian random number: $Cr = N(Crm, 0.15)$. If a trial vector created with this Cr dominates its parent, such a value is recorded. On that basis, after each learning period, Crm is recalculated as the average of these successful Cr values. It must be noted that once Crm is computed, Cr records are emptied to avoid possible side-effects accumulated in the previous learning periods.

Similar updating process proposed for the scaling factor F did not bring satisfactory results, so ultimately, after testing several ideas, a very simple idea was chosen: F will change randomly before every mutation operation as $F = N(0.5, 0.15)$. According to the Three-Sigma Rule, with a very high probability, F will then take values in the range $(0, 1)$. Values greater than one are accepted, but if a negative number is drawn, then its absolute value is used.

6.5 Numerical Results

In this part, the results of numerical experiments are demonstrated and discussed. Two algorithms are compared, namely DEMOSA and DEMO, the latter of which uses fixed parameters values and only one mutation scheme (*DE/rand/1*). The following paragraphs describe the benchmark problems, efficiency measures and show differences in computational cost and quality of nondominated solutions sets found by the two algorithms. Also, the dynamics of parameters in the self-adaptation learning process is presented.

6.5.1 Test Problems and Parameter Settings

DEMOSA was tested on seven unconstrained bi-objective test functions, named DEB, KUR, ZDT1, ZDT2, ZDT3, ZDT4, and ZDT6, which are among the most popular and broadly used multiobjective benchmark problems. However, in this chapter, numerical results for only four of them are presented (Table 6.1), as most depictive ones. As for the initial parameter settings, the following values were used:

Table 6.1 Test problems analyzed in the chapter

test problem	D	variable space	objective functions
ZDT1	30	$\mathbf{x} \in [0, 1]$	$f_1(\mathbf{x}) = x_1$ $f_2(\mathbf{x}) = g(\mathbf{x})\left(1 - \sqrt{x_1/g(\mathbf{x})}\right)$ $g(\mathbf{x}) = 1 + \frac{9}{D-1}\sum_{i=2}^{D} x_i$
ZDT2	30	$\mathbf{x} \in [0, 1]$	$f_1(\mathbf{x}) = x_1$ $f_2(\mathbf{x}) = g(\mathbf{x})\left(1 - (x_1/g(\mathbf{x}))^2\right)$ $g(\mathbf{x}) = 1 + \frac{9}{D-1}\sum_{i=2}^{D} x_i$
ZDT3	30	$\mathbf{x} \in [0, 1]$	$f_1(\mathbf{x}) = x_1$ $f_2(\mathbf{x}) = g(\mathbf{x})\left(1 - \sqrt{x_1/g(\mathbf{x})} - \frac{x_1}{g(\mathbf{x})}\sin(10\pi x_1)\right)$ $g(\mathbf{x}) = 1 + \frac{9}{D-1}\sum_{i=2}^{D} x_i$
ZDT4	10	$x_1 \in [0, 1]$, $x_i \in [-5, 5]$, for $i = 2, \ldots, D$	$f_1(\mathbf{x}) = x_1$ $f_2(\mathbf{x}) = g(\mathbf{x})\left(1 - \sqrt{x_1/g(\mathbf{x})}\right)$ $g(\mathbf{x}) = 1 + 10(D-1) + \sum_{i=2}^{D}\left(x_i^2 - 10\cos(4\pi x_i)\right)$

- number of generations $GEN = 200$,
- number of individuals in the population $NP = 100$,
- learning period $LP = 25$.

6.5.2 *Efficiency metrics*

The suitable definitions of efficiency metrics are crucial to reliably asset the quality of obtained solutions set, or to compare the performance of different multiobjective algorithms. Defining such metrics is not a simple task, though, because the result of an evolutionary multiobjective optimization algorithm is not a single scalar value, but a set of compromising solutions. Fortunately, various metrics for nondominated sets have been suggested [5, 26]. Four of them are described here and used in the numerical experiments. The metrics measure the accuracy of the solutions (IGD, HV) and uniformity of spread along the identified Pareto Front (Δ, SP). Moreover, two of the following metrics (IGD and Δ) require a reference set of nondominated solution, while the remaining two (HV and SP) can be calculated without such a set.

Inverted Generational Distance (IGD): This measure estimates how far are the vectors in the Pareto Front produced by the algorithm (PF_{known}) from those in the true Pareto Front of the problem (PF_{true}), which serves as a reference set of solutions. It is defined as:

$$IGD = \frac{\sqrt{\sum_{i=1}^{N} d_i^2}}{N}, \tag{6.17}$$

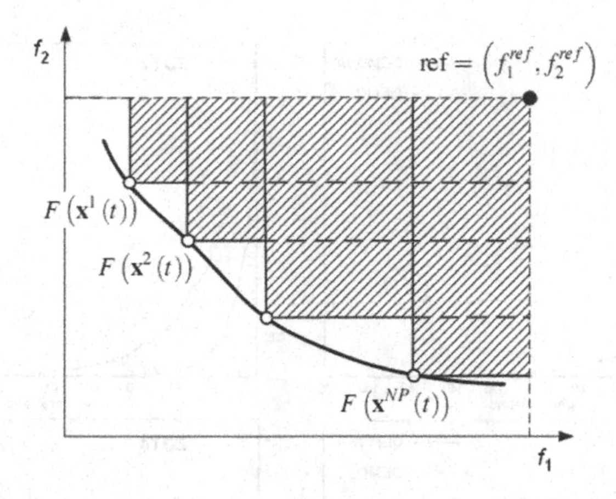

Fig. 6.1 Graphical illustration of the Hypervolume metric for a two-objective MOP

where N is the number of vectors in PF_{true}, and d_i is Euclidean distance between the i-th reference vector in PF_{true} and *the closest* solution vector in PF_{known}. It should be clear that $IGD = 0$ is the best possible value and it indicates that all solutions found by the algorithm are in the true Pareto Front.

Hypervolume (HV): For a two-objective MOP, this metric equals to the summation (union) of all rectangular areas bounded by some reference point and solution vectors from PF_{known} (Fig. 6.1). Mathematically it is defined as:

$$HV = \left\{ \bigcup_i vol_i \,|\, vec_i \in PF_{known} \right\}. \tag{6.18}$$

Δ (Delta): This metrics measures the extent of spread among nondominated solutions found by the algorithm:

$$\Delta = \frac{d_f + d_l + \sum_{i=1}^{N-1} |d_i - \overline{d}|}{d_f + d_l + (N-1)\overline{d}}. \tag{6.19}$$

Here, d_f and d_l denote the Euclidean distances between the boundary solutions from PF_{true} and boundary solution obtained by the algorithm. The parameter $\overline{d}$ is the average of all distances d_i between the consecutive solutions from PF_{known}. N is the size of PF_{known}. The most uniform distribution of solutions gives $d_f = d_l = 0$ and $|d_i - \overline{d}| = 0$. In that case, Δ also equals zero, and it is the best possible value for this metric.

Spacing (SP): Similarly to Δ, SP allows to asset the distribution of solutions along the Pareto Front. This metric is defined as:

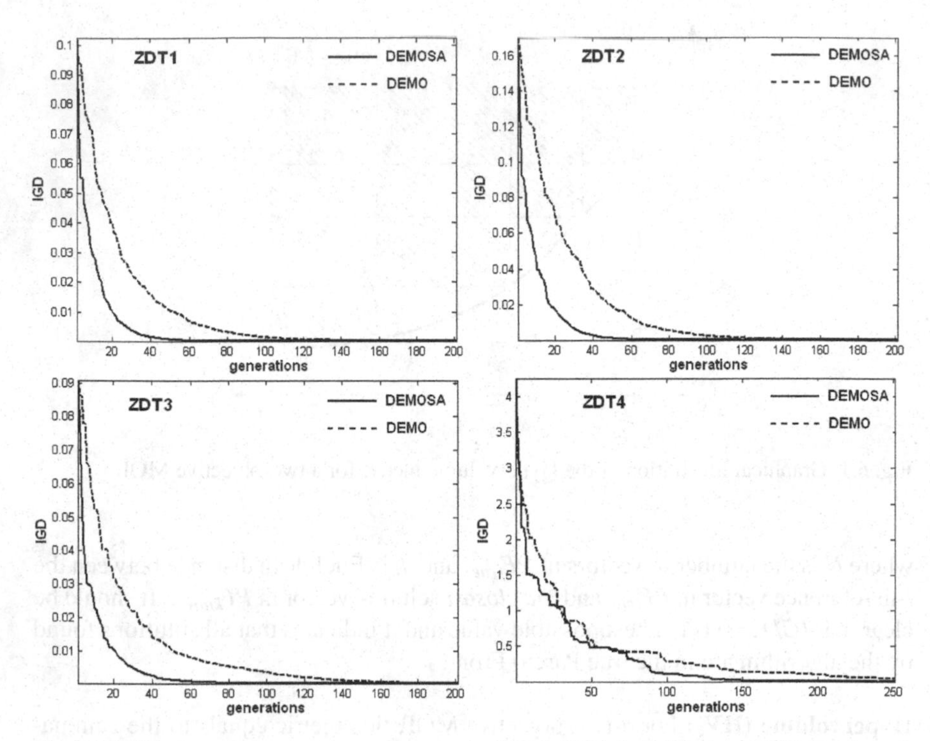

Fig. 6.2 Differences between convergence rates of DEMOSA (solid line) and DEMO (dashed line) for test problems ZDT1, ZDT2, ZDT3 and ZDT4

$$SP = \sqrt{\frac{1}{N-1}\sum_{i=1}^{N}\left(\overline{d}-d_i\right)^2}. \tag{6.20}$$

Again, N is the numbers of nondominated solutions found by the algorithm, but d_i denotes the Manhattan distance between the i-th solution and *the closest* j-th solution from PF_{known}: $d_i = \min_j \left(\left|f_1\left(\mathbf{x}^i(t)\right) - f_1\left(\mathbf{x}^j(t)\right)\right| + \left|f_2\left(\mathbf{x}^i(t)\right) - f_2\left(\mathbf{x}^j(t)\right)\right|\right)$, $i, j = 1,\ldots,N$, $i \neq j$. The parameter $\overline{d}$ is the average of all such distances.

6.5.3 Convergence Rate

To analyze and compare the convergence speed of DEMO and DEMOSA, the Inverted Generational Distance metric was used. The tests were repeated ten times and the mean values are plotted in Fig. 6.2.

Fig. 6.2 presents differences in convergence speed between DEMO and DEMOSA algorithms. DEMO is the core algorithm working with fixed parameter values ($Cr = 0.4$ and $F = 0.5$) and with only one offspring-creation scheme: *DE/rand/*1. It is clear that by using self adaptation and suitable DE strategies, one can achieve significant improvement to the convergence rates for all investigated

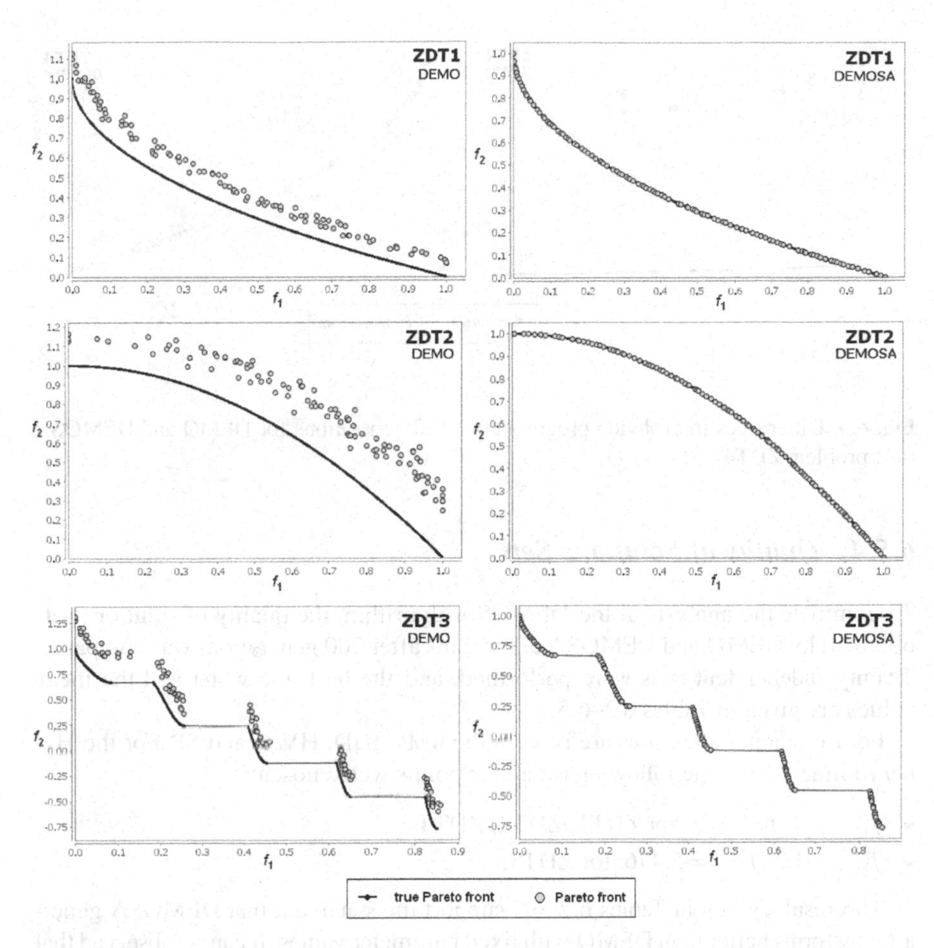

Fig. 6.3 Differences in evolution progress after 80 generations for DEMO and DEMOSA (test problems ZDT1, ZDT2 and ZDT3)

problems. These differences become even more visible when the evolution progress after a certain number of generations is depicted (Fig. 6.3 and Fig. 6.4).

Differences in the results accomplished by the two algorithms are most outstanding for the test problem ZDT4 (Fig. 6.4). It is the most difficult of the analyzed test problems, having 21^9 local fronts. While DEMOSA finds (after 130 generations) a very good approximation of the true Pareto Front, concerning both accuracy and diversity of the solutions, DEMO still needs dozens of iterations before it can achieve a solutions set of comparable quality.

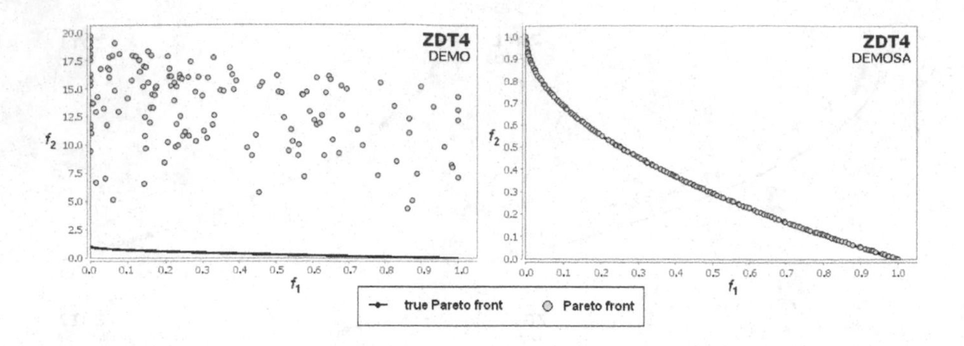

Fig. 6.4 Differences in evolution progress after 130 generations for DEMO and DEMOSA (test problem ZDT4)

6.5.4 *Quality of Solutions Sets*

To complete the analysis of the DEMOSA algorithm, the quality of solutions sets obtained by DEMO and DEMOSA algorithms after 200 generations was compared. Twenty independent runs were performed, and the best, the worst and the mean values are given in Tables 6.2–6.5.

Four efficiency measures are used in the tests: IGD, HV, Δ and SP. For the Hypervolume metric, the following reference points were chosen:

- $f_1^{ref} = 3$, $f_2^{ref} = 3$: for ZDT1, ZDT2, ZDT4,
- $f_1^{ref} = 1.2$, $f_2^{ref} = 5.446$: for ZDT3.

The results given in Tables 6.2–6.5 support the statement that DEMOSA generally performs better than DEMO with fixed parameter values. It can be observed that DEMOSA finds solutions closer to the true Pareto Front (metrics IGD and HV) but sometimes the distribution of solutions is a bit worse (e.g., SP and Δ on ZDT1). The HV value equal to 0 for ZDT4 means that the accuracy of the identified solutions was so poor (as in Fig. 6.4) that at least one solution vector exceeded the bounds defined by the reference point. It also explains the values of the remaining metrics for DEMO on that test problem. In fact, the values of DEMO/ZDT4 are not reliable in this particular case, because performance metrics should not be calculated for such disordered solutions sets.

6.5.5 *Dynamics of Parameters in the Learning Process*

This paragraph presents how probabilities p_1, p_2 and the mean value of Cr change during the evolving procedure. The following plots show the dynamics of these parameters recorded for each of the test problems for two different values of the learning period LP.

Table 6.2 Results for efficiency measures on test function ZDT1

efficiency metric		DEMO	DEMOSA
IGD	best	2.42E-04	**2.28E-04**
	mean	2.58 E-04	**2.40E-04**
	worst	2.70E-04	2.58E-04
	std	9.00E-06	11.00E-06
HV	best	8.6567	**8.6613**
	mean	8.6552	**8.6610**
	worst	8.6533	8.6604
	std	10.59E-04	2.60E-04
Δ	best	**0.2524**	0.2866
	mean	**0.2876**	0.3272
	worst	0.3231	0.3776
	std	0.0209	0.0294
SP	best	**5.18E-03**	5.89E-03
	mean	**5.88E-03**	6.64E-03
	worst	7.03E-03	7.43E-03
	std	5.76E-04	4.19E-04

Table 6.3 Results for efficiency measures on test function ZDT2

efficiency metric		DEMO	DEMOSA
IGD	best	2.75E-04	**2.35E-04**
	mean	2.86E-04	**2.46E-04**
	worst	3.04E-04	2.54E-04
	std	9.00E-06	6.00E-06
HV	best	8.3186	**8.3279**
	mean	8.3151	**8.3277**
	worst	8.3118	8.3271
	std	2.39E-03	2.32E-04
Δ	best	0.2664	**0.2642**
	mean	**0.3189**	0.3227
	worst	0.3543	0.3502
	std	0.0225	0.0247
SP	best	5.27E-03	**4.77E-03**
	mean	6.33E-03	**6.26E-03**
	worst	7.44E-03	6.88E-03
	std	6.65E-04	6.42E-04

Table 6.4 Results for efficiency measures on test function ZDT3

efficiency metric		DEMO	DEMOSA
IGD	best	2.90E-04	**2.50E-04**
	mean	3.47E-04	**2.65E-04**
	worst	3.85E-04	2.77E-04
	std	3.10E-05	1.00E-05
HV	best	10.5680	**10.5881**
	mean	10.5574	**10.5875**
	worst	10.5506	10.5872
	std	5.00E-03	2.72E-03
Δ	best	0.5066	**0.4741**
	mean	0.5263	**0.5108**
	worst	0.5401	0.5605
	std	0.0105	0.0262
SP	best	**5.29E-03**	6.01E-03
	mean	**6.43E-03**	6.90E-03
	worst	7.55E-03	8.06E-03
	std	7.92E-04	5.95E-04

Table 6.5 Results for efficiency measures on test function ZDT4

efficiency metric		DEMO	DEMOSA
IGD	best	9.01E-02	**2.27E-04**
	mean	1.21E-01	**2.50E-04**
	worst	1.47E-01	2.99E-04
	std	1.92E-02	2.40E-05
HV	best	0.00	**6.1965**
	mean	0.00	**6.1957**
	worst	0.00	6.1954
	std	0.00	3.57E-04
Δ	best	0.6298	**0.2829**
	mean	0.6894	**0.3284**
	worst	0.7476	0.4116
	std	0.0428	0.0365
SP	best	7.58E-02	**5.48E-03**
	mean	1.23E-01	**6.62E-03**
	worst	2.30E-01	8.23E-03
	std	4.14E-02	9.42E-04

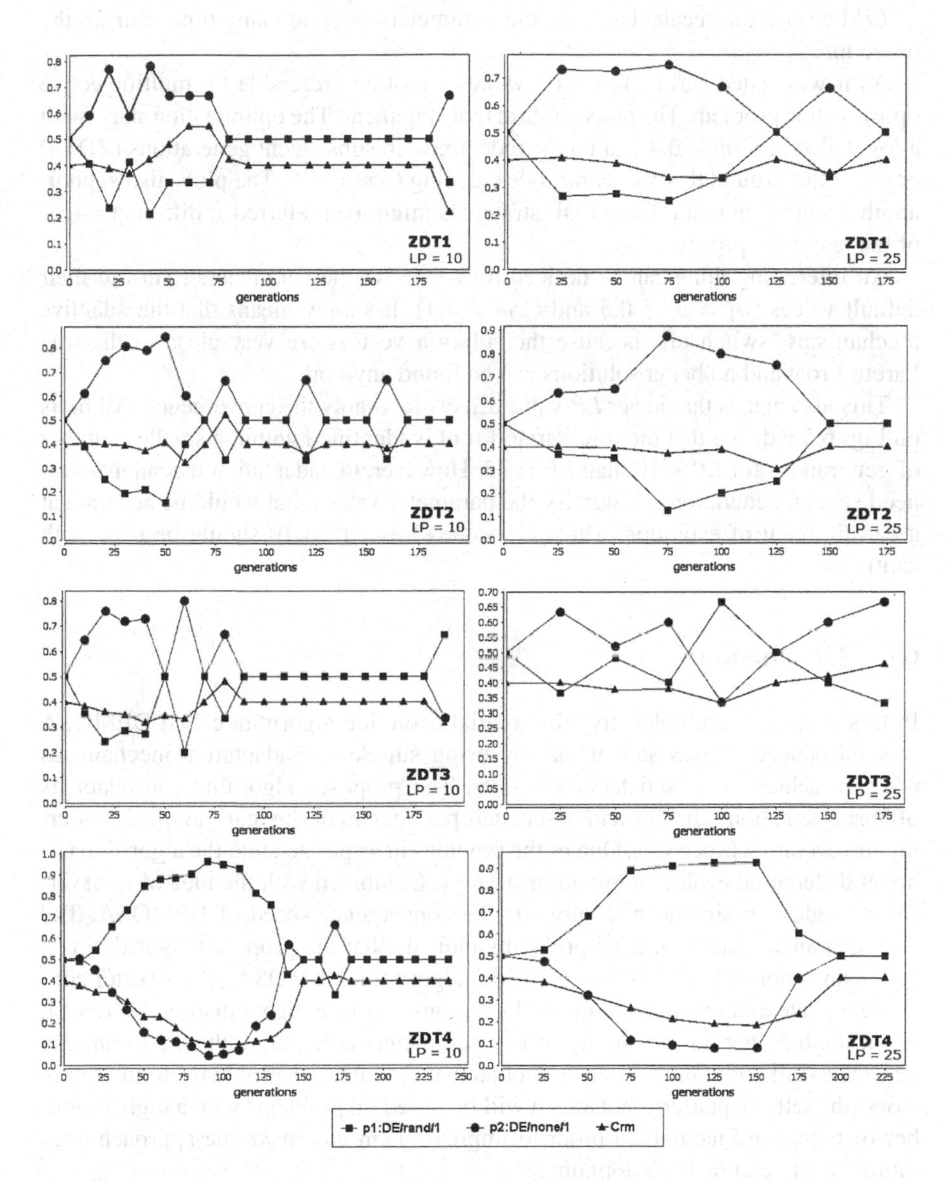

Fig. 6.5 Examples of parameters dynamics: p_1, p_2 and the mean value of Cr for $LP = 10$ and $LP = 25$ (test problems ZDT1, ZDT2, ZDT3 and ZDT4)

Plots on Fig. 6.5 confirm that the dynamics of parameters is rather high and different for each problem. Clearly, it is noticeable especially for the lower values of *LP* because the recalculation of the parameters is done many times during the procedure.

As it was stated earlier, lower *Cr* values are often preferable for multiobjective optimization problem. The plots confirm that statement. The optimization starts with a low value of $Crm = 0.4$ and tends to decrease in subsequent generations (ZDT4) or oscillates around that value, rarely exceeding $Crm = 0.5$. The plots also support another statement that different DE strategies might be preferred at different stages of the evolution process.

An interesting thing can be noticed for $LP = 10$: the parameters return to their default values ($p_1 = p_2 = 0.5$ and $Crm = 0.4$). It simply means that the adaptive mechanisms 'switch off' because the solution vectors are very close to the true Pareto Front and no better solutions can be found anymore.

This also means that lower *LP* value affects favorably the convergence. All plots on Fig. 6.5 indicate that the true Pareto Front is identified within a smaller number of generations for $LP = 10$ than $LP = 25$. However, the adaptation mechanisms *do* need several generations to identify the parameter values that would be accurate at a certain point of evolution. Thus, *LP* values lower than 10 should be used with caution.

6.6 Conclusion

In this chapter, a multiobjective differential evolution algorithm called DEMOSA was introduced. It was shown that by using simple self-adaptation mechanisms one can achieve very satisfactory results. The proposed algorithm can adapt its offspring-creation schemes and associated parameters in the learning process during the evolution procedure. One of the schemes incorporated into the algorithm is a novel differential evolution mutation strategy. Combined with the idea of transversal individual, it significantly supports the convergence speed of DEMOSA. The computational results for tests problems indicate that the proposed algorithm is a promising approach. Compared to DEMO algorithm, DEMOSA shows better convergence rate and finds nondominated solutions sets of a higher quality. Moreover, given the fact that no fine-tuning of key parameters is required, the algorithm can be successfully used on various multiobjective optimization problems. In the future work, the self-adaptation mechanism will be tested on problems with a higher number of objective functions in order to improve them and make the approach even more attractive in real-life domains.

References

1. Abbass, H.A., Sarker, R., Newton, C.: PDE: A Pareto-frontier Differential Evolution Approach for Multi-objective Optimization Problems. IEEE Congr. Evol. Comput. 2, 971–978 (2001)
2. Abbass, H.A.: The Self-Adaptive Pareto Differential Evolution Algorithm. In: IEEE Congr. Evol. Comput., CEC 2002, vol. 1, pp. 831–836 (2002)
3. Brest, J., Greiner, S., Bošković, B., Mernik, M.: Self-Adapting Control Parameters in Differential Evolution: A Comparative Study on Numerical Benchmark Problems. IEEE Trans. Evol. Comput. 10(6) (2006)
4. Cichoń, A., Kotowski, J.F., Szlachcic, E.: Differential evolution for multi-objective optimization with self-adaptation. In: IEEE Int. Conf. Intell. Eng. Syst., Las Palmas of Gran Canaria, Spain, pp. 165–169 (2010)
5. Coello Coello, C.A., Lamont, G.B., Van Veldhuizen, D.A.: Evolutionary algorithms for solving multi-objective problems, 2nd edn. Genetic and Evolutionary Computation Series. Springer Science+Business Media, New York (2007)
6. Das, S., Abraham, A., Konar, A.: Particle swarm optimization and differential evolution algorithms: Technical analysis, applications and hybridization perspectives. J. Stud. Comput. Intell. 116, 1–38 (2008)
7. Deb, K.: Multi-objective genetic algorithms: problem difficulties and construction of tests problems. Evol. Comput. 7(3), 205–230 (1999)
8. Deb, K., Pratap, A., Agarwal, S., Meyarivan, T.: A fast and elitist multi-objective genetic algorithm: NSGA-II. IEEE Trans. Evol. Comput. 6(2) (2001)
9. Feoktistov, V.: Differential Evolution. In: Search of Solutions. Springer Science + Business Media, New York (2006)
10. Gaemperle, R., Mueller, S.D., Koumoutsakos, P.: A parameter study for differential evolution. In: Gmerla, A., Mastorakis, N.E. (eds.) Advances in intelligent systems, fuzzy systems, evolutionary computation, pp. 293–298. WSEAS Press (2002)
11. Knowles, J.D., Corne, D.W.: Approximating the Non-dominated Front Using the Pareto Archived Evolution Strategy. Evol. Comput. 8(2), 149–172 (2000)
12. Konak, A., Coit, D., Smith, A.: Multi-objective optimization using genetic algorithms: A tutorial. Reliability Engineering and System Safety, 992–1007 (2006)
13. Kukkonen, S., Lampinen, J.: An extension of generalized differential evolution for multi-objective optimization with constraints. In: Yao, X., Burke, E.K., Lozano, J.A., Smith, J., Merelo-Guervós, J.J., Bullinaria, J.A., Rowe, J.E., Tiňo, P., Kabán, A., Schwefel, H.-P. (eds.) PPSN 2004. LNCS, vol. 3242, pp. 752–761. Springer, Heidelberg (2004)
14. Lai, J.C.Y., Leung, F.H.F., Ling, S.H.: A New Differential Evolution with Wavelet Theory Based Mutation Operation. In: IEEE Congr. Evol. Comput (CEC 2009), pp. 1116–1122 (2009)
15. Mezura-Montes, E., Reyes-Sierra, M., Coello Coello, C.A.: Multi-objective optimization using differential evolution: a survey of the state-of-the-art. In: Chakraborty, U.K. (ed.) Advances in Differential Evolution, pp. 173–196. Springer, Heidelberg (2008)
16. Neri, F., Tirronen, V., Rossi, T.: Enhancing Differential Evolution Frameworks by Scale Factor Local Search – Part I. In: IEEE Congr. Evol. Comput (CEC 2009), pp. 94–101 (2009)
17. Neri, F., Tirronen, V., Käkkäinen, T.: Enhancing Differential Evolution Frameworks by Scale Factor Local Search – Part II. In: IEEE Congr. Evol. Comput., pp. 118–125 (2009)
18. Pant, M., Thangaraj, R., Abraham, A., Grosan, C.: Differential Evolution with Laplace Mutation Operator. In: IEEE Congr. Evol. Comput., pp. 2841–2849 (2009)

19. Price, K.V., Storn, R.: Differential Evolution – a simple evolution strategy for fast optimization. Dr. Dobb's Journal 22, 18–24 (1997)
20. Price, K.V., Storn, R., Lampinen, J.A.: Differential Evolution. A Practical Approach to Global Optimization. Springer, Heidelberg (2005)
21. Qin, A.K., Suganthan, P.N.: Self-adaptive Differential Evolution Algorithm for Numerical Optimization. In: IEEE Congr. Evol. Comput., vol. 2, pp. 1785–1791 (2005)
22. Ali, M., Pant, M., Singh, V.P.: A Modified Differential Evolution Algorithm with Cauchy Mutation for Global Optimization. In: Ranka, S., Aluru, S., Buyya, R., Chung, Y.-C., Dua, S., Grama, A., Gupta, S.K.S., Kumar, R., Phoha, V.V. (eds.) IC3 2009. Communications in Computer and Information Science, vol. 40, pp. 127–137. Springer, Heidelberg (2009)
23. Robič, T., Filipič, B.: DEMO: Differential evolution for multiobjective optimization. In: Coello Coello, C.A., Hernández Aguirre, A., Zitzler, E. (eds.) EMO 2005. LNCS, vol. 3410, pp. 520–533. Springer, Heidelberg (2005)
24. Santana-Quintero, L.V., Hernandez-Diaz, A.G., Molina, J., Coello Coello, C.A.: DEMORS: A hybrid multi-objective optimization algorithm using differential evolution and rough set theory for constrained problems. Computers & Operations Research 37, 470–480 (2010)
25. Thiele, L., Miettinen, K., Korhonen, P., Molina, J.: A preference-based evolutionary algorithm for multi-objective optimization. J. Evol. Comput. 17(3), 411–436 (2007)
26. Villalobos Arias, M.A.: Análisis de Heurísticas de Optimización para Problemas Multiobjetivo, PhD Thesis, Centro de Investigación y de Estudios Avanzados del Instituto Politécnico Naciónal, Departamento de Matemáticas, México (2005)
27. Zitzler, E., Deb, K., Thiele, L.: Comparison of multi-objective evolutionary algorithms: Empirical results. Evol. Comput. 8(2), 173–195 (2000)
28. Zitzler, E., Laumans, M., Thiele, L.: Improving the strength Pareto evolutionary algorithm for multi-objective optimization. In: Evolutionary methods for design, optimization and control with application to industrial problems, Barcelona, pp. 95–100 (2002)

Chapter 7
Model Complexity of Neural Networks in High-Dimensional Approximation

Věra Kůrková

Abstract. The role of dimensionality in approximation by neural networks is investigated. Methods from nonlinear approximation theory are used to describe sets of functions which can be approximated by neural networks with a polynomial dependence of model complexity on the input dimension. The results are illustrated by examples of Gaussian radial networks.

7.1 Introduction

Experimental results have shown that networks with various types of computational units (perceptrons, radial and kernel units) obtain good performances in high-dimensional tasks. Typically, input-output functions of such networks have the form of linear combinations of functions from certain "dictionaries" containing functions computable by units of various types. The number of terms in such a linear combination corresponding to the number of computational units in so called "hidden layer" plays the role of *model complexity*.

Several authors investigated speed of growth of the number of computational units needed for increasing accuracy in approximation of functions from various classes. Such estimates can be obtained from inspection of upper bounds on errors in approximation of multivariable functions by input-output functions of networks with n hidden units. Often, the estimates have been formulated using "big O" notation. For example, in the form $O(\kappa(n))$, with $\kappa(n) = n^{-1/2}$ [2], $\kappa(n) = n^{-1/p}$ [3] or $\kappa(n) = n^{-1/2} (\log n)^{1/2}$ [5]. In these estimates, the focus was only on the dependence of approximation errors on model complexity n, while the dependence on the number d of variables (which is equal to the dimension of input data) was hidden

Věra Kůrková
Institute of Computer Science, Academy of Sciences of the Czech Republic
Pod Vodárenskou věží 2, 18207 Prague, Czech Republic
e-mail: vera@cs.cas.cz

J. Fodor et al. (Eds.): Recent Advances in Intelligent Engineering Systems, SCI 378, pp. 151–160.
springerlink.com

in the "big O". In [16, 13], dependence of model complexity on the input dimension d was studied for perceptron networks, it was shown that in some cases this dependence can be even exponentional.

In this chapter, we investigate the role of the input dimension in neurocomputing. We investigate sets of multivariable functions which can be approximated by neural networks with a polynomial dependence of approximation errors on the input dimension d. Our approach is based on estimates of worst-case errors formulated in terms of certain norms of multivariable functions to be approximated. We exploit norms tailored to computational units and explore their relationships to norms defined in terms of magnitudes of derivatives and smoothness properties.

The chapter is organized as follows. In Section 7.2, basic concepts concerning approximation from a dictionary are introduced and in Section 7.3, norms induced by dictionaries are defined. In Section 7.4, sets of functions which can be approximated with a polynomial dependence on input dimension are described in terms of these norms. In Section 7.5, these norms are estimated using more common norms defined in terms of derivatives and the results are illustrated by examples of Gaussian radial networks. Section 7.6 is a brief conclusion.

7.2 Approximation from a Dictionary

Let $(\mathscr{X}_d, \|.\|_{\mathscr{X}_d})$ be a normed linear space of functions or equivalence classes of functions of d variables and let A_d, G_d be two nonempty subsets of $\mathscr{X}_d$. Functions in A_d are to be approximated by linear combinations of n elements of G_d, i.e., by functions from the set

$$\mathrm{span}_n G_d := \left\{ \sum_{i=1}^{n} w_i g_i \,|\, w_i \in \mathbb{R}, g_i \in G_d \right\}.$$

The set G_d is sometimes called a *dictionary*. With suitable choices of dictionaries, this approximation scheme includes sets of functions computable by one-hidden layer neural networks, radial-basis functions, kernel models, splines with free nodes, trigonometric polynomials with free frequencies, Hermite functions, etc. The integer n can be interpreted as the *model complexity* measured by the number of computational units.

Worst-case errors measured by the distance defined by the norm $\|.\|_{\mathscr{X}_d}$ in approximation of functions from A_d by elements of $\mathrm{span}_n G_d$ can be formally described in terms of the *deviation*

$$\delta(A_d, \mathrm{span}_n G_d)_{\mathscr{X}_d} := \sup_{f \in A_d} \inf_{g \in \mathrm{span}_n G_d} \|f - g\|_{\mathscr{X}_d}.$$

We investigate upper bounds of the form

$$\delta(A_d, \mathrm{span}_n G_d)_{\mathscr{X}_d} \leq \psi(d,n), \tag{7.1}$$

with a special emphasis on the case when the upper bound is in the factorized form

$$\psi(d,n) = \xi(d)\,\kappa(n),$$

where $\xi : \mathbb{N} \to \mathbb{R}_+$ is a function of the number d of variables of functions in $\mathscr{X}_d$ and $\kappa : \mathbb{N} \to \mathbb{R}_+$ is a non increasing function of the model complexity n.

Model complexity n sufficient for approximation of all functions from A_d within a prescribed accuracy $\varepsilon = \xi(d)\,\kappa(n)$ can be estimated from the bound (7.1) as a function of the dimension d. For example, for $\kappa(n) = n^{-1/2}$, we get

$$n \geq \frac{\xi(d)^2}{\varepsilon^2}.$$

To describe cases when model complexity grows only polynomially with the dimension d, in [10] we introduced a concept of tractability in approximation from a dictionary. Approximation of d-variable functions from sets A_d by elements of $\mathrm{span}_n\, G_d$ is called *tractable with respect to d in the worst case* or simply *tractable* if an upper bound in the form

$$\delta(A_d, \mathrm{span}_n\, G_d)_{\mathscr{X}_d} \leq \xi(d)\,\kappa(n), \tag{7.2}$$

holds with a polynomial $\xi(d)$ in the number d of variables. Note that tractability with respect to other quantities and in different settings has been investigated; see, e.g., [22, 17].

7.3 Norms Induced by Dictionaries

Description of sets of functions which can be tractably approximated by networks with various types of units can provide some understanding to the role of input dimension in neurocomputing. It is desirable to describe such tractable sets as sets of functions satisfying suitable quantitative constrains on their derivatives or smoothness. The constrains we consider in this chapter can be formulated in terms of various norms. Recall that a norm is a functional assigning to a function a non negative number, i.e., a mapping $\|.\| : \mathscr{X} \to \mathbb{R}_+$ such that $\|cf\| = |c|\|f\|$ and $\|f+g\| \leq \|f\| + \|g\|$ for all $c \in \mathbb{R}$ and all $f, g \in \mathscr{X}$.

In this section we present upper bounds on rates of approximation in the form

$$\delta(B_{r_d}(\|.\|_F), \mathrm{span}_n\, G_d)_{\mathscr{X}_d} \leq \psi(d,n) = \xi(d)\,\kappa(n), \tag{7.3}$$

where $B_{r_d}(\|.\|_F)$ denotes the ball of radius r_d centered at zero in a suitable norm $\|.\|_F$ defined on a subspace of $\mathscr{X}_d$. Note that $\|.\|_F$ is usually a different norm than the ambient space norm $\|.\|_{\mathscr{X}_d}$ in which approximation errors are measured. For example $\|.\|_{\mathscr{X}_d}$ can be $\|.\|_{\mathscr{L}^2}$-norm, while $\|.\|_F$ can be some of Sobolev norms which are defined as maxima or weighted averages of iterated partial derivatives up to certain orders.

We will use upper bounds in the form (7.3) to derive estimates of network complexity and description of tractable sets of d-variable functions in approximation by $\mathrm{span}_n G_d$ for various dictionaries G_d. Inspection of bounds of the form (7.3) can lead to a description of conditions on radii r_d which guarantee tractability of approximation of functions from balls $B_{r_d}(\|.\|_F)$, i.e., which guarantee that $\xi(d)$ in the upper bound (7.3) is a polynomial.

We first describe tractable sets in the form of balls in norms tailored to computational units from quite general dictionaries. Recall that a norm is uniquely determined by its unit ball. For any nonempty bounded subset G of a normed linear space $(\mathscr{X}, \|.\|_{\mathscr{X}})$, the symmetric convex closure $\mathrm{cl}_{\mathscr{X}} \mathrm{conv}(G \cup -G)$ defines a norm for which it forms the unit ball. The norm is called *G-variation* and is denoted $\|.\|_{G,\mathscr{X}}$ (or only $\|.\|_G$ when $\mathscr{X}$ is clear from the context), i.e.,

$$\|f\|_G := \inf\{c > 0 \,|\, c^{-1} f \in \mathrm{cl}_{\mathscr{X}} \mathrm{conv}(G \cup -G)\}. \tag{7.4}$$

The term "variation" is motivated by the special case of the dictionary formed by functions computable by Heaviside perceptron networks for which in the one-variable case the variational norm is equal (up to a constant) to the concept of total variation from integration theory. The concept of G-variation was defined by Barron [1] for sets of characteristic functions and extended to general sets by Kůrková [8]. The class of variational norms also includes as a special case the ℓ_1-norm (for G orthonormal). For dictionaries G corresponding to Heaviside perceptrons and Gaussian radial units, G-variation can be estimated using Sobolev and Bessel norms [11, 9].

7.4 Tractability for Balls in Variational Norms

The following theorem gives estimates of worst-case errors of functions in balls in G-variation in approximation by connectionistic computational models of the form $\mathrm{span}_n G$ with quite general dictionaries G. The estimates are reformulations of results by Maurey [20], Jones [7], Barron [2], and Darken et. al. [3] in terms of G-variation (see [12]).

For approximation errors measured by Hilbert space norms, these upper bounds are formulated in terms of G_d-variation of the function to be approximated. For errors measured by the supremum norm $\|.\|_{\sup}$, the upper bounds also depend on the co-VC-dimension of the dictionary G_d. We recall its definition. By $\chi_S : \Omega \to \{0,1\}$ denote the *characteristic function* of $S \subseteq \Omega$, i.e., $\chi_S(x) = 1$ if $x \in S$, otherwise $\chi_S(x) = 0$. Let $\mathscr{F}$ be any family of characteristic functions of subsets of Ω and $\mathscr{S}_{\mathscr{F}} = \{S \subseteq \Omega \,|\, \chi_S \in \mathscr{F}\}$ be the family of the corresponding subsets of Ω. Then a subset A of Ω is said to be *shattered* by $\mathscr{F}$ if $\{S \cap A \,|\, S \in \mathscr{S}_{\mathscr{F}}\}$ is the whole power set of A. Recall that the *VC-dimension* of $\mathscr{F}$ is the largest cardinality of any subset

A which is shattered by $\mathscr{F}$. The *co-VC-dimension* of $\mathscr{F}$ is the VC-dimension of the set $\mathscr{F}' := \{ev_x \,|\, x \in \Omega\}$, where the *evaluation* $ev_x : \mathscr{F} \to \{0,1\}$ is defined for every $\chi_S \in \mathscr{F}$ as $ev_x(\chi_S) = \chi_S(x)$.

Theorem 7.1. *Let d be a positive integer, $(\mathscr{X}_d, \|.\|_{\mathscr{X}_d})$ be a Banach space of d-variable functions, G_d its bounded subset with $s_{G_d} = \sup_{g \in G_d} \|g\|_{\mathscr{X}_d}$, $r_d > 0$ and n be a positive integer. Then*
(i) for $(\mathscr{X}_d, \|.\|_{\mathscr{X}_d})$ a Hilbert space,

$$\delta(B_{r_d}(\|.\|_{G_d}), \mathrm{span}_n\, G_d)_{\mathscr{X}_d} \le s_{G_d}\, r_d\, n^{-1/2},$$

(ii) for $(\mathscr{X}_d, \|.\|_{\mathscr{X}_d}) = (\mathscr{L}^p(\Omega_d, \rho), \|.\|_{\mathscr{L}^p})$ with $p \in (1,\infty)$ and ρ a σ-finite measure on $\Omega_d \subseteq \mathbb{R}^d$,

$$\delta(B_{r_d}(\|.\|_{G_d}), \mathrm{span}_n\, G_d)_{\mathscr{L}^p} \le 2^{1+1/\bar{p}}\, s_{G_d}\, r_d\, n^{-1/\bar{q}},$$

where $\frac{1}{p} + \frac{1}{q} = 1$, $\bar{p} = \min(p,q)$, and $\bar{q} = \max(p,q)$;
(iii) for $(\mathscr{X}_d, \|.\|_{\mathscr{X}_d}) = (\mathscr{M}(\Omega_d), \|\cdot\|_{\sup})$ the space of bounded measurable functions on $\Omega_d \subseteq \mathbb{R}^d$ with the supremum norm and G_d a subset of the set of characteristic functions on Ω_d such that the co-VC-dimension $h^*_{G_d}$ of G_d is finite,

$$\delta(B_{r_d}(\|.\|_{G_d}), \mathrm{span}_n G_d)_{\sup} \le 6\sqrt{3}\, \left(h^*_{G_d}\right)^{1/2} r_d\, (\log n)^{1/2}\, n^{-1/2}.$$

In the upper bounds from Theorem 7.1, the functions $\xi(d)$ are of the forms

$$\xi(d) = c\, s_{G_d}\, r_d \quad \text{and} \quad \xi(d) = c\left(h^*_{G_d}\right)^{1/2} r_d,$$

where c is an absolute constant, while s_{G_d}, r_d and $h^*_{G_d}$ depend on the input dimension d. So Theorem 7.1 implies tractability when $s_{G_d}\, r_d$, and $h^*_{G_d}\, r_d$ grow at most polynomially with d increasing. Note that s_{G_d} and $h^*_{G_d}$ are determined by the choice of the dictionary G_d, so the only flexible term is the radius r_d. We can guarantee tractability of the ball $B_{r_d}(\|\cdot\|_{G_d})$ by choosing its radius r_d in such a way that $\xi(d)$ is a polynomial.

A condition on a radius r_d which by Theorem 7.1 guarantees tractability of the ball $B_{r_d}(\|.\|_{G_d})$ can be illustrated by an example of the dictionary G_d formed by Gaussian kernel units. Let $\gamma_{d,b} : \mathbb{R}^d \to \mathbb{R}$ denote the *d-dimensional Gaussian function of the width $b > 0$*, defined as

$$\gamma_{d,b}(x) = e^{-b\|x\|^2}.$$

We consider the dictionary $G^{\gamma}_d(b)$ formed by *Gaussian d-variable functions with a fixed width b and varying centers*, i.e.,

$$G^{\gamma}_d(b) = \left\{\gamma_{d,b}(.-y) \,|\, y \in \mathbb{R}^d\right\} = \left\{e^{-b\|.-y\|^2} \,|\, y \in \mathbb{R}^d\right\}.$$

A simple calculation shows that for any $b > 0$,

$$\|\gamma_{d,b}\|_{\mathscr{L}^2} = \left(\int_{\mathbb{R}^d} e^{-2b\|x\|^2} \, dx \right)^{1/2} = (\pi/2b)^{d/4}.$$

Thus for every $b > 0$, $\sup_{g \in G_d^\gamma(b)} \|g\|_{\mathscr{L}^2} = (\pi/2b)^{d/4}$ and so the set $G_d^\gamma(b)$ of Gaussians with a fixed width b is bounded in $(\mathscr{L}^2(\mathbb{R}^d), \|.\|_{\mathscr{L}^2})$ and thus $G_d^\gamma(b)$-variation is defined. Applying Theorem 7.1(i) to approximation by linear combinations of translates of Gaussians with a fixed width b we get the upper bound

$$\delta(B_{r_d}(\|.\|_{G_d^\gamma(b)}), \operatorname{span}_n G_d^\gamma(b))_{\mathscr{L}^2} \leq r_d \left(\frac{\pi}{2b}\right)^{d/4} n^{-1/2}. \tag{7.5}$$

In this upper bound, $\xi(d) = (\pi/2b)^{d/4} r_d$. So for $b = \pi/2$, (7.5) implies tractability of balls $B_{r_d}(\|.\|_{G_d^\gamma(b)})$ with radii r_d growing with d polynomially, for $b > \pi/2$, it implies tractability even when r_d are increasing exponentially fast, while for $b < \pi/2$, it merely implies tractability when r_d are decreasing exponentially fast. Hence, the width b of Gaussians has a strong impact on the size of radii r_d of balls in $G_d^\gamma(b)$-variation for which $\xi(d)$ is a polynomial. The sharper the Gaussians (i.e., the larger widths b they have), the larger the balls for which (7.5) implies tractability.

7.5 Tractability for Sets of Smooth Functions

For perceptron and Gaussian radial functions, variational norms can be estimated using more common norms (Sobolev and Bessel) defined in terms of various properties of derivatives and smoothness. Thus also tractability results for balls in these norms can be obtained from Theorem 7.1.

In this section we describe relationships between variational norms and norms defined in terms of smoothness. We derive such relationships from estimates of variational norms holding for dictionaries formed by parameterized sets of functions.

Computational units are functions of two vector variables: an input vector and a parameter vector. So formally they compute functions

$$\phi : \Omega \times A \to \mathbb{R},$$

where $\Omega \subseteq \mathbb{R}^d$ is a domain of input vectors and $A \subseteq \mathbb{R}^k$ is a set of parameters.

For example, *perceptrons with an activation function* $\sigma : \mathbb{R} \to \mathbb{R}$ can be described by a mapping $\phi_\sigma : \mathbb{R}^d \times \mathbb{R}^{d+1}$ defined for $(v,b) \in \mathbb{R}^d \times \mathbb{R} = \mathbb{R}^{d+1}$ as

$$\phi_\sigma(x,v,b) := \sigma(v \cdot x + b). \tag{7.6}$$

An important type of an activation function is the *Heaviside function* $\vartheta : \mathbb{R} \to \mathbb{R}$ defined as $\vartheta(t) = 0$ for $t < 0$ and $\vartheta(t) = 1$ for $t \geq 0$.

Similarly, *radial units* with a radial function $\beta : \mathbb{R} \to \mathbb{R}$ can be described by a mapping $\phi_\beta : \mathbb{R}^d \times \mathbb{R}^{d+1}$ defined for $(v,b) \in \mathbb{R}^d \times \mathbb{R} = \mathbb{R}^{d+1}$ as

$$\phi_\beta(x,v,b) := \beta(b\|x-v\|). \tag{7.7}$$

Kernel units with a symmetric positive semidefinite kernel $K : \Omega \times \Omega \to \mathbb{R}$ are described as

$$\phi(x,a) = K(x,a).$$

Computational units generate dictionaries formed by parameterized sets of functions

$$G = G_\phi(A) = \{\phi(.,a) \,|\, a \in A\}.$$

Functions from large classes can be represented in the form of "infinite neural networks", i.e., as

$$f(x) = \int_A w_f(a)\phi(x,a)da. \tag{7.8}$$

where the "output-weight" functions w_f depend on the represented functions f (see, e.g., [6, 4, 15, 9]). Such representations have been used to derive universal approximation properties [6, 18, 19].

The next theorem is a corollary of a more general result from [14]. It shows that for functions representable as infinite networks in the form (7.8), G_ϕ-variation is bounded from above by the $\mathscr{L}^1$-norm of the output-weight function w_f from the infinite network.

Theorem 7.2. *Let $\Omega \subseteq \mathbb{R}^d, A \subseteq \mathbb{R}^k$, $\phi : \Omega \times A \to \mathbb{R}$ be a mapping such that $G_\phi(A) = \{\phi(.,a) \,|\, a \in A\}$ is a bounded subset of $(\mathscr{L}^q(\Omega), \|.\|_{\mathscr{L}^q})$ with $q \in [1,\infty)$, and $s_\phi = \sup_{a\in A} \|\phi(.,a)\|_{\mathscr{L}^q}$. Let $f \in \mathscr{L}^q(\Omega)$ be such that for some $w_f \in \mathscr{L}^1(A)$, $f(x) = \int_A w_f(a)\phi(x,a)da$. Then*

$$\|f\|_{G_\phi(A),\mathscr{L}^q} \leq \|w_f\|_{\mathscr{L}^1(A)}.$$

For various special cases including perceptron and Gaussian radial networks, $\mathscr{L}^1$-norms of output-weight functions w_f can be farther estimated in terms of Sobolev and Bessel norms defined in terms of derivatives and smoothness. Theorem 7.2 (and its versions holding for other function spaces [14]) combined with various integral representations in the form of networks with infinitely many units can be applied to obtain descriptions of sets of tractable functions.

Here, we illustrate applications of Theorem 7.2 by an example of the dictionary formed by Gaussian radial units with varying widths and centers. We use an integral representation in the form of an infinite network of normalized Gaussians

$$\gamma^o_{d,b} = \frac{\gamma_{d,b}}{\|\gamma_{d,b}\|_{\mathscr{L}^2}}$$

from [9]. This representation holds for all functions in Sobolev spaces.

Recall that for a positive integer s and $q \in [1,\infty)$, the *Sobolev space* $\mathscr{W}^{q,s}(\mathbb{R}^d)$ is formed by all functions having t-th order partial derivatives in $\mathscr{L}^q(\mathbb{R}^d)$ for all $t \leq s$ and the norm $\|.\|_{\mathscr{W}^{q,s}}$ is defined as $\|f\|_{\mathscr{W}^{q,s}} = \left(\sum_{|\alpha|\leq s} \|D^\alpha f\|^q_{\mathscr{L}^2}\right)^{1/q}$, where α denotes a multi-index (i.e., a vector of non-negative integers), $|\alpha| = \alpha_1 + \cdots + \alpha_d$, and D^α is the corresponding partial derivative operator.

For $s > 0$, the *Bessel potential* of the order s on $\mathbb{R}^d$, is the function $\beta_{d,s}$ with the Fourier transform

$$\hat{\beta}_{d,s}(\omega) = (1 + \|\omega\|^2)^{-s/2}.$$

Every function f in the Sobolev space $\mathscr{W}^{q,s}(\mathbb{R}^d)$ can be expressed as a convolution of a function $w \in \mathscr{L}^q(\mathbb{R}^d)$ with the Bessel potential $\beta_{d,s}$. Moreover, the Sobolev norm $\|.\|_{\mathscr{W}^{q,s}}$ is equivalent to the Bessel norm $\|.\|_{L^{q,s}}$ which is defined for $f = w * \beta_{d,s}$ as

$$\|f\|_{L^{q,s}} = \|w\|_{\mathscr{L}^q}.$$

The following integral representation of smooth functions as infinite networks with Gaussian units follows from results from [9]. By $\mathbb{R}_+$ is denoted the set of positive real numbers and $\Gamma(z) = \int_{\mathbb{R}_+} t^{z-1} e^{-t}\, dt$ is the Gamma function.

Theorem 7.3. *Let s,d be positive integers such that $s > d/q$. Then every $f \in \mathscr{W}^{q,s}(\mathbb{R}^d)$ can be represented as*

$$f(x) = \int_{\mathbb{R}^d} \int_{\mathbb{R}_+} w_f(y)\, v_s(t)\, \gamma^o_{d,\pi/t}(y - x)\, dt\, dy,$$

where $v_s(t) = c_1(s,d)\ 2^{-d/4}\ e^{-t/4\pi}\ t^{-d/4+s/2-1}$ with $c_1(s,d) = (2\pi)^{d/2}(4\pi)^{-s/2}/\Gamma(s/2)$, and w_f is the unique function in $\mathscr{L}^q(\mathbb{R}^d)$ such that $f = \beta_{d,s} * w_f$.

Combining this representation with Theorem 7.1(i), we obtain the following estimate of worst-case errors in approximation by Gaussian radial networks of functions in balls in the Bessel norm $\|.\|_{L^{1,s}}$ which is equivalent to the Sobolev norm $\|.\|_{\mathscr{W}^{1,s}}$ [21].

Corollary 7.1. *Let d be a positive integer, $s > d/2$, and $r_d > 0$. Then in $(\mathscr{L}^2(\mathbb{R}^d), \|.\|_{\mathscr{L}^2})$ for all n*

$$\delta(B_{r_d}(\|.\|_{L^{1,s}}) \cap \mathscr{L}^2(\mathbb{R}^d), \operatorname{span}_n G^\gamma_d)_{\mathscr{L}^2} \leq \left(\frac{\pi}{2}\right)^{d/4} \frac{\Gamma(s/2 - d/4)}{\Gamma(s/2)}\, r_d\, n^{-1/2}.$$

The upper bound on the worst-case errors in approximation by Gaussian radial functions from Corollary 7.1 is of the factorized form $\xi(d)\kappa(n)$. The function $\kappa(n)$ is equal to $n^{-1/2}$, while $\xi(d)$ is the product of the radius r_d of the ball in the Bessel space and the function $\left(\frac{\pi}{2}\right)^{d/4} \frac{\Gamma(s/2-d/4)}{\Gamma(s/2)}$.

When the degree od smoothness $s = s(d) = d$, then we get $\left(\frac{\pi}{2}\right)^{d/4} \frac{\Gamma(d/2-d/4)}{\Gamma(d/2)}$, which is going to zero exponentially fast with d increasing. This shows that in the case when the degree $s(d)$ of derivatives which are controlled is increasing with d linearly, even an exponential growth of radii r_d can be allowed as it is compensated by the factor $\left(\frac{\pi}{2}\right)^{d/4} \frac{\Gamma(d/2-d/4)}{\Gamma(d/2)}$. Thus rather large balls of functions in Bessel norms can be tractably approximated by Gaussian radial networks.

7.6 Conclusion

We have investigated the role of an input dimension in approximation of functions by one-hidden layer networks with units from various dictionaries. We have analyzed several upper bounds on rates of approximation of multivariable functions by one-hidden-layer networks to obtain formulations with explicitly stated dependence on input dimension d. Using these upper bounds, we derived characterizations of sets of multivariable functions which can be tractably approximated by networks with units from various dictionaries, i.e., their approximation errors grow with increasing input dimension polynomially. In particular, for Gaussian radial networks, we characterized sets which can be tractably approximated in terms of suitable norms defined by constraints on magnitudes of derivatives.

Acknowledgements. The author thanks to Prof. L. Kóczy for fruitful discussions. This work was supported by MŠMT of the Czech Republic under the project COST INTELLI OC10047 and the project MEB 040901 of the program KONTAKT of Czech-Hungarian collaboration and the Institutional Research Plan AV0Z10300504.

References

1. Barron, A.R.: Neural net approximation. In: Narendra, K. (ed.) Proc. 7th Yale Workshop on Adaptive and Learning Systems. Yale University Press, London (1992)
2. Barron, A.R.: Universal approximation bounds for superpositions of a sigmoidal function. IEEE Transactions on Information Theory 39, 930–945 (1993)
3. Darken, C., Donahue, M., Gurvits, L., Sontag, E.: Rate of approximation results motivated by robust neural network learning. In: Proceedings of the Sixth Annual ACM Conference on Computational Learning Theory, pp. 303–309. The Association for Computing Machinery, New York (1993)
4. Girosi, F.: Approximation error bounds that use VC- bounds. In: Proceedings of ICANN 1995, Paris, pp. 295–302 (1995)
5. Gurvits, L., Koiran, P.: Approximation and learning of convex superpositions. J. of Computer and System Sciences 55, 161–170 (1997)
6. Ito, Y.: Representation of functions by superpositions of a step or sigmoid function and their applications to neural network theory. Neural Networks 4, 385–394 (1991)
7. Jones, L.K.: A simple lemma on greedy approximation in Hilbert space and convergence rates for projection pursuit regression and neural network training. Annals of Statistics 20, 608–613 (1992)
8. Kůrková, V.: Dimension-independent rates of approximation by neural networks. In: K. Warwick, M. Kárný (eds.) Computer-Intensive Methods in Control and Signal Processing. The Curse of Dimensionality, pp. 261–270. Birkhäuser (1997)

9. Kainen, P.C., Kůrková, V., Sanguineti, M.: Complexity of Gaussian radial basis networks approximating smooth functions. J. of Complexity 25, 63–74 (2009)
10. Kainen, P.C., Kůrková, V., Sanguineti, M.: On tractability of neural-network approximation. In: Kolehmainen, M., Toivanen, P., Beliczynski, B. (eds.) ICANNGA 2009. LNCS, vol. 5495, pp. 11–21. Springer, Heidelberg (2009)
11. Kainen, P.C., Kůrková, V., Vogt, A.: A Sobolev-type upper bound for rates of approximation by linear combinations of Heaviside plane waves. J. of Approximation Theory 147, 1–10 (2007)
12. Kůrková, V.: High-dimensional approximation and optimization by neural networks. In: Suykens, J., Horváth, G., Basu, S., Micchelli, C., Vandewalle, J. (eds.) Advances in Learning Theory: Methods, Models and Applications. ch. 4, pp. 69–88. IOS Press, Amsterdam (2003)
13. Kůrková, V.: Minimization of error functionals over perceptron networks. Neural Computation 20, 252–270 (2008)
14. Kůrková, V.: Model complexity of neural networks and integral transforms. In: Alippi, C., Polycarpou, M., Panayiotou, C., Ellinas, G. (eds.) ICANN 2009. LNCS, vol. 5768, pp. 708–717. Springer, Heidelberg (2009)
15. Kůrková, V., Kainen, P.C., Kreinovich, V.: Estimates of the number of hidden units and variation with respect to half-spaces. Neural Networks 10, 1061–1068 (1997)
16. Kůrková, V., Savický, P., Hlaváčková, K.: Representations and rates of approximation of real–valued Boolean functions by neural networks. Neural Networks 11, 651–659 (1998)
17. Mhaskar, H.N.: On the tractability of multivariate integration and approximation by neural networks. J. of Complexity 20, 561–590 (2004)
18. Park, J., Sandberg, I.: Universal approximation using radial–basis–function networks. Neural Computation 3, 246–257 (1991)
19. Park, J., Sandberg, I.: Approximation and radial basis function networks. Neural Computation 5, 305–316 (1993)
20. Pisier, G.: Remarques sur un résultat non publié de B. Maurey. In: Séminaire d'Analyse Fonctionnelle 1981, vol. I (12). École Polytechnique, Centre de Mathématiques, Palaiseau, France (1981)
21. Stein, E.M.: Singular Integrals and Differentiability Properties of Functions. Princeton University Press, Princeton (1970)
22. Traub, J.F., Werschulz, A.G.: Complexity and Information. Cambridge University Press, Cambridge (1999)

Chapter 8
A Comprehensive Survey on Fitness Landscape Analysis

Erik Pitzer and Michael Affenzeller

Abstract. In the past, the notion of fitness landscapes has found widespread adoption. Many different methods have been developed that provide a general and abstract framework applicable to any optimization problem. We formally define fitness landscapes, provide an in-depth look at basic properties and give detailed explanations and examples of existing fitness landscape analysis techniques. Moreover, several common test problems or model fitness landscapes that are frequently used to benchmark algorithms or analysis methods are examined and explained and previous results are consolidated and summarized. Finally, we point out current limitations and open problems pertaining to the subject of fitness landscape analysis.

8.1 Introduction

The notion of a *fitness landscape* dates back to Sewall Wright, who, while not calling it a "fitness landscape" himself, is often cited as the originator of the idea [103]. The analogy to real landscapes seems to give an intuitive understanding of how and where heuristic algorithms operate. In practice, however, this notion is too imprecise and can be misleading, especially in higher-dimensional spaces [6]. In [30] and [82] a widely accepted formal definition is given. It is emphasized that a fitness *function*, an assignment of a fitness value to individuals, is not enough, as it does not relate these individuals to one another. Therefore, the notion of connectedness is included which can either be a set of neighbors, a distance, or even a more complex structure to associate solution candidates with each other. Following this definition, every operator induces its own neighborhood and therefore its own landscape.

Erik Pitzer · Michael Affenzeller
Josef Ressel Center "Heureka!", School of Informatics, Communications and Media
Upper Austria University of Applied Sciences, Campus Hagenberg, Softwarepark 11,
4232 Hagenberg, Austria
e-mail: {erik.pitzer,michael.affenzeller}@fh-hagenberg.at

J. Fodor et al. (Eds.): Recent Advances in Intelligent Engineering Systems, SCI 378, pp. 161–191.
springerlink.com

This perspective simplifies basic landscape analysis. Every landscape is well connected and no large jumps occur at all. However, often enough, different operators are employed simultaneously, which would mean that the "landscape" *changes*, possibly several times, during a single step of the algorithm.

In the light of algorithm analysis, as opposed to operator analysis, this perspective causes more problems than it solves. Moreover, in many cases, the operators are based on simple metrics, like hamming distance for a binary problem encoding or geometric distances in the real-valued case. As shown in [51], the neighborhood defined by an operator is strongly related or even equivalent to an underlying distance function. Even crossover operators can be used to induce a distance function [78, 97]. Therefore, it should be sufficient to study a fitness landscape defined by a fitness function and one or several induced distances.

8.1.1 Definition

We will use the following simple model of a fitness landscape: Given the space of possible solutions or search space $\mathbf{S}$ we need a function $\mathrm{enc} : \mathbf{S} \to \mathscr{S}$ that encodes a proposed solution candidate into a form that can be processed by an algorithm, according to the chosen *representation* of solution candidates. This will be something like a bit-vector ($\mathscr{S} = C(n)$), a real vector ($\mathscr{S} = \mathbb{R}^n$) or a permutation ($\mathscr{S} = P_n$) or even a more complex structure. Sometimes, encoded solution and the actual solution can be identified, making this first step superfluous. In the following we will always refer to an already encoded solution candidate $x \in \mathscr{S}$.

The next step is to assign a *fitness value* to the encoded solution candidate. The *fitness function* $f : \mathscr{S} \to \mathbb{R}$ is often sufficient. For some studies, however, this projection is further broken down into a *fitness mapping* $f_m : \mathscr{S} \to \Phi$ from the encoded solution candidate space to a phenotype space Φ or sometimes also directly from the original solution candidate space, i.e., $f_m : \mathbf{S} \to \Phi$. And only in a second step with $f_v : \Phi \to \mathbb{R}$ fitness values are assigned to phenotypes. Again, however, we can obtain a composed function $f = f_v \circ f_m$.

Throughout this chapter we will be using terminology like "phenotype" to refer to the proposed solution candidate and "genotype" to refer to the encoded solution and other terms from genetics as the notion of a fitness landscape was initially coined in the field of evolutionary biology and many parallels have been drawn. It has to be noted, however, that this does not imply the restriction to only evolutionary methods.

We are chiefly interested in the fitness function $f : \mathscr{S} \to \mathbb{R}$ that assigns a fitness *value* to our chosen *solution candidate*. However, this is not yet a fitness *landscape*. What we need is a notion of connectedness between different solution candidates. In contrast to many others, we do not define a neighborhood function $N : \mathscr{S} \to 2^{\mathscr{S}}$ (where $2^{\mathscr{S}}$ is the power set of $\mathscr{S}$) but instead use the underlying or induced distance $d : \mathscr{S} \times \mathscr{S} \to \mathbb{R}$ which usually forms a metric.

In summary we define a fitness landscape $\mathscr{F}$ as a set of the two functions f and d that define the fitness value and the distances between (encoded) solutions in $\mathscr{S}$ similar to [67].

$$\mathscr{F} = (\mathscr{S}, f, d) \tag{8.1}$$

Once we have this basic definition, notions like peaks and valleys start to become tangible objects. For example, a solution candidate x is a singular peak in the landscape if all other solution candidates n that are close, have smaller fitness as shown in Eq. (8.2). We will usually talk about fitness maximization, therefore, unless stated otherwise an optimum will be assumed to be a maximum.

$$\text{local optimum}(x) :\Leftrightarrow (\exists \varepsilon > 0) \ \left(\forall n \in N_\varepsilon^+(x)\right) \ f(x) > f(n) \tag{8.2}$$

For this definition we need the notion of an ε-neighborhood of x: $N_\varepsilon^+(x) = \{n \mid n \in \mathscr{S}, n \neq x, d(x,n) \leq \varepsilon\}$. Our definition of a neighborhood makes a detour over a distance to become more general than a neighborhood directly induced by an operator.

By requiring a distance measure in addition to the fitness value, or, in other words, by requiring a notion of connectedness, the fitness *landscape* is not only dependent upon the problem but also strongly linked to the choice of representation and its connectedness using certain operators for moving between or recombining solution candidates.

8.1.2 Purpose

The most common motivation for fitness landscape analysis is to gain a better understanding of algorithm performance on a related set of problem instances. Phrases, relating to this apparent *landscape* and its "geological" shapes can be found in many articles about heuristic optimization. This aims at creating an intuitive understanding of how a heuristic algorithm performs and progresses "through the landscape".

The use of a more formal approach like the one given in Section 8.1.1 is most often used to understand the performance of heuristic algorithms on a certain problem class e.g. graph drawing in [41], Graph Bi-Partitioning in [48] or Timetabling Problems in [59].

A common approach is to derive problem or problem class-dependent measures of difficulty or recommended parameter settings. While all of this can be done without resorting to the notion of a fitness landscape this has often helped in getting a deeper understanding of the underlying problem class [49, 17].

An important question is the usefulness of such a landscape analysis: Why not better spend the time and resources solving the problem itself? It is a valid claim that fitness landscape analysis is usually much more resource intensive than just solving a given problem, however, the resulting insight is much more profound as well. Therefore, one should probably not perform a landscape analysis, if the task is to solve a single problem instance, but one should employ this heavy machinery when a deeper understanding of a whole problem class is desired and to elucidate the commonalities and differences between other problem classes.

The ultimate goal of such a rigorous analysis is to gain an understanding of the characteristic landscape properties and its variation over different problem instances. Once the rough "image" of such a landscape over a problem class has been determined, one can investigate problem dependent and simpler measures that reveal characteristics of a single problem instance.

This task is at the core of heuristic research. The universal notion of an underlying fitness landscape across many different scenarios can help to create a common vocabulary and an analysis tool set to better understand the performance of different heuristics.

8.2 Fitness Landscape Analysis

8.2.1 Local Optima and Plateaus: Modality

One of the relatively apparent features of a fitness landscape is its set of local optima. Studying local optima is often one of the first investigations performed on fitness landscapes. The term modality stems from statistics, where a distribution with a single most likely value (the mode) is called unimodal while a distribution with several likely values is called multimodal.

A simple definition for a singular local optimum was already given in Eq. (8.2), where a local optimum is defined as an isolated solution candidate $o \in \mathscr{S}$ with higher fitness than all neighboring points for a certain ε-neighborhood.

In many interesting cases, however, local optima are not singular points but ridges or plateaus of equal fitness. We start with a *connected set* C of solution candidates. If $(\forall x,y \in C)\, f(x) = f(y)$, i.e., all points have equal fitness, this set can either be a generalized local optimum or a generalized saddle point depending on the fitness distribution of the bordering solution candidates $B(C) := \{b \in \mathscr{S} \mid b \notin C, (\exists x \in C)\, b \in N(x)\}$. If they are all of lower fitness we have a generalized local optimum and otherwise a saddle point.

Depending on the search space $\mathscr{S}$ the notion of connectedness will differ. In a discrete space two solution candidates can be connected if they are neighbors. In a continuous space we can use the notion of a topologically connected set, i.e., an open set that cannot be represented as the union of two or more disjoint nonempty open subsets [53]. As a side note we observe that the global optimum needs no connectedness as it is simply a value o where $(\forall x \in \mathscr{S})\, f(x) \leq f(o)$.

A first and obvious measure is the *number* of local optima n_o and the frequency of local optima with respect to the size of the solution space which is often referred to as *modality* of a fitness landscape.

Studying the distribution of local optima can provide important information about the fitness landscape. In [86] the average distance between local optima is examined. We can extend this idea by looking at the normalized distribution of distances between local optima. Based on the set of all local optima O and the diameter of the search space $\text{diam}(\mathscr{S}) := \max\{d(x,y) \mid x,y \in \mathscr{S}\}$, we can analyze the set $\{d(x,y)/\text{diam}(\mathscr{S}) \mid x,y \in \mathscr{O}\}$.

Keeping our focus on just the local optima alone, we can derive another set of measures comparing their "location" and their fitness value. Are better optima close together? Can we argue that they form some kind of succession from lower fitness to the global optimum? How much worse are local optima compared to the global optimum? It may be sufficient to find a local optimum to get a *good enough* solution. These basic questions can lead to simplified problem statements and reduced computation time. We can extend the simple set of distances to include the fitness value as well. Analyzing this extended set $\{(d(x,y), |f(x) - f(y)|) \mid x,y \in \mathscr{O}\}$ gives us the basis to answer these questions.

8.2.2 Basins of Attraction

Another structure derived from local optima are basins of attraction. Simply put, they are the areas around the optima that lead directly to the optimum when going downhill (assuming a minimization problem). This intuitively clear description, however, bears some subtleties that are not immediately obvious as described in [63]: One important question is what we should do if we have several choices of going downhill. This will have a profound influence on the extent of our basins of attraction. In general we can distinguish between a basin of *unconditional* or *strong* attraction where any form of descent will lead to a single optimum, and a bigger basin of *conditional* or *weak* attraction within which we can reach any of a number of different optima depending on the choice of algorithm or operators.

To begin studying basins of attraction we need a notion of convergence to an optimum. For this reason we will define a *downward path* $p_\downarrow$ between x_0 and x_n as a connected sequence of solution candidates $\{x_i\}_{i=0}^n$ where $(\forall i < j)\ x_i \geq x_j$, $x_0 > x_n$, and $x_{i+1} \in N(x_i)$.

Once we have the definition of a downward path, we can use it to collect all possible incoming trajectories of a local optimum which results in the definition of a weak basin of attraction $b(o) := \{x \mid x \in \mathscr{S}, p_\downarrow(x,o)\}$.

If we further restrict this area to those solution candidates that exclusively converge to a certain optimum, we arrive at the definition of a strong basin of attraction $\hat{b}(o) := \{x \mid x \in b(o), (\nexists\, o' \neq o \in \mathscr{O})\ x \in b(o')\}$. To determine exclusivity, however, we need again, the set of all local optima $\mathscr{O}$ which is usually not easy to obtain.

Several different measures can be defined: Starting with the *basin extent*, the fraction of solution candidates that falls within the weak and strong basins, followed by the minimum, maximum and average "diameter" and most importantly the relation between basin size and depth. Figure 8.1 shows a visualization of several properties.

Jones' *reverse hill climber* (assuming a maximization problem) in [30] can be used to exhaustively explore the basins of attraction for discrete spaces. It can be

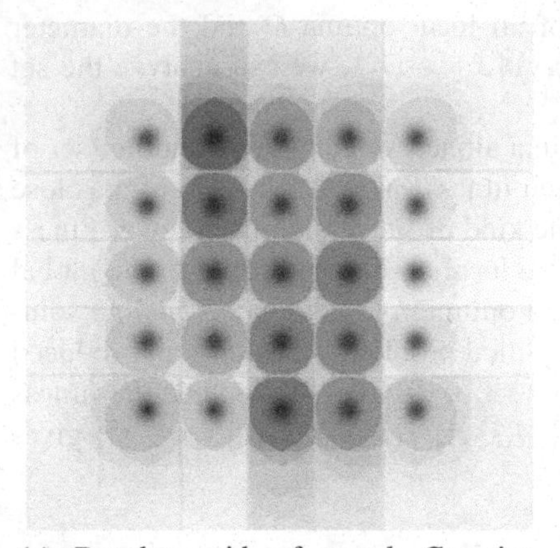

(a) Regular grid of equal Gaussian functions.

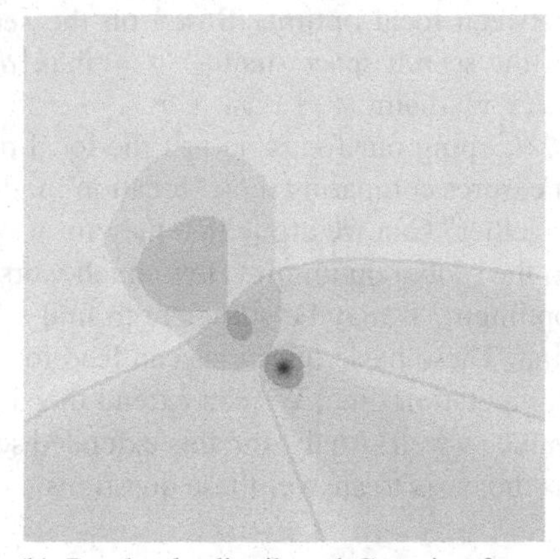

(b) Randomly distributed Gaussian functions with different variances.

Fig. 8.1 2D visualization of basins of attraction for a test function that is a superposition of several normal distributions. Every local optimum receives a distinct color and is separated from the other basins by an extra line that shows a shift in dominant convergence. The color intensity shows the probability of converging to a certain optimum (see [63] for details). In B&W you can still see the different shades and the smoothly changing intensities around the optima.

seen as a breadth-first exploration starting from an optimum to all points that would have been chosen to be uphill given a certain algorithm's choice.

We further define the probabilistic *convergence volume* for every basin of a local optimum. First we define the convergence probability $p_c(x,o) := \sum_{n \in N_\varepsilon(x)} p(n|x) \cdot p_c(n,o)$ which gives the probability of reaching a certain optimum o starting at an arbitrary candidate x. The probability $p(n|x)$ is the probability of continuing a search to solution candidate n from solution candidate x. With $p_c(o,o) := 1$ we can initiate a dynamic programming scheme to efficiently calculate these values recursively [63]. Now we can calculate the convergence volume $V_c(o) := \sum_{s \in \mathscr{S}} p_c(s,o)$.

A clever combined analysis of local optima and landscape walks (see Section 8.2.4) was conceived in [47]. Instead of performing a truly random walk from a random starting point to random final point, a directed walk is conducted starting at a local optimum to another local optimum. The fitness trajectory of this walk is then compared to the fitness trajectory of a random walk starting at the same local optimum ending somewhere on the landscape. This can give a clue about whether "intermediate" points between two optima provide a more valuable alternative as random points at the same distance and decide whether crossover might be useful.

8.2.3 Barriers

Directly derived from the concepts of local optima and basins of attraction are *barriers* as described in [82]. This terminology stems from the use in physics describing energy barriers that need to be crossed in order to reach one metastable (locally optimal) state from another. The same terminology can be generalized for fitness landscapes as described in [82]. As the notion of a barrier stems from physical equilibrium states that are usually the result of energy minimization, they will be defined here in terms of minimization problems.

A *fitness barrier* is the minimum fitness value necessary to reach one optimum from another through an arbitrary path. As shown in Eq. (8.3) a barrier is defined between two solution candidates or, most interestingly, between two local minimums.

$$B_f(x,y) := \min\{\max\{f(x) \mid x \in p\} \mid p \text{ path from } x \text{ to } y\} \tag{8.3}$$

An important question that remains unanswered in this definition is the difference and relative fitness difference that is required to go from one optimum to another. It might seem easy in terms of activation energy to reach one optimum from another, however, these optima might be separated by a great distance and, therefore, difficult to reach from each other in reality. Moreover, if the fitness values of these optima differ significantly, it might be much easier to go from one optimum to the other but not the other way around. Therefore, it remains essential to also look at the measures previously introduced in Section 8.2.1.

It can already be seen by considering relatively simple aspects of fitness landscapes that all "perspectives" of the fitness landscape remain incomplete. Therefore, one should always consider several different analysis methods to obtain a complete overview of a fitness landscape.

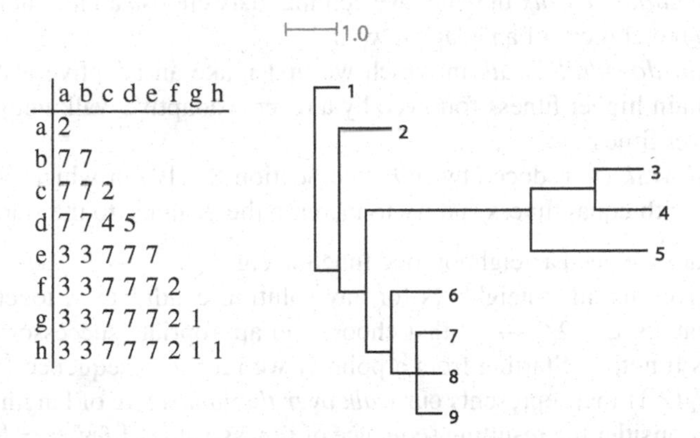

Fig. 8.2 Distance matrix between local optima and resulting barrier tree created from the symmetric distance matrix. These ultrametric distances have been transformed into a tree using the Neighbor-Joining method described in [74].

Fitness barriers between optima form an ultrametric [22], i.e, a metric that satisfies the strong triangle inequality $d(x,z) \leq \max\{d(x,y), d(y,z)\}$. It can serve as an alternate distance measure between local optima that captures the difficulty of moving between two optima. In [82] it has been used to construct a *barrier tree* that creates a hierarchy of local optima, separated by the minimum fitness needed to reach another optimum as shown in Figure 8.2. Using this analysis it is possible to detect clusters of related optima separated by "activation energy" as opposed to the previous analyzes using only fitness and distances differences.

8.2.4 Landscape Walks

One essential tool for fitness landscape analysis is the notion of a *walk*. If we imagine a landscape analogous to a real-world landscape the basic principle is clear: We continuously move a small distance from one solution candidate to a nearby one while keeping a close eye on the development of the fitness.

For discrete spaces, with a finite choice of neighboring solution candidates with minimal distance ε, the definition of a walk is straightforward. For a continuous walk, however, we have to explicitly choose a step size ε or probability distribution of neighbors. In other words we have the choice of what *kind* of walk we want to perform. For every solution candidate there may be several other solution candidates with a distance of ε. Depending on the selection criterion we might be conducting any of a number of different walks, including:

- A *random walk* where we arbitrarily choose any of the neighbors.
- An *adaptive walk* in which we choose a fitter neighbor. Here we have again a choice of which fitter neighbor we choose resulting in e.g. *any ascent, steepest ascent, minimum ascent* or similar adaptive walks.
- A *reverse adaptive walk* in which we continuously choose a more unfit neighbor reversing the choices of an adaptive walk.
- An *"uphill-downhill" walk* in which we first make an adaptive walk until we cannot attain higher fitness followed by a reverse adaptive walk until we cannot attain lower fitness.
- A *neutral walk* (introduced by [69], see Section 8.2.10) in which we choose a neighbor with equal fitness and try to increase the distance to the starting point.

In summary we need a neighborhood function e.g. $N_\varepsilon : \mathcal{S} \to 2^{\mathcal{S}}, x \mapsto \{n \mid d(x,n) = \varepsilon\}$ that gives us all ε-neighbors for any solution candidate x, together with a selection strategy $\zeta_f : 2^{\mathcal{S}} \to \mathcal{S}$ that chooses an appropriate successor based on a given fitness function. Starting from a point x_1 we receive a sequence $\{x_i\}_{i=1}^n$ with $x_{i+1} = \zeta_f(N_\varepsilon(x_i))$ that represents our *walk over the landscape* of length n. For this analysis we consider the resulting sequence of fitness values $\{f_i\}_{i=1}^n = \{f(x_i)\}_{i=1}^n$.

While a random walk explores a greater area, an adaptive walk explores a more "interesting" area. Therefore, it is not a single type of walk that should be used exclusively to obtain a comprehensive picture of a fitness landscape. Variations, combining aspects of both, like a Metropolis random walk [24], or a walk-based distance function d_n that is the reciprocal of the probability of reaching one point through a random walk starting at another are all important tools for fitness landscape analysis.

8.2.4.1 Population-Based "Walks"

A variation of a walk is a population transition where one evolutionary step is performed on every individual of a usually large population to obtain a second generation. From these populations, the average progression characteristics are then derived (e.g. [43, 42]). This is already closely connected to the notion of evolvability discussed in Section 8.2.9.

An extension of this idea is to use a population-based algorithm such as a genetic algorithm [25] or evolution strategy [66] with parameter settings geared towards exploration. This can provide a relatively quick overview while concentrating on the more "interesting" parts of the landscape by employing a slow convergence and large populations instead of random sampling. With these settings, a population can be thought to slowly raise to higher fitness maintaining high dispersion over the landscape. Using this technique, fitness landscapes can be explored with a bias towards the fitter regions by sacrificing details in lower quality regions. However, as with any selective sampling technique, using just a subset of samples to generalize over a whole fitness landscape for an unknown problem is dangerous. If sampling is the only option, however, this probably provides a good complementary sample selection technique that focuses more on global properties as compared to widely adopted sampling technique of trajectory-based walks that can be used to inspect more local features.

8.2.5 *Ruggedness*

The notion of *ruggedness* was among the first measures of problem difficulty. In essence, it can be described as the frequency of changes in slope from "up-hill" to "down-hill" or simply the number of local optima. However, for different encodings and different distance measures, there is a wide array of concrete measures that all measure a certain variation of ruggedness.

Edward Weinberger was probably the first to define a measure of ruggedness in [100] where he describes *autocorrelation* and *correlation length.* These two ideas have later been generalized in [30]. Autocorrelation is the correlation of neighboring fitness values along a random walk in the landscape. It can thus be defined as the average correlation of fitness values at distance ε as shown in Eq. (8.4). An example of an autocorrelation analysis of the Rosenbrock test function is shown in Figure 8.3.

$$\rho(\varepsilon) = \frac{\mathrm{E}(f_t \cdot f_{t+\varepsilon}) - \mathrm{E}(f_i)E(f_{t+\varepsilon})}{\mathrm{Var}(f_t)} \tag{8.4}$$

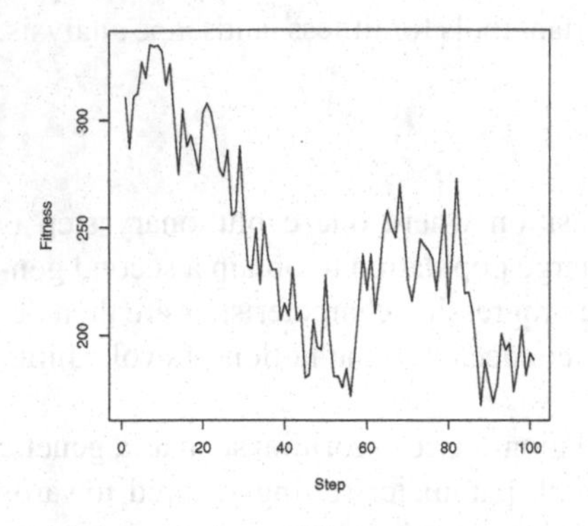

(a) Fitness values of consecutive steps of a random walk on the Rosenbrock test function

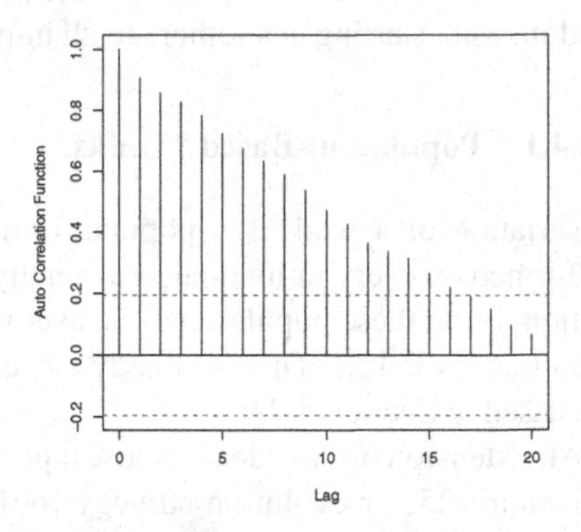

(b) Auto correlation function of the fitness values of the random walk in Figure 3(a)

Fig. 8.3 The autocorrelation function plot in Figure 3(b) shows a high autocorrelation. Even after 15 steps the fitness values are still significantly correlated (dotted horizontal line).

Assuming isotropy and a large enough sample, it has been shown that this relatively simple measure can be used as an indicator of problem difficulty for certain problem classes e.g. in [43, 13, 49].

In addition, *correlation length* is defined as the average distance between points until they become "uncorrelated". There are several different definitions of correlation lengths. The most common formulation in [100] is based on the assumption of a stationary stochastic process or an elementary landscape (see Section 8.2.6) which yields an exponential decay of the autocorrelation function. Thus, the correlation length τ is simply $1/\rho(1)$.

Hordijk, on the other hand, who later extended the model of autocorrelation to a statistically more robust time series analysis gives a more conservative definition, where correlation is only significant while it exceeds the two standard-error-bound. "So the statistical definition of the correlation length τ of a time series is one less than the first time lag i for which the estimated autocorrelation ρ_i falls inside the region $(-2/\sqrt{T}, +2/\sqrt{T})$ [T being the length of the time series], and thus becomes (statistically) equal to zero." [27]

Analogous to the autocorrelation function that gives the correlation for different distances, one can create a variance-corrected correlation coefficient as given in [43] in Eq. (8.5).

$$\rho(f_t, f_{t+\varepsilon}) = \frac{\mathrm{Cov}(f_t, f_{t+\varepsilon})}{\sqrt{\mathrm{Var}(f_t)\mathrm{Var}(f_{t+\varepsilon})}} \tag{8.5}$$

Despite their apparent usefulness (applied in e.g. [87, 99]), these measures are basic and give only a local view of the fitness landscape. Therefore, it has been shown that they are not always useful to predict problem hardness [46].

In [27] this correlation analysis is enhanced by perceiving the random walk as time series of fitness measurement values. Hence, the Box-Jenkins time series analysis [8] is used to get a fuller and statistically sounder view on the landscape which was only *sampled* through a random walk. This approach is used to create an autoregressive moving-average process (ARMA). It consists of an AR(p) process with references to p past values and an MA(q) process that averages over q members of a white noise series. However, the result is not a single number, but a whole model that, therefore, provides a richer picture of the fitness landscape.

These three measures (autocorrelation, correlation length and the Box-Jenkins model) do not take into account the number of neighbors at each step and have only been examined for *random* walks on discrete landscapes. Even though a random walk could appear to be highly uncorrelated, an adaptive walk following along a continuous improvement could present a much smoother landscape.

8.2.6 Spectral Landscape Analysis

A different perspective on the fitness landscapes is provided by perceiving it as a graph. Peter Stadler and co-workers have analyzed many popular problems using an approach that can be described as "Fourier Analysis" of fitness landscapes [79, 80, 81, 85, 82, 70]. This idea is originally described in [101] as Fourier and Taylor series on fitness landscapes and requires a relatively strict assumption of isotropy. It can also be seen as a graph theoretic generalization of Holland's "hyperplane transform" [26] which is based on Holland's widely known schema theory [25].

The basic idea is to transform the fitness landscape in a way that allows easier analysis. Eq. (8.6) shows the general scheme of such a transformation, where the original fitness function f is transformed into a weighted sum of component fitness functions φ_i.

$$f(x) = \sum_i a_i \varphi_i(x) \tag{8.6}$$

This is already reminiscent of a Fourier analysis that determines coefficients for a set of standard base functions. Indeed, this is also the basis for the hyperplane transform already described in [26] which is further extended in [101] to commutative groups and later in [79] to graphs.

First the graph Laplacian $-\Delta$ is derived from a graph's adjacency matrix $\mathbf{A}$ as defined in Eq. (8.7). For simplicity, a d-regular graph is assumed, i.e., all vertexes in the graph have the same number of edges.

$$-\Delta := d\mathbf{I} - \mathbf{A} \tag{8.7}$$

In spectral graph theory, the graph Laplacian is often preferred over the adjacency matrix to calculate eigenvalues and eigenvectors of a graph. To understand spectral landscape theory, it is important to understand the meaning of an eigenvector e of a graph with $-\Delta e = \lambda e$. Looking at this construct from a graph centric perspective, every row represents the connected neighbors of a vertex in the graph. Hence, once the eigenvectors have been obtained for each vertex they can be used as weights of the vertexes' fitness function. As these "weights" are actually values of an eigenvector e Eq. (8.8) is equivalent to the previous definition, however, it uses graph parlance, where $N(v)$ are the neighbors of vertex v or in other words the non-zero entries in the adjacency matrix. The subscript index i in the eigenvector e is the i-th component in the vector e and the function $I : \mathscr{S} \to \mathbb{N}$ gives vertex index of a certain node in the graph.

$$\sum_{n \in N(v)} e_I(n) = \lambda e_i \tag{8.8}$$

As the adjacency matrix of an undirected regular graph is symmetric, the eigenvectors form an orthogonal basis. Hence, we have produced a transformation of the original graph into a form that might be easier to analyze. Because of Eq. (8.8) this transformed landscape satisfies the condition in Eq. (8.9), and is called an *elementary landscape*. In other words, the fitness of a vertex in the transformed landscape graph can be calculated as a weighted sum of its neighbors. In Eq. (8.9), Δf is the graph Laplacian applied to the fitness function, or in other words a linear combination of the neighbor's fitness values, and λ and f^* are scalar constants.

$$\Delta f + \lambda (f - f^* \mathbf{1}) = 0 \tag{8.9}$$

This enables us to transform the graph of a fitness landscape into the form shown in Eq. (8.6) where every φ_i is an eigenvector of the graph Laplacian. Hence, the landscape is decomposed into a set of elementary landscapes which are easier to analyze [81]. Many landscapes of combinatorial optimization problems are elementary or easily decomposable into a small set of elementary landscapes [27].

Even crossover landscapes have been successfully analyzed with this method in [78, 97] with the help of so-called P-structures where a graph is defined through an incidence matrix over the crossover "neighborhood". This matrix can than be treated similarly to the incidence matrix of a graph and allows similar analyzes. However, mating is simulated with a random second parent which is not the case in a population based heuristic where the second parent is usually drawn from the same population. It resembles the "headless chicken" test proposed in [30] which is significantly different to how crossover actually works in most algorithms.

In summary, spectral landscape analysis, provides a powerful basis for many analytical methods that can provide more exact estimates than mere sampling. However, due to the high complexity of this method and its theoretic nature, deriving results for new landscapes is less accessible through automatic means that other analysis methods. Nevertheless, it remains a powerful complementary tool to all other methods presented in this review.

8.2.7 *Information Analysis*

Related to the concept of ruggedness is the information analysis method proposed in [94]. It is based on the idea of information content of a system as a measure of the difficulty to describe this system [10]. The authors assume statistical isotropy of the landscape and define several measures. These measures are derived from a time series obtained by a random walk on the landscape, similar to measures of ruggedness.

The idea is to analyze the amount of information that is necessary to fully describe the random walk. The more information is needed, the more difficult the landscape is. It can be be seen as trying to compress the information of the random walk into as few bits as possible or as a function of the distribution of elements over the states of a system (Shannon Entropy in [75]).

However, instead of directly using the fitness values of the random walk $\{f_t\}_{t=1}^n$, this information is first differentiated $\Delta\{f_t\}_{t=1}^n := \{f_t - f_{t-1}\}_{t=2}^n$ and quantized using a relaxed sign function that is either -1, 0 or 1 depending whether x is less than $-\varepsilon$, in the interval $(-\varepsilon, \varepsilon)$ or greater than ε. Depending on the choice ε we obtain different views of the landscape. Please note that this parameter ε is completely different from the ε that we used as step size to obtain the random walk in the continuous case. While the step size defines the precision with which we sample the landscape locations, the quantization defines how much of a change in fitness value we want to consider being "smooth". The obtained sequence is called $S(\varepsilon) = \widetilde{\text{sign}}_\varepsilon(\Delta\{f_t\}_{t=0}^n)$ and regarded as a sequence of *symbols* that are further analyzed as follows.

1. *Information content* is designed to capture the "variety of *shapes*", i.e., the number and frequency of different combinations of consecutive symbols:
$$H(\varepsilon) := -\sum_{p \neq q} P_{[pq]} \log_6 P_{[pq]} \tag{8.10}$$
where $P_{[pq]}$ is the relative frequency of consecutive symbols $p, q \in \{-1, 0, 1\}$ in the sequence $S(\varepsilon)$. So, $H(\varepsilon)$ is an entropy measure of the number of consecutive symbols that are not equal.
2. *Partial information content* measures the number of slope changes. Therefore, the sequence $S(\varepsilon)$ is filtered into $S'(\varepsilon)$ removing all zeros and all symbols that equal their preceding symbol. The length of this string is designated as μ and its relation to the length of the walk is defined as the partial information content:
$$M(\varepsilon) := \frac{\mu}{n} \tag{8.11}$$
where n is the length of the $S(\varepsilon)$. In other words it is the relative number of changes in slope direction as compared to the total number of steps in the walk.
3. *Information stability* is simply the highest fitness difference between neighbors in the time series. It can be measured as the smallest ε for which the landscape appears completely flat, i.e., $M(\varepsilon) = 0$ and is called ε^*.

4. Finally the authors define the *density-basin information* for analysis of the smooth areas:

$$h(\varepsilon) := -\sum_{p=q} P_{[pq]} \log_3 P_{[pq]} \tag{8.12}$$

where $P_{[pq]}$ is again the probability of two consecutive symbols. Note that this time sections with equal slope are considered.

By varying the parameter ε one can "zoom in and out" to view the same walk at different levels of detail. This analysis can provide additional insight as compared to looking at the measures of ruggedness by going towards the analysis of the distribution of rugged and smooth parts of the landscape.

A related idea is explored in [7] where an attempt is made to measure problem hardness through algorithmic entropy. Here, the number of possibilities to continue a search is perceived as the entropy. The more choices an algorithm has to continue, i.e., the less guidance can be obtained from the current search state, the more difficult it is to find the optimum.

Entropy or information analysis provides again a different perspective of a fitness landscape that is related but dissimilar to ruggedness, and can also be extended to population analysis as demonstrated in [7].

8.2.8 Fitness Distance Correlation

While ruggedness, local optima analysis and information analysis all provide insights into the connectedness and local structure of fitness landscapes, the fitness distance correlation (FDC) was designed to obtain a much more global view. In the original definition of fitness distance correlation in [30, 31], the knowledge of the global optimum is a prerequisite. Therefore, much more can be said about the structure of the landscape leading to this optimum. At the same time, knowledge of the global optimum is often not easy to achieve and can, therefore, also be considered a weakness of this approach.

The underlying idea is to examine the correlation between fitness value and distance to the global optimum. Ideally, this should be done with a representative sample of the landscape and knowledge of the global optimum. The original definition is simply the correlation coefficient of a joint distribution of fitness and distance values.

What we need is a global optimum $x_{\max}$ with $(\forall x \in \mathscr{S})\, f(x_{\max}) \geq f(x)$ and a representative sample of encoded solution candidates $\{x_i\}_{i=1}^n, x \in \mathscr{S}$. For these solution candidates we compute the fitness values $\{f_i\}_{i=1}^n = \{f(x_i)\}_{i=1}^n$ and the distances to the global optimum $\{d_i\}_{i=1}^n = \{d(x_i, x_{\max})\}_{i=1}^n$. Then, we can compute the means $\overline{f}, \overline{d}$ and standard deviations σ_F, σ_D of these sequences and finally compute their correlation coefficient.

$$\text{FDC} = \frac{\frac{1}{n}\sum_{i=1}^n (f_i - \overline{f})(d_i - \overline{d})}{\sigma_F \sigma_D} \tag{8.13}$$

Due to its simplicity and discriminatory power in many real-world scenarios this has been one of the most popular measures used in the past [47, 13, 59, 87, 99, 17, 90, 54, 40].

With the FDC *coefficient* alone, however, even during its inception in [30] several cases were noted where it fails to correlate to problem difficulty [55]. This is due to the same problems that any correlation coefficient faces in the light of differently-distributed variables. Moreover, compressing the information of a large multidimensional space into a single number naturally incurs information loss. It is not surprising that this method has its limits. What is surprising, however, is that this number reflects problem difficulty for a large array of model problems [30].

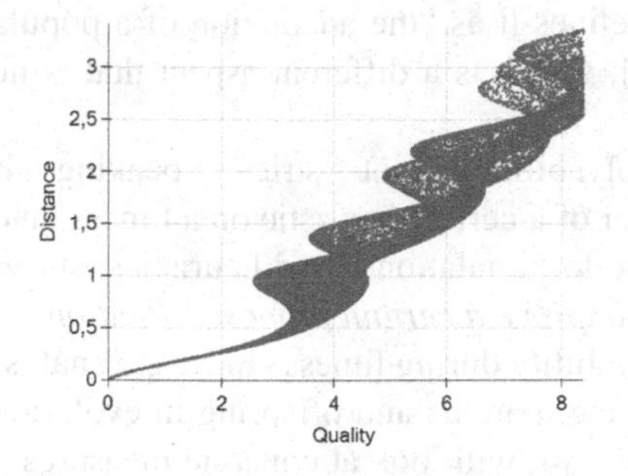

(a) Close-up of the fitness-distance correlation plot of the two-dimensional Ackley function [1]: While this function has a clear correlation of fitness vs. distance in the lower quality regions (not shown), the difficulty stemming from local optima near the global optimum can be clearly identified.

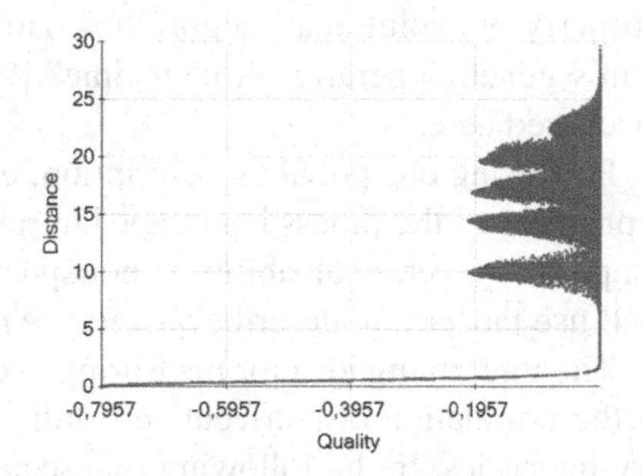

(b) FDC plot of a multi-normal test function described in [63]. This plot shows the clearly deceptive nature of this test function leading to large local optima that can easily trap a search algorithm.

Fig. 8.4 Examples of fitness distance correlation plots, generated by randomly sampling the solution space with a preference to areas with higher fitness as described in Section 8.2.4.1

Additionally, for the cases where the single coefficient alone is inconclusive or wrong a look at the *distribution* of fitness vs. distance can provide remarkable insight as shown in Figures 4(a) and 4(b). However, due to the projection of distance to a single number, points with similar distance to the optimum do not necessarily have to be close to each other at all.

While FDC plots can greatly facilitate the understanding of a fitness landscape it is important to be reminded that it represents a severe simplification. All dimensions are reduced to a single number—the distance to the optimum. While it seems to work nicely for many scenarios e.g. [30, 47, 87] it has also been shown in [4, 32, 65] that it is easy to construct pathological cases where the insights of an FDC plot analysis can be inconclusive or even misleading.

8.2.9 *Evolvability*

Evolvability or the *ability to evolve* is a population's chance to improve. The term is used both in evolutionary biology as well as evolutionary computation. Altenberg defines it as "the ability of a population to produce variants fitter than any yet existing' [3]. While it is not yet clear why evolvability is a trait of natural evolution [44] it has been shown that it is a desirable long-term trait of an evolutionary algorithm [89]. Continued evolvability is often ensured through *variability* [98] which is achieved through e.g. keeping diversity in a population-based heuristics, random restarts in local search or temporarily allowing inferior solutions in simulated annealing.

The term evolvability has also been used by Grefenstette to describe a different property of evolutionary algorithms. He defines it as "the adaptation of a population's genetic operator set over time" [23], which is a different aspect that is not discussed here.

Following our previous description, evolvability cannot—strictly speaking—be a property of the fitness landscape but rather of a certain heuristic on a fitness landscape. However, evolvability is not specific to population based heuristics and we will use the term to describe *the chance to improve a certain proposed solution*.

The underlying idea for measuring evolvability during fitness landscape analysis is the correlation of consecutive solutions e.g. parents and offspring in evolutionary heuristics. In the following two sections, we will look at concrete measures of evolvability that have been proposed previously.

8.2.9.1 Evolvability Portraits

One approach to quantify evolvability was given in [77] termed "Fitness Evolvability Portraits". The authors show how evolvability portraits relate to both ruggedness and neutrality (Section 8.2.10). The underlying function for all measures is the evolvability of a single solution candidate x. From a fitness landscape perspective, this reduces to having neighbors of equal or higher fitness and can thus be formulated as

$$\mathscr{E}(x) = \frac{|N^*(x)|}{|N(x)|} \qquad \text{or} \qquad \mathscr{E}(x) = \int_{f(n) \geq f(x)} p(n|x)\, \mathrm{d}n \tag{8.14}$$

where $N^*(x) = \{n \mid n \in N(x), f(n) \geq f(x)\}$, which is simply the number of non-deleterious neighbors over the number of neighbors. Alternatively, we can define this measure for a specific algorithm as the probability of choosing a fitter or equally fit offspring where $p(n|x)$ is the probability of a certain algorithm choosing a particular neighbor n of x. Since $\int p(n|x)\mathrm{d}n$ sums to unity, this is a probability density function.

Using this definition we can derive the proposed evolvability metrics for every solution candidate x:

- $E_a(x) = \mathscr{E}(x)$, probability of non-deleterious mutation.
- $E_b(x) = \int p(n|x)f(x)\mathrm{d}n$, the average expected offspring fitness,
- $E_c(x) = \frac{100}{C}\int_{f(n)\geq f_c(x)} p(n|x)f(x)\mathrm{d}n$ where $f_c(x)$ is defined implicitly through $\int_{f(n)\geq f_c(x)} p(n|x)\mathrm{d}n = \frac{C}{100}$. Hence, E_c is the top C-th percentile of the expected offspring fitness.
- $E_d(x) = \frac{100}{C}\int_{f(n)\leq f_d(x)} p(n|x)f(x)\mathrm{d}n$ where $f_d(x)$ is implicitly defined through $\int_{f(n)\leq f_d(x)} p(n|x)\mathrm{d}n = \frac{C}{100}$. Hence, E_d is the bottom C-th percentile of the expected offspring fitness.

To extend this to a population based analysis, the values, that are defined for every single solution candidate, can now be averaged over a whole set of solution candidates with the same fitness. Evolvability portraits capture an important piece of information and provide a new perspective as compared to the previously discussed measures. However, by design, evolvability portraits alone do not provide enough information to allow meaningful conclusions. They have been devised as a complementary addition to the arsenal of fitness landscape analysis tools.

8.2.9.2 The "Fitness Cloud"

Another, related analysis proposed in [11] is a simple scatter plot of parent fitness vs. offspring fitness. In this approach every solution candidate is considered to have only one unique offspring which is determined by the algorithm or operator. The fitness of this offspring is called the *bordering fitness*. Using the pair of fitness value $f(x)$ and bordering fitness value $\widetilde{f}(x)$ as coordinates we obtain a scatter plot which is called a *fitness cloud* by the authors as shown in Figure 8.5.

The authors further define the set of minimum, mean and maximum bordering fitness for every fitness value as $\mathrm{FC}_{\mathrm{min}}$, $\mathrm{FC}_{\mathrm{mean}}$, and $\mathrm{FC}_{\mathrm{max}}$ respectively and define the "intersection" of these sets with the diagonal (i.e., $f(x) = \widetilde{f}(x)$) as α, β and γ respectively. These mark the boundaries between strictly advantageous, advantageous, deleterious and strictly deleterious areas of parent fitness for a given algorithm.

Local optima are points under the diagonal (for a maximization problem), i.e., all their offspring's fitness values are less than their own fitness. This kind of analysis facilitates finding the achievable fitness for a certain algorithm as it clearly shows easily attainable fitness values for a certain heuristic. The β level is assumed to be a "barrier of fitness" that cannot be easily overcome for a certain algorithm. Again, this analysis method has been devised as a complement to other methods to provide a further view of fitness landscapes.

Based on the definition of a fitness cloud the authors have extracted a singular number as a measure for problem hardness in [91]. This measure was developed and tested for genetic programming but it seems that the concept is general enough to be easily applicable to other heuristic methods. Using the fitness cloud, a set of fitness and neighbor fitness pairs (f,g) is divided into consecutive groups based

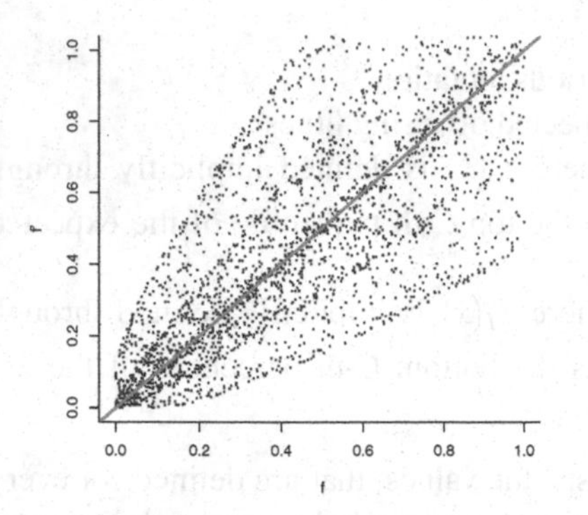

(a) Bordering fitness (ordinate) is determined by choosing a neighbor at **random**

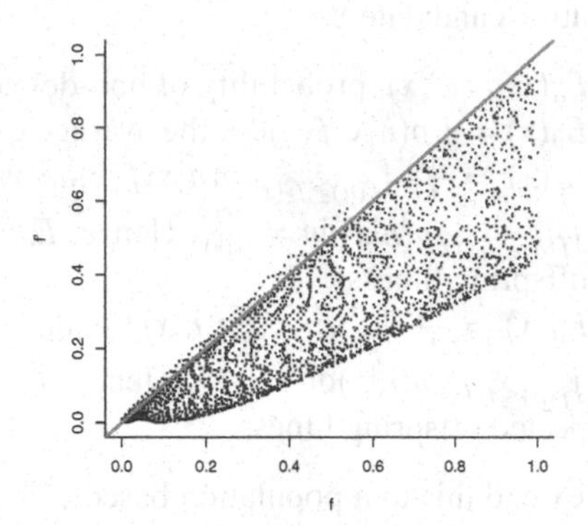

(b) Bordering fitness (ordinate) is determined by choosing the **best** neighbor

Fig. 8.5 Two examples of fitness clouds for the Rosenbrock test function close to the optimum: The abscissa represents possible parent fitness values while the ordinate has all possible offspring fitness values. Please note that the Rosenbrock function is a minimization problem, hence, better neighbors are below the diagonal.

on their fitness value (f). For each of these groups the mean is calculated $(\overline{f_i}, \overline{g_i})$, i.e., the mean of fitness and mean of neighbor fitness within each group. Between consecutive groups, the slope S_i between the respective mean points is calculated as $S_i = (\overline{g_{i+1}} - \overline{g_i})/(\overline{f_{i+1}} - \overline{f_i})$. The sum of all negative slopes then defines the *negative slope coefficient* $\text{nsc} = \sum_i \min(S_i, 0)$.

Later, in [92], it was pointed out that a more elaborate segmentation into fitness groups improves the predictive performance of this measure. Finally, it should be noted that this measure currently lacks a normalizing term that would make it comparable across different problems. The negative slope coefficient designates an "easy" problem if it is zero and increasingly negative values describe increasingly difficult problems, however, without a scale. In a follow-up study [93], this method has been applied to the popular NK landscapes (Section 8.3.1) where it is successful in correlating the negative slope coefficient to increased ruggedness of the NK landscapes. Moreover, it is confirmed that sampling NK landscapes produces comparative results to an exhaustive analysis which probably pertains to the NK landscape's high isotropy.

A similar idea to the fitness cloud is presented in [9], where the authors quantize fitness values into several classes. However, this so-called "correlation analysis" comes pretty close to what we have termed evolvability analysis here.

Evolvability analysis is yet another viewpoint of the fitness landscape that enables a better understanding of problem difficulty and suitability of a certain heuristic.

8.2.10 Neutrality

Simply put, neutrality is the degree to which a landscape contains *connected areas of equal fitness*. It was first introduced in biological evolution theory [37] and immediately stirred some controversy as it was initially considered a complete alternative to Darwinian selection. As neutral mutation causes no fitness change it cannot be selected against and therefore, makes it possible to explore a much greater genotype space. Later, the notion of neutrality was toned down to form the theory of *nearly-neutral networks* [60, 61, 62]. Here, it also becomes plausible that even slightly deleterious mutations are not selected against. Upon a changed environment, however, these mutations and, hence, a population's increased diversity, might provide the basis for a mutation with a significant fitness advantage at which time it will be selected for. Therefore, it was proposed to work as a complement to Darwinian selection [96]. A crucial insight was that a "neutral mutation does not change one aspect of a biological system's function in a specific environment and genetic background," but might have a profound influence later on. The difficulty of coming to this conclusion within biological evolution was the difficulty to quantify fitness and, hence, the difficulty to measure neutrality [95].

For natural evolution, neutrality plays an important role: "[D]ue to the high-dimensionality of sequence space, networks of frequent structures penetrate each other so that each frequent structure is almost always realized within a small distance of any random sequence" [29]. "Fisher, for example, argued that local optima may not exist in a large class of high-dimensional spaces; the probability that a solution is optimal in every single dimension simultaneously is negligible [64, 274]" [77]. Wright's 3-dimensional landscape metaphor [103] has often been criticized as a model for natural evolution, especially in the light of neutrality: "peaks are largely irrelevant because populations are never able to climb there [...], and valleys are largely irrelevant because selection quickly moves populations away from there."[19].

As described in the previous two paragraphs, neutrality plays an important role in natural evolution. For this reason, the idea of neutrality has also found its way into heuristic optimization. For discrete spaces, the definition is obvious: A so called "neutral network" is a set of n solution candidates S_n and their corresponding fitness values forming a connected network of solution candidates with equal fitness.

To extend the notion of neutral networks to neutral *areas* in continuous spaces, we can replace all direct neighborhood functions through neighborhood ε-discs for an arbitrary ε. Furthermore, for a fitness function that is continuous in the codomain, one might consider relaxing the equality criterion and look at *almost* neutral areas.

An important property of evolution in neutral areas is constant innovation [28, 77] which ensures that despite a constant fitness value, the population is actively exploring the neutral area. A poorly designed algorithm might simply jump back and forth between equally fit solutions, ceasing to explore the landscape and, hence, being stuck.

The first measure of neutrality was proposed in [69]: Neutral areas are explored by neutral walks, a variation of random walks that continue only within a neutral area trying to continuously increase the distance to the starting point. The maximum distance can then be used as an indicator of neutrality. In [34, 33] Nei's standard genetic distance operator [57] is used to gain more insight into neutrality. Later, Barnett defined more measures of neutrality in [5, 6].

In summary we can define the following measures for every (almost) neutral area.

- The volume or size, i.e., the number of neighboring solution candidates with equal or nearly equal fitness.
- The number of different fitness values *surrounding* the neutral area. Called the "percolation index" in [5]. In addition we can also include the number of different phenotypes and genotypes, as different phenotypes might still have the same fitness values if a landscape has a high degree of neutrality.
- The maximum distance between two members [69].
- The average distance between two members.

The distributions of these values can then provide an additional complementary perspective of the fitness landscape. Neutrality plays an important role in the performance of heuristic algorithms [5, 6, 69, 94] and cannot be subsumed by other measures [69, 77].

8.2.11 Epistasis

The biological definition of epistasis is the impact one gene has on the effect of another gene, or in other words, the non-linearity or degree of concerted effect of several genes. Simple examples are one gene inhibiting or enhancing the phenotype expression of another gene. Already in the biological sense as described in [12] there is ambiguity about the concrete definition of the term.

When it comes to heuristic algorithms, several different measures for epistasis have been proposed. Davidor's *epistasis variance* [14] is an often cited attempt to quantify epistasis: We start with a linear model of the encoded solution candidates. With no epistasis, the fitness function can be reformulated as a linear combination of the elements x_i of the solution candidate x (e.g. the bits in a binary vector) as $f(x) = \sum_{i \in I} f_i(x_i)$. With maximum epistasis, the fitness function contains all possible combinations of interactions between elements of the solution candidate i.e. $f(x) = \sum_{i \in 2^I} f_i(x_i)$ where I is the index set of all components in the solution candidate. For this new formulation of the fitness function we have to estimate the parameters using a large enough sample from the solution space.

After parameters for a linear model have been estimated, an analysis of variance between the model and the actual fitness function reveals how much variance is explained by the linear model. The remaining variance is assumed to be due to epistatic interactions across different elements of the solution candidates (i.e., different loci in genetic parlance). Epistasis variance can the then be defined as $(\sigma - \sigma_l)/\sigma$ where σ_l is the variance explained by the linear model and σ is the total variance of the fitness function. While this works for some cases it fails for several others and has been heavily criticized in the past [68].

In [54] epistasis variance and other measure are used to measure hardness for a genetic algorithm and are shown to be insufficient for several cases. In [56] a closer look to a more complete definition of epistasis (in contrast to epistasis variance) is compared to deceptiveness and they conclude that epistasis is indeed related to deceptiveness. However, for these results a full analysis of the solution space is required.

As already noted by Davidor, epistasis variance cannot differentiate between different orders of non-linearity. Therefore, in [71] *graded epistasis* and *graded epistasis correlation* are introduced. The first idea is to model more complex combinations including some orders of non-linearity. The second idea is to measure correlation with the original fitness function instead of an analysis of variance.

Using our notation, graded epistasis can be described as a model with at most g interacting genes as $f^g(x) = \sum_{i \in G(g)} f_i(x_i)$ where the multi index i from the power set of all indexes is limited to g simultaneous elements through $G(g) = \{i \mid i \in 2^I, |i| \leq g\}$.

However, even this more elaborate model fails to establish a clear relationship between measured epistasis and problem difficulty. An interesting experiment is carried out in [72] where the encoding is changed to reduce the epistasis, however, the problem does not become easier for their genetic algorithm.

In [55] it is pointed out that both epistasis variance and epistasis correlation, although set out to measure the extent of epistasis only measure the absence of epistasis. Another review of epistasis can be found in [67], where several other difficulties of this approach are pointed out.

While the concept of epistasis itself is indeed important to understand problem difficulty or deceptiveness for heuristic algorithms and despite extensive research on this topic, so far, we have yet to find a good model that is both easy to compute and has enough predictive power. In light of the shortcomings of other indicators, the previously mentioned epistasis measures can also be useful as one of several perspectives that are considered in a more complete analysis of a fitness landscape.

8.3 Popular Problem Classes

For the derivation and analysis of fitness landscapes a whole series of test problems and model landscapes have evolved. The most popular benchmark problem classes and findings of various fitness landscape analysis techniques are described in the following sections.

8.3.1 Additive Random Fields

Several specialization of random fields are frequently used as model landscapes when developing or testing new fitness landscape analysis methods. An additive random field has the general form shown in Eq. (8.15).

$$F(x) = \sum_i C_i F_i(x) \tag{8.15}$$

where every C_i is a random variable and every $F_i : \mathscr{S} \to \mathbb{R}$ is an arbitrary function [69].

An example of a random field is the energy function of spin glasses [76, 15] where only neighboring spins are interacting. Here, the constants C_i correspond to the neighborhood relation and are either 0 or 1, while the function F_i makes a positive fitness contribution based on whether neighboring spins are equal.

The *NK landscapes* and their extensions are used almost universally for testing new fitness landscape analysis methods. (It has to be noted that various spellings of NK landscapes can be found in the literature e.g. "NK", "Nk" or even "n-k". For consistency, we will exclusively spell them as "NK".) NK fitness landscapes were introduced by Kauffman in [35, 36]. It describes a model fitness function based on the idea of epistasis or overlapping fitness contribution of sequence elements. It contains two parameters: N, the number of loci, and K, the number of epistatic interactions. Every locus can have one of two possible alleles e.g. 0 or 1. There are two different models of epistasis that are proposed: The simple *neighborhood* model assumes that neighboring loci epistatically interact with each other. The *random* model assumes that epistatic interactions occur at random (but with a fixed number of $K+1$ interaction partners). The final fitness value for a sequence is the average of all fitness contributions for all loci including the epistatic interactions.

$$NK(s) := \frac{1}{n} \sum_{i=1}^{n} f_i(n_i(s)) \tag{8.16}$$

with $n_i : C(n) \to C(K+1)$, $f_i : C(K+1) \to \mathbb{R}$ where $C(n)$ is the set of binary sequences of length n, f_i is the fitness contribution of the i-th epistatic component and $n_i(s)$ is the value of this component of $s \in \mathscr{S}$. For example, the simple neighborhood variant with $K=2$, $n_i(s)$ could be defined as $n_i(s) := (s_{i-1 \bmod |s|}, s_i, s_{i+1 \bmod |s|})$ where s_i denotes the i-th element in the sequence s. The fitness value contributions f_i can be a simple table with a fitness for every value of every epistatic component.

NK landscapes, especially with higher values of K, contain a substantial amount of variability and complexity. One important simple property, however, is their isotropy. This is in stark contrast to many fitness landscapes of real-world problems that have a tendency towards non-uniform structures. The most important property of NK landscapes, however, is their tunable ruggedness, which is achieved by varying the parameter K [100, 27, 77, 94, 47]. In [30] it was shown that NK landscapes with increasing K loose any structure and resemble purely random landscapes.

In [6] an extension to NK landscapes was proposed that introduces a third parameter p which allows tuning the neutrality of a landscape. The parameter p is the probability that a certain allele combination is zero, i.e., that it makes no fitness contribution. As shown in [6] this gives rise to increased neutrality of the landscape, while it does not significantly influence ruggedness.

A similar modification is achieved with NKq landscapes, where the possible contribution of a certain allele combination is quantized to integers between 0 and

q [58]. In [20] NKq and NKp landscapes are compared and found to be quite different in how they achieve neutrality and, therefore, should both be considered valuable additions to the library of model fitness landscapes.

In [4], the NK model is re-examined and a more general *block model* is proposed. While Kauffman's NK model identifies loci with fitness components, Altenberg introduces an arbitrary map between fitness components and genes:

$$F(x) := \frac{1}{n} \sum_{i=1}^{n} F_i(x_{j_1(i)}, x_{j_2(i)}, \ldots, x_{j_{p_i}}) \tag{8.17}$$

where n is the number of fitness components and F_i is the fitness function for each fitness component and the indexes j_i are a selection of loci that are the inputs for a certain fitness component.

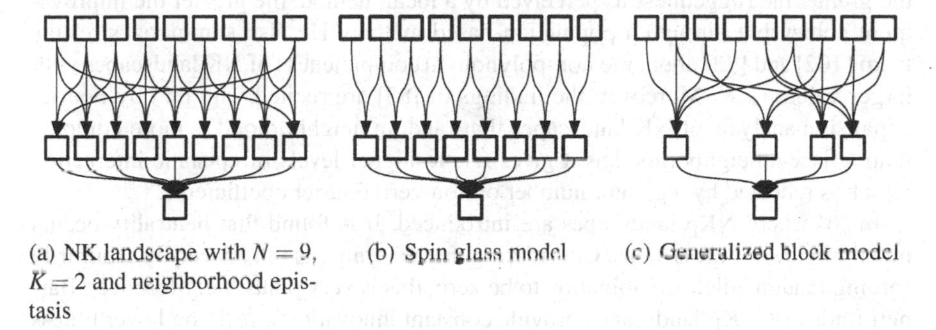

(a) NK landscape with $N = 9$, $K = 2$ and neighborhood epistasis (b) Spin glass model (c) Generalized block model

Fig. 8.6 Random field landscapes: The upper row is the sequence of loci or genes that is represented in the encoding. The middle row represents the fitness components or features and the bottom box is the final fitness value, typically the sum or average of the fitness components. Every fitness component computes an arbitrary function over its inputs.

The example of an NK landscape in Figure 6(a) has nine genes and two epistatically linked genes ($K = 2$). Thus, the input values for every feature are obtained from three different genes. Figure 6(b) shows the links for a spin glass example. Here the "interaction" of two spins or two neighboring genes can also be modeled by a feature function that analyzes the correspondence of neighboring spins. Finally, Figure 6(c) shows the generalization proposed by Altenberg, where the number of features is independent of the number of loci and gives rise to a great variety of model problems.

Many problems can be recast as additive random fields, which makes this model relatively generic. On the other hand, this kind of representation easily becomes "frustrating" [4], i.e., hard to solve, since it possesses—by design—a large potential for epistasis. Therefore, even though additive random fields provide a remarkably insightful playground for fitness landscape analysis methods it has to be examined how they relate to real-world problems.

The family of additive random fields, in particular NK and NKp landscapes have been subject to many different analyzes in the past. Often a newly-introduced landscape measure is first applied to NK landscapes e.g. [77, 94, 93, 27, 47]. It was found that NK landscapes serve as a good model for molecular evolution e.g. in [2]. In [4] a summary of the findings of [36], [101] and [18] can be found. It describes the properties of a one-mutant adaptive walk as found in populations in an equilibrium state. For $K = 0$, there is a single globally attractive genotype, the average number of steps is $N/2$ and the landscape is highly correlated. For the maximum number of epistatic interactions ($K = N - 1$) the expected number of local optima grows to $(2^N)/(N+1)$ while the expected fitness of a local optimum from a random starting point decreases towards the mean fitness of the entire landscape. Kauffman called this the "complexity catastrophe". For low values of K, the highest local optima share many alleles with each other.

In [4] results for full population analysis are also reviewed which conclude that the greater the ruggedness as perceived by a local method, the greater the improvement achievable through a population-based method. He also summarizes results from [102] and [88] about the non-polynomial completeness of NK landscapes with larger values of K. Moreover, the findings of [83] are recited who find by Fourier expansion analysis of NK landscapes that random neighborhood is more complex than adjacent neighborhood as it gives rise to higher levels of epistatic interaction which is reflected by a greater number of non-zero Fourier coefficients.

In [6] where NKp landscapes are introduced. It is found that neutrality occurs mostly in lower fitness areas. Considering that neutrality is introduced by arbitrarily forcing random allele combination to be zero, this is very plausible. Moreover, Barnett finds that NKp landscapes provide constant innovation mostly on lower fitness regions, which is also consistent with a higher level of neutrality in these regions.

When compared to other optimization problems, NK problems are by far more difficult in comparison as pointed out in [7], where it is suggested that the difficulty is close to random landscapes.

8.3.2 *Combinatorial Problems*

Together with real-valued problems, combinatorial problems are the primary target of heuristic algorithms. Therefore, they have frequently been subjected to fitness landscape analysis as well as served as benchmark problems for heuristic algorithms. In the following paragraphs we will give a brief overview of what characteristics of specific combinatorial problems have been determined by dedicated fitness landscape analysis methods in the past.

8.3.2.1 Quadratic Assignment Problem

The quadratic assignment problem (QAP) is defined as the assignment of n facilities to n locations as originally described in [38]. Between every pair of facilities (i, j) we have an associated flow f_{ij} and between every pair of locations we have an associated distance d_{ij}, which are both encoded as matrices. The task is to minimize cost that is defined as the product of flow and distance by optimally assigning locations to facilities encoded in the permutation π as defined in Eq. (8.18).

$$C(\pi) = \sum_{i=1}^{n} \sum_{j=1}^{n} f_{ij} \, d_{\pi(i)\pi(j)} \tag{8.18}$$

In [49] we can find a detailed analysis of the fitness landscapes of several instances of the quadratic assignment problem. They analyze local optima, ruggedness, fitness distance correlation as well as a problem specific measure as an estimate for epistasis. They also include instances of the traveling salesman problem (see Section 8.3.2.2) which they cast into quadratic assignment problems and specifically crafted instances with varying epistasis. They conclude that no single measure alone is able to make any prediction of problem hardness. However, taken together, they used epistasis, ruggedness and fitness distance correlation to classify QAP instances into a few classes of varying difficulty for a memetic algorithm.

An application of spectral landscape analysis to the QAP can be found in [73], where they show that it can be decomposed into a handful elementary landscapes implying a strong internal structure of this problem class.

8.3.2.2 Traveling Salesman Problem

The objective in a traveling salesman problem (TSP) is to find the order in which to visit a series of n cities to minimize total distances traveled [39]. Eq. (8.19) shows a simple definition. Again the distances are encoded in a matrix d where d_{ij} is the distance from city i to city j. The solution is, again, represented as a permutation π as shown in Eq. (8.19).

$$C(\pi) = d_{\pi(n)\pi(1)} + \sum_{i=1}^{n-1} d_{\pi(i)\pi(i+1)} \tag{8.19}$$

TSP instances were one of the first model problems to be subject to fitness landscape analysis. In [84] it was found that Euclidean traveling salesman instances stemming from actual city maps have low ruggedness and that ruggedness is usually independent of a certain instance but dependent upon the number of cities. Later, in [50] it was found that local optima are frequently close to each other and also close to the global optimum termed a "massif central" reminiscent of the mountains and plateaus in south-central France. Moreover, as also stated in [17] it was argued that especially small changes during mutation, i.e., a 2-opt provide the smoothest and hence the "easiest" landscape for optimization with recombination. In [70] it is argued that depending on the algorithm, different mutation operators provide better performances, hence, emphasizing the situation-dependent interplay of problem, encoding, algorithm and its operators. In [45] problem-dependent measures are used in addition to a correlation analysis to augment the prediction of operator performance.

8.4 Conclusion

As elaborated in the previous sections fitness landscape analysis has been successfully used to produce a deeper understanding of problem and algorithm characteristics. However, there remain several unsolved problems.

One important question is the choice of encoding which cannot be solved by fitness landscape analysis alone. Different encodings with different operators can still create the same fitness landscape as shown in [67]. Moreover, as discussed in Section 8.2.10 it can be advantageous to have a high degree of neutrality which can be facilitated by an appropriate encoding. Goldberg's "Messy" genetic algorithms are an intriguing example [21] of the power of degeneracy and neutrality.

Another open question is how to efficiently model landscapes of recombination algorithms. Several approaches have been investigated in [30, 51, 52, 78, 97], however, no satisfying solution has been found so far. One interesting approach is the "headless chicken" test in [30] which compares crossover of two individuals from the population with crossover of one randomly generated parent to discern between macro-mutation and true genetic evolution.

Fitness landscape analysis is a powerful tool for conducting repeatable and standardized tests on problems and problem classes. Previous approaches have tried to produce a difficulty measure as a single number which is too much of a simplification. However, what fitness landscape analysis can accomplish is reduce the dimensionality of a problem to several key numbers that describe the landscape's characteristics and create a rough image of its structure. In general this can only be achieved through several complementary analysis methods that provide different perspective and, therefore, a more complete picture.

As stated in [16] there might not be a *free lunch*, but at least, there might be some *free appetizer* that can be reaped more easily using fitness landscape analysis techniques.

Acknowledgements. This work was supported by a sabbatical grant from the Upper Austria University of Applied Sciences to EP. Special thanks go to Franz Winkler, Stephan Winkler and Andreas Beham for insightful discussions and help with experimental setups.

References

1. Ackley, D.H.: A connectionist machine for genetic hillclimbing. Kluwer, Dordrecht (1987)
2. Aita, T., Hayashi, Y., Toyota, H., Husimi, Y., Urabe, I., Yomo, T.: Extracting characteristic properties of fitness landscape from in vitro molecular evolution: A case study on infectivity of fd phage to E.coli. J. Theor. Biol. 246, 538–550 (2007)
3. Altenberg, L.: The Evolution of Evolvability in Genetic Programming. In: Advances in Genetic Programming, pp. 47–74. MIT Press, Cambridge (1994)
4. Altenberg, L.: NK Fitness Landscapes. In: The Handbook of Evoluationary Computation, pp. B2.7.2:1–11. Oxford University Press, Oxford (1997)

5. Barnett, L.: Tangled Webs: Evolutionary Dynamics on Fitness Landscapes with Neutrality. MSc dissertation, School of Cognitive Sciences, University of East Sussex, Brighton (1997)
6. Barnett, L.: Ruggedness and neutrality—the NKp family of fitness landscapes. In: ALIFE, pp. 18–27. MIT Press, Cambridge (1998)
7. Borenstein, Y., Poli, R.: Decomposition of fitness functions in random heuristic search. In: Stephens, C.R., Toussaint, M., Whitley, L.D., Stadler, P.F. (eds.) FOGA 2007. LNCS, vol. 4436, pp. 123–137. Springer, Heidelberg (2007)
8. Box, G.E.P., Jenkins, G.M., Reinsel, G.: Time Series Analysis: Forecasting & Control. Holden Day (1970)
9. Brandt, H., Dieckmann, U.: Correlation Analysis of Fitness Landscapes. Working papers, International Institute for Applied Systems Analysis (1999)
10. Chaitin, G.J.: Information, Randomness & Incompleteness. World Scientific, Singapore (1987)
11. Collard, P., Verel, S., Clergue, M.: Local search heuristics: Fitness Cloud versus Fitness Landscape. In: ECAI, pp. 973–974. IOS Press, Amsterdam (2004)
12. Cordell, H.J.: Epistasis: what it means, what it doesn't mean, and statistical methods to detect it in humans. Hum. Mol. Genet 11, 2463–2468 (2002)
13. Czech, Z.: Statistical measures of a fitness landscape for the vehicle routing problem. In: IPDPS, pp. 1–8 (2008)
14. Davidor, Y.: Epistasis variance: suitability of a representation to genetic algorithms. Complex Systems 4, 369–383 (1990)
15. Derrida, B.: Random energy model: Limit of a family of disordered models. Phys. Rev. Lett. 45, 79–82 (1980)
16. Droste, S., Jansen, T., Wegener, I.: Perhaps Not a Free Lunch But At Least a Free Appetizer. In: GECCO, July 13-17, vol. 1, pp. 833–839. Morgan Kaufmann, San Francisco (1999)
17. Fonlupt, C., Robilliard, D., Preux, P.: Fitness Landscape and the Behavior of Heuristics. In: Monmarché, N., Talbi, E.-G., Collet, P., Schoenauer, M., Lutton, E. (eds.) EA 2007. LNCS, vol. 4926, pp. 321–329. Springer, Heidelberg (2008)
18. Fontana, W., Stadler, P.F., Bornberg Bauer, E.G., Griesmacher, T., Hofacker, I.L., Tacker, M., Tarazona, P., Weinberger, E.D., Schuster, P.: RNA folding and combinatory landscapes. Phys. Rev. E 47(3), 2083–2099 (1993)
19. Gavrilets, S.: Evolution and Specitation in a Hyperspace: The Roles of Neutrality, Selection, Mutation and Random Drift. In: Evolutionary Dymaics: Exploring the Interplay of Selection, Accident, Neutrality and Function, pp. 135–162. Oxford University Press, Oxford (1999)
20. Geared, N., Wiles, J., Hallinan, J., Tonkes, B., Skellet, B.: A Comparison of Neutral Landscapes – NK, NKp and NKq. In: CEC, pp. 205–210. IEEE Press, Los Alamitos (2002)
21. Goldberg, D.E., Deb, K., Kargupta, H., Harik, G.: Rapid, Accurate Optimization of Difficult Problems Using Fast Messy Genetic Algorithms. In: ICGA, pp. 56–64. Morgan Kaufmann, San Francisco (1993)
22. Gouvêa, F.Q.: p-adic Numbers: An Introduction. Springer, Heidelberg (1993)
23. Grefenstette, J.J.: Evolvability in Dynamic Fitness Landscapes: A Genetic Algorithm Approach. In: CEC, pp. 2031–2038 (1999)
24. Hastings, W.K.: Monte Carlo sampling methods using Markov chains and their applications. Biometrika 57, 97–109 (1970)
25. Holland, J.H.: Adaptation in Natural and Artificial Systems. University of Michigan Press, Ann Arbor (1975)

26. Holland, J.H.: Evolution, Learning and Cognition, pp. 111–127. World Scientific, Singapore (1988)
27. Hordijk, W.: A measure of landscapes. Evol. Comput. 4(4), 335–360 (1996)
28. Huynen, M.A.: Exploring phenotype space through neutral evolution. J. Mol. Evol. 43, 165–169 (1996)
29. Huynen, M.A., Stadler, P.F., Fontana, W.: Smoothness within ruggedness: The role of neutrality in adaptation. P. Natl. Acad. Sci. USA 93, 397–401 (1996)
30. Jones, T.: Evolutionary Algorithms, Fitness Landscapes and Search. Ph.D. thesis, University of New Mexico, Albuquerque, New Mexico (1995)
31. Jones, T., Forrest, S.: Fitness Distance Correlation as a Measure of Problem Difficulty for Genetic Algorithms. In: ICGA, pp. 184–192. Morgan Kaufmann, San Francisco (1995)
32. Kallel, L., Naudts, B., Schoenauer, M.: On functions with a given fitness-distance relation. In: CEC, vol. 3, pp. 1910–1916 (1999)
33. Katada, Y.: Estimating the Degree of Neutrality in Fitness Landscapes by the Nei's Standard Genetic Distance – An Application to Evolutionary Robotics. In: CEC, pp. 483–490 (2006)
34. Katada, Y., Ohkura, K., Ueda, K.: The Nei's standard genetic distance in artificial evolution. In: CEC, pp. 1233–1239 (2004)
35. Kauffman, S.: Adaptation on rugged fitness landscapes. Lectures in the Sciences of Complexity, vol. 1, pp. 527–618. Addison-Wesley Longman, Amsterdam (1989)
36. Kauffman, S.A.: The Origins of Order. Oxford University Press, Oxford (1993)
37. Kimura, M.: Evolutionary Rate at the Molecular Level. Nature 217, 624–626 (1968)
38. Koopmans, T.C., Beckmann, M.: Assignment Problems and the Location of Economic Activities. Econometrica 25(1), 53–76 (1957)
39. Lawler, E.L., Lenstra, J.K., Kan, A.H.G.R., Shmoys, D.B. (eds.): The Traveling Salesman Problem: A Guided Tour of Combinatorial Optimization. Wiley, Chichester (1985)
40. Le, M.N., Ong, Y.-S., Jin, Y., Sendhoff, B.: Lamarckian memetic algorithms: local optimum and connectivity structure analysis. Memetic Comp. 1, 175–190 (2009)
41. Lehn, R., Kuntz, P.: A contribution to the study of the fitness landscape for a graph drawing problem. In: Boers, E.J.W., Gottlieb, J., Lanzi, P.L., Smith, R.E., Cagnoni, S., Hart, E., Raidl, G.R., Tijink, H. (eds.) EvoWorkshops 2001. LNCS, vol. 2037, p. 172. Springer, Heidelberg (2001)
42. Lipsitch, M.: Adaptation on rugged landscapes generated by iterated local interactions of neighboring genes. In: ICGA, vol. 135, pp. 128–135 (1991)
43. Manderick, B., de Weger, M.K., Spiessens, P.: The Genetic Algorithm and the Structure of the Fitness Landscape. In: ICGA, pp. 143–150 (1991)
44. Marrow, P.: Evolvability: Evolution, Computation, Biology. In: GECCO, pp. 30–33 (1999)
45. Mathias, K., Whitley, L.D.: Genetic operators, the fitness landscape and the traveling salesman problem. In: Parallel Problem Solving From Nature, vol. 2, pp. 219–228 (1992)
46. Mattfeld, D.C., Bierwirth, C.: A Search Space Analysis of the Job Shop Scheduling Problem. Ann. Oper. Res. 86, 441–453 (1996)
47. Merz, P.: Advanced fitness landscape analysis and the performance of memetic algorithms. Evol. Comput. 12(3), 303–325 (2004)
48. Merz, P., Freisleben, B.: Memetic Algorithms and the Fitness Landscape of the Graph Bi-Partitioning Problem. In: Eiben, A.E., Bäck, T., Schoenauer, M., Schwefel, H.-P. (eds.) PPSN 1998. LNCS, vol. 1498, pp. 765–774. Springer, Heidelberg (1998)

49. Merz, P., Freisleben, B.: Fitness landscape analysis and memetic algorithms for the quadratic assignment problem. IEEE T. Evolut. Comput. 4(4), 337–352 (2000)
50. Merz, P., Freisleben, B.: Memetic Algorithms for the Traveling Salesmen Problem. Complex Systems 13(4), 297–345 (2001)
51. Moraglio, A., Poli, R.: Topological interpretation of crossover. In: Deb, K., et al. (eds.) GECCO 2004. LNCS, vol. 3102, pp. 1377–1388. Springer, Heidelberg (2004)
52. Moraglio, A., Poli, R.: Topological Crossover for the Permutation Representation. In: GECCO 2005, pp. 332–338 (2005)
53. Munkres, J.R.: Topology. Prentice-Hall, Englewood Cliffs (2000)
54. Naudts, B.: Measuring GA-hardness. Ph.D. thesis, University of Antwerp (June 1998)
55. Naudts, B., Kallel, L.: A Comparison of Predictive Measures of Problem Difficulty in Evolutionary Algorithms. IEEE T. Evolut. Comput. 4(1), 1–15 (2000)
56. Naudts, B., Verschoren, A.: Epistasis and Deceptivity. B Belg. Math. Soc-Sim 6(1), 147–154 (1999)
57. Nei, M.: Genetic distance between populations. Am. Nat. 106, 283–292 (1972)
58. Newman, M.E.J., Engelhardt, R.: Effects of selective neutrality on the evolution of molecular species. Proc. R Soc. Lond B 265, 1333–1338 (1998)
59. Ochoa, G., Qu, R., Burke, E.K.: Analyzing the landscape of a graph based hyper-heuristic for timetabling problems. In: GECCO 2009, pp. 341–348. ACM, New York (2009)
60. Ohta, T.: The Nearly Neutral Theory of Molecular Evolution. Annu. Rev. Ecol. Syst. 23, 263–286 (1992)
61. Ohta, T.: The neutral theory is dead. The current significance and standing of neutral and nearly neutral theories. BioEssays 18(8), 673–677 (1996)
62. Ohta, T.: Evolution by nearly-neutral mutations. Genetica 102/103, 89–90 (1998)
63. Pitzer, E., Affenzeller, M., Beham, A.: A Closer Look Down the Basins of Attraction. In: UKCI, Colchester, UK, pp. 1–6 (2010)
64. Provine, W.B.: Sewall Wright and Evolutionary Biology. University of Chicago Press, Chicago (1989)
65. Quick, R.J., Rayward-Smith, V.J., Smith, G.D.: Fitness Distance Correlation and Ridge Functions. In: Eiben, A.E., Bäck, T., Schoenauer, M., Schwefel, H.-P. (eds.) PPSN 1998. LNCS, vol. 1498, pp. 77–86. Springer, Heidelberg (1998)
66. Rechenberg, I.: Evolutionsstrategie – Optimierung technischer Systeme nach Prinzipien der biologischen Evolution. Frommann-Holzboog (1973)
67. Reeves, C.R., Rowe, J.E.: Genetic Algorithms—Principles and Perspectives. Kluwer, Dordrecht (2003)
68. Reeves, C.R., Wright, C.C.: Epistasis in Genetic Algorithms: An Experimental Design Perspective. In: ICGA, pp. 217–224. Morgan Kaufmann, San Francisco (1995)
69. Reidys, C.M., Stadler, P.F.: Neutrality in fitness landscapes. Appl. Math. Comput. 117(2-3), 321–350 (1998)
70. Reidys, C.M., Stadler, P.F.: Combinatorial Fitness Landscapes. SIAM Rev. 44, 3–54 (2002)
71. Rochet, S.: Epistasis in genetic algorithms revisited. Inform. Sciences 102(1-4), 133–155 (1997)
72. Rochet, S., Venturini, G., Slimane, M., El Kharoubi, E.M.: A Critical and Empirical Study of Epistasis Measures for Predicting GA Performances: A Summary. In: Hao, J.-K., Lutton, E., Ronald, E., Schoenauer, M., Snyers, D. (eds.) AE 1997. LNCS, vol. 1363, pp. 275–286. Springer, Heidelberg (1998)
73. Rockmore, D., Kostelec, P., Hordijk, W., Stadler, P.F.: Fast Fourier Transform for Fitness Landscapes. Appl. Comput. Harmon A 12(1), 57–76 (2002)

74. Saitou, N., Nei, M.: The Neighbor-joining Method: A New Method for Reconstructing Phylogenetic Trees. Mol. Biol. Evol. 4(4), 406–425 (1987)
75. Shannon, C.E.: A Mathematical Theory of Communication. AT&T Tech. J. 27, 379–423, 623–656 (1948)
76. Sherrington, D., Kirkpatrick, S.: Solvable model of a spin-glass. Phys. Rev. Lett. 35(26), 1792–1795 (1975)
77. Smith, T., Husbands, P., Layzell, P., O'Shea, M.: Fitness landscapes and evolvability. Evol. Comput. 10(1), 1–34 (2002)
78. Stadler, P., Wagner, G.: The Algebraic Theory of Recombination Spaces. Evol. Comput. 5, 241–275 (1998)
79. Stadler, P.F.: Linear Operators on Correlated Landscapes. J. Phys. I 4, 681–696 (1994)
80. Stadler, P.F.: Towards a Theory of Landscapes. Lecture Notes in Physics, vol. 461, pp. 78–163. Springer, Heidelberg (1995)
81. Stadler, P.F.: Landscapes and Their Correlation Functions. J. Math. Chem. 20, 1–45 (1996)
82. Stadler, P.F.: Fitness Landscapes. In: Biological Evolution and Statistical Physics, pp. 187–207. Springer, Heidelberg (2002)
83. Stadler, P.F., Happel, R.: Random field models for fitness landscapes. Working Papers 95-07-069, Santa Fe Institute, Santa Fe, NM (1995)
84. Stadler, P.F., Schnabl, W.: The Landscape of the Travelling Salesmand Problem. Phys. Lett. A 161, 337–344 (1992)
85. Stadler, P.F., Seitz, R., Wagner, G.P.: Evolvability of Complex Characters: Population Dependent Fourier Decomposition of Fitness Landscapes over Recombination Spaces. B Math. Biol. 62, 399–428 (2000)
86. Talbi, E.-G.: Metaheuristics: From Design to Implementation. Wiley, Chichester (2009)
87. Tavares, J., Pereira, F., Costa, E.: Multidimensional Knapsack Problem: A Fitness Landscape Analysis. IEEE T. Syst. Man Cy. B 38(3), 604–616 (2008)
88. Thompson, R.K., Wright, A.H.: Additively Decomposable Fitness Functions. Tech. rep., University of Montan, Computer Science Dept., Missoula, MT, USA (1996)
89. Turney, P.: Increasing evolvability considered as a large-scale trend in evolution. In: GECCO, pp. 43–46 (1999)
90. Uludağ, G., Uyar, A.Ş.: Fitness Landscape Analysis of Differential Evolution Algorithms. In: ICSCCW (2009)
91. Vanneschi, L., Clergue, M., Collard, P., Tomassini, M., Vérel, S.: Fitness Clouds and Problem Hardness in Genetic Programming. In: Deb, K., et al. (eds.) GECCO 2004. LNCS, vol. 3103, pp. 690–701. Springer, Heidelberg (2004)
92. Vanneschi, L., Tomassini, M., Collard, P., Vérel, S.: Negative slope coefficient: A measure to characterize genetic programming fitness landscapes. In: Collet, P., Tomassini, M., Ebner, M., Gustafson, S., Ekárt, A. (eds.) EuroGP 2006. LNCS, vol. 3905, pp. 178–189. Springer, Heidelberg (2006)
93. Vanneschi, L., Verel, S., Tomassini, M., Collard, P.: NK Landscapes Difficulty and Negative Slope Coefficient: How Sampling Influences the Results. In: Giacobini, M., Brabazon, A., Cagnoni, S., Di Caro, G.A., Ekárt, A., Esparcia-Alcázar, A.I., Farooq, M., Fink, A., Machado, P. (eds.) EvoWorkshops 2009. LNCS, vol. 5484, pp. 645–654. Springer, Heidelberg (2009)
94. Vassilev, V.K., Fogarty, T.C., Miller, J.F.: Information Characteristics and the Structure of Landscapes. Evol. Comput. 8(1), 31–60 (2000)
95. Wagner, A.: Robustness, evolvability, and neutrality. FEBS Lett. 579, 1772–1778 (2005)

96. Wagner, A.: Robustness and evolvability: a paradox resolved. Proc. R Soc. B 275, 91–100 (2008)
97. Wagner, G., Stadler, P.: Complex Adaptations and the Structure of Recombination Spaces. In: Algebraic Engineering, pp. 96–115. World Scientific, Singapore (1998)
98. Wagner, G.P., Altenberg, L.: Complex Adaptations and the Evoluation of Evolvability. Evolution 50(3), 967–976 (1996)
99. Wang, S., Zhu, Q., Kang, L.: Landscape Properties and Hybrid Evolutionary Algorithm for Optimum Multiuser Detection Problem. In: ICCSA, pp. 340–347 (2006)
100. Weinberger, E.: Correlated and uncorrelated fitness landscapes and how to tell the difference. Biol. Cybern. 63(5), 325–336 (1990)
101. Weinberger, E.D.: Local properties of Kauffman's N-k model, a tuneably rugged energy landscape. Phys. Rev. A 44(10), 6399–6413 (1991)
102. Weinberger, E.D.: NP Completeness of Kauffman's N-k Model, A Tuneable Rugged Fitness Landscape. Working Papers 96-02-003, Santa Fe Institute (1996)
103. Wright, S.: The Roles of Mutation, Inbreeding, Crossbreeding and Selection in Evolution. In: 6th International Congress of Genetics, vol. 1, pp. 356–366 (1932)

Part II
Intelligent Computation in Networks

Chapter 9
Self-Organizing Maps for Early Detection of Denial of Service Attacks

Miguel Ángel Pérez del Pino, Patricio García Báez, Pablo Fernández López, and Carmen Paz Suárez Araujo

Abstract. Detection and early alert of Denial of Service (DoS) attacks are very important actions to make appropriate decisions in order to minimize their negative impact. DoS attacks have been catalogued as of high-catastrophic index and hard to defend against. Our study presents advances in the area of computer security against DoS attacks. In this chapter, a flexible method is presented, capable of effectively tackling and overcoming the challenge of DoS (and distributed DoS) attacks using a CISDAD (Computer Intelligent System for DoS Attacks Detection). It is a hybrid intelligent system with a modular structure: a pre-processing module (non neural) and a processing module based on Kohonen Self-Organizing artificial neural networks. The proposed system introduces an automatic differential detection of several Normal Traffic and several Toxic Traffics, clustering them upon its Transport-Layer-Protocol behavior. Two computational studies of CISDAD working with real networking traffic will be described, showing a high level of effectiveness in the CISDAD detection process. Finally, in this chapter, the possibility for specific adaptation to the Healthcare environment that CISDAD can offer is introduced.

9.1 Introduction

During the last decade, the growth of computers connected to the Internet has risen in an exponential way. In 1998, globally speaking, just 4 countries had a connection index higher than 31%. Ten years later, in 2008, all European countries exceed

Miguel Ángel Pérez del Pino · Pablo Fernández López · Carmen Paz Suárez Araujo
Instituto Universitario de Ciencias y Tecnologas Cibernéticas
Universidad de Las Palmas de Gran Canaria, Las Palmas de Gran Canaria,
Canary Islands, Spain
e-mail: miguel.perez107@alu.ulpgc.es,
{pfernandez,cpsuarez}@dis.ulpgc.es

Patricio García Báez
Departamento de Estadística, Investigación Operativa y Computacón,
Universidad de La Laguna, La Laguna, Tenerife, Spain
e-mail: patricio@etsii.ull.es

J. Fodor et al. (Eds.): Recent Advances in Intelligent Engineering Systems, SCI 378, pp. 195–219.
springerlink.com

the mentioned percentage widely, together with many countries such as Australia, Brazil, Japan or the United States of America, among others. In particular, Spain had approximately 1.700.000 computers connected to the Internet in 1998, while 2008 data revealed a connectivity index over 26.000.000, [3]. This growth has favoured the provision of services of socio-economical interest to the network users; many of them involving critical considerations, such as banking, medical, administrative procedures, etc. Virtualization of services and resources has provided a simpler way to develop and consume applications in the Cloud. On the other hand, the number of infractions which took advantage of security vulnerabilities in systems and computer networks has hugely increased, causing serious loss at the companies and institutions who suffer them. Recently, well-known businesses in the technological sector have been subject to serious threats, preventing users to make use of the offered services, [34].

The attempt to make a computing resource unavailable to its intended users is known as a Denial of Service (DoS) attack. These attacks consist of concerted efforts to prevent a system or a service from functioning efficiently, causing it a temporary or permanent interruption. When several systems participate together in order to consume the resources or bandwidth of a target system (victim), it is said a distributed DoS attack is being performed. One common way to undertake these attacks is saturating the victim machine with connection requests, making slow (or impossible) for it to answer to legitimate traffic. These actions provoke an overload on resources causing system and/or network congestion, in such a way it is difficult for the victim (or its perimetral partners) to defend under attack. In last instance, the victim can no longer provide its intended service; that is to say, legitimate users can no longer communicate adequately with it and vice-versa. New security concerns have arisen out of Cloud Computing. The Internet architecture underlies the Cloud Computing scenario; so do its communication, control and management protocols. If determined critical systems were attacked, entire Cloud applications could misfunction. The *2010 State of Enterprise Security Survey*, performed by Symantec, [32], showed that 75% of the respondents experienced cyberattacks in the last 12 months, considering them highly effective. In the same line, 40% of the respondents indicated their organizations were currently using Cloud Computing applications and services. Because of the closed context of the original ARPANET and NSFNet, no consideration was given to these threats in the original Internet Architecture. As a result, almost all Internet services are vulnerable to DoS attacks of sufficient scale, [27, 17]. DoS attacks have become a real threat, being considered as one of the three more frequent attacks by several computer security laboratories and catalogued as of high-catastrophic index and hard to defend against. Because of this, they constitute violations of the laws of several individual nations around the globe.

Until recently, monitoring and auditing of computer systems were based on individual and independent machines. However, the most common organization needs to defend a distributed set of systems in interconnected communication networks. Although it is possible to increase computer security by using isolated and independent monitoring systems, experience has shown that cooperation and coordination among these systems guarantee a more effective defense, [35, 29]. Thus, a set of monitoring, auditing and data analysis systems oriented towards the defense of computer systems have appeared, all of which should be capable of detecting any illegal action against a protected system, minimizing possible damage and guaranteeing the continuity of offered services.

Different models have been proposed for detection, early alert and fast decision-making against DoS attacks using different approaches: expert, probabilistic and heuristics systems, Petri networks, automata and artificial neural networks, the last ones usually under some type of supervision, [15, 16, 14, 4, 1]. Most of these proposals present weaknesses in the efficiency required for a complex implementation and a difficult level of maintenance, [15, 16]. There are two main reasons for this situation: first one, the difficulty in defining precise behavior patterns for these attacks; second one, the frequent appearance of new attacking techniques. Moreover, it is rather difficult to determine when a DoS attack is taking place in real time. A follow-up analysis of the captured data is required to exactly determine what happened; a process that also requires great computational and human resources.

Every process (connection, flow control, data sharing, disconnection) undertaken on the Internet, either normal or abnormal, relies on the IP protocol, mainly on the upper-level TCP (Transport Control Protocol) and UDP (User Datagram Protocol) protocols. In some cases, the ICMP protocol (Internet Control Message Protocol) appears as a reporting state mechanism interacting in the communication flow. DoS attacks are constructed using methods and techniques based on these three protocols.

Regarding this, we think the key to detect and early alert DoS attacks is to discover how they behave at this networking level, trying to detect abnormal behavior patterns (toxic traffic) which differ from what should be considered as normal behavior (legitimate normal traffic) in the networking area being watched. Moreover, discovering new behavior patterns and being able to classify them in such a way it can be analyzed their similarity or dissimilarity to previously known ones would be desirable. We propose to tackle this problem of computer security using an unsupervised neural computation approach. Unsupervised Artificial Neural Networks are suitable for complex data analysis in real time, where information overlapping is present, in addition to data noise and the absence of explicit knowledge of the situation, like the one studied in this research.

In this work, a flexible method is presented, capable of effectively tackling and overcoming the challenge of DoS (and distributed DoS) attacks using a CISDAD (Computer Intelligent System for DoS Attacks Detection), [17]. It can be defined as a hybrid intelligent system with a modular structure: a pre-processing module (non-neural) and a processing module based on unsupervised artificial neural networks, concretely Kohonen Self-Organizing Maps , [12, 20]. One of the strong points of our proposed CISDAD is that it introduces an automatic non bi-modal detection, quite different than the majority of the ones that are available today, [15, 16, 14, 4, 1]. This architecture has been designed to detect Toxic Traffics (TT), distinguishing them from legitimate Normal Traffic (NT) and performing clusterization upon the behavior of transactions by taking approach of the tendency of determined features at the Transport-Layer Protocols employed.

Obtained results in the first stage of CISDAD show a high level of effectiveness in the detection process, with a low rate of false negatives, being probably the most desirable feature in any security system. In addition, it offers a high level of performance, based on low resource consumption and a fast real-time processing and analysis. Our CISDAD has the capability to recognize, at an early stage, the presence of any unusual flooding flow that is produced inside and outside the organization. This feature helps managers to consider contingency plans, increasing their knowledge in real time about transactions that affect their managed networks and, especially important, aiding in the response, in time, to potential DoS attacks against certain threatened information systems. Medical care environments, such as EDEVITALZH, [19, 18], a web clinical virtual platform with features of clinical workstation, where systems avalaibility is highly critical, is one of the fields of application where our CISDAD promises great successful results.

9.2 Fingerprints of DoS Attacks

DoS attacks are featured by an explicit attempt to prevent legitimate users from using a service provided by the victim systems. Symptoms *during* and *after* the execution of these threats include, [34]:

- *Network Congestion*, that is, slow network performance.
- *Resources Congestion*, by means of increasing resources consumption by the device.

In a short period of time, these signs will benefit the appearance of difficulties in accessing the offered services by the victim/s, or in last instance, their complete unavailability. Because of the closed context of the original ARPANET and NSFNet, no consideration was given to these threats in the original Internet Architecture. As a result, almost all Internet services, including those ones based on the Cloud Computing Cloud Computingparadigm, are vulnerable to DoS attacks of sufficient scale, [27, 17]. In general, any network device is susceptible to be attacked using DoS techniques: end systems (HTTP servers, DNS servers, etc.), links, routers and gateways, firewalls and IDS systems.

Two general forms of DoS attacks have been catalogued, [7]: (1) the ones that *flood* services and (2) the ones that *crash* services. DoS attacks have evolved in such a way they combine different techniques to highly increase caused damage in short amounts of time. DoS techniques can be classified according to [27]:

- Disruption of device configuration data; i.e. routing information
- Disruption of state information; i.e. reset of TCP sessions
- Interfere the media to cause damage in communications between the victim and its intended users
- Excessive consumption of computational resources; i.e. CPU time, disk space, bandwidth
- Disruption of physical components

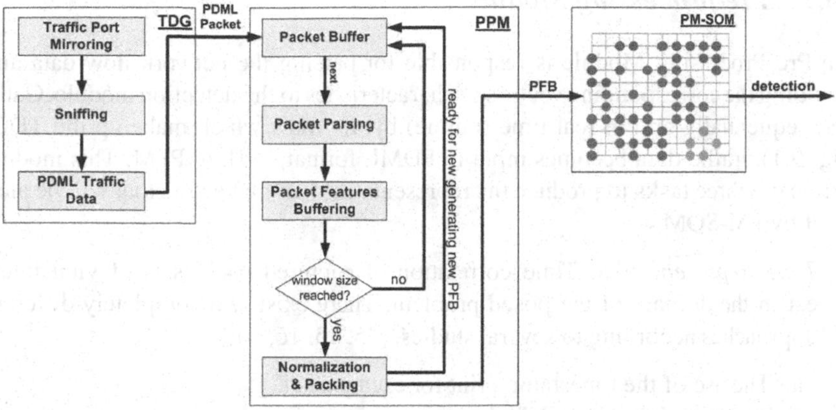

Fig. 9.1 CISDAD Modular Structure. Traffic Data Gathering (TDG), Pre-Processing Module (PPM), Processing Module based on Kohonen Self-Organizing Maps, (PM-SOM).

In the end, it becomes a must to distinguish whether congestion is provoked by exceeded use of the provided services by legitimate users (so that it is a task for network admins to analyze, plan and set up network configuration and management to fit the actual needs) or it has appeared because the network (or one of its devices) is under attack.

9.3 SOM-Based System for DoS Attacks Detection (CISDAD)

The system proposed to tackle this problem of computer security consists of two modules; the first one, in charge of parsing the network flow data and giving structure to the information which the second module, able to perform detection of depending-on-upper-IP-protocol toxic types of traffic, will handle. Thus, our CISDAD is a hybrid intelligent system with a modular structure, (Fig. 9.1), made up of a Pre-Processing Module (PPM), which parses network traffic, generating and providing feature vectors as input to the Processing Module (PM-SOM), which is based on Kohonen Self Organizing Maps, [12, 13]. This artificial neural map is responsible for detecting any potential DoS attacks flowing over the communication network, [17].

The information environment employed in our study consists of real networking traffic data, gathered by means of a sniffer tool and related techniques (TDG) from two scenarios: first one, a benchmark local network which services web applications to its users; second one, a High Performance Computing facility where services such as HTTP, NFS, NIS+, SSH, FTP, SNMP, RDP operate. The connectivity scheme allows wired and wireless connections of allowed clients (legitimate users) in both scenarios. Access to provided services can be performed using LAN connections (local network) or WAN connections (extranet).

9.3.1 Pre-processing Module

The Pre-Processing Module is responsible for parsing the network-flow data and providing the corresponding vectors of characteristics to the detection module. Gathered sequentially and in real time (online) by the tools which make up the TDG, (Fig. 9.1), traffic data becomes input in PDML format, [24], to PPM. This module carries out three tasks to produce the representative feature vectors that will be analyzed by PM-SOM:

1. *Time Representation*: Time correlation of captured packets is of vital interest in the domain of the posed problem. There exist two completely different approaches according to several studies, [35, 15, 16, 14, 1]:

   a. The use of the timestamp print for each packet.
   b. To determine a Time Window.

   Our PPM development implements approach (*b*). Our time window is represented by a FIFO queue and a maximum number of packets threshold to deal with the set. Let M be the size of the window from a time representation point of view. We use the term Packet Flow Block (PFB) to name the set of M packets that, based on order of arrival, are parsed together by PPM. It was empirically determined that the optimal value for M was 50. For values of M lower than 20, the amount of traffic concentration was not significant enough. For values of M higher than 50, gathered results were not significantly better and there was a much higher consumption of computational resources than when using the finally established empirical threshold $M=50$.

2. *Data Representation*: PPM of our system presents important changes in the information representation regarding previous studies, [15, 16, 14, 4, 1], with the objective of defining patterns that can indicate the presence of potential flooding attacks on the network. Our proposed representation does not include anything related to the packet payload (data content), [4]. However, our study was focused on the management of specific network session data, given that the management of complete packet data (*packet headers + data content*) does not present relevant semantics to improve the detection of possible DoS attacks. Reasons for such decision are:

   - The information about communication flow and control resides in the headers of protocols, specifically for the faced problem, in the ones above the Network Level; that is to say, mainly, TCP, UDP, ICMP. These protocols suppose approximately 99% of all the Internet transactions. Table 9.1 shows sample statistics for an online server machine connected to the tested networking environment as reference.
   - A DoS attack, by definition, reflects an abuse of specific actions by the attacking systems to the victim systems. Consequently, and exactly opposite

Table 9.1 IP Traffic Statistics for a Sample Server Machine (5-hours snapshot)

Protocol	In Packets	Out Packets	Sum I/O	Traffic %
TCP	7.388.892	3.680.102	11.068.994	91,13%
UDP	785.669	159.345	945.014	7,78%
ICMP	53	7.094	7.147	0,05%
TCP+UDP+ICMP	8.174.614	3.846.541	12.021.155	98,98%
Other Protocols	121.996	2.766	124.762	1,02%
All IP Traffic	8.296.610	3.849.307	12.145.917	100%

of what occurs with other vulnerabilities (virus, trojans, exploitation of software services, etc.), data content of packets is not relevant for the posed problem.

Several studies consider that the IP addresses represent valuable information regarding potential attacking systems [15], or protected systems, [1]. It must be faced that a DoS attack can be carried out by any system that is connected to a network (not just to the WAN), even inside a protected one. In addition, the attack can be accompanied by an identity hijacking of the sender source address in sent packets (IP address spoofing), [5], or even a distributed attack method could be employed. Thus, our proposal considers that the IP addresses representation does not provide additional information to the detector. Conversely, it will be useful to temporarily store these addresses during the pre-processing and processing stages. The goal is to gather, and then provide, information to the network traffic processing systems (routers, firewalls, etc.) in the early stage of a potential flooding attack with the intention of minimizing the damage that could take place.

Next step in our research was to analyze and select which features of packets could be relevant for early detection of potential DoS attacks. There were analyzed the 25 leading indicators, referring to network flow and control, related to the TCP, UDP and ICMP protocols. Performed studies show that:

- In the cases of UDP and ICMP flooding, the increase in the number of these types of packets was immediately detected as apparently abnormal regarding the usual network flow.
- In the case of TCP-based flooding attacks, and due to its connection orientation, it was observed that, although the amount of packets during an attack grew up highly, it was not enough information to decide whether the received packets should be considered as abnormal or not. On the other hand, it

was observed that several of the TCP featuring flags (ACK, PSH, RST, SYN, URG) registered a change in their usual tendencies.

Our focus proposes an analysis of the tendency of these featuring packet-flags because it is this tendency that could indicate a potential current flooding attack based on exploiting the mechanisms of the TCP protocol.

3. *Data Normalization and Packing*: Normalization is needed to place information attributes on the same scale and to prevent some of them with a large original scale from biasing the solution. It also minimizes the likelihood of overflows and/or underflows in the original data to be exploded. PPM extracts the previously commented featuring indicators from each input packet $\overrightarrow{P_i} = \{c_0, c_1, c_2, c_3, c_4, c_5, c_6, c_7, c_8\}$, where the c_0 component stores the protocol indicator according to expression (9.1).

$$c_0 = \begin{cases} -1 \; if \; Protocol = 'UDP' \\ \;\;0 \; if \; Protocol = 'TCP' \\ \;\;1 \; if \; Protocol = 'ICMP' \end{cases} \tag{9.1}$$

Every TCP segment contains an 8-element array of indication flags, sizing 1 bit per element, which handles the transmission flow between two peers, the sender and the receiver. These flags, which control the actual dialog state in such a way the sender or the receiver asks for or performs any action needed to regulate the transmission state from the beginning to the end of the communication event, are briefly described below, [30]:

$$TCPSegmentFlags = \begin{cases} c_1 = ACK \\ c_2 = CWR \\ c_3 = ECN \\ c_4 = FIN \\ c_5 = PSH \\ c_6 = RST \\ c_7 = SYN \\ c_8 = URG \end{cases} \tag{9.2}$$

- ACK: This flag indicates the TCP Acknowledgment field is significant, according to the TCP handshaking and communication flow mechanisms.
- CWR: This flag stands for Congestion Window Reduced. It is set by the sending host to indicate that it received a TCP segment with the ECN flag set and had responded in Congestion Control Mechanism.

- ECN: This flags works in conjunction with the SYN flag. If SYN flag is set, the TCP peer is ECN capable. In case the SYN flag is unset but ECN is, it indicates that a packet with Congestion Experienced flag set in the IP header was received during normal transmission.
- FIN: This flag indicates transmission end. No more data to be transmitted.
- PSH: This flag indicates the peer to push the buffered data to the receiving application.
- RST: This flag indicates the connection must be resetted.
- SYN: This flag indicates synchronization of sequence numbers. Only the first packet sent from each end should have this flag set. Some other flags change meaning based on this flag, for example, the ECN flag. Some are only valid for when it is set; other ones when it is unset.
- URG: This flag indicates urgency in the data being transmitted.

The $\{c_1..c_8\}$ components store the PFB normalized activation state value for each one of the TCP flags, according to expression (9.2).

Once the time window size M is reached, that is to say, when M packets have been buffered in order of arrival, PPM will carry out a normalization and packing process of data according to expression (9.3). Results of this stage will constitute the information space of PM-SOM.

$$\overrightarrow{PFB} = \frac{1}{M} \sum_{i=0}^{M-1} \overrightarrow{P_i} \qquad (9.3)$$

PPM has been implemented using Perl programming language, [25], to take approach of its fast data-string processing capabilities. The developed module works online, in such a way that it receives in real time a detailed structure in PDML format, [24], for every packet received or sent through the network interface. This allows a fast information processing seeking those packet sections (headers) considered as interesting in this research. It is not necessary to store input neither output packets because the information processing is performed online. This method improves the handling of data regarding DoS attacks, tweaking tools such as [21, 2, 22]. It can be optionally stored, according to whether the global operating system mode is training or production, the corresponding generated PFB when the time window size is reached.

Thus, PPM of the proposed system allows processing packets in real time upon their order of arrival and generating the corresponding feature input vectors (PFBs) to be next analyzed by PM-SOM, improving computing time by avoiding the storage of monitored information by means of semantics compression.

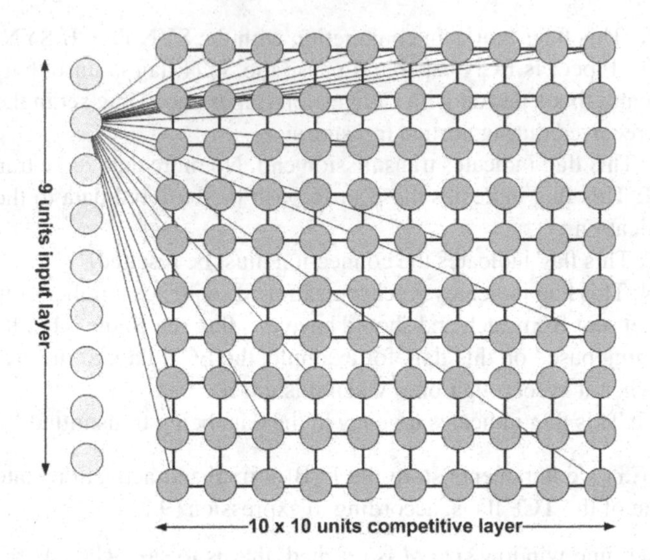

Fig. 9.2 PM-SOM Structure. Self-Organizing Map consisting of a 9-neuron input layer and a 100-neuron competitive layer, using a fully-connected topology scheme.

9.3.2 Processing Module

The Processing Module, core of the detection process, is responsible for clustering networking traffics. This module is based on a Self-Organizing Map (SOM), a neuro-computational architecture which carries out a non-linear mapping of a n-dimensional input space to a *two*-dimensional space through competitive and unsupervised learning, [12, 13]. This architecture distinguishes from other artificial neural networks because of the creation of feature maps that preserve the topological properties of the input space by means of a neighbourhood function, approximating its probability distribution producing feature detectors. SOM is also a first step in the search for meaning and structure in both input data and the classification process, [8, 31].

SOM consists of a two-dimensional Competitive Layer of cells or neurons (CL), which are linked to n inputs (IL) by a total-connectivity topology, but also to their CL relatives using lateral connections. As a competitive and unsupervised learning neural network, SOM performs an inhibition of the activation of other neuron cells in the CL when a certain cell responds to a determined pattern while, at the same time, reinforcing this last one. In the end, this means that neurons are specialized in recognizing one kind of pattern. The comparative process between existing entries and detectors can use different similarity measures; the more noteworthy include Euclidean and Scalar Product distances. The Euclidean one, employed in this work, is defined according to expression (9.4), where x is the input feature vector and m_l is the weights vector between the input and detector l.

$$\|x - m_l\| = \sqrt{\sum_i (x_i - m_{li})^2} \tag{9.4}$$

The training process allows a global ordering so that presented patterns are used to organize the SOM under topological constraints of the input space; this way, SOM acts as a quantifying system. In this competitive process, the winner unit will be the detector with the best adjustment to the input; that is, the one that is closer to this pattern, or in other words, the detector whose distance to the input vector is minimal. The winner unit, and those that are located topographically close, will learn something from the input vector x, increasing the degree of pairing with it. A mapping between the input data space and the network space is built, resulting that two close patterns will stimulate two close cells of the SOM. Equation (9.5) describes the variance motivated in SOM weights and is referred to as the SOM architecture learning law, where $\varepsilon(t) = 1/t$ is the learning rate, usually implemented by monotonically decreasing functions of time (although other kind of functions are also possible to be used), and $V_c(l)$ is a function defining the neighbourhood centered in c, between the winner unit c and any other unit l. $V_c(l)$ can be thought as a function that decreases as long as l is further from c, and at the same time, this decrease is highlighted according to the elapsing learning time.

$$\Delta m_{li} = \varepsilon(t) V_c(l)(x_i - m_{li}) \tag{9.5}$$

The proposed structure for SOM integrated into CISDAD is made up of a 9-unit IL and a two-dimensional CL of 10x10 neurons, being topologically fully-connected, (Fig. 9.2). As previously mentioned, it uses Euclidean distance as similarity measure according to expression (9.4). In order to carry out the quantification of obtained results during the training process, Average Quantization Error Q, also known as Reconstruction Error, is used, [8]. Q reflects the average dissimilarity between every pattern x_i and the weights corresponding to unit $c(i)$, winner unit for a determined input pattern, according to expression (9.6); thus, it expresses the measure in which patterns can be represented by the map units.

$$Q = \frac{1}{n} \sum_{i=1}^{n} \|x_i - m_{c(i)}\| \tag{9.6}$$

9.3.3 Application Fields

The spread of broadband connection for users has favoured the expansion of the Internet, and with it, the provision of services to users over the Internet, many of them involving critical considerations such as banking, financial or medical. During the last months, several enterprises and governments around the globe have suffered serious loss because of DoS attacks causing failure to their systems. No one is immune to DoS attacks.

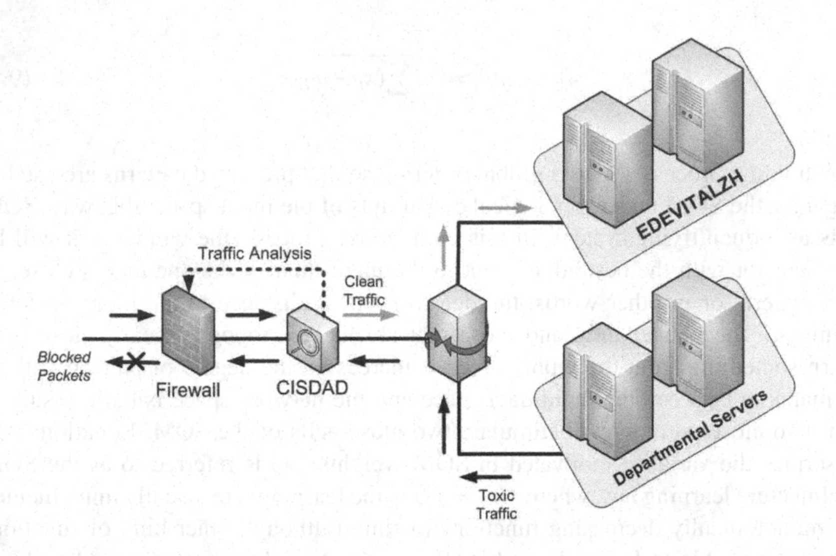

Fig. 9.3 Proposed scenario where CISDAD collaborates hand-in-hand with firewalls in inspecting traffic and aiding in decision-making. Insider toxic traffic can also be controlled so proprietary network devices are prevented from being misused for DoS attacks to the outside networks.

The medical case is highly critical due to the role of such systems in service to patients. Current healthcare information systems centralize patients data in a HIS, the central warehouse of the electronic clinical/medical records, and encourage information sharing between the HIS and the so-called departmental systems, servers running special software applications/services according to the medical unit where they are being used. A sample of departmental system is EDEVITALZH, [19, 18], a web virtual platform with features of clinical workstation, developed to aid clinicians in the diagnosis and prognosis of Alzheimer's Disease and other dementias using intelligent computational assistants and systems, that takes advantage of the Internet allowing the appropriate tools and patients medical records to be available for any primary care, specialized care or hospital in a virtual context. This scheme allows EDEVITALZH components and clients to be geographically separated but interacting together assuring a good availability and quality of service.

Different protocols for information sharing have been proposed for the healthcare environment, [9, 6], all of them supported by the IP protocol. In the case of EDEVITALZH, it has been developed based on a distributed architecture, empowering data sharing between the systems which make it up. Thus, communication between these systems becomes a critical spot, since reliability and security in transmissions between the involved components is a must. The wide variety of data managed and stored by systems such as EDEVITALZH are always subject to potential computer attacks, being one of the most common the DoS ones. Their consequences can be catastrophic in a so sensitive area such as this one.

According to data from the U.S. consulting firm SecureWorks, in the first nine months of the year 2009 there were approximately 6.500 registered attacks per day on hospital information systems, while this same tendency doubled in the last three months of the same year, exceeding 13.400 registered attacks per day. What is more revealing is that other industries that are also protected by the same firm did not report any change in the tendency of these security incidents. According to SecureWorks, it can be said that the worst attacks that can be produced in a clinical setting are those whose finality results in a denial of service. In addition, their study reveals that more than 50% of all large hospitals in 2009 were subject to security breaches of these characteristics.

It is necessary to have available tools and policies to protect systems that store sensitive data, such as medical are, guaranteeing services availability 24/7. The proposed system in this study is capable of solving this type of problems, incorporating it to network topologies and data communications, (Fig. 9.3). Our CISDAD has the capability to alert at an early stage of any unusual flooding congestion that is produced inside and outside of the organization. This feature helps managers to consider contingency plans such as traffic redirection schemes over other routes, increasing security levels in their firewalls, knowledge in real time and about transactions that affect their networks, avoid the use of their resources as part of distributed attacks to outsider machines and, especially important, to respond in time to potential attacks against their information systems.

9.4 CISDAD Computational Analysis

Presented studies A and B are described by means of the gathered and used Training (TS) and Validation (VS) datasets, and the five best obtained configurations for PM-SOM, according to Q_t and Q_v values.

Real networking traffic was used in the performed studys. Different TS and VS were gathered for each study; thus, four real networking data sets were collected in total: 2 TS and 2 VS. It must be highlighted that, in each study, VS was made up of completely different data than the ones constituting its corresponding TS. No data sanitization was needed because, as described in section 3.1.2, no packet payload was inspected; just headers of the Network and Transport levels were used. Gathered toxic traffics were generated by using the *hping* tool, [10], a command-line oriented TCP-IP packet assembler.

In what relates to the PM-SOM implementation, *SOM_PACK*, [28], a public domain software package was used. This software contains all the necessary applications to employ Self-Organizing Maps with experimental data. It was compiled for Linux x86_64 platform. To reduce the needed time to gather results, several instances of the same problem, according to the SOM CL dimension parameter, were parallel- and simultaneously executed using multiple computing nodes. Regarding the training process, on both studies, every SOM configuration was first initialized using random numbers, [28].

The following PM-SOM parameters were varied on each executed dimension-based instance to accomplish all possible combinations: map topology type,

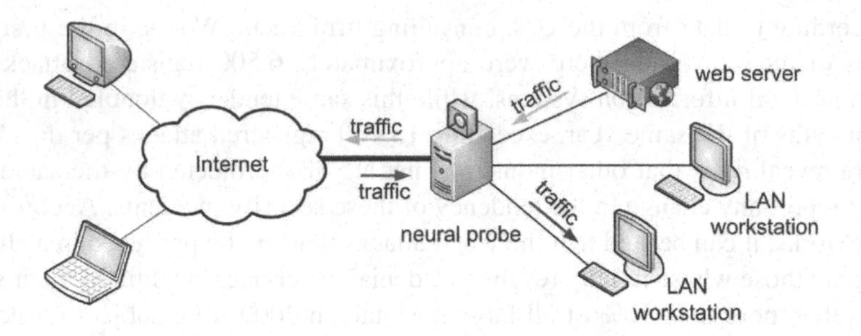

Fig. 9.4 Study A. The scenario where the tests were perfomed is a benchmark local network where all the traffic flows transparently over a network probe with IPv4 Forwarding and Network Address Translation enabled. The network offers web pages services to users.

neighbourhood function, neighbourhood ratio, the training rate α (decreasing its value in the range of $[0,09,0,01]$) and the number of training steps. 54.000 PM-SOM configurations were trained and studied for each performed analysis, corresponding to *20x20*, *15x15*, *10x10*, *9x9*, *8x8*, *7x7*, *6x6*, *5x5*, *4x4* map dimensions.

In order to perform the validation process, one of the most clarifying validation functions for clustering quality was used: the Confusion Matrix (CM), [11]. CM is considered to be a visualization tool in which each column of the matrix represents the instances in a *predicted* class, while each row represents the instances in an *actual* class. CM benefits checking if the developed system is mislabeling or confusing two classes in the classification process.

9.4.1 Study A

Study A was undertaken in a benchmark local network, (Fig. 9.4). A web server was placed into the local network as well as several client workstations. The local network was connected to the Internet by means of a Linux Gateway (LG) with IPv4 forwarding with Network Address Translation enabled. This way, the web server could be accessed from the inner and the outer network environments. The interfaces of the LG were sniffed in promiscuous mode, capturing all data packets flowing over the communication network.

Traffic capture was performed by gathering the four described types of real traffic data in a controlled manner; that is to say, Normal Traffic (NT), TCP Toxic Traffic (TT-TCP), UDP Toxic Traffic (TT-UDP) and ICMP Toxic Traffic (TT-ICMP). The four gathered types of traffic were presented to the trained PM-SOM networks during their training stages.

Regarding the constitution of the training and validation data sets for this study, TS was made up of 63.200 data packets, of which 38.200 were considered NT, 10.000 were TCP SYN-type-based flooding attacks, 5.000 were UDP-type flooding attacks and the last 10.000 were ICMP-type flooding attacks. In what relates to VS, it was made up of 4.550 data packets, divided as follows: 2.150 NT type,

1050 TT-TCP type, 700 TT-UDP type and 650 TT-ICMP type. Table 9.2 shows the volume breakdown of captured packets versus obtained PFBs using the proposed time window threshold M=50.

Table 9.2 Study A. Constitution of the TS and VS

Traffic Type	Training		Validation	
	Packets	PFBs	Packets	PFBs
Normal Traffic (NT)	38.200	764	2.150	43
TCP Toxic Traffic (TT-TCP)	10.000	200	1.050	21
UDP Toxic Traffic (TT-UDP)	5.000	100	700	14
ICMP Toxic Traffic (TT-ICMP)	10.000	200	650	13
Total	63200	1264	4550	91

Table 9.3 shows the five best obtained configurations for PM-SOM by means of the Average Quantization Errors in the training and validation stages. According to these Q_t and Q_v values, network configuration $Net = 2$ reveals as the best map for this study. Network configuration $Net = 1$ offers a Q_t value which is slightly lesser than the same value for the $Net = 2$ configuration, not following its Q_v this tendency.

Table 9.3 Study A. Best 5 Configurations for PM-SOM according to Q_t and Q_v

Net	Dimension	Topology	Neighbourhood	α	Radius	Steps	Q_t	Q_v
1	10x10	hexagonal	rectangular	0,05	3	85000	0,0491	0,0815
2	10x10	hexagonal	rectangular	0,05	3	75000	0,0495	0,0707
3	10x10	hexagonal	rectangular	0,05	3	45000	0,0500	0,0730
4	10x10	hexagonal	rectangular	0,05	3	30000	0,0503	0,0722
5	10x10	hexagonal	rectangular	0,08	2	80000	0,0511	0,0807

9.4.2 Study B

This study was performed in a High Performance Computing facility (HPC), offering planning and execution services for user jobs by means of a distributed resource management implementation which manages and assigns system resources to user tasks according to the available compute capabilities, [23]. This computing cluster scheme is based on the Internet architecture over the IP protocol to make communications possible between the heterogeneous cluster components. All servers were connected in the same network segment. In order to collect all network traffic, port-mirrored switching, [26], was configured in a D-Link DGS-1216T switching device.

Traffic flowing through all switch ports was mirrored to one specific interface. This switch interface was linked to a probe system (TDG), which received and gathered all the network traffic by means of a software sniffer tool, [33], (Figs. 9.1, 9.5). Under normal conditions, most of the legitimate network-flowing traffic (normal traffic) was expected to be made up of encapsulated data of Application-Layer Protocols such as HTTP, SSH and SFTP, NFS, NIS+, SMB, and SNMP, among others. Toxic traffics for the mentioned protocols were generated by using again the *hping* tool. This tool was compiled under several Linux operating system machines, which acted as attacking systems inside and outside the HPC networking environment. The procedure for gathering data was the same in Study B as in Study A: real networking traffic data was collected in a controlled manner. In this case, instead of four types of traffic, five types were gathered:

- First, a new collection of the four previously described types was gathered: NT, TT-TCP, TT-UDP and TT-ICMP.
- Secondly, a new data collection based on a DoS-flavoured SYN attacking procedure which contains source IP addresses spoofing (TT-SPF), [5], was harvested while perfoming the attack using the *hping* tool, [10]. This type of attack is featured by a huge set of asking-for-connection queries which never accomplish associated to invalid source addresses to lie to the victim system.

4.847.500 NT-type data packets (96950 PFBs) were collected for Study B, related to sniffing during 20 hours (the highest day-time workload hours). The most frequent PFBs were statistically analyzed, observing that around 73.5% of NT-type PFBs repeated themselves more than 50 times each. Moreover, there were some PFBs which repeated themselves more than 4.900 times. There were selected those PFBs which repeated themselves more than 50 times and a new Selective Training Dataset (S-TS) set was generated. Same as with S-TS, a new Selective Validation Dataset (S-VS) was created from the resting 26.5%, different than the ones corresponding to S-TS. Same procedure was followed for creating new selective datasets for the other traffic types: TT-TCP, TT-UDP, TT-ICMP and TT-SPF. Table 9.4 shows a summary of the collected traffic data.

Regarding the constitution of the selective training and validation data sets in this study, S-TS was made up of the four basic types: NT, TT-TCP, TT-UDP and TT-ICMP. Only these four types of patterns were presented to PM-SOM; that is to say, TT-SPF type patterns were not a part of the TS. S-TS was formed by 16700 data packets, of which 11500 were considered NT, 3100 were TCP SYN-type-based flooding attacks, 50 were UDP-type flooding attacks and the last 2050 were ICMP-type flooding attacks. In what relates to S-VS, it was made up of the four basic traffic data types plus the new TT-SPF type. It was pretended to test the system reaction to a new type of attack (one which the neural network did not see during its training stage). The S-VS set was constituted by 3.100 data packets, divided as follows: 1.100 NT type, 600 TT-TCP type, 150 TT-UDP type, 450 TT-ICMP type and 800 TT-SPF type. Table 9.5 shows the volume breakdown of captured packets versus obtained PFBs using the proposed time window threshold M=50.

Table 9.4 Study B. Gathered Traffic-Type Statistics

Traffic Type	Collected Packets	Generated PFBs
Normal Traffic (NT)	4.847.500	96950
TCP Toxic Traffic (TT-TCP)	274.500	5.490
SPOOF Toxic Traffic (TT-SPF)	112.500	2.260
ICMP Toxic Traffic (TT-ICMP)	76.100	1.522
UDP Toxic Traffic (TT-UDP)	60.150	1.203
Total	5.370.750	107.425

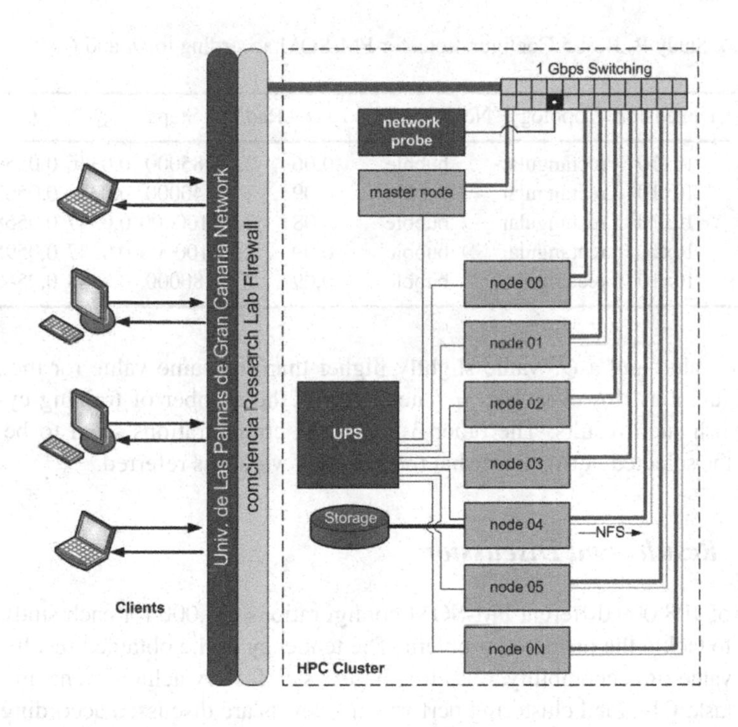

Fig. 9.5 Study B. The scenario where the test is performed is a High Performance Computing facility offering job dispatching/processing queues and several over-IP services. A network probe taking approach of port mirroring is placed transparently inside the network.

Table 9.6 shows the five best obtained configurations for PM-SOM by means of the Average Quantization Errors in the training and validation stages for Study B. According to these Q_t and Q_v values respectively, network configuration $Net = 6$ reveals as the best map for this study. Network configuration $Net = 7$ offers the

Table 9.5 Study B. Constitution of the Selective TS and VS

Traffic Type	Training		Validation	
	Packets	PFBs	Packets	PFBs
Normal Traffic (NT)	11500	230	1100	22
TCP Toxic Traffic (TT-TCP)	3100	62	600	12
UDP Toxic Traffic (TT-UDP)	50	1	150	3
ICMP Toxic Traffic (TT-ICMP)	2050	41	450	9
SPOOF Toxic Traffic (TT-SPF)	-	-	800	16
Total	16700	334	3100	62

Table 9.6 Study B. Best 5 Configurations for PM-SOM according to Q_t and Q_v

Net	Dimension	Topology	Neighbourhood	α	Radius	Steps	Q_t	Q_v
6	10x10	rectangular	bubble	0,06	2	85000	0,0346	0,0559
7	10x10	rectangular	bubble	0,09	2	40000	0,0346	0,0560
8	10x10	rectangular	bubble	0,08	2	100000	0,0347	0,0568
9	10x10	rectangular	bubble	0,09	2	100000	0,0347	0,0595
10	10x10	rectangular	bubble	0,09	2	80000	0,0348	0,0560

same Q_t value and a Q_v value slightly higher than the same value for the $Net =$ 6 configuration. However, $Net = 7$ needed half the number of training cycles to accomplish such results. The other 3 PM-SOM configurations seem to be really near to the selected network in what to Q_t and Q_v values is referred.

9.4.3 Results and Discussion

A total of 108.000 different PM-SOM configurations (54.000 for each study) were trained to tackle the proposed problem. The tendency of the obtained results shows a great value of acceptability, revealing highly satisfactory achievements in the detection task. Obtained clustering performance levels are discussed according to the Q_v and Q_t quantification errors respectively, presented in Table 9.3 and Table 9.6 as well as through the CM for each scenario.

Regarding Study A, once the training period was finished, the selected-as-best network configuration $Net = 2$ properly detected four different clusters in the input TS, according to the four pattern types presented, (Fig. 9.6, (a)). Obtained results in the validation process reveal that PM-SOM was able to detect the four different existing clusters in the analyzed traffic VS with a high level of accuracy by each class, (Fig. 9.6, (b)). Table 9.7 shows a 100% detection accuracy for NT events. Such accuracy is also gathered for TT-UDP and TT-ICMP type flooding episodes. In what relates to TT-TCP traffic, PM-SOM was able to perform an accurate detection in the 95.24% of the TT-TCP attacking episodes (only 1 of the 21 validation PFBs

Table 9.7 Study A. Confusion Matrix for PM-SOM ($Net = 2$)

	NT	TT-TCP	TT-ICMP	TT-UDP
NT	43	0	0	0
TT-TCP	1	20	0	0
TT-ICMP	0	0	13	0
TT-UDP	0	0	0	14

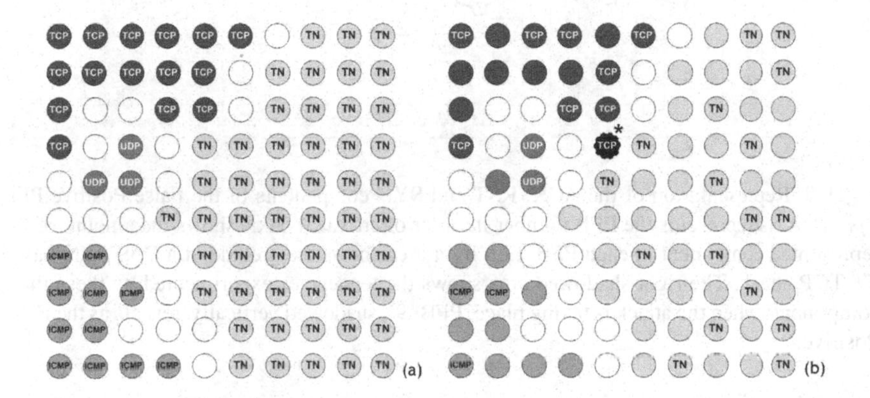

Fig. 9.6 Study A. $Net = 2$. (a) Four clusters detection in training dataset. (b) Classification of Validation PFBs. False Positive located in neuron $fp_0(3,4)$.

was misclassified). The winner neuron that identified the misclassified validation PFB was supposed to react to NT patterns, while the presented PFB was supposed to be of TT-TCP type, (Fig. 9.6 (b)).

Once the components of the studied traffic feature vector were in-order represented and the incorrectly classified PFB was placed, it was observed that this pattern was located in the prelude of a perfomed DoS SYN-based TCP attack. This event, false positive, showed up a high level of ACKs, which indicates acknowledgments by one of the peers during a established communication, being very common in NT data transactions. Analyzing in-order the prior and subsequent PFBs, it was revealed that this concrete PFB had the highest ACK rate of all the traffic analyzed before and during the performed DoS attack, (Fig. 9.7). The values of ACK and SYN suggest that this PFB contained a combination of NT and TT-TCP, where there was greater tendency for indicators of NT packets than usual in pure TT-TCP traffic.

Study A has shown how a prelude of DoS attack could be misclassified as Normal Traffic. Traffic toxicity concentration was not enough to be considered as Toxic Traffic in the inspected PFB_n: NT traffic concentration was higher than the expected to be in a PFB which was supposed to be identified as toxic TCP traffic. However, it must be highlighted that the following PFB_{n+1} was clearly classified as Toxic Traffic, which means that, in about 100 transmitted packets, just 2 PFBs, the DoS attack was correctly detected by CISDAD. A higher value of M, the time window

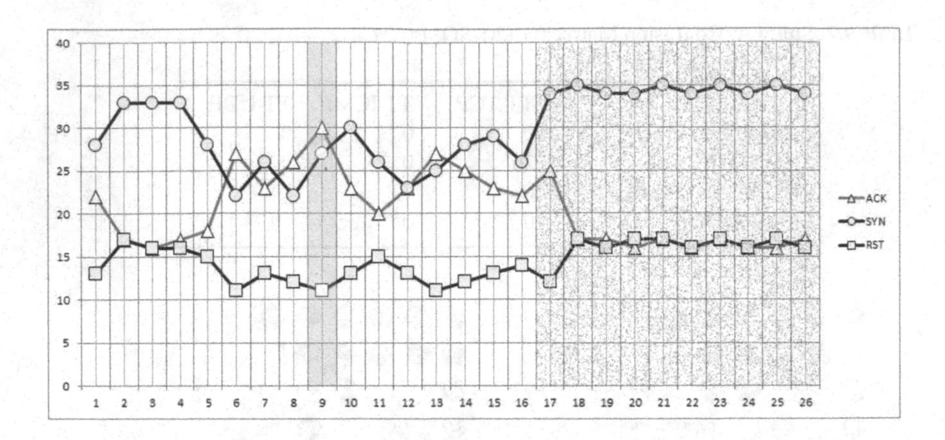

Fig. 9.7 Representation of the ACK, RST and SYN components of the False Positive PFB fp_0. X-Axis represents the PFB number in order of arrival. Y-Axis shows the amount of the represented component in each PFB. Left light area shows the prelude of a DoS SYN-based TT-TCP attack. The right shadowed area shows the tendencies experimented by these three components when the attack is taking place. PFB '9', shadowed vertically, reveals as the False Positive.

size, could maybe help reduce this kind of false positive, because by increasing the number of toxic packets per PFB, the concentration of toxic traffic also increases in the subject PFB. Therefore, the described event that originates the false positive fp_0 could be minimized, improving the detection.

Regarding Study B, the selected-as-best network configuration $Net = 6$ also detected four different clusters in the input TS, according to the four pattern types presented, (Fig. 9.8, (a)). In this case, obtained results in the validation process were certainly interesting. Table 9.8 shows the resulting CM for this study.

Relating to NT, VS contained 23 patterns for validation tasks of which 20 were correctly classified as NT while the resting 3 ones were classified as TT-ICMP. Although these 3 PFBs packets were gathered surrounded by dozens of NT packets in a normal-traffic state of the network environment, PM-SOM classified them as toxic ICMP traffic. In fact, after inspecting the corresponding patterns, a higher index of ICMP packets than normal was observed, advising some source workstation that the port it was trying to connect with was unreachable. Despite the index of ICMP packets was not high enough to be considered as a possible flooding attack, the concentration of ICMP packets in such PFBs was sufficient to reduce the amount of other packets generated by other normal transactions and to appear similar to excessive ICMP PFBs. However, PM-SOM detected that the kind of traffic being properly inspected was not purely normal and denoted it as ICMP-excessive. This situation provoked a false positive which should not have been considered as a toxic traffic by the system. Nevertheless, it must be taken into account that the constitution of these PFBs make them kind of special event, so they should be deeply analyzed with the intention of increasing PM-SOM accuracy.

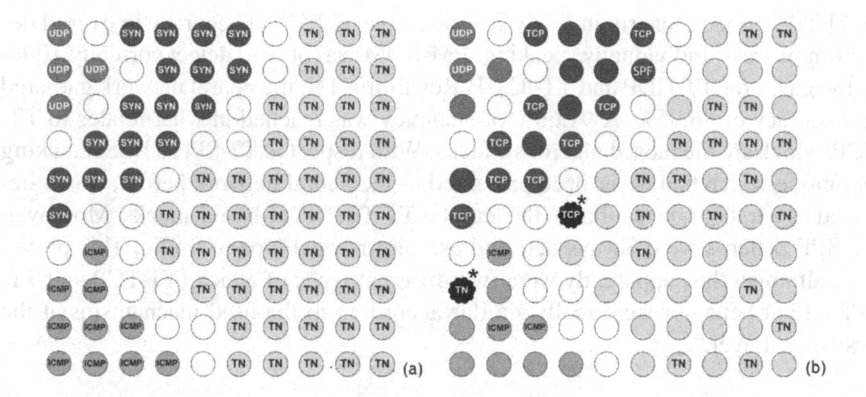

Fig. 9.8 Study B. $Net = 6$. (a) Four clusters detection in training dataset. (b) Validation PFBs classification. False Positives located in neuron $fp_1(7,0)$ and $fp_2(5,3)$. TT-SPF patterns detected by neuron(5,1).

Regarding the TT-TCP validation patterns, CM shows a misclassification of 1 of the 12 patterns as NT, instead of being detected as TT-TCP as expected. The false positive from Study A repeated again in this case. Inspecting the PFB, the same situation re-appeared: PFB was made up of a combination of NT and TT-TCP packets just at the beginning of a DoS SYN-based TT-TCP episode. Once again, it was described by a high index of ACK in company of a mix of SYN, RST and PSH indicators. Toxic concentration was not enough to classify this PFB as TT-TCP. However, next PFB in order of sequence indeed was.

TT-SPF PFBs were all recognized by neuron (1,5), (Fig. 9.8, (b)), which was specialized in detecting SYN-based TT-TCP flavour attacks. The performed attack follows the phylosophy of SYN-based flooding attacks but it differs in performing a massive set of asking-for-connection queries which never accomplish and no reset of connection is provided by the attacker, neither priority PSH or URG flags are activated, unlike the classic SYN-flooding attack used to train PM-SOM. So in the end, the general tendency of flag-featuring indicators present little dissimilarities regarding the classic attack presented to PM-SOM during its training. The neural network detected that the input data was likely to previously TT-TCP seen attacks.

Table 9.8 Study B. Confusion Matrix for PM-SOM ($Net = 6$)

	NT	TT-TCP	TT-UDP	TT-ICMP	TT-SPF
NT	20	0	0	3	0
TT-TCP	1	11	0	0	0
TT-UDP	0	0	3	0	0
TT-ICMP	0	0	0	5	0
TT-SPF	0	16	0	0	0

The described scenario in Study B shows up a 93,55% of accuracy in overall detection process. Individually speaking, PM-SOM was able to detect correctly 100% of the cases for TT-UDP and TT-ICMP. Regarding TN, the neural network gathered an accuracy of 86,95%. A 91,66% of accuracy was reached in what relates to TT-TCP, which are the most dangerous attacks. With respect to TT-SPF, a DoS attacking technique which had never been presented to the neural network before, it was detected as a traffic type similar to the learned TT-TCP SYN-based attacks. Moreover, TT-SPF patterns were always detected as a not-normal type of traffic. This proves that, although they apparently were two different types of attack (TT-TCP and TT-SPF), their behavior were really similar according to the used mechanisms at the Transport Layer.

9.5 Conclusion

In this chapter, a hybrid, intelligent and modular networking traffic analyzer which parses features of protocols that operate over the Network Layer has been presented. We have proposed a solution based on unsupervised neural networks, especifically on Kohonen Self-Organizing Maps , which is capable of detecting between legitimate and several types of toxic traffics. The presented representation scheme for information provides concrete features to distinguish with accuracy the most dangerous DoS and DDoS attacks, the TCP-based flooding attacks. It is also capable of identifying other flooding events based on the ICMP and UDP protocols.

Two studies have been performed. The first one, Study A, performed under a benchmark networking environment, allowed to locate and study a critical spot that arises in the early stage of a TCP SYN-based DoS attack, symptom which seems to be directly related to the traffic concentration levels in our featuring vector, the PFB. The second one, Study B, gathered traffic activity in a HPC facility and described the behavior of the neural detector when a new type of attack was presented. Although DoS attacks can manifest different results according to the victims reactions, mechanisms employed to produce flooding or crashing of certain systems are always based on communication protocols overlaying the IP protocol architecture. Thus, Study B showed that a TCP SYN-based distributed DoS attack with spoofing of source IP addresses was similar to general toxic TCP DoS attacks, and moreover, it was definetely clustered as not-normal traffic. Obtained results in both studies show the effectiveness of CISDAD managing traffic in highly distributed networking environments, under different workload levels and circumstances.

Another important conclusion from our developments is the capability of CISDAD for detecting traffic toxicity anomalies in short time. As soon as the toxicity concentration level was enough to be potentially dangerous in just one PFB, that is to say, only 50 IP packets, CISDAD is able to detect both incoming and outgoing DoS procedures. This system shows high accuracy and reliability in detection. We have shown that in between 50 and 100 transmitted packets, (that is to say, just 1 or 2 PFBs), DoS attacks were detected by CISDAD. Considering that a DoS attack can generate thousands of millions of packets, CISDAD detection can be considered as highly accurate. Regarding this, CISDAD early detection capabilities will allow to apply security and prevention policies in an early stage reducing unnecesary resources congestion and minizing damage on victims.

CISDAD can offer specific adaptation to the any environment where data sharing exists, such as Cloud ComputingCloud Computing applications, satisfying the underlying need to protect systems against unwanted attacks and preventing these systems from being used to attack relatives, assisting in the creation of security procedures with minimum investment in human resources while providing a high level of security. With special interest in critical scopes such as medical care, advantages of embedding CISDAD into the networking environments have been described. Concretely, its integration into a clinical workstation system, EDEVITALZH, has been reviewed.

Future work will be centered into two lines. First one, increasing the neural network knowledge about already existing attacks to study their relationships, based on the underlying mechanisms of TCP, UDP and ICMP. This will help in the study and proposal of prevention, contingency and early decision-making plans against DoS. Moreover, it will assist in gathering a better understading of the techniques, tricks and intentions supporting these attacks. Second, the clustering improvement in what relates to mixes of traffic data, with more effective neural architectures in blind clustering processes and with greater capabilities to deal with noise, such as HUMANN, in the Processing Module.

Last, but not least, our study presents advances in the area of computer security against denial of service attacks. The development of reliable, fast and low-cost strategies against DoS attacks is one of the desired achievements of this proposal. In the near future, and thanks to simplicity and accuracy, CISDAD could be integrated into traffic processing systems with great ease, such as routers, firewalls and other detection systems, improving decision-making and efficiency of such devices against DoS events.

Acknowledgements. We would like to thank Canary Islands Government, the Ministry of Science and Innovation of the Spanish Government and EU Funds (FEDER) for their support under Research Projects SolSubC200801000347 and TIN2009-13891 respectively.

References

1. Amini, M., Jalili, R., Shahriari, H.R.: RT-UNNID: A Practical Solution to Real-time Network-based Intrusion Detection Using Unsupervised Neural Networks. Computers & Security 25-6, 321–354 (2006)
2. Argus: Auditing Network Activity, `http://www.qosient.com/argus` (cited January 11, 2011)
3. BBC News. Visualizing the Internet, `http://news.bbc.co.uk/2/hi/8552410.stm` (cited January 31, 2011)
4. Bivens, A., Palagiri, C., Smith, R., Szymanski, B.K., Embrechts, M.: Network Based Intrusion Detection Using Neural Network. In: Intelligent Engineering Systems through Artificial Neural Networks: Proceedings of ANNIE, vol. 12 (2002)

5. Ali, F.: IP Spoofing. The Internet Protocol Journal 10-4, 2–9 (2007)
6. Digital Imaging and Communications in Medicine Standard, http://medical.nema.org/ (cited February15, 2011)
7. Erikson, J.: HACKING the art of exploitation, 2nd edn. No Starch Press, San Francisco; ISBN: 1-59327-144-1
8. García Báez, P.: HUMANN: Una Nueva Red Neuronal Artificial Adaptativa, No Supervisada, Modular y Jerárquica. Aplicaciones en Neurociencia y Medioambiente (Ph.D. Thesis). University of Las Palmas de Gran Canaria (2005)
9. Health Level 7 International, http://www.hl7.org/ (cited February 15, 2011)
10. hping. Salvatore Sanfilippo, http://www.hping.org/ (cited January 23, 2011)
11. Kohavi, R., Provost, F.: Glossary of Terms. Machine Learning 30-2,3, 271–274 (1998)
12. Kohonen, T.: Self-Organization and Associative Memory, 3rd edn. Springer Series in Information Sciences, pp. 3–540 (1989); ISBN: 3-540-51387-6
13. Kohonen, T.: Self-Organizating Maps, 2nd edn. Springer Series in Information Sciences (1997); ISBN: 3-540-62017-6
14. Labib, K., Vemuri, R.: NSOM: A Real-Time Network-Based Intrusion Detection System Using Self-Organazing Maps (2002)
15. Lichodzijewski, P., Nur Zincir-Heywood, A., Heywood, M.I.: Dynamic Intrusion Detection Using Self-Organizing Maps. In: Proceedings of the 14th Annual CITASS (2002)
16. Lichodzijewski, P., Nur Zincir-Heywood, A., Heywood, M.I.: Host-Based Intrusion Detection Using Self-Organizing Maps. In: Proceedings of the 14th Annual CITASS (2002)
17. Pérez-del-Pino, M.A., García Báez, P., Fernández López, P., Suárez Araujo, C.P.: Towards Self-Organizing Maps based Computational Intelligent System for Denial of Service Attacks Detection. In: 14th International Conference on Intelligent Engineering Systems (INES), pp. 978–971 (2010); ISBN: 978-1-4244-7650-3
18. Pérez-del-Pino, M.A., Suárez Araujo, C.P., García Báez, P., Fernández López, P.: EDEVITALZH: an e-Health Solution for Application in the Medical Fields of Geriatrics and Neurology. In: 13th International Conference on Computer Aided Systems Theory, EUROCAST 2011 (2011)
19. Suárez Araujo, C.P., Pérez-del-Pino, M.A., García Báez, P., Fernández López, P.: Clinical Web Environment to Assist the Diagnosis of Alzheimers Disease and other Dementias. WSEAS Transactions on Computers 6, 2083–2088 (2004); ISSN: 1109-2750
20. Matsopoulos, G.K.: Self-Organizing Maps.In: InTech. ISBN: 978-953-307-074-2
21. NetFlow by Cisco Systems, http://en.wikipedia.org/wiki/Netflow(cited December 12, 2010)
22. Network Grep, http://ngrep.sourceforge.net/ (cited January11, 2011)
23. OGE: Oracle Grid Engine, http://www.oracle.com/us/products/tools/oracle-grid-engine-075549.html (cited January 21, 2011)
24. Packet Details Markup Language Specification, http://gd.tuwien.ac.at/.vhost/analyzer.polito.it/docs/dissectors/PDMLSpec.htm (cited January 15, 2011)
25. Perl Programming Language, http://www.perl.org (cited December 14, 2010)
26. Port Mirroring. Wikipedia, http://en.wikipedia.org/wiki/Port_mirroring (cited January 21, 2011)
27. RFC 4732: Internet Denial-of-Service Considerations, http://tools.ietf.org/html/rfc4732 (cited November 21, 2010)

28. SOM_PACK. Dept. of Information and Computer Science, Helsinki University of Technology,
http://www.cis.hut.fi/research/som-research/nnrc-programs.shtml (cited January 21, 2011)
29. Stalling, W.: Network Security Essentials. Applications and Standards. Prentice Hall, Englewood Cliffs (2007); ISBN: 0-13-238033-1
30. Stalling, W.: Comunicaciones y Redes de Computadores, 6th edn. Prentice Hall, Englewood Cliffs (2000); ISBN: 84-205-2986-9
31. Suárez Araujo, C.P., García Báez, P., Hernández Trujillo, Y.: Neural Computation Methods in the Determination of Fungicides. Fungicides, 471–496 (2010); ISBN: 978-953-307-266-1
32. Symantec State of Enterprise Security Survey (2010),
http://www.symantec.com/content/en/us/about/presskits/SES_report_Feb2010.pdf
(cited March 25, 2011)
33. TShark: The Wireshark Network Analyzer. Documentation,
http://man-wiki.net/index.php/1:tshark (cited January 21, 2011)
34. Denial-of-Service Attacks, Incidents. Wikipedia,
http://en.wikipedia.org/wiki/Denial-of-service_attack
(cited January 02, 2011)
35. Zanero, S.: Analyzing TCP Traffic Patterns Using Self Organizing Maps. In: Roli, F., Vitulano, S. (eds.) ICIAP 2005. LNCS, vol. 3617, pp. 83–90. Springer, Heidelberg (2005),
http://man-wiki.net/index.php/1:tshark (cited January 21, 2011)

Chapter 10
Combating Security Threats via Immunity and Adaptability in Cognitive Radio Networks

Jan Nikodem, Zenon Chaczko, Maciej Nikodem, Ryszard Klempous, and Ruckshan Wickramasooriya

Abstract. In this chapter we shall consider security, immunity and adaptability aspects of Cognitive Radio (CR) networks and its applications. We shall cover design of a immunity/adaptability and security simulation model for Cognitive Radio and discuss results of conducted experiments using Matlab simulation tools and Crossbow's XMesh using MoteWorks software platform. The main goal of this chapter is to provide an overview of various applications of CR as well as methods of combating security threats faced when applying the CR technology. The immunity/adaptability functions, their benefits and applications in CR are analyzed, along with the challenges faced. We shall discuss in detail how the proposed immunity and adaptability model can mitigate security threats faced by CR and carry out research on a range of selected techniques that can help to mitigate malicious attacks and provide examples of simulation experiments.

10.1 Introduction

A Cognitive Radio is a device capable of receiving and transmitting signal that can detect changes in its operating environment and adapt accordingly [6]. Cognitive Radio Networks are a field that is still in its infancy but it is developing fast. Basically, a CR has sufficient intelligence to detect user communications needs and can

Jan Nikodem · Maciej Nikodem · Ryszard Klempous
Institute of Computer Engineering, Control and Robotics, Wroclaw University of Technology
Wybrzeze Wyspianskiego 27, 50-370 Wrocław, Poland
e-mail: {jan.nikodem,maciej.nikodem,ryszard.klempous}@pwr.wroc.pl

Zenon Chaczko · Ruckshan Wickramasooriya
Faculty of Engineering and IT, University of Technology Sydney (UTS),
Bld1, Broadway, Ultimo 2007 NSW, Australia
e-mail: {zenon.chaczko,ruckshan.wickramasooriya}@uts.edu.au

J. Fodor et al. (Eds.): Recent Advances in Intelligent Engineering Systems, SCI 378, pp. 221–242.
springerlink.com

provide suitable resources to satisfy those needs. In other words, it is its operating environment which decides on which transmitting parameters to change. A variety of devices can be considered CRs. Simple CRs can detect interference and avoid it by switching to another frequency. A more complex CR can reconfigure itself to interact with any nearby networks as per user requirements. Other CR devices can adapt to comply with changing network and regulatory policies. A common theme among them is their adaptability to a changing environment.

10.1.1 Benefits of Cognitive Radio

Studies have shown that several frequency bands are sometimes not utilized to their maximum potential. These under-utilized areas are known as spectrum holes or white spaces. CR offers the opportunity for these white spaces to be exploited and keep them from going to waste. This offers the following benefits:

- Increases the efficiency of the frequency bands.
- Assists new business models that are not strictly tied to the availability of spectra, as well as facilitating macro and micro level spectrum trading. Spectrum trading is the practice of allowing the user to change the use of the spectrum from its initial requirement and still allow the user to maintain his right to use the spectrum [6].
- It will introduce more bandwidth that can be used in emergencies.
- Users will be able to enjoy higher data rates.
- Enhanced coverage.
- Improved and more extensive roaming.
- Decreased requirement for centralized spectrum management.

10.1.2 Applications of Cognitive Radio

Cognitive Radio has several potential applications. Among most common are:

- Non-RT applications such as email, mobile internet, etc.
- Broadband wireless networking in WiFi hotspots.
- Localized wireless multimedia distribution networks.
- Emergency communications with high priority.

Since CR is about devices making maximum use of available bandwidth, means of sharing bandwidth will have to be devised. The below list includes bands that can be shared. Analogue land mobile radio operates in the 148 MHz - 470 MHz range. The advantage is that this band currently has low activity and has ample space for more signals. It uses local spectrum sensing to avoid interference and the locations of its base stations are already known [6].

Television broadcast is similar to analogue land mobile radio. Its base stations are well known and it uses local spectrum sensing to counter interference. Given the nature of television, a large topographic area can be covered [6].

Radar is another possibility since it consumes a large part of the spectrum and therefore can accommodate more extra bandwidth when it is not used for its primary function. Its bandwidth can be used when the radar is pointing in a different direction. An advanced mechanism is required for signal sensing and synchronization. The radar locations are also required [6].

10.1.3 Challenges to Cognitive Radio

As with any emerging technology, Cognitive Radio faces several technical challenges that must be solved. These challenges are related to such areas as:

- Protecting primary users and guaranteeing they would have a positive experience with CR networks.
- Control mechanisms.
- Spectrum mobility.
- Regulatory policies.

Protecting primary users is about avoiding consequences of the hidden node problem. For example, in a scenario setting of 3 users, users A and B cannot hear user C and therefore they presume it is OK to transmit using Cs frequency. Naturally, C experiences interference at his intended receiver because of this. The solution to this problem is to have restrictions on location, using expertise on the transmitted signal to improve transmitter detection, cooperative detection of nodes and to have distinct distributed spectra sensing networks [6].

Control mechanisms introduce several issues that need to be taken into consideration in this regard. Among the most important issues [6] are such problem as:

- Should control be centralized or distributed?
- What are the optimal communications parameters?
- What are the common control channels that can negotiate initial bands?
- Should the spectrum sharing approach be overlay or underlay?
- Ensure unlicensed bands get resources fairly allocated.
- Access cost negotiation and micro-trading for licensed bands.
- How to deal with different propagation losses in different bands?

To ensure spectrum mobility CR users frequently perform changes to frequency of their operations. This may involve such aspects as [6]:

- Fast detection of primary users.
- Fast negotiation of new band and spectrum handover.

- Ability to use multiple simultaneous bands .
- Managing higher layer protocols during spectrum handover.

However, there are some forbidden bands that should be avoided. The ideal scenario is for the user to switch to a fallback channel when the main channel cannot be used.

Last but not least, new technology such as CR, requires regulations [6] to utilize its full potential and to keep it from being exploited. The challenges it faces are as follows:

- The devices should be flexible and able to operate in various frequency bands.
- They should be able to operate even without a fixed infrastructure.
- Naturally, frequency bands vary from country to country.

Among possible solutions for the above listed challenges is software with components ensuring that:

- Nodes are locally aware and have access to a central database that regulates them.
- Comply with internationally standard bands to operate in.
- Provide local beacons that broadcast regulatory information.

There are a few projects aimed at tackling the current challenges involving CR [6]. The key areas are spectrum and equipment. IEEE 1900 is working on standard definitions, interference and coexistence analysis, Software Defined Radio (SDR) conformance, air interface and device certification. IEEE 802.22 is a set of standards that are being devised for the use of white space in the television band. This was recently demonstrated by DARPA Xg when the technology was integrated into several US military projects. The Federal Communications Commission of the USA is bringing in regulations that will accommodate SDR and CR, as well as enable dynamic frequency selection in radar bands. The International Telecommunication Union (ITU) has also pledged to consider the relevant regulations at the upcoming World Radio Communications Conference in 2011.

10.2 Security Threats in Cognitive Radio Systems

Cognitive Radio based systems have two main components: Artificial Intelligence (AI) and Dynamic Spectrum Access (DSA) [14]. AI involves reasoning and learning. This gives CR its intelligent characteristic and allows it to learn about its changing environment. DSA is the processes involved in getting a CR to detect and occupy a vacant spectrum. It involves spectrum sensing, spectrum management, spectrum mobility and spectrum sharing [14]. Spectrum sensing detects holes in the spectrum and ensures the spectrum is shared between devices without interfering with each other. Spectrum management makes sure that the best available channels are selected. Spectrum mobility allows transition between spectra without any problem.

The task of spectrum sharing is to see that users coexist with each other in one channel [14]. CR applications face the following threats:

- Attackers can alter sensory input statistics.
- This in turn can lead to belief manipulation.
- The manipulated information can be spread throughout a CR network.
- Behavior algorithms based on manipulated information can result in faulty performance.

To counter these threats, CR devices need to perform the following:

- Always assume that sensory input statistics are noise and at constant risk of manipulation.
- Include a common sense software routines to validate learned beliefs.
- Compare and validate learned beliefs with other devices in the CR network.
- Perform adaptive learning in environment that is known to be secure.
- Let learned beliefs to be expired after some period of time in order to prevent long-term effects of malicious attacks.

Two main types of attackers can exist in the Cognitive Radio network environment: *off-path* and *on-path*. Off-path attackers can insert false data into a data stream and spoof other network devices but they cannot observe traffic while it is transmitted. Therefore, they can be neutralized by implementing protocols which allow devices to participate only if they can see the traffic [3]. On-path attackers are more ominous as they can observe as well as insert new data elements into a data stream in real time. Therefore, they can also launch Denial of Service (DoS) attacks. For protection against other types of attacks, mutual authentication, data integrity protection and data encryption should be used in combination [3]. On-path adversaries find wireless protocols relatively easy to attack because the link from devices to the network is more exposed than in a conventional wired, switched network and all devices can see traffic from its counterparts in the network [3]. Furthermore, the many physical links between servers and clients means that attackers can easily spoof network devices. Transmitting a simple jamming signal prompts signal degradation and eventual deletion, facilitating packets to be spoofed and DoS attacks. Therefore, wireless networks always assume that on-path attackers are present and have a link-layer security protocol built in, typically IEEE 802.11i. are two classes of radios: policy radios and learning radios. A policy radio has a reasoning engine while a learning radio has both reasoning and learning engine. A reasoning engine is a software module that has the ability to surmise logical consequences from a preprogrammed set of logical inference rules [3]. A learning engine has no preprogrammed policy and can experiment with various radio configurations to gauge system response radios, therefore they are more flexible. A typical cognition cycle is Observe, Orient, Plan, Decide, Act [11]. If it is a learning radio, a new stage called Learn is added whenever a new operating stage is encountered, thus allowing it to learn about the new stage. This system, however, is also exposed to a wide range of attackers.

10.2.1 CR Threat Types

10.2.1.1 Threats to Policy Radios

The primary threat is that an attacker could spoof faulty sensor information and lead the radio to select a below par configuration. An attacker can cause this by manipulating the radio frequency seen by the radio. This form of attack is called the sensory manipulation type of an attack as the attacker needs to understand complex logic and should know the type of input to provide [3].

10.2.1.2 Threats to Learning Radios

Learning radios face the same threats as policy radios, but the threats they face are much stronger as it develops long-term behavior over time. For example, an attacker can send a jamming signal when a policy radio switches to a faster modulation rate, resulting in link degradation for the duration of the attack (lower data rates and link speeds), but a learning radio may infer that higher modulation rates are always associated with lower data rates and always use lower data rates. Such attacks are called belief manipulation attacks [3].

10.2.1.3 Self-propagating Behavior

This is a very powerful attack where, for example, a certain state S on radio 1 induces state S on radio 2, which in turn induces state S on radio 3, eventually propagating throughout all the radios in the area. It is like a Cognitive Radio virus [3], [8]. They spread among non-cooperative radios that do not have direct protocol interaction.

10.2.2 Classes of Attacks

10.2.2.1 Dynamic Spectrum Access Attacks

This is also known as a Primary User Emulation (PUE) attack and is of concern in Dynamic Spectrum Access environments [3]. In such a scenario, a primary user (PU) has the licence to use a particular frequency band and allows secondary user (SU) to use the band when the PU is idle. These SUs utilize spectrum sensing algorithms to detect when the PU is inactive. An attacker can create a waveform that is similar enough to the PU and create a false positive in the SUs algorithm, leading the SUs to believe that the PU is still active. This enables the attacker to have exclusive access to the empty frequency band. But once the attacker vacates the frequency band the SU can use it as they can sense that the band is once again available. There are more powerful DSA attacks, some of which gather information from PUs to calculate when the band will be free based on the PUs current and past behavior. This is a belief manipulation attack as the attacker spoofs the PUs waveforms [3]. In the presence of a Time Division Multiple Access (TDMA) PU, an attacker can even completely deny access to a SU by making the PUs access pattern appear random instead of periodic while the SU is in its learning phase. This renders the SU unable to gather much information on when the band is available and it is a long term problem.

10.2.2.2 Objective Function Attacks

CRs have cognitive engines that manipulate radio parameters in order to maximize its service functions. Objective function attacks target learning algorithms that use such functions and are therefore belief manipulation attacks [3]. Input parameters could be center frequency, bandwidth, power, modulation type, coding rate, channel access protocol, encryption type, and frame size.

Typical radios have three goals: low power, high rate and secure communication. These three goals have different weights depending on the application being run. Low power and secure communication are defined by the system inputs while high rate is defined by the system outputs. An attacker can determine if high rate communication is achieved by attacking the channel. For example, he can do this by jamming the channel to increase to decrease the rate, ensuring that the system cannot achieve a higher security state [3]. This kind of attack is feasible in a static system where human configuration is involved. If the attacker jams the network whenever it is activated with encryption enabled, the engineer might assume that there is a crypto-related obstacle to using encryption and run it without encrypting, presenting a security risk [3]. The IEEE 802.11u standards have defined weaker security for WLANs to support 911 calls. This can be exploited with a learning engine. Attackers can also make some bandwidths, frequencies and modulation types seem undesirable for CRs. These attacks work only when CRs are engaging in on-line learning, not off-line, because on-line learning involves constant on-line optimization.

10.2.2.3 Malicious Behavior Attacks

These are an extension of objective function attacks and teach CRs to become malicious. For example, consider the case where a CR is made into a jammer. In a system where the PU is accessing a channel sporadically, SUs have channel sensing algorithms that detect when the band is available. Their objective function seeks to maximize throughput and minimize interference. If an attacker sends a jamming wave that cannot be detected by the SUs algorithms, throughput will be artificially increased when the PU is idle. The CR will then believe that communication can be achieved only when the PU is active [3]. Thus the CR can become a jammer. This can be done using commercial waveforms.

10.2.3 Counter-Measures against Attacks

A range of techniques are available to foil attackers in CR systems.

10.2.3.1 Robust Sensory Input

Sensor input is of great importance in CRs so that any improvement to the quality of sensor input can increase security [3]. Radios that can distinguish between interference and noise can also distinguish between natural and artificial RF signals. CRs like that can also seek out hostile signals that have malicious intent. Sensor data can be used in a distributed environment to more accurately check if a signal is from an attacker or just noise by running cross-correlation algorithms [3]. Actually, all sensory input initially needs to be considered noise because errors can occur even with no attacker present.

10.2.3.2 Mitigation in Individual Radios

In individual radios, security threats mitigation and prevention requires some common sense to be inserted into radios [3]. It is hard to protect individual policy radios from attack because they decide on their course of action based on their policy and the individual environment. An attacker familiar with the policy can simply provide false sensor inputs to carry out a malicious attack. If an attacker does not know the policy, he can easily infer it by trying out various techniques. Cognitive Radio policies must be cautiously chosen to protect against sensor input from attackers. A bad state can be made unreachable by calculating all possible states, defining a state transition framework to be overlaid on the state space and applying formal state-space validation. Nevertheless, attackers can work around this measure by carrying out PUE attacks. Neutralizing such attacks requires improved sensing algorithms with lower false-positive rates [3]. Such an algorithm can identify an attacker from an authentic PU. Providing security to individual radios is also a challenge but it can be done. Their beliefs need to be under constant supervision and a feedback loop needs to be updating learned relationships between CR inputs and outputs. If this is impractical, the learning phases need to be done in a well controlled environment with outside auditing to make sure that no undesired signals are present. An alternative method is to infuse logic that is able to identify learned and potentially dangerous actions. However this might be a rather challenging undertaking.

10.2.3.3 Mitigation in Networks

In the CR security threats mitigation environment, several AI agents exist, each trying to perform at their optimum level [2], [3], [12]. Swarm intelligence can be used in security measures as there is a lot of devices communicating. Swarm intelligence is a collection of algorithms that imitate biological system behavior. Biomimetic algorithms such as Immune-computing [2], [10], Self Organising Maps (SOM) [9] or Particle Swarm Optimization (PSO) [3] algorithms can be used, where each element of CR network is modeled as a particle or immune network component with its own presumptions about how to perform best in a certain situation. The chosen behavior is a weighted average of all the hypotheses in a network The PSO model can be used against PUE attacks. The group decision is the decision that is common to most devices. The risk with such group decision-making is that an attacker might be able to affect (infect) the entire neighborhood [2],[8], [12].

10.3 Immunity and Adaptability in CR Applications

10.3.1 Basic Ideas and the Rationale

The immunity and adaptability model presented in this chapter assumes that the Cognitive Radio network consists of a set of relatively homogeneous network elements interacting with each other. CR network is not organized centrally, but rather in a distributed manner. Each individual element has finite capabilities (i.e. resources such as: energy, hardware, software and communication range) hence at the individual level it is not able to realize overall system tasks. Due to these limitations, elements are required to some degree strictly follow an integrated cooperation between them. This activity is realized predominately in the information domain, as it is essential for the CR dependability to provide a robust communication capability.

From one point of view, each element works in its precinct (vicinity) autonomously, interacting with environment stimulus; And from the other point of view, elements must communicate with each other, therefore communication channels are crucial elements of the CR network architecture. In general, elements are simple, unsophisticated technical components performing tasks determined by programmers and engineers. Because of that the risk of unauthorized access or even a possible destruction of communication links by outsiders can be an increasing threat. External attackers are often responsible for causing communication and routing disruptions including the breach of security. Frequently, however, it is the internal event which may contribute to serious decrease in efficiency of communication channels.

10.3.2 Mathematical Model

When describing the Cognitive Radio activity, we will be discussing the concepts of actions and behavior. *Action* should be considered the property of every network element such as: an element, a router, a switch or a node. The *Behavior*, on the other hand is an external attribute which can be considered either as an outcome of actions performed by the whole CR network or its subset (i.e. cluster, tree, element field, vicinity). *Action* is a ternary relation which can be defined as:

$$Act : Nodes \times States \rightarrow States. \tag{10.1}$$

Based on this we can construct the quotient set Behavior, elements of which are called equivalence classes linked to the relation $\mathscr{R}$ and here denoted as:

$$Beh : Act/\mathscr{R} = \{act_{[x]} \in Act \mid act_{[x]} \mathscr{R} x\} \tag{10.2}$$

10.3.2.1 Neighborhood Abstraction

Let us define $Map(X,Y)$ as a set of mapping functions from X onto Y (surjection), and $Sub(X)$ is defined as a family of all X subsets and where we define the neighborhood $\mathscr{N}$ as follows:

$$\mathscr{N} \in Map(Nodes, Sub(Nodes)). \tag{10.3}$$

Thus, $\mathscr{N}(k)$ is the neighborhood of node k , and $\mathscr{N}(C)$ is the neighborhood of C (set of nodes) defined as:

$$\mathscr{N}(k)_{|k \in Nodes} := \{y \in Nodes \mid y \; \mathscr{R}_{\mathscr{N}} \; k\}, \tag{10.4}$$

$$\mathscr{N}(C)_{|C \subset Nodes} := \{y \in Nodes \mid (\exists x \in C)(y \; \mathscr{R}_{\mathscr{N}} \; x)\}. \tag{10.5}$$

The formal view emerging from the above discussion will project a construction of the extended Cognitive Radio behavioral model and related to the challenges of individual tasks versus collective activities of network elements. By Individual it is meant that the node/element improves its goal-reaching activity, interacting with its neighborhood $\mathscr{N}(k)$ while collective activity relates to $\mathscr{N}(C)$ neighborhood.

10.3.2.2 Relational Attempt to Network Activity

In 1978 J. Jaroń has developed an original methodology for the systemic cybernetics [7] which describes three basic relations between systems components such as: subordination (π), tolerance (ϑ) and collision ($\varkappa$). We find these basic relations very useful in describing activities and qualitative relations between components of the Cognitive Radio. The concepts developed in this chapter and the discussion, on the immune functions in communication activities, require to define the following three key relations:

$$Subordination \qquad \pi = \{<x,y>; x,y \in Act \mid x \, \pi \, y\}. \tag{10.6}$$

The expression $x\pi y$ defines the action x which is subordinated to the action y or action y dominate over action x.

$$Tolerance \qquad \vartheta = \{<x,y>; x,y \in Act \mid x \vartheta y\}. \tag{10.7}$$

The expression $x\vartheta y$, states that the actions x and y tolerate each other,

$$Collision \qquad \varkappa = \{<x,y>; x,y \in Act \mid x \varkappa y\}, \tag{10.8}$$

and finally $x\varkappa y$ means the actions x and y are in collision to one another. The basic properties of mentioned above relations could be formulated succinctly as follows [7]:

$$\pi \cup \vartheta \cup \varkappa \subset Act \times Act \neq \emptyset, \tag{10.9}$$

$$\iota \cup (\pi \cdot \pi) \subset \pi, \tag{10.10}$$

where $\iota \subset Act \times Act$ is the identity on the set *Action*. Moreover:

$$\pi \cup \vartheta^{-1} \cup (\vartheta \cdot \pi) \subset \vartheta, \tag{10.11}$$

where ϑ^{-1} is the converse of ϑ. That is:

$$\vartheta^{-1} = \{<x,y> \in X \times Y \mid y \vartheta x\}. \tag{10.12}$$

For collision,

$$\varkappa^{-1} \cup \{\pi \cdot \varkappa\} \subset \varkappa \subset \vartheta', \tag{10.13}$$

where ϑ' is the complement of ϑ i.e.:

$$\vartheta' = \{<x,y> \in X \times Y \mid <x,y> \notin \vartheta\}. \tag{10.14}$$

The formula (10.9) indicates that all three relations are binary on nonempty set *Actions*. The formula (10.10) describes fundamental properties of subordination relation which is reflexive and transitive. Therefore it is also ordering relation on the set *Actions*. The formula (10.11) states that subordination implies the tolerance. Hence we can obtain:

$$\{\forall x,y \in Act \mid x \pi y \Rightarrow x \vartheta y\}, \tag{10.15}$$

and subordinated actions must tolerate all actions tolerated by dominants

$$\{\forall x,y,z \in Act \mid \{x \pi y \wedge y \vartheta z\} \Rightarrow x \vartheta z\}. \tag{10.16}$$

10.3.3 Immunity and Adaptability Functions

Immunity is the distributed rather than the global property of a complex system. The complexity of the immune function, the absence of simple feedback loops, a high complexity of tasks and activities as well as collective (not central) rather than individual regulatory processes all results in serious challenges for attempts to define systems immune competences. There are a number of fundamental works corresponding to this domain and related to artificial systems. Several authors have drawn inspiration from the biological immune system, incorporated a lot of properties from autonomous immune systems [4], [5], [10], [11], [13]. In this article, we discuss a novel relational approach to modeling of the system immunity [4]. In our attempt we are considering the immunity not as a set of particular mechanisms like clonal selection, affinity maturation or elements (anti-gens, antibodies, idiotopes, paratopes, etc.). In the proposed approach the phenomenon of immunity appears to be a result of inter-relations among members of community or neighborhood [1], [12]. Let us consider for the node k, its neighborhood $\mathscr{N}(k)$. Any communication activity act_k that is performed by node k relates to some members of $\mathscr{N}(k)$ and the set of actions act_k within neighborhood $\mathscr{N}(k)$ can be defined as follows:

$$Act_{\mathscr{N}}(k) := \{act_k \in Act \mid (\exists x \in \mathscr{N}(k))(act_x \mathscr{R}\, act_k)\}. \tag{10.17}$$

If n is a number of actions act_k within neighborhood $\mathscr{N}(k)$, then it can be expressed as cardinality $Card(Act_{\mathscr{N}}(k))$. The collection of all subsets of $Act_{\mathscr{N}}(k)$ is determined as power set $Pow(Act_{\mathscr{N}}(k))$ with cardinality 2^n. Finally, any subset of that power set is called a family $Fam(Act_{\mathscr{N}}(k))$ of communication activities within neighborhood $\mathscr{N}(k)$. The Cartesian product defined as:

$$IS_k := Act_{\mathscr{N}}(k) \times Act_{\mathscr{N}}(k) \subseteq \pi \cup \vartheta \cup \varkappa, \tag{10.18}$$

describes interaction space IS_k within $\mathscr{N}(k)$. Let us now consider a set of possible interactions fulfilled relation $\mathscr{R}$ within neighborhood $\mathscr{N}(k)$ which can be expressed as:

$$\mathscr{R}_k := \{y \in Act_{\mathscr{N}}(k) \mid < k,y > \in IS_k \wedge k\mathscr{R}y\}. \tag{10.19}$$

Thus, for a given relation $\mathscr{R}$ we define intensity quotient within neighborhood $\mathscr{N}(k)$ as follows

$$IR_k = Card(\mathscr{R}_k)/Card(IS_k). \tag{10.20}$$

The rationale behind our choice of the relational approach to modeling immune and adaptability functions is the fact that interactions are factual and a very relevant aspect of elements of the immune system. The immune system itself can be defined as a very large complex network with a layered and hierarchical architecture. Relationships between components of this architecture can be described by the subordination relations (π) . Furthermore, a positive response of the immune functions would result from the collision relation ($\varkappa$) while a negative response would be attributed to tolerance (ϑ) relation. An immune system is adaptive and autonomous in nature in the sense that its positive or negative responses should be adequate to environment stimulus. Since resources are limited, the allocation rule predicts that an increased investment in system adaptability will come at a cost to investment in immunity (and vice versa). Growing adaptability decreasing system immunity and extends possibility of adversarys attacks. On the other hand, growing immunity barrier tends to adaptability reduction. Bellow we define immunity and adaptability as follows;

$$Immunity := \{< x,y >; x,y \in Act \mid < x,y > \in \vartheta' \times \varkappa\} \tag{10.21}$$

and

$$Adaptability := \{< x,y >; x,y \in Act \mid < x,y > \in \vartheta \times \pi\} \tag{10.22}$$

Therefore, the scope of immunity is determined by decreasing tolerance and growing collision relations while the extend of adaptation is determined by expanding tolerance and subordination. Additionally, based on (10.21), (10.22) it is possible to model adaptation-immunity characteristics of a system and consider it as a process of finding a fine homeostatic balance between them (Fig.10.1). In our case the set of feasible solutions is modeled as a line segment (Fig.10.1) linking strong immunity point SI= $(\varkappa, \pi, \vartheta) = (1,0,0)$ and strong adaptability point SA= $(\varkappa, \pi, \vartheta) = (0,1,1)$. This line is obtained as the intersection of a plane [SI, (0,0,1), SA, (1,1,0)] (immunity) with a plane [SA, (0,0,0), SI, (1,1,1)] (adaptation). In the matter of fact, three points are enough to determine plane but inside the cube we have intersection of two rectangles. In such approach it is practically impossible to determine balance point $b \in [SI;SA]$ which corresponds to both global and any local situations. Based on relational approach (10.21), (10.22), it is evident (Fig.10.1) that the issues of keeping the balance between adaptation and immunity is far more refined then simply finding a point on line segment $[SI;SA]$. This line segment is a canonical projection of topological space subset (tetrahedral [SA, (0,0,0), SI, (0,0,1)] on Fig.10.1). Any point at this tetrahedral structure represents one of many, feasible solution which can fulfill the quality requirements of the network system. Choosing the point $b \in [SI;SA]$ we determine globally the desirable adaptability/immunity homeostatic level(s). However, for each node, owing to its neighborhood state, it is necessary to determine an individual point of balance. The set of such points has to be determined with reference to equivalent class (triangle abc in (Fig.10.1) that corresponds exactly to one point b (Fig.10.1) on a line segment $[SI;SA]$.

10.4 Modeling and Simulation Results

10.4.1 Practical Implementation in Matlab

In order to illustrate the modeling immunity and adaptability results with a concrete model simulation, we consider Cognitive Radio with 10 network elements and one gateway/base station (*BS*). To determine neighborhood $\mathcal{N}(k)$ assignment we use radio link range. Therefore, coming back to formula (10.3) and (10.4) we create subsets $\mathcal{N}(k)$,k=1, 2.., 10. Each cell n of row k in the binary matrix N (Fig.10.4) represents a membership $n \in \mathcal{N}(k)$. It is worth mentioning that a neighborhood always relate

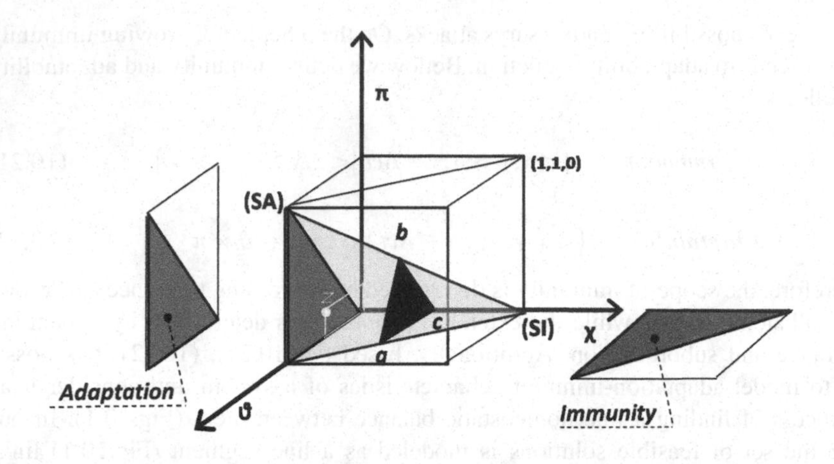

Fig. 10.1 Modeling immunity and adaptability based on π, ϑ and $\varkappa$ relations.

to communication range, but very rarely corresponds with a structure of clusters. Such constructed neighborhood subsets are unambiguous and stable as long as a communication range is fixed.

In the context of Cognitive Radio activity, we focus our attention on routing aspects. Therefore, in further consideration, from all action sets $Act_{\mathcal{N}}(k)$ within neighborhood k we select only these, performed routing activity. In such way we obtain a family $Fam(Act_{\mathcal{N}}(k))$ of routing activities. Considering the distance from the base station (BS) to a node position, a routing activity within $\mathcal{N}(k)$ is partially ordered ($\leq$). Additional preferences were given to cluster heads (CH) and those nodes belonging to the routing tree. Firstly, a cluster heads on routing path that is the nearest BS, next other CHs, finally the regular nodes within $\mathcal{N}(k)$ (Fig.10.4) .

In order to proceed any further, it is necessary to combine the Cognitive Radio spatial structure (topology) and the Cognitive Radio activity (Fig.10.2). To facilitate this process we are required to construct a product that describes the interaction space IS_k within $\mathcal{N}(k)$ as defined in (10.18). Each element of the interaction space (regardless of which neighborhood we consider), according to the right side of eq.(10.18) is mapping (as a point) to (may be not injection) the 3D relational space $[\varkappa,\pi,\vartheta]$.

In order to model $[\varkappa,\pi,\vartheta]$ space, we exploit three additional matrices (Fig.10.5). The real number in any cell of these matrices expresses an intensity quotient of relation. Elements $[r_{k,k}]$ represent intensity quotient within $\mathcal{N}(k)$, while $[r_{i,j}]$ related to particular $i-j$ nodes interactions. Henceforth, we are ready to model and simulate the dichotomies of the Cognitive Radio characteristics - immunity and adaptability (as shown on Fig.10.2 and Fig.10.3). The initial step is the determination of the global (for the whole network) strategy GS_{CR} for the network activity. For this aim, we construct a subset of relational space (10.9) as a conjunction of (10.10), (10.11), (10.13) with (10.21), (10.22). This subset is a solid object (tetrahedral structure on Fig.10.1) that contains all feasible actions. The derived set constitutes a global

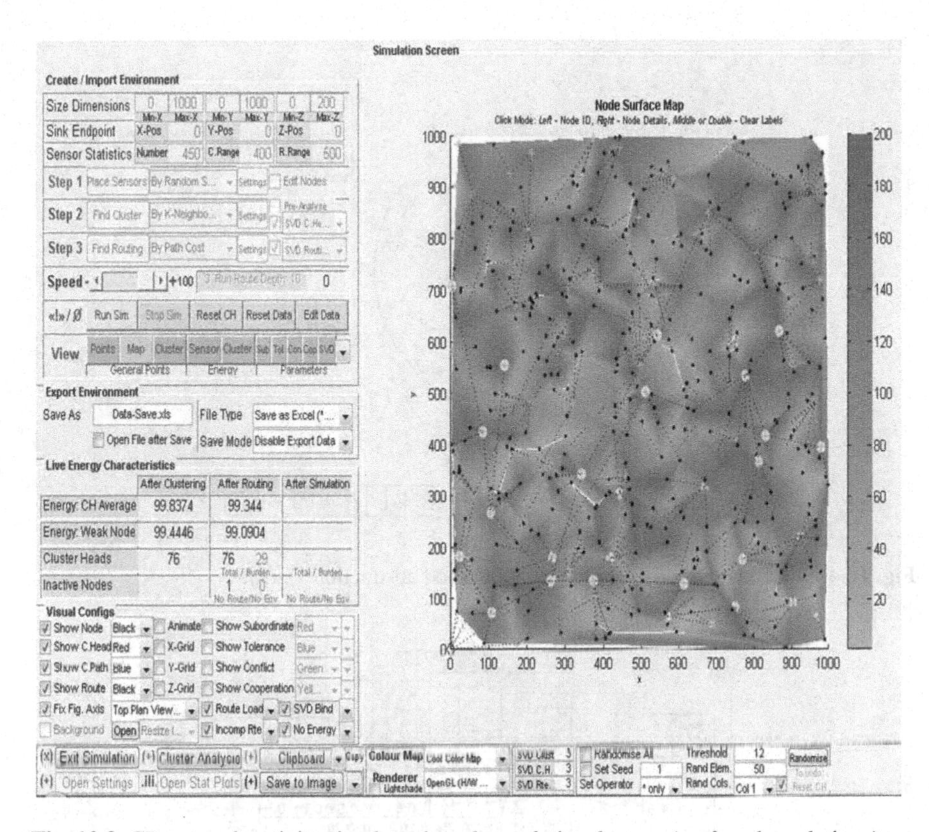

Fig. 10.2 CR network activity simulator based on relational space (π, ϑ and $\varkappa$ relations).

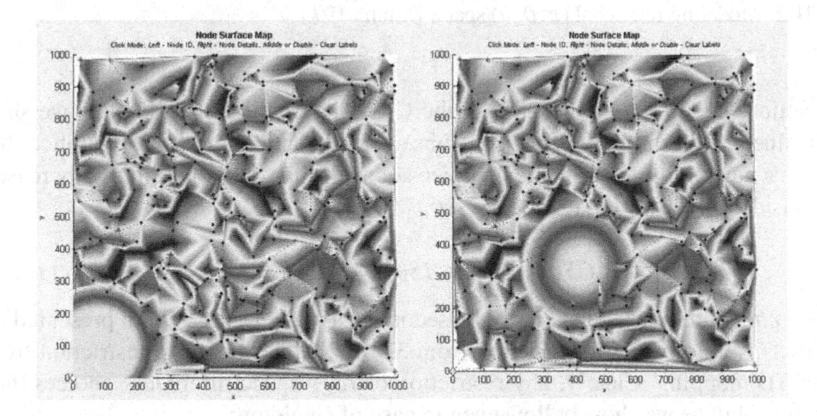

Fig. 10.3 Simulation of CR activity - setting a balance of π, ϑ and $\varkappa$ relations in $\mathcal{N}(k)$.

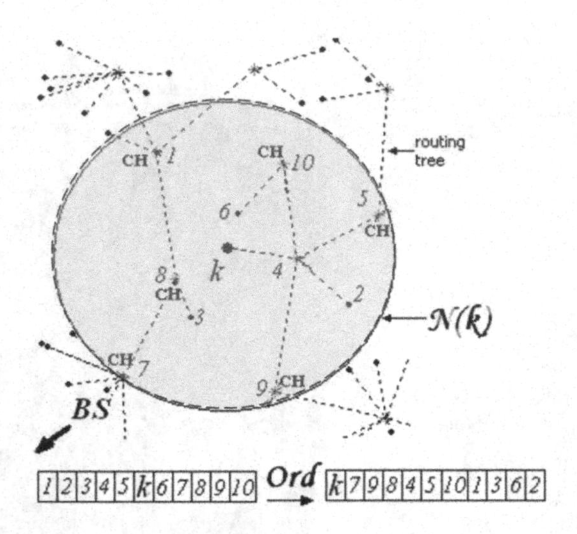

Fig. 10.4 Modeling partially ordered neighborhood abstraction $\mathcal{N}(k)$.

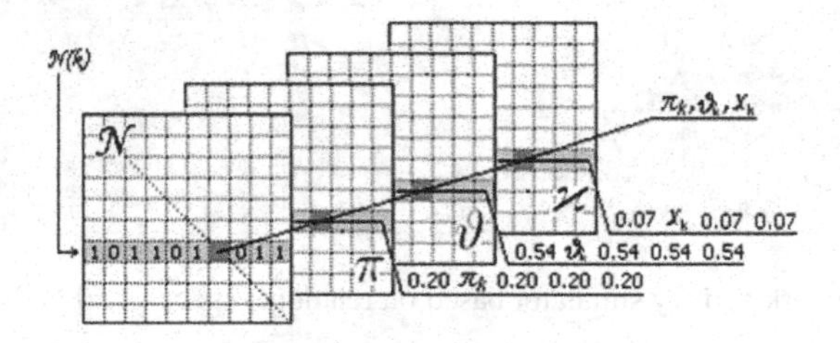

Fig. 10.5 Modeling relational $(\pi,\vartheta,\varkappa)$ space within $\mathcal{N}(k)$.

interaction space IS_{CR} (10.18) within the Cognitive Radio network. Now we shall determine global strategy GS_{CR} as a subset of interaction space IS_{CR}. Notice that there is a huge number of different choices of such subset, but for simplicity reason we consider only two. First is the subset:

$$GS^1_{WSN} = \{y \in IS_{WSN} \mid y \in abc\}, \tag{10.23}$$

where, *abc* represents a black triangle, second is a white point Z (both presented on Fig.10.1). While first is a restriction from 3D to 2D, the second is restriction from 3D to 1D mapping. Clearly, any restriction of IS_{CR} space offers less choices then the whole, but as we show bellow even in case of singleton:

$$\{Z\} = \{< \pi, \vartheta, \varkappa >\} = \{< 0.2, 0.54, 0.07 >\} = GS^2_{WSN}, \tag{10.24}$$

the spectrum of possible activities is rather wide. In the following step we shall determine local strategy for each neighborhood $\mathcal{N}(k)$,k=1, 2.., 10. In order to deal with the case where the global strategy is determined by (10.24) the local strategy for neighborhood activity need to be identical. Hence for each node k its interaction space IS_k is a singleton $Z = < \pi_k, \vartheta_k, \varkappa_k >$ exactly the same like (10.24) and the real numbers $\pi_k, \vartheta_k, \varkappa_k$ occupied main diagonal cells of relational matrices $\pi, \vartheta, \varkappa$ respectively (Fig.10.5).

An adequate level of local activity (within neighborhood $\mathcal{N}(k)$) can be accomplished in accordance with the global strategy. Therefore, instead of forcing the local activity at each node, we are rather formulating requirements for the nature and intensity of relations within $\mathcal{N}(k)$). The identification and actual fulfillment of requirements result in desired global strategy. In the last step of preparing the simulation process we shall determine non-diagonal elements of relational matrices (Fig.10.5). In order to facilitate the representation of nodes interactions (point-to-point) we assign them a real numbers exactly the same like on diagonal. The starting point of simulation determined in such way is feasible of course (holds both local and global strategies).

The relational state for any $\mathcal{N}(k)$ is now identified with ordered sequence of node indexes as follows:

$$< \pi_k, \vartheta_k, \varkappa_k >= ceiling(< 0.2, 0.54, 0.07 >) =< 2, 6, 1 > . \qquad (10.25)$$

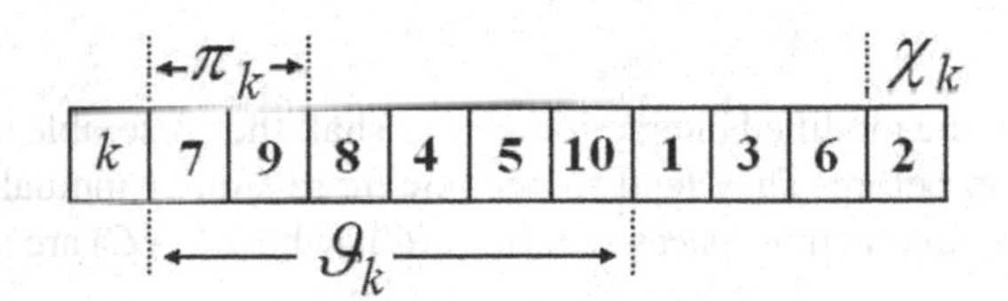

Fig. 10.6 Neighborhood vector for $\mathcal{N}(k)$ space.

Considering that a given Cognitive Radio consists of 10 network elements, we say there is ten different ordered sequences, which all together constitute the CR relational State. In each iteration, we choose randomly both the sender s (node sending information to BS) and a position in the relational space $\mathcal{N}(s)$. It determines the next-hop path forwards to BS (the receiver node) and type of relation describing this hop. In the following, for chosen receiver r, we will make a decision taking into account relational State for $sender - receiver$ interaction (cells $\pi_{s,r}$, $\vartheta_{s,r}$, $\varkappa_{s,r}$). According to the obtained results we needed to modify the intensity quotients for sender-receiver interaction and repeat this process until information reaches the base station. Modeling CR network lifetime activity we modify repeatedly matrices π, ϑ, $\varkappa$ but diagonal elements remains the same.

It is apparent that the spatial distribution of intensity quotients fulfills global requirements, but is has captured the essential local relationships within any neighborhood. The method meets other important requirements for immunity and adaptability. The interaction space within $\mathcal{N}(k)$ shall be modified only in the local vicinity of each

Algorithm 10.1 Rules of component wiring

```
configuration Security_Threats { }
implementation {
  components Main, Security_Threats, TimerC, LedsC, PhotoTemp, GenericComm as Comm,
             XMeshNodeM, MULTIHOPROUTER;
    Main.StdControl -> TimerC.StdControl;
    Main.StdControl -> Security_ThreatsM.StdControl;
    Main.StdControl -> Comm.Control;
    Main.StdControl -> XMeshNodeM.StdControl;
    Main.StdControl -> MULTIHOPROUTER.StdControl;

    Security_Threats.Timer -> TimerC.Timer[unique("Timer1")];
    Security_Threats.TimerRouting -> TimerC.Timer[unique("Timer2")];
    Security_Threats.TimerSecurity -> TimerC.Timer[unique("Timer3")];
    Security_Threats.Leds -> LedsC.Leds;
    Security_Threats.MhopSend -> MULTIHOPROUTER.MhopSend[10];
    Security_Threats.health_packet -> MULTIHOPROUTER;

    Security_Threats.SendNeighbor -> Comm.SendMsg[AM_XSXMSG];
    Security_Threats.ReceiveNeighbor->Comm.ReceiveMsg[AM_XSXMSG];

    Security_Threats.SendNetwork -> Comm.SendMsg[AM_XSXCMD];
    Security_Threats.ReceiveNetwork->Comm.ReceiveMsg[AM_XSXCMD];
    ...
}
```

node k and that all the modified interaction spaces shall then resemble the prevailing activity better than before. They tend to become more similar mutually i.e. differences between any interaction spaces within $\mathcal{N}(C)$ (where $k \in C$) are smoothed.

10.4.2 Implementation of Relational Model for Crossbow's *XMesh* Platform

The proposed relational approach refers to the ability of the nodes to dynamically seek new routes for delivering packets when parts of the network go off-line due to security threats, radio interference or power heavy-working. In order to observe how relational approach expands routing and supports combating security threats in CR network, we extend *XMesh* platform. Using Crossbow's MoteWorks software platform we employ *XMesh* procedures which routes data from nodes to a base station (upstream) or downstream to individual nodes.

In our modification, each node additionally can also broadcast (Alg.3) within a neighborhood area of range. On the other side, a node continuously listens to radio traffic (Alg.2) in its neighborhood and collects an information to fill in a security evaluation matrix *SecurMatrix*. It estimates how secure it can use a neighbor as a re-transmitter.

In order to better understand how it works, we first present some fragments of *NesC* codes (sometimes, for clarity, we left type descriptors). Algorithm 1 defines rules of components wiring, interface and structures. Interfaces; *SendNeighbor* and

Algorithm 10.2 Interfaces and structures definition

```
module Security_ThreatsM {
    provides {
     interface StdControl;
    }
    uses {
     interface Timer;   interface Leds;
     interface Timer as TimerRouting;   interface Timer as TimerSecurity;
     interface SendNeighbor;   interface ReceiveNeighbor;
     interface SendNetwork;   interface ReceiveNetwork;
     interface MhopSend;
     ...
     }
    }
implementation {
    uint8_t NeighMatrix[1][21], SecurMatrix[21][21];
    typedef struct NodeData {
     uint16_t vref, thermistor, light, mic, accelX, accelY, magX, magY;
     uint16_t Spare[12];
    }__attribute__ ((packed)) NodeData;
    typedef struct NeighborsData {
     uint8_t NodeNo, NeighborsAmount, NeighborsEvalData[40],
    } __attribute__ ((packed)) NeighborsData;
    typedef struct XSensorHeader{
     uint8_t board_id, packet_id, node_id, rsvd;
    } __attribute__ ((packed)) XSensorHeader;
    typedef struct XDataMsg {
     XSensorHeader xSensorHeader;
     union {
       NodeData NodeDat;   NeighborsData NeighEval;
     }xData;
    } __attribute__ ((packed)) XDataMsg;
    ...
}
```

ReceiveNeighbor are most important due to local (within neighborhood) communication/broadcasting while *NeighMatrix* and *SecurMatrix* are crucial for neighborhood description and neighbors evaluation. Due to the *TinyOS* packet size each route update message will contain at most 5 neighbors, but we extend the neighborhood's size up to 20.

10.5 Conclusion

The primary goal of this chapter was to provide an overview of available CR models, threats and their mitigation patterns that fit the category of CR modeling and

Algorithm 10.3 Receiving evaluation data from neighbor

```
eventTOS_MsgPtr ReceiveNeighbor.receive(TOS_MsgPtr m) {
  ...
  if (received_neigh_packet==TRUE) break;
  atomic {
    neigh_buffer_rec = *m; received_neigh_packet = TRUE;
  }
  if (LedsSignalCommunication) call Leds.greenOn();
  temp = neigh_buffer_rec -> xData. NeighEval.NodeNo;
  done = FALSE;
  for(j=1; j<20; j++){                    // there is my NodeNo in [0]
   if(SecurMatrix[0][j]==temp){
     temp1 = neigh_buffer_rec -> xData. NeighEval.NeighborsAmount;
     for(k=0;k<temp1-1 ; k++){
       temp2 = neigh_buffer_rec -> xData.NeighEval.NeighborsEvalData[2*k];
       for(i=0; i<20 ; i++){
         if (SecurMatrix[0][i]== temp2){
           temp3=neigh_buffer_rec -> xData. NeighEval.NeighborsEvalData[2*k+1]
           SecurMatrix[(i+1)][j]=temp3
         }
       }
     }
     done = TRUE;
   }
   if (done) return m;
  }
  return m;
}
```

integration facilitators. The vision of future demands for CR network that is characterized by immunity,adaptability, scalability and extensibility was also presented.

Dependability of a complex distributed system closely relies on its immunity and adaptation abilities for information exchange. The immunity and adaptability of communication channels are essential to our approach. The proposed novel relational method, allows reconciling two, often dichotomous, points of view: immunity and adaptability to neighborhood. Thanks to the global network strategy, each node has an adequate level of immunity and adaptation functions that guarantee sufficient communication services. Additionally, nodes using these resources provide a suitable level of the CR adaptability and immunity towards the desired common network security level. Management of complex system in such environment yields in growing both network adaptability and immunity. By modeling Cognitive Radio network activities using relational approach we have managed to accurately describe the complex characteristics of the network interactions, whilst at the same time we have eliminated many different and distributed variables (the Cognitive Radio network parameters).

Algorithm 10.4 Broadcasting evaluation data within neighborhood

```
bool broadcastEval(TOS_Msg *EvalToSend, bool *SendFlag){
  ...
  XDataMsg *pack;
  if (sending_neigh_packet==TRUE) return FALSE;
  if (LedsSignalCommunication) call Leds.redOn();
  atomic {
    sending_packet = TRUE;  EvalToSendFlag = SendFlag;
    pack = (XDataMsg *)(EvalToSend->data);
  }
  pack->xSensorHeader.rsvd == 0xFF;
  pack->xSensorHeader.node_id = TOS_LOCAL_ADDRESS;
  pack->xData. NeighEval. NodeNo = TOS_LOCAL_ADDRESS;
  k=0;
  for (j=1; j<20; j++){
    if (SecurMatrix[0][j]==0xFF) break;
    pack->xData. NeighEval. NeighborsEvalData[(2*k)] = SecurMatrix[0][j];
    pack->xData. NeighEval. NeighborsEvalData[(2*k)+1] = SecurMatrix[j][0];
    k=k+1;
  }
  pack->xData. NeighEval. NeighborsAmount =k;
  call SendNeighbor.send(TOS_BCAST_ADDR, sizeof(XDataMsg),&EvalToSend);
  return TRUE;
}
```

This reduction can be seen as crucial for system simulation. Hence, with the presented relational approach for modeling immunity and adaptation functions we are able to scale the complexity of interactions and model with much higher precision various behavioral aspects of Cognitive Radio networks.

The main challenge with this project was adapting the *XMesh* framework to model CR interactions and making the software to work in a completely new way. We have encountered quite a few difficulties trying to adapt it and then operate it. However, overall we are satisfied with what was achieved in our experimentation, particularly since *XMesh* framework, Cognitive Radio and immunity to adaptivity functions and are not entirely compatible areas. We were able represent biomimetic relations in CR to assess threats and plan mitigation strategies to counter-measure various threats.

Acknowledgements. This chapter has been partially supported by the project entitled: Detectors and sensors for measuring factors hazardous to environment modeling and monitoring of threats. The project financed by the European Union via the European Regional Development Fund and the Polish State budget, within the framework of the Operational Programme Innovative Economy 2007-2013. The contract for refinancing No. POIG.01.03.01-02-002/08-00.

References

1. de Castro, L.N., Timmis, J.: Artificial Immune Systems: A new computational intelligence approach. Springer, Heidelberg (2002)
2. Chaczko, Z.: Towards epistemic autonomy in adaptive biomimetic middleware for cooperative sensornets. Ph.D. thesis, University of Technology Sydney (2009)
3. Clancy, T., Goergen, N.: Security in cognitive radio networks: Threats and mitigation. In: Third International Conference on Cognitive Radio Oriented Wireless Networks and Communications (2008)
4. Dasgupta, D.: Immunity-based intrusion detection system: A general framework. In: Proceedings of the 22th International Information System Security Conference (1999)
5. Hofmeyr, S.: Architecture for an artificial immune system. Evolutionary Computation 8(3), 443–473 (2000)
6. ICT Regulation Toolkit: What is spectrum trading. Tech. rep., (2010), `http://www.ictregulationtoolkit.org`
7. Jaroń, J.: Systemic prolegomena to theoretical cybernetics. In: Scientific Papers of Institute of Technical Cybernetics, vol. (45). Wroclaw University of Technology (1978)
8. Jungwon, K., Bentley, P.: Towards an artificial immune system for network intrusion detection: an investigation of clonal selection with a negative selection operator. In: Proceedings of the Congress on Evolutionary Computation (2001)
9. Kohonen, T.: Self Organizing Maps. Springer, Heidelberg (2001)
10. Luther, K., Bye, R., Alpcan, T., Muller, A., Albayrak, S.A.: Cooperative ais framework for intrusion detection. In: Proceedings of the IEEE International Conference on Communications ICC 2007 (2007)
11. Mitola, J.: Cognitive radio: An integrated agent architecture for software defined radio. Ph.D. thesis, KTH Royal Institute of Technology (2000)
12. Nikodem, J.: Autonomy and cooperation as factors of dependability in wireless sensor network. In: Proceedings of the Conference in Dependability of Computer Systems (2008)
13. Timmis, J.I.: Artificial immune systems: A novel data analysis technique inspired by the immune network theory. Ph.D. thesis, University of Kent (2000)
14. Zhang, Y., Xu, G., Geng, X.: Security threats in cognitive radio networks. Tech. rep., Department of Computer Science and Technology. Jilin University, China (2008)

Chapter 11
Novel Modifications in WSN Network Design for Improved SNR and Reliability

Kamil Staniec and Grzegorz Debita

Abstract. The chapter addresses the issue of Wireless Sensor Network planning (in fact - well applicable to any mesh network design) for achieving a network structure of minimized intra-network interference and reliability. The former feature can be attained by means of directional antennas, which - in combination with a planar topology - leads to a considerable increase in the average SNIR level. The latter feature is to provide a structure of sensor nodes resistant to single-node failures, deployed with the minimum-pathloss criterion for better radio performance.

11.1 Introduction

Scattered wireless sensor networks (WSN) are a simplest example of mesh networks (Fig. 11.1). In such networks, however, there exists the same problems associated with creating topology as with broadband networks. For this reason the solution to the procedure of establishing a topology satisfying the reliability criterion, can be demonstrated with the use of any mesh network.

Scattered networks can be divided into two kinds with respect to the topology establishment procedures:

1. Networks with a PAN (Personal Area Network) coordinator. In this type of networks one element or a group of its elements are responsible for signaling control;
2. Fully scattered networks without a coordinator. In this type of networks all network elements possess information on the network state and store their own packet routing tables.

Kamil Staniec · Grzegorz Debita
Wroclaw University of Technology, Faculty of Electronics, Wroclaw, Poland
e-mail: {kamil.staniec,grzegorz.debita}@pwr.wroc.pl

J. Fodor et al. (Eds.): Recent Advances in Intelligent Engineering Systems, SCI 378, pp. 243–259.
springerlink.com

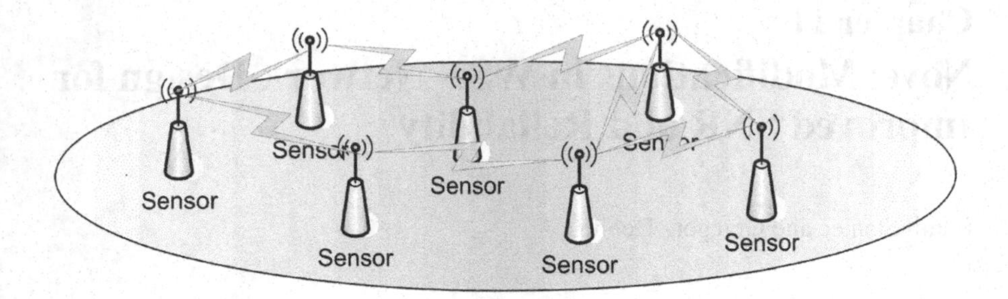

Fig. 11.1 Sensor nodes connected in a scattered - mesh network

11.2 Sensor Networks with a Coordinator

Sensor networks with a coordinator are characterized by having at least one of their elements held accountable for managing all the other nodes. The coordinator has an assigned task to manage the network in such a manner that all the nodes are interconnected, either directly or indirectly, to form a coherent network. The method of creating a topology by the coordinator will be referred to in the rest of this chapter as the network spanning. Creating a network with the use of a coordinator has many common features irrespective of the system that utilizes it. In the current work investigations concern the ZigBee (IEEE 802.15.4) system. The network spanning routing consists of two stages:

1. a broadcast stage (Fig. 11.2);
2. a distribution of the routing tables calculated by the coordinator (Fig. 11.3).

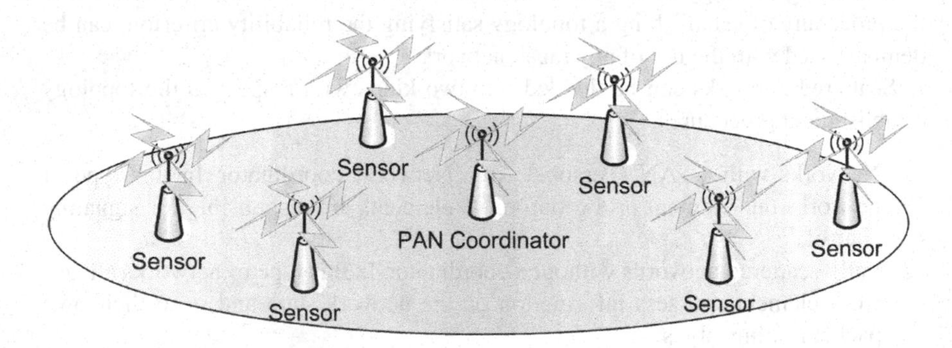

Fig. 11.2 Network topology formation stages - the broadcast phase

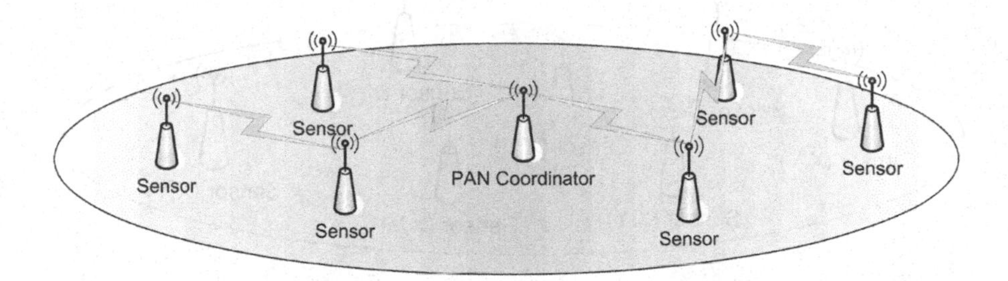

Fig. 11.3 Network topology formation stages - the resultant (unreliable) network topology

11.2.1 Fully Scattered Networks without Coordinator

The topology-creating process in the fully scattered networks consist of a few following steps. It is assumed the position (location) of each node in the initial network (Fig. 11.4) is known; be it due to the intentionally placing them in predefined locations (in which a factor to be sensed is expected to appear) or because the particular node has distributed information on its localization to other nodes.

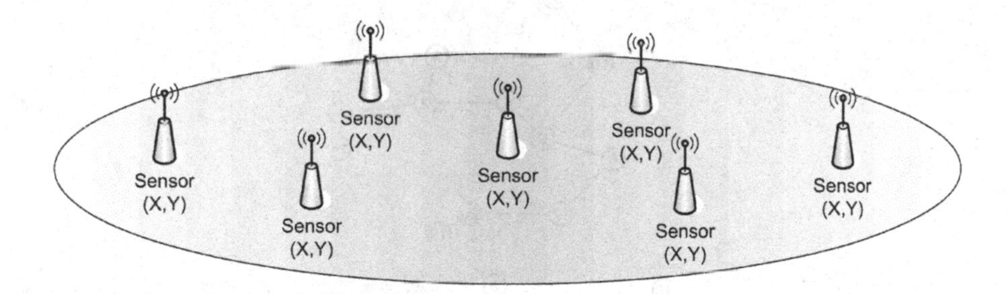

Fig. 11.4 Sensors deployment phase

In either case the exact positions of each node is finally known throughout the network. For each node a network designer (a network management entity) determines a routing table in such a way that at least two alternative paths exist for each node, to account for possible node failures along the transmission route (Fig. 11.5). Moreover, all nodes implement mechanisms of accommodating new nodes (provided that the latter are also aware of their localization) and routing/topology reorganization in the case of failure of one the network nodes.

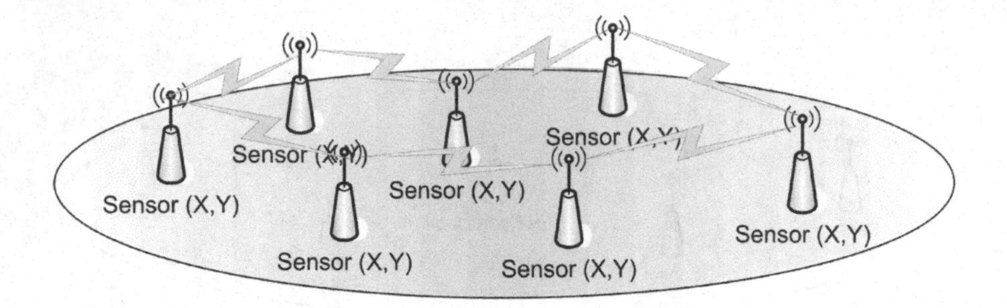

Fig. 11.5 Wireless sensor network formation stage

11.3 Neighbor-Searching Routine in WSN

This chapter will cover some of the basic aspects of mathematical apparatus involved in the wireless sensor network spanning process. An algorithm will be described with indication of how it accounts for the nature of the physical medium which these links exist in.

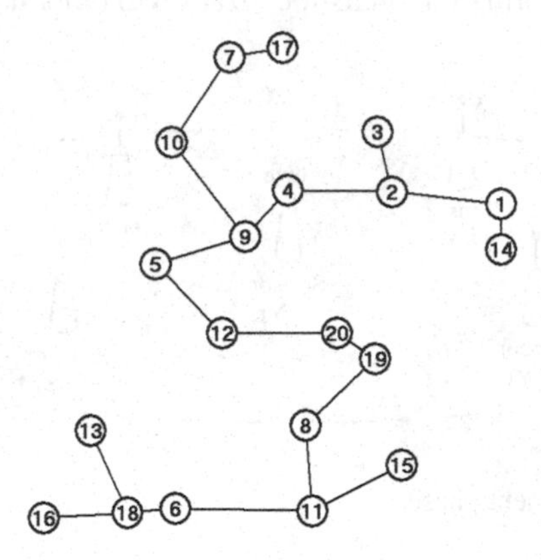

Fig. 11.6 A WSN formed with a Stojmenovic algorithm at the listen window set to basic

Out of multiple concepts of modeling and building the WSN topology some of the most popular ones are those described by Stojmenovič [1], initially intended for military applications. The Minimum Spanning Tree, MST, algorithm was chosen as the most relevant to WSN design due to its outstanding energetic efficiency (a strategic parameter in wireless sensor networks). The MST algorithm starts with the PAN coordinator (refer to [1]) broadcasting a beacon message announcing a new network formation, which is to announce that a new network is about to be created. Soon after this broadcast the PAN coordinator enters the listening mode to hear incoming response calls from nodes within the radio audible range. The successfully received calls are being collected for the duration of the, so called, listening window, which depending on the manufacturer lasts for a few up to several seconds (values experimentally measured by the authors with the use of the Daintree Sensor Network Analyzer Software 100A, Prof. Edition a commercial software for ZigBee networks analysis).

Now, after forming connections to the PAN coordinator each connected node performs the same searching routine to find nodes of its own neighbor set (further away from the PAN coordinator) and which have not connected to the PAN coordinator upon the primary broadcast. Those nodes that respond within the listen window are recruited as neighbors. The procedure then propagates along all the nodes on the area until eventually the minimum (global) spanning tree is formed (Fig 11.6). It is assumed that the nodes which respond the fastest will be those that lie the closest to the broadcasting node, thus they may be considered as the minimum-pathloss connections as compared to the calls that arrived after the listen window. Algorithms with a small (basic) window form an network initial structure. It is obvious that if the listen window is prolonged into a multiple of the basic listen window duration, more nodes will join in forming redundant connections. Based on this assumption, the topology-creating algorithms can be divide into two groups:

- with a small listen window;
- with a large listen window.

The multiple by which the basic listen window is to be extended is heuristic since neither distances between network nodes are known, nor are their geographical locations. In [1] it is shown that the listen window size may be extended up to three timc the basic size in order to achieve satisfactory level of redundancy (and the network reliability and fault tolerance). It is shown in Fig. 11.7 for the unchanged network scenario as in Fig. 11.6 with a doubled basic window.

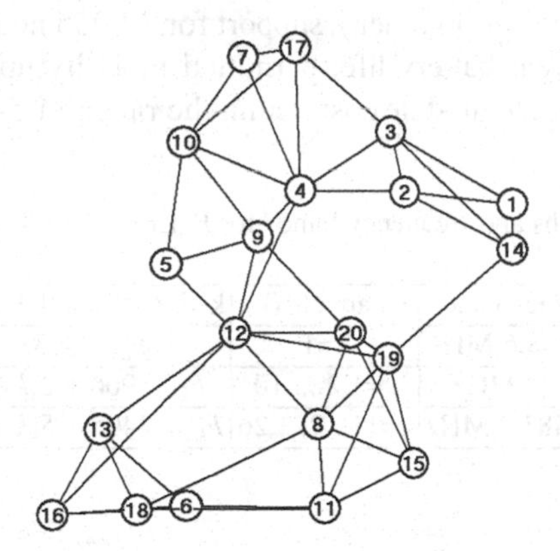

Fig. 11.7 A WSN formed with a Stojmenovic algorithm with the listen window set to extended duration

It should, however, be also noticed that there is a particular feature that may be potentially threatening. Namely, the algorithm enables connections that intersect other connections, e.g. pairs (1;2) and (3;14) in Fig. 11.7. In radio communications avoidance of such a situation is a strongly recommended practise. Another issue worth mentioning is that since now, with an extended listen window, most nodes have redundant connections, it is a purely random matter which of them will be chosen for transmission of a current packet. For example, the node 16 in Fig. 11.7 may choose to transmit via nodes 12, 13 or 18. The greatest transmit power would, of course, be necessary if the link 12 were selected and this may happen with a 33% probability since all links in the neighbor set are equally probable.

11.4 ZigBee System

The system under investigation is based on IEEE 802.15.4 specification [2] known under its commercial name as ZigBee and defining its physical and Medium Access Control (MAC) layers. It is an ad-hoc networking technology for low-rate Wireless Personal Area Networks, LR-WPAN intended for 2.45 GHz, 868 MHz and 915 MHz bands, characterized by being low in cost, complexity and power consumption as compared to competing technologies. It allows for data rates up to 250 kb/s for 2.45 GHz, 40 kb/s for 915 MHz and 20 kb/s for 868 MHz band which are speeds just adequate of transmissions of data from sensing or control modules. As a system it has been designed to respond to a few requirements defined for a new family of devices intended not for user data transmissions but rather for sending volume-limited readings from control and sensor simple appliances. These requirements included: global license free ISM band operation, unrestricted geographic use, RF penetration through walls and ceilings, automatic/semi-automatic installation, ability to add or remove devices, possible voice support. The just enumerated needs posed the following demands on the target system: 10 kb/s-115.2 kb/s of data throughput, 10-100 m coverage range (home, yard/garden), support for 32-255 nodes, 4-100 co-located networks, up to two- year battery life, permitted mobility up to 5 m/s (18 km/h) mobility and eventually the module cost within the range $1.5-$2.5 in 2004/5.

Table 11.1 Channel widths and frequency bands for IEEE 802.15.4 (ZigBee) system

Frequency band	Channels (k)	k-th channel definition
868.0-868.6 MHz	k=1	$F_k = 863.3$
902-928 MHz	k=1,2,...,10	$F_k = 906 + 2(k-1)$
2400-2483.5 MHz	k=11,12,...,26	$F_k = 2405 + 5(k-11)$

Depending on the chosen band, ZigBee offers from one up to 15 distinct channels defined as in tab. 11.1. Depending on the application requirements, an IEEE 802.15.4 LR-WPAN may operate in either of two topologies [2]: the star topology or the peer-to-peer topology. In the star topology communication is established between devices and a single central controller, called the PAN coordinator. The peer-to-peer topology also has a PAN coordinator; however, it differs from the star topology in that any device may communicate with any other device as long as they are in range of one another.

11.5 Interference Origin in WSN

As will be treated in detail in later sections, the peer-to-peer topology, beside offering excellent network scaling and reliability/fault tolerance advantages, is quite prone to generating mutual radio interference between network nodes (see [4],[5]). As for the reliability provision, ZigBee offers a built-in capability assuring that each node holds at least two other nodes as candidates for connections. As was shown in [3] this native fault-tolerance mechanism is not optimal from the radio interference-mitigation viewpoint since each time a connection is to be established it is picked at random from the neighbor set. Such a selection has an inherent weakness in that it does not favor the least-pathloss candidates.

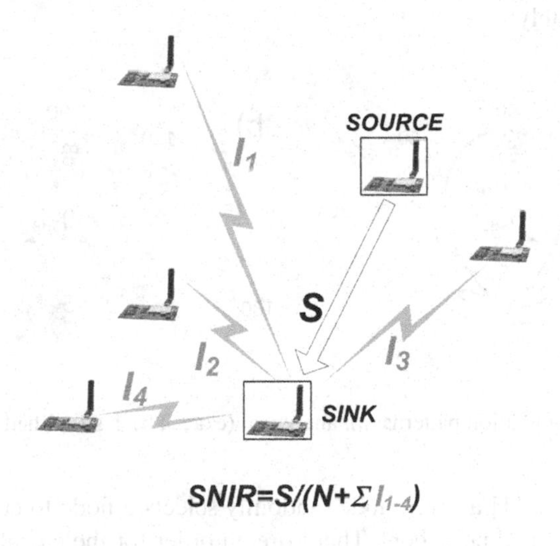

Fig. 11.8 A visualization of the signal-to-noise interference origin in WSN

A basic measure of the system performance is the Signal-to-Noise and Interference, SNIR, a factor defined by (11.1), where S is the desired signal power in [W], kB - Boltzmans constant in [J/K], T - ambient temperature in [K], BW - channel bandwidth in [Hz], NF - Noise Factor [-], M - the total number of network nodes, whereas the last term in the denominator denotes the sum of radiations from all other nodes in the network (excluding, of course, the radiation from the k-th node for which SNIR is being calculated and the l-th node to which the node k is transmitting, see Fig. 11.8). The desired signal power S as well as that of all interfering signals, can be found from one of the available pathloss models, e.g. the two-ray model which assumes up to some distance (a breakpoint) the power decays as $\sim 1/r^2$ and for further distances as $\sim 1/r^4$ (refer to [3]). The model is quite fit for low-positioned antennas in the absence of other reflectors than the ground - the received power is assumed to be the phasor sum of the direct and the reflected wave.

$$SINR_k[dB] = 10log(\frac{S}{N + \sum\limits_{M/(k,l)} I_m}) = 10log(\frac{S}{k_B \cdot T \cdot BW \cdot NF + \sum\limits_{M/(k,l)} I_m}) \quad (11.1)$$

11.6 Antenna Beamwidth Control

Based on the neighbor-searching procedure according to Stojmenovič algorithm, a simulator has been developed that, for a given number M of nodes (M varied from 10 to 90) randomly scattered over a square area of 500x500 m, calculates SNIR for each node separately.

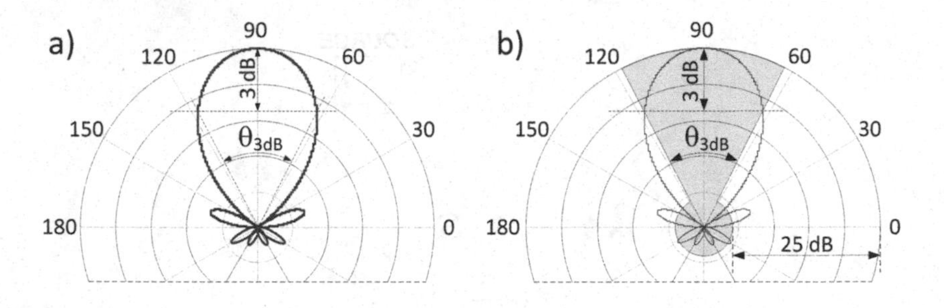

Fig. 11.9 Antenna radiation patterns: a). analytical (exact); b). a simplified model.

As was stated in [1] a given node randomly selects a node to connect to out of a set of predetermined neighbors. Therefore, in order for the calculations of SNIR to acquire statistical robustness, the final value of SNIR was averaged over 15 such neighbor selections. Such a procedure was to prevent SNIR from flickering due to different choices the nodes make at every instance.

As for the directional antennas, a simplified model of the Antenna Radiation Pattern, ARP, was used, as shown in Fig. 11.9. What will be considered as a beamwidth in [3] is in fact a 3-dB angle θ_{3dB} (a boresight) of the exact analytical ARP. An immediate expected advantage from the application of directional antennas is the reduction in interference observed from transmissions on angles other than the beamwidth as in Fig. 11.10. In this example out of the two interfering sessions, i.e. (3;4) and (5;6), only the first one will affect the desired (1;2) transmission. In real-life solutions, however, omnidirectional antennas are most commonly used in sensor

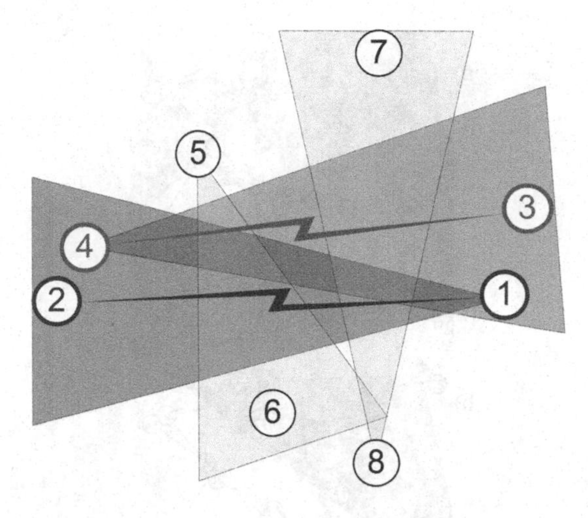

Fig. 11.10 An example scene of interference mitigation by means of directional antennas.

applications (particularly in ZigBee modules with a short monopole antenna). As an example of interference mitigation by means of directive antennas in radio modules, a visual example is given in Fig. 11.10 of how narrowing the beamwidth influences the number of effective interferers (i.e. those nodes which radiation pattern falls within the beamwidth window of the interfered node as in Fig. 11.10). In the example red ARPs are radiated from interferers; their number is clearly decreasing as the beamwidth narrows from 150° through 100° and 50°, respectively.

11.7 Results of Simulations

The results cited and discussed in this section (see also [6],[7]) have been obtained provided the network operates at 100% duty cycle which, in fact, constitutes the blackest of possible scenarios. All M nodes transmit simultaneously and hence pose an interfering threat to each other. While this assumption is obviously unrealistic to last for a longer time of the network functioning, one may imagine such discrete instances (rather than continuous long periods), where most if not all of the network nodes will transmit to their peers at the same time.

An example scenario is when a critical event is sensed over a large area (e.g. fire) and causes multiple nodes to transfer this information to the sink. It should be mentioned that while there exists a regulatory limit on the duty cycle for devices in 868 band (it cannot exceed 1% [8]), no such constraint exists in the analyzed 2.45 GHz ISM band.

The purpose of the investigations was to find a limiting bound on the directional antenna beamwidth that still allows to meet the minimum value of SNIR recommended by IEEE 802.15.4 specification i.e. 5-6 dB [2] (in the presented calculations the lower value was chosen as the threshold SNIRthr).

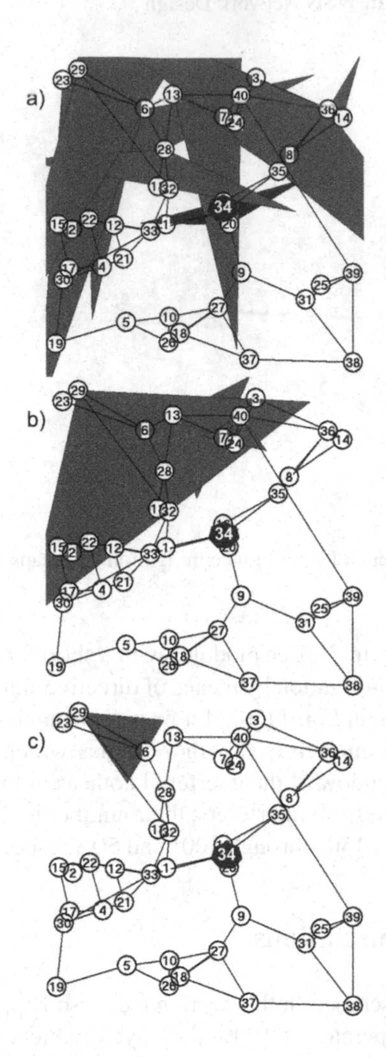

Fig. 11.11 A visualization of the number of Effective interferers experienced by the 34th node with the half-power angle of : a) 150°; b) 100°; c) 50°

Fig. 11.12,11.13 summarize the obtained results for the small and large window, respectively, of the Stojmenovič algorithm (refer to [1]). With the dashed line representing the 5 dB SNIR_{min} it can be directly observed that regardless of the considered algorithm version (i.e. with a small or large listen window), in the best case (with just ten nodes) the maximum beamwidth max assuring SNIR_{min} equals 180°. As can be also noticed min was most severely affected at the first increments in M: the receive beamwidth would have to be decreased by as much as 60° and 55°, respectively, for the small and large window, as the number of nodes rises from 10 to 20. Further increase of M (from 30 up to 100 nodes) resulted in min reduction by merely 35° and 30°, respectively. This conclusion is best demonstrated in 11.14.

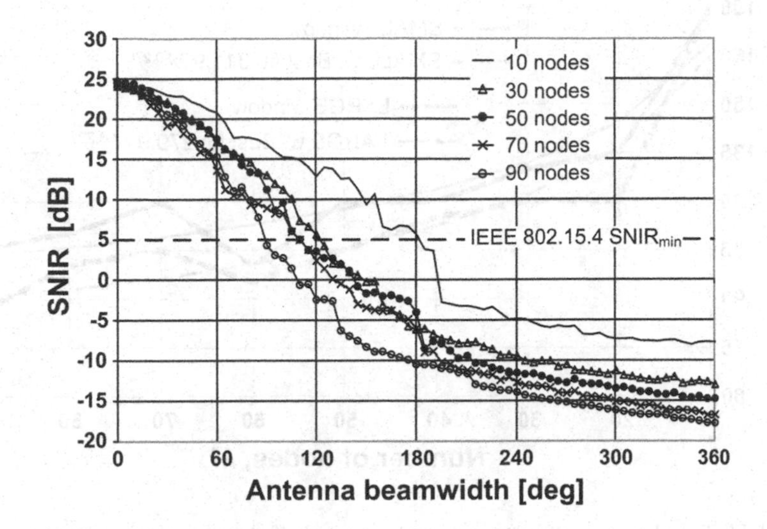

Fig. 11.12 Signal-to-noise/interference as a function of the half-power angle. Assumed "small window" in the MST algorithm.

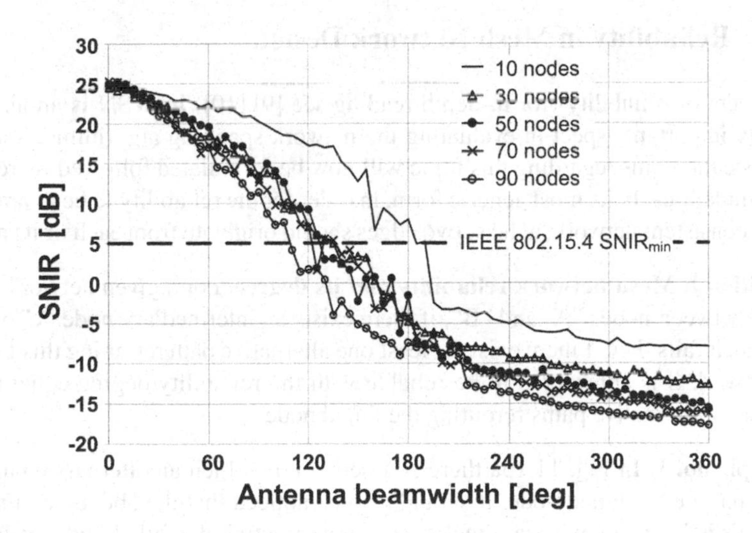

Fig. 11.13 Signal-to-noise/interference as a function of the half-power angle. Assumed "large window" in the MST algorithm.

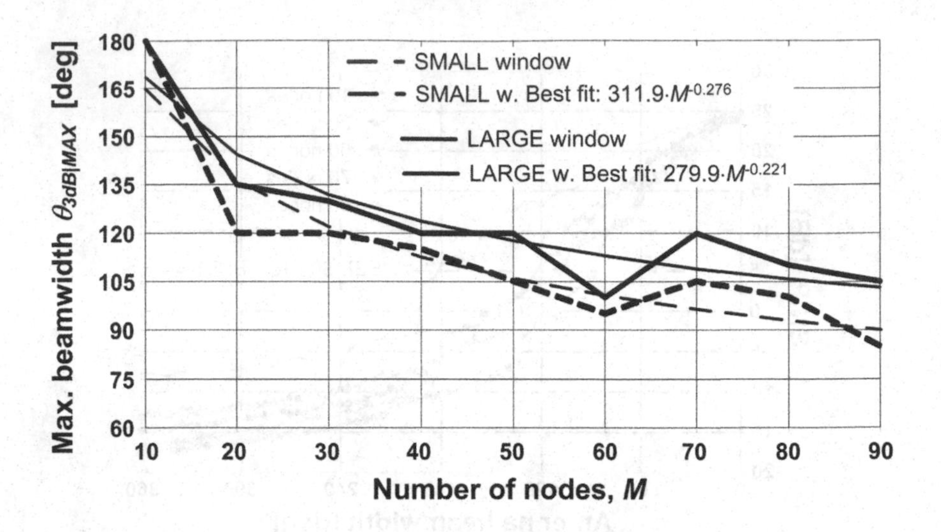

Fig. 11.14 The plot of a maximum required antenna half-power angle as a function of the number of nodes. M.

11.8 Reliability in Mesh Network Design

A concept of reliability (for in-depth reading see [9],[10]) in WSN is another extremely important aspect in evaluating the network spanning algorithms. Some of the basic theorems regarding this topic will now be formulated followed by respective simulations. In its most generic form, in order for the reliability to be guaranteed in any consistent network, at least two edges should originate from each of its nodes.

Definition I. Mesh network reliability and its degree. For a given network S and a flow between nodes "A" and "B", if there exists an intermediary node "C" on the path and it fails then if there exists at least one alternative path rerouting this failure, the network S is considered to be reliable with the reliability degree equal to the number of alternative paths rerouting the failed node.

Example no. 1. In Fig. 11.15a there is a network in which an alternative path exists to reroute the failed node. It should also be noticed that the above definition is also satisfied if there exists a direct connection rerouting the failed node, as in Fig. 11.15b.

It is crucial to remember that the reliability degree is not dependent of the number of outputs from the transmit node unless these are inputs to the failed node. Moreover, reliability is not determined by the number of intermediate nodes either. It should also be kept in mind that a particular node failure causes the damage of at least two network edges satisfying the reliability criterion cited in the above definition.

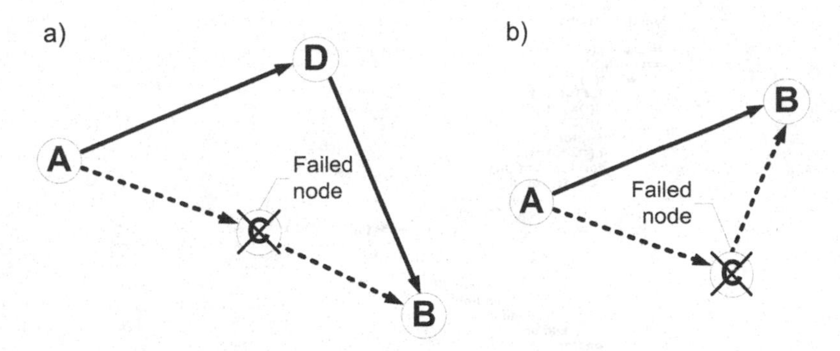

Fig. 11.15 An example network satisfying the reliability theorem

Taking the above considerations into account one may formulate a corollary that the reliability is determined by the value of the maximum flow between network nodes, assuming that the throughput between nodes equals unity (i.e. all network edges are assigned values of one).

Theorem of reliability. In a given network S all connections are assigned with unit values. Now let the maximum flow be calculated for all nodes (with any known method). If the flow value in all cases is greater than two the network is regarded as having at least two alternative routes in the case of failure of one of the intermediate nodes, whereby satisfying the reliability theorem.

The maximum flow value of a network dimensioned in a manner just described will be referred to as the network reliability degree.

Example no. 2. An example realization of the reliability theorem can be easily implemented in the form of an OCTAVE script in Fig. 11.16.

```
function [F]=reliability(network)
[n,k]=size(siec);                          % Counts the number of nodes
        for i=(1:n)                        % For all nodes
            for j=(1:k)
                if i!=j
                [f]=maxflow(network,i,j); % Calculates the maximum flow
                F(i,j)=sum(f(i,:));        % Write the maximum flow value
                else

                F(i,j)=0;                  % Insert zeros on the matrix diagonal

                end
            end
        end
    end
```

Fig. 11.16 OCTAVE implementation of the reliability theorem

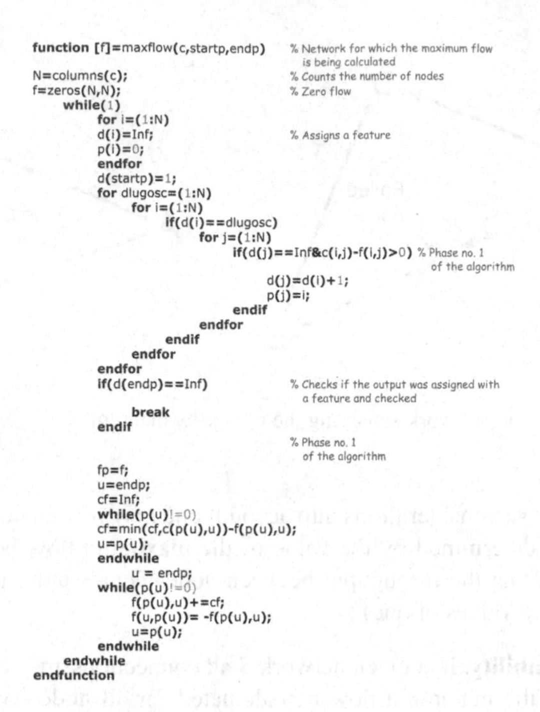

```
function [f]=maxflow(c,startp,endp)   % Network for which the maximum flow
                                        is being calculated
N=columns(c);                         % Counts the number of nodes
f=zeros(N,N);                         % Zero flow
    while(1)
        for i=(1:N)
        d(i)=Inf;                     % Assigns a feature
        p(i)=0;
        endfor
        d(startp)=1;
        for dlugosc=(1:N)
            for i=(1:N)
                if(d(i)==dlugosc)
                    for j=(1:N)
                        if(d(j)==Inf&c(i,j)-f(i,j)>0) % Phase no. 1
                                                        of the algorithm
                            d(j)=d(i)+1;
                            p(j)=i;
                        endif
                    endfor
                endif
            endfor
        endfor
        if(d(endp)==Inf)              % Checks if the output was assigned with
                                        a feature and checked
            break
        endif
                                      % Phase no. 1
                                        of the algorithm
        fp=f;
        u=endp;
        cf=Inf;
        while(p(u)!=0)
        cf=min(cf,c(p(u),u))-f(p(u),u);
        u=p(u);
        endwhile
            u = endp;
        while(p(u)!=0)
            f(p(u),u)+=cf;
            f(u,p(u))= -f(p(u),u);
            u=p(u);
        endwhile
    endwhile
endfunction
```

Fig. 11.17 OCTAVE implementation of the maximum flow algorithm

The procedure is to be performed with the following settings:

- for nodes incidental with each other the transition (connections) with a unitary weight is assumed;
- for the remaining nodes the transition weight is equal zero provided that no node is incident with itself.

The maxflow procedure can be implemented in OCTAVE as an algorithm for calculating the maximum flow in a fashion shown in Fig. 11.17.

A number of computations have been carried out for different scenarios in order to verify whether the proposed topologies meet the assumption of the network reliability. By definition it is known that topologies based on MST or LMST do not satisfy this condition in the small window mode since only structures with increased number of connections can be seriously considered for fulfillment of the above reliability conditions. Therefore the authors have also verified whether despite increased number of outputs from WSN nodes in the large window mode the redundancy will be also increased to the point that the reliability criterion is satisfied. The scenarios studied here are the same as those for which the interference analyses were performed in Section 6. As a result it turned out the increase in the listening window size does not always assure connection redundancy in the event of failure of one node. Some sample results are presented in Fig. 11.18 and Fig. 11.19.

It turns out that both methods do not always span the network that has two alternative paths for each connection. It is also evident that there may exist a bottleneck effect at some parts of the network, which is caused by the nature of MST and LMST algorithms (that despite different spanning principle yield quite similar final effects). conditions do not allow for a uniform RSSI distribution across the network.

It is easy to notice that in Stojmenovič case this effect appears between nodes 40 and 39 whereas in ZigBee case - between nodes 21 and 12 (Fig. 11.18). It occurs because both algorithms are based on the Received Signal Strength Indicator (RSSI) in the network. The examples show that the algorithms tends to work well for dense networks. However, they reveal some deficiencies if the number of nodes at a given area is low and the propagation.

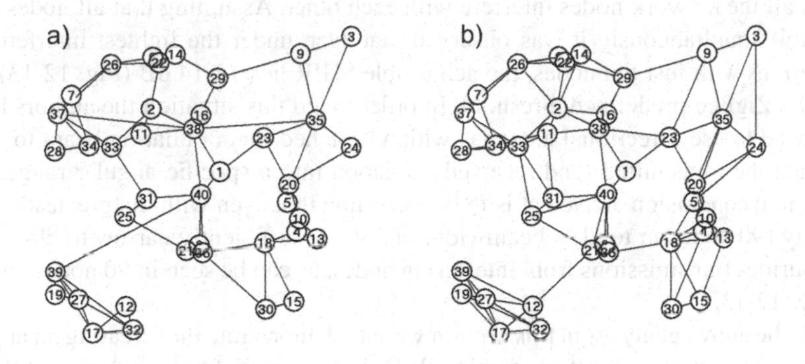

Fig. 11.18 A mesh network spanned with (a) Stojmenovič MST and (b) ZigBee LMST method. Both not satisfying the reliability criterion - example connections

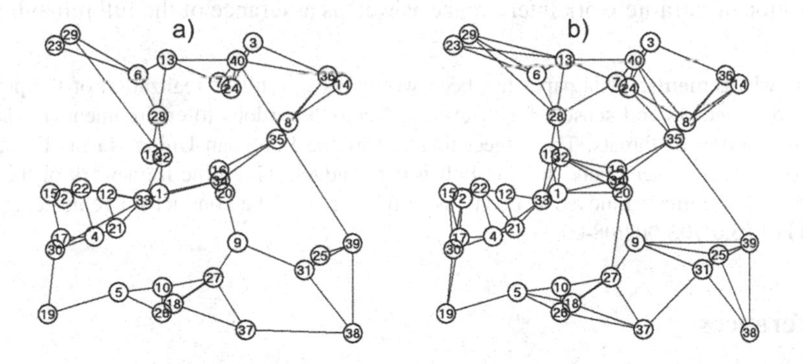

Fig. 11.19 A mesh network spanned with (a) Stojmenovič MST and (b) ZigBee LMST method. Both not satisfying the reliability criterion - example connections

11.9 Conclusion

The chapter presents simulated results of a wireless sensor network operating in 2.45 GHz ISM band in an interference-intensive environment. It has been demonstrated that the major source of this susceptibility to interference lies, on one hand, in the imperfect neighbor-searching algorithm which does not account for radio interference issues by allowing the existence of intersecting links. On the other hand, connections are determined with negligence of the minimum-pathloss criterion. Both issues are responsible for the achievable values of SNIR far below the ZigBee-defined minimum in the generated worst-case scenarios. The authors have developed a simulating tool for calculating ZigBee performance as a function of the number of nodes to verify how the average SNIR on the network would behave when all the network nodes interfere with each other. Assuming that all nodes will transmit simultaneously it was observed that even under the lightest interference conditions with just ten nodes, the achievable SNIR lies 12-14 dB (Fig. 12-13) below the ZigBee predefined threshold. In order to aid this situation the authors have proposed to use directional antennas with a switched main radiation beam to concentrate the transmitted (and received) radiation into a specific angular range. An important conclusion arrived at is the observation that even with the greatest nodes density (90 nodes in total) a beamwidth of 180° is sufficiently narrow to eliminate all spurious transmissions from interfering nodes, as can be seen in 90 nodes curves in Fig. 12-13.

All the above analyses in this section were to demonstrate that creating an appropriate redundancy in a self-organizing WSN is not a trivial task. It turns out that a mere knowledge of the electromagnetic power level received from neighbors is not a sufficient indicator. The factor that could aid the situation is the information of the nodes locations and, perhaps, some support from a network designer (which will, however, limit the self-organizing effect). The simulations presented in the paper show that self-organizing networks still leave a lot to be solved as concerns the minimization of intra-network interference as well as assurance of the full reliability.

Acknowledgements. This paper has been written as a result of realization of the project entitled: Detectors and sensors for measuring factors hazardous to environment modeling and monitoring of threats. The project financed by the European Union via the European Regional Development Fund and the Polish state budget, within the framework of the Operational Programme Innovative Economy 2007 ÷ 2013. The contract for refinancing No. POIG.01.03.01-02-002/08-00.

References

[1] Stojmenovič: Handbook of Sensor Networks - Algorithms and Architectures. Wiley and Sons, Ottawa (2004)

[2] IEEE, IEEE Std 802.15.4-2006, Part 15.4: Wireless Medium Access Control (MAC) and Physical Layer (PHY) Specifications for Low-Rate Wireless Personal Area networks (WP ANs)

[3] Debita, G., Staniec, K.: Reliable mesh network planning with minimization of intra-system interference. In: Proc. of Broadband Communication, Information Technology & Biomedical Applications, BROADBANDCOM 2009, Wroclaw, Poland, July 15-18, pp. 243–244 (2009)
[4] Xu, W., Trappe, W., Zhang, Y.: Defending Wireless Sensor Networks from Radio Interference through Channel Adaptation. Proceedings of ACM Trans. on Sensor Network 4(4), article 18, 18.1–18.34 (2008)
[5] Vakil, S., Liang, B.: Balancing cooperation and interference in wireless sensor networks. In: Proceedings of 3rd IEEE Annual IEEE Communications Society Conference on Sensor, Mesh and Ad Hoc Communications and Networks (SECON), vol. 1, pp. 198–206 (2006)
[6] Staniec, K., Debita, G.: Interference mitigation in WSN by means of directional antennas and duty cycle control. In: Wireless Communications and Mobile Computing; doi: 10.1002/wcm.1089
[7] Staniec, K., Debita, G.: Methodology of the wireless networks planning in mesh topology. In: Proc. Third International Conference on Dependability of Computer Systems, DepCos-RELCOMEX 2008, Szklarska Poreba, Poland, June 26-28, pp. 375–382 (2008)
[8] CEPT, ERC Recommendation 70-03 Relating to the Use of Short Range Devices (SRD), Annex 1: Non-specific Short Range Devices. Version of (October 16, 2009)
[9] Frank, H., Frisch, I.: Analysis and design of survivable networks. IEEE Transactions on Communication Technology 18(5), 501–518 (1970)
[10] Ford, L.R., Fulkerson, D.R.: Maximal flow through a network. Canadian Journal of Mathematics 8, 399–404 (1956)

Chapter 12
A Conceptual Framework for the Design of Audio Based Cognitive Infocommunication Channels

Ádám Csapó and Péter Baranyi

Abstract. Cognitive infocommunication channels are abstract channels which use a combination of sensory substitution and sensorimotor extension to convey structured information via any number of sensory modalities. Our goal is to develop engineering systems which are capable of using cognitive infocommunication channels in order to convey feedback information in novel and meaningful ways. Such applications could help reduce the cognitive load experienced by the user on the one hand, and help alleviate the undesirable effects of hidden parameters on the other. We describe the main challenge behind the development of cognitive infocommunication channels as a two-part problem which consists of the design of a synthesis algorithm and the design of a parameter-generating function for the synthesis algorithm. We use formal concept algebra to describe the kinds of synthesis algorithms which are capable of reflecting realistic forms of interaction between the user and the information which is to be communicated. Using this model, we describe an application example in which auditory signals are used to convey information on tactile percepts. Through an experimental evaluation of the application, we demonstrate that our approach can be used successfully for the design of cognitive infocommunication channels.

12.1 Introduction

Feedback devices facilitate teleoperation because they help reproduce events which occur in the slave environment on the master side. Instead of exclusively using our feedback devices to try to synthesize replicas of remote realities, perhaps better

Ádám Csapó
MTA SZTAKI, Kende u. 11-13, 1111 Budapest
e-mail: csapo.adam@gmail.com

Péter Baranyi
Budapest Univ. of Technology and Economics, Egry u. 18, 1111 Budapest
e-mail: baranyi@sztaki.hu

J. Fodor et al. (Eds.): Recent Advances in Intelligent Engineering Systems, SCI 378, pp. 261–281.
springerlink.com

results can be achieved if we rely on the plasticity of the human brain. Sensory substitution involves conveying data through sensory modalities (the *substituting modalities*) that are different from the ones normally used in the given application (the *substituted modalities*). If the substituted information is sufficiently different from the kinds of information afforded by the substituting modality, it can be argued that a completely new sensory modality emerges, and we may speak of sensorimotor extension [2]. There are considerable number of studies in the scientific literature which demonstrate that various biological sensory systems are amenable to successful sensory substitution and sensorimotor extension [2, 18, 19, 1, 3].

From an engineering point of view, a system which uses alternative forms of communication can have several merits. First, the cognitive load experienced by the user can be relieved if feedback information is distributed among a variety of different channels instead of being conveyed through a single channel [21, 22]. Second, the effect of control instabilities can be mitigated if the control loop is opened and the user can obtain feedback information from devices that are different from the actuator devices [16]. Third, and equally important is the fact that hidden physical parameters arise naturally in remote or virtual environments, and the transfer of these hidden parameters through alternative means of communication could lead to a tighter coupling between the user and the remote environment. We refer to communication channels which use sensory substitution and sensorimotor extension to provide structured feedback information as cognitive infocommunication channels.

In this chapter, we give a general overview of the two major aspects which arise during the creation of cognitive infocommunication channels. The first aspect focuses on the design of synthesis algorithms. The output of this design stage is a set of algorithms that will produce output signals for the substituting modality. The second aspect focuses on converting the descriptors of the transmitted information into input parameters for the synthesis algorithms. In the second half of the chapter, we demonstrate the concepts we introduce through an application example in which audio signals are used to communicate information on tactile percepts. Based on the proposed synthesis algorithm design approach, we use abstract sounds in general, not specific sonifications which arise as a natural byproduct of some physical phenomenon. We argue that the model developed can be used successfully in engineering systems requiring remote teleoperation. In such environments, hidden physical parameters arise naturally and their transfer through alternative means of communication can lead to a tighter coupling between the user and the remote environment.

The chapter is structured as follows. In section 12.2, we describe the problem of designing cognitive infocommunication channels in general. In section 12.3, we introduce our interaction-based model for the design of synthesis algorithms. In section 12.4, we describe the mathematical foundations for the parameter generating method we propose, which will allow the user to link the substituted information with input parameters to the synthesis algorithms. Finally, in section 12.5, we describe the application example along with test results.

12.2 Problem Description

A typical sensory substitution application involves two sensory modalities: a *substituted* modality and a *substituting* modality. The goal of the application is to create signals which are able to map percepts that are normally perceived through the substituted modality onto signals that can be perceived by the substituting modality, and then interpreted by the human brain. We use this terminology in the treatment of cognitive infocommunication channels, even in more extreme cases where the application is used to convey more abstract information types which are not normally perceived by any physical modality (i.e., in cases where there is no substituted modality).

A key component of any cognitive infocommunication channel is a synthesis algorithm, $\mathscr{A}$, which produces sensory substitution signals for the substituting modality based on a set of parameters. These parameters are a function of the synthesis algorithm itself, and they do not necessarily have much to do with the application. For example, when designing audio-based sensory substitution applications, simple examples of these parameters may include the frequency or the amplitude of the various kinds of sounds that are used.

We assume that the parameters used to generate the sensory substitution signals can be calculated as a function, F, of a set of descriptors which define the percept that is normally perceived by the substituted modality. Thus, the output signal can be expressed as:

$$sig = \mathscr{A}(F(\mathbf{x})) \tag{12.1}$$

where $\mathbf{x}$ is a vector that describes the percept being communicated by the application. For example, in a touch-to-audio application, $\mathbf{x}$ might contain parameters such as the roughness, softness, stickiness, or temperature of the surface. Function F would convert these descriptors into a set of parameters which can be supplied to the synthesis algorithm, $\mathscr{A}$.

Sections 12.3 and 12.4 of this chapter give a formal overview of the proposed interactive approach for the design of synthesis algorithms and parameter-generating functions.

12.3 Interaction-Based Model for Synthesis Algorithm Design

When dealing with semiotic systems, the concepts of iconic and abstract (symbolic) signals emerge in practically every domain [12, 7, 5, 4]. For example, in earlier research on audio interfaces, the dual concepts of **auditory icons** and **earcons** have emerged. Auditory icons were first proposed by Gaver as natural, everyday sounds that can be used to represent actions or objects within an interface [12, 11]. Auditory icons are natural sounds that occur as byproducts of physical processes in real-world environments. In contrast, earcons – developed by Blattner, Sumikawa and Greenberg – were defined as non-verbal audio messages that do not necessarily create an intuitive set of associations with the objects or events that they represent [6].

Earcons are therefore abstract sequences of sounds which are usually temporal in nature. There is a tradeoff involved in the decision to use auditory icons or earcons: auditory icons are more easily learned by users because they appeal to common, everyday physical associations, while earcons are more difficult to learn due to their abstract nature; on the other hand, the environments in which auditory icons may be used are more limited because they are constrained to the associations they help create, while earcons can be used in virtually any context.

In our approach, we do not distinguish between these two aspects, because both iconic and symbolic signals can effectively contribute to cognitive infocommunication channels. We formally describe our methodology for the design of synthesis algorithms in terms of concept algebra. This will be useful for future work because concept algebra affords a natural way of producing automated inferences on taxonomies.

12.3.1 Basic Definitions

In order to introduce our framework, it is necessary to provide a few definitions in advance. We adopt these definitions from [23, 10].

Definition 12.1 (Formal context). A (formal) context Θ can be defined as a 3-tuple, consisting of objects, attributes and a set of relationships between them:

$$\Theta = (\mathbb{O}, \mathbb{A}, \mathscr{R}) \tag{12.2}$$

where $\mathscr{R}$ is a set of relations between objects and attributes.

$$\mathscr{R} : \mathbb{O} \to \mathbb{O} | \mathbb{O} \to \mathbb{A} | \mathbb{A} \to \mathbb{O} | \mathbb{A} \to \mathbb{A} \tag{12.3}$$

Objects are instantiations of concrete entities and / or abstract concepts (defined later), while attributes are subconcepts which are used to characterize the properties of a given concept.

Definition 12.2 (Abstract concept). An (abstract) concept c in a context Θ can be defined as a 5-tuple:

$$c = (O, A, \mathscr{R}^c, \mathscr{R}^i, \mathscr{R}^o) \tag{12.4}$$

where

$$O \subseteq \mathbb{O}^*, A \subseteq \mathbb{A}^*, R^c \subseteq O \times A, R^i \subseteq C' \times C, R^o \subseteq C \times C' \tag{12.5}$$

and $\mathscr{R}^c$, $\mathscr{R}^i$ and $\mathscr{R}^o$ are a set of internal, input and output relations, respectively.

Definition 12.3 (Intension.). The intension of a concept is the intersect of the sets of attributes of all the objects in the concept. More formally:

$$c^*(c) = c^*(O,A,R^c,R^i,R^o) = \bigcap_{j=1}^{\#O} (A_{o_j}) \tag{12.6}$$

where $\#O$ is the number of elements in O.

12.3.2 Perceptual Icons, Body Icons and Interactive Icons

Based on the above definitions, we define the following icons:

Definition 12.4 (Perceptual icons of modality x). Perceptual icons of a modality x are a set of concepts, $\mathbb{PI}$, within the context of percepts which can be obtained through a sensory modality x, that have a single attribute in their intension. Formally:

$$\mathbb{PI}(x) = \left\{ c \,\middle|\, ctx(c) = (percepts(x), A, R), \exists! a : a \in c^*(c) \right\} \tag{12.7}$$

where $ctx(c)$ is a function that returns the context of a concept, c, and $percepts(x)$ is a function that returns the set of all sensory signals which can be perceived through sensory modality x.

Remark 12.1. Some examples of perceptual icons of the tactile modality (also referred to as "tactile icons") are the concepts of "bumpiness", "roughness", "stickiness", "softness", or "temperature". Examples of perceptual icons of the auditory modality (also referred to as "auditory icons")auditory icons may include the concepts of "tonality", "consonance", "dissonance", "raspiness", or "brightness".

Definition 12.5 (Evaluation functions of perceptual icons). Perceptual icons of any modality x have an evaluation function whose output value represents the degree to which the given perceptual icon represents a sensory signal obtained by sensory modality x:

$$\begin{aligned} eval\_fun : percepts(x) \times \mathbb{PI}(x) \rightarrow (0,1] \\ \mathbb{O}^* \backslash percepts(x) \times \mathbb{PI}(x) \equiv 0 \end{aligned} \tag{12.8}$$

An output value of 0 is only permitted for objects (percepts) which cannot be perceived using the modality of the perceptual icon because a non-existent quality of a sensory percept cannot be used to describe the sensory percept.

Definition 12.6 (Body icons). Body icons are a set of concepts, $\mathbb{BI}$, within the context of the human body, that have a single attribute in their intension. Formally:

$$\mathbb{BI} = \left\{ c \,\middle|\, ctx(c) = (BodyParts, A, R), \exists! a : a \in c^*(c) \right\} \tag{12.9}$$

where *BodyParts* is the set of all body parts.

Remark 12.2. Examples of body icons include the degree a finger is bent, the height of the position of a finger, or the pulse rate at which the heart is beating.

Definition 12.7 (Impact functions of body icons). For every 3-tuple of perceptual icons, sensory signals obtained through the modality of the perceptual icons and body icons, there exists an impact function which returns a numerical value representing the state of the attribute within the body icon intension after the sensory signal has been perceived:

$$impact\_fun : \mathbb{PI}(x) \times percepts(x) \times \mathbb{BI} \rightarrow [0,1] \tag{12.10}$$

Definition 12.8 (Interactive icons of perceptual icon *pi*). Interactive icons of a perceptual icon *pi* are a set of body icons, $\mathbb{II}$, that have an attribute whose value changes during the course of interaction to a degree that is proportional to the value of the perceptual icon's evaluation function. More formally:

$$\mathbb{II}(pi(x)) = \left\{ bi \middle| \begin{array}{l} bi \in \mathbb{BI}, \forall sig_1, sig_2 \in percepts(x), sig_1 \neq sig_2, \\ \frac{impact\_fun(pi(x),sig_1,bi)}{impact\_fun(pi(x),sig_2,bi)} \propto \frac{eval\_fun(pi(x),sig_1)}{eval\_fun(pi(x),sig_2)} \end{array} \right\} \tag{12.11}$$

Definition 12.9 (Related interactive and perceptual icons). A perceptual icon of modality *x*, $pi_1(x)$, and an interactive icon of another perceptual icon, $\mathbb{II}(pi_2(y))$ are related if the attribute in the intension of $\mathbb{II}(pi_2(y))$ is also linked (through $\mathscr{R}$) with at least one object, *o* within the perceptual icon $pi_1(x)$. More formally:

$$\begin{aligned} r(pi_1(x), \mathbb{II}(pi_2(y))) \Leftrightarrow pi_1(x) = (O, A, \mathscr{R}^{pi_1(x)}, \mathscr{R}^i, \mathscr{R}^o), \exists a \in A, o \in O : \\ (o,a) \in \mathscr{R}^{pi_1(x)}, c^*(\mathbb{II}(pi_2(y))) = a \end{aligned} \tag{12.12}$$

The concepts defined above are demonstrated in figure 12.1. On the top left, we see all concepts in the context of tactile percepts as a subset of all concepts. Two examples are given: the concepts of stickiness and roughness. These are perceptual icons because the only common attribute of all percepts within these concepts are a degree of stickiness and a degree of roughness, respectively. Similarly, on the top right side we see all concepts in the context of auditory percepts as a subset of all concepts. The two concepts given here as examples are raspiness and pitchiness. The evaluation function of stickiness would return how sticky a tactile percept is, and the evaluation function of pitchiness might return how high or low the pitch of an auditory percept is.

The concepts in the figure are in turn connected with all objects that are included in the concepts (it is clear that not all audible sounds are raspy to at least a certain degree, and not all audible sounds have a pitch, but some sounds do, and for some sounds it is both true that they are raspy to at least some degree and that they have a

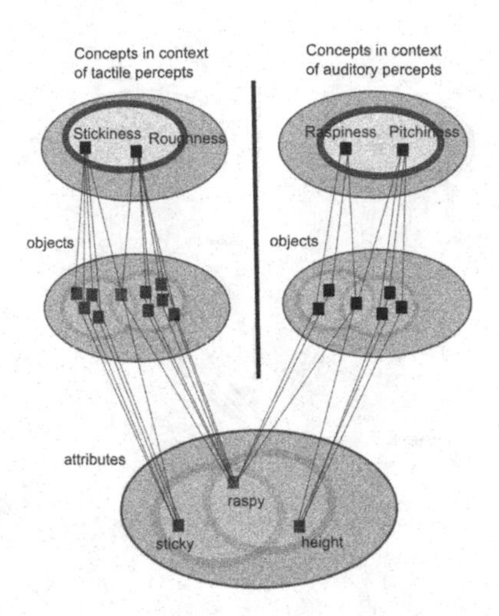

Fig. 12.1 The relationship between auditory and tactile icons.

pitch). Finally, at the third level we have the attributes that tactile and auditory percepts can assume. Whenever the intensions of a concept consist of a single attribute, the concept becomes an icon.

12.3.3 Methodology for the Design of Synthesis Algorithms

When we view perceptual icons as sets of concepts, our goal is to find a matching between attributes in the intensions of perceptual icons of the substitute and substituted modalities. An example of such a matching can be seen in figure 12.1, where the raspy attribute is present in both the tactile icon of roughness and the auditory icon of raspiness. If such a matching exists, meaningful iconic sounds can be created, because the substitute percepts can give rise to the same interpretations as the substituted percepts.

In reality, however, more often than not it is impossible to find attributes that form the intension of a perceptual icon of both the substituted and the substitute modalities. In this case, we have no choice but to use abstract signals and a set of attributed meanings to describe the perceptual icons of the substituted modality. In our research, we have found that cognitive infocommunication channels can be tailored to suit the human cognitive system to a greater extent if we can find an interactive icon of the substituted modality and a perceptual icon of the substituting modality which are related to each other, and use the attribute of the perceptual icon in cognitive infocommunication channels.

Figure 12.2 demonstrates this case. The figure shows an example when we would like to convey the stickiness of a surface through sound. Because there is no generally accepted concept of stickiness for sounds, we need to use an interactive icon (i.e., a symbol of stickiness). We see that the contact time between a person's finger

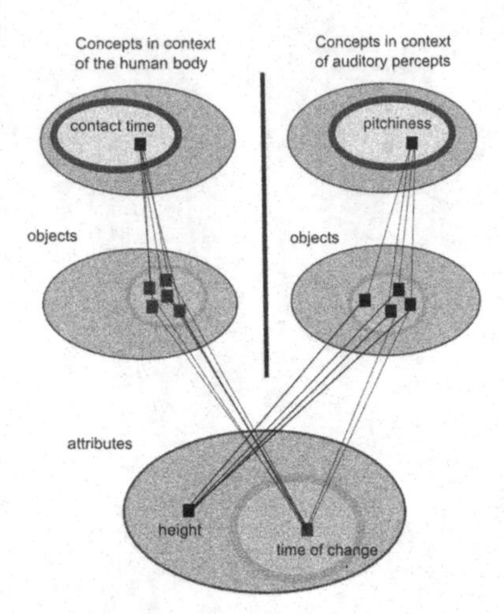

Fig. 12.2 When there is no direct link between the perceptual icon of the substituted modality we would like to represent and the perceptual icons of the substituting modality, the goal is to find an interactive icon of the substituted modality that is related to a perceptual icon of the substituting modality.

and a surface is a body icon whose attribute value changes depending on how sticky the surface is. Therefore, contact time is an interactive icon for sticky surfaces. Incidentally, there are at least some audio signals that are pitchy to a certain extent and whose pitchiness changes through time. Thus, the "time of change" attribute, which is also in the intension of contact time, ensures that the interactive icon of contact time is related to the audio icon of pitchiness, and changes in the time of a change in pitchiness can be used to represent different degrees of stickiness.

12.4 Interaction-Based Model for Parameter Generating Function Design

In this section, we turn our attention to the second aspect of designing cognitive infocommunication channels: the creation of parameter-generating functions which transform the input parameter space into input parameters for the synthesis algorithms (function F in section 12.2).

Our goals can be summarized as follows:

1. We would like to create a compact, canonical and granular representation of function F.
2. We would like to be able to locally manipulate the output values of F in specific points of its input domain. The ultimate goal of these manipulations will be to change the output signal of the synthesis algorithm in only a specific point of the percpetual space that is defined by the substituted modality. We will refer to this manipulation as *local tuning*.

12.4.1 *Mathematical Background*

12.4.1.1 Bounded Discreteized Multivariate Functions and Their Tensor Product Form

We begin our discussion by providing the definition of *bounded discretized multivariate vector functions* and their *tensor product form*.

Definition 12.10 (Bounded multivariate vector functions). Let $F(\mathbf{x},\mathbf{y}) \in \mathbb{R}^H$ be a multivariate vector function, where $\mathbf{x} \in \mathbb{R}^M$, $\mathbf{y} \in \mathbb{N}^L$, $M+L=N$, and $H,N<\infty$ ($\mathbf{x}$ is an element of a continuous, and $\mathbf{y}$ is an element of a discrete subspace of the domain of function F). If input vector $\mathbf{x}$ is taken from a bounded hypercube of real numbers, and input vector $\mathbf{y}$ is taken from a bounded hypercube of natural numbers ($\mathbf{x} \subset \mathbb{R}^{[a_1,b_1]\times[a_2,b_2],\ldots,\times[a_M,b_M]}$, and $\mathbf{y} \subset \mathbb{N}^{[c_1,d_1]\times[c_2,d_2],\ldots,\times[c_L,d_L]}$), the function is a *bounded multivariate vector function.*

Definition 12.11 (Bounded discretized multivariate vector functions). A bounded multivariate vector function $F(\mathbf{x},\mathbf{y}) \in \mathbb{R}^H$ is *discretized* if it is defined only for a discrete set of points within the input space defined by hyper-rectangular grid $G = \{g_{p_1,p_2,\ldots,p_N} \in \mathbb{R}^{[a_1,b_1]\times[a_2,b_2],\ldots,\times[a_M,b_M]} \times \mathbb{N}^{[c_1,d_1]\times[c_2,d_2],\ldots,\times[c_L,d_L]}\}_{p_n=1}^{P_n}$ (where P_n is the number of unique discretization points along the n-th dimension, and $n=1..N$). Such functions can be represented by an $(N+1)$-dimensional tensor $\mathcal{F}^{D(G)}$ of size $P_1 \times \ldots \times P_N \times H$ such that

$$\mathcal{F}^{D(G)}_{p_1,\ldots,p_N} = F(g_{p_1,p_2,\ldots,p_N}) \tag{12.13}$$

where $F(g_{p_1,p_2,\ldots,p_N})$ is the output vector of F corresponding to the input vector in point $g_{p_1,p_2,..p_N}$ on grid G.

Definition 12.12 (Bounded multivariate TP functions). There exists a subset of all bounded multivariate vector functions which can be written in the following tensor product form:

$$F(\mathbf{x},\mathbf{y}) = \mathcal{S} \boxtimes_{m=1}^{M} \mathbf{w}_m(x_m) \boxtimes_{l=1}^{L} \mathbf{u}_l[y_l] \tag{12.14}$$

where $\mathcal{S} \in \mathbb{R}^{I_1\times\ldots\times I_N\times H}$ is a core tensor with finite dimensions, each $\mathbf{w}_m(x_m) = [w_{m,1}(x_m), w_{m,2}(x_m), \ldots, w_{m,I_m}(x_m)]$ is a vector of continuous univariate weighting functions, and each $\mathbf{u}_l[y_l] = [u_{l,1}[y_l], u_{l,2}[y_l], \ldots, u_{l,I_l}[y_l]]$ is a vector of discrete univariate weighting functions. x_m denotes the input on the m-th continuous input dimension, and y_l denotes the input on the l-th discrete input dimension. Bounded multivariate vector functions which can be written in this form are referred to as *bounded multivariate tensor product (TP) functions.*

Theorem 12.1. *(Higher-order singular value based canonical form of discretized multivariate functions). Every discretized bounded multivariate vector function, $\mathscr{F}^{D(G)}$ can be written as the product:*

$$\mathscr{F}^{D(G)} = \mathscr{S} \boxtimes_{m=1}^{M} \mathbf{W}_m \boxtimes_{l=1}^{L} \mathbf{U}_l \times_{N+1} \mathbf{U}_{L+1} = \mathscr{S} \boxtimes_{n=1}^{N+1} \mathbf{X}_n \tag{12.15}$$

in which:

1. $\mathbf{W}_m = (\mathbf{w}_1^{(m)}, ..., \mathbf{w}_{I_m}^{(m)})$, $m = 1..M$ *is a unitary matrix of size* $(P_m \times I_m)$
2. $\mathbf{U}_l = (\mathbf{u}_1^{(l)}, ..., \mathbf{u}_{I_l}^{(l)})$, $l = 1..L$ *is a unitary matrix of size* $(P_l \times I_l)$
3. $\mathbf{U}_{L+1} = (\mathbf{u}_1^{(L+1)}, ..., \mathbf{u}_{I_{L+1}}^{(L+1)})$ *is a unitary matrix of size* $(H \times I_{L+1})$
4. $\mathscr{S}$ *is a real tensor of size* $I_1 \times ... \times I_N \times H$*, the subtensors* $\mathscr{S}_{i_n=\alpha}$ *of which have the following properties:*
   - *all-orthogonality: any pair of the subtensors of* $\mathscr{S}$ *are orthogonal, i.e. for all possible values of* n, α *and* β *subject to* $\alpha \neq \beta$:

$$< \mathscr{S}_{i_n} = \alpha, \mathscr{S}_{i_n} = \beta >= 0 \tag{12.16}$$

   - *ordering: All of the subtensors of* $\mathscr{S}$ *along any given dimension n are ordered according to their Frobenius norm, i.e.* $\forall n = 1..N+1$:

$$||\mathscr{S}_{i_n=1}|| \geq ||\mathscr{S}_{i_n=2}|| \geq ... \geq ||\mathscr{S}_{i_n=I_n}|| \geq 0 \tag{12.17}$$

Proof. The HOSVD (higher-order singular value decomposition) of any N-dimensional tensor with real values was introduced by de Lathauwer in [15]. The fact that discretized multivariate functions can be stored in such tensors, as demonstrated in equation 12.13 proves the theorem.

12.4.2 Tuning Model for Discretized Multivariate Functions

In the previous theorem, we established that any discretized multivariate function can be expressed in the form shown in equation 12.15. We begin this section by stating a trivial but very important corollary to this theorem.

Corollary 12.1. *(Local tuning.) If we change the values in just the* p_k*-th row of any* $\mathbf{X}_k$ *in equation 12.15 (that is, the values of* $\mathbf{x}_{k,p_k}$*), then only those output values of function* $\mathscr{F}^{D(G)}$ *will be changed which belong to the* p_k*-th discretization point along* *the k-th dimension of hyper-rectangular grid G. This can be easily seen if we express a single element of $\mathscr{F}^{D(G)}$ as follows:*

$$\mathscr{F}^{D(G)}(g_{p_1,\dots,p_N}) = \mathscr{S} \mathop{\boxtimes}_{\substack{n=1,\\ n\neq k}}^{N+1} \mathbf{x}_{n,p_n} \times_k \mathbf{x}_{k,p_k} \tag{12.18}$$

and it is obvious that if any point on hyper-rectangular grid G is chosen in which the value of the k-th dimension is not the p_k-th discretization point, then the output value of the function will be unchanged. For this reason, we refer to the changing of values within vector $\mathbf{x}_{k,p_k}$ as a local tuning of $\mathscr{F}^{\mathscr{D}(\mathscr{G})}$ along the k-th input dimension. The values in vector $\mathbf{x}_{k,p_k}$ in turn are referred to as tuning weights.

Regarding the uniqueness of the weights used to tune the output, we have the following theorem:

Theorem 12.2. *Let us consider the HOSVD-based form of a discretized multivariate function $\mathscr{F}^{D(G)} = \mathscr{S} \mathop{\boxtimes}_{n=1}^{N+1} \mathbf{X}_n \in \mathbb{R}^H$. Assuming that the values in $\mathbf{X}_{n\neq k}$ are kept fixed, $\mathbf{x}_{k,p_k}$ is unique $\forall n = 1..N, k, p_k$ if the number of outputs, $H \geq I_k$.*

To facilitate the proof of this theorem, we define the n-mode unfolding of a tensor based on [15] as follows:

Definition 12.13. Let the *n-mode unfolding* of tensor $\mathscr{A}^{(I_1 \times I_2 \times \dots \times I_N)}$ be a matrix $\mathscr{A}_{(n)} = \mathbf{A}_{(n)} \in \mathbb{R}^{I_n \times (I_{n+1} \dots I_N I_1 I_2 \dots I_{n-1})}$ such that the value of $a_{i_1,i_2,\dots,i_N}$ is stored in the i_n-th row, and r_n-th column of $A_{(n)}$, where

$$\begin{aligned} r_n = {} & (i_{n+1}-1)I_{n+2}\dots I_N I_1 \dots I_{n-1} + \\ & + (i_{n+2}-1)I_{n+3}\dots I_N I_1 \dots I_{n-1} + \dots + \\ & + (i_N - 1)I_1 \dots I_{n-1} + (i_1 - 1)I_2 \dots I_{n-1} + \dots \\ & \dots + i_n - 1 \end{aligned} \tag{12.19}$$

Proof. The system of linear equations which has to be satisfied by $\mathbf{x}_{k,p_k}$ can be written as:

$$\begin{aligned} \mathscr{F}^{D(G)}(g_{p_1,\dots,p_N}) &= \mathbf{x}_{k,p_k}(\mathscr{S} \mathop{\boxtimes}_{n=1,n\neq k}^{N+1} \mathbf{X}_n)_{(k)} = \\ &= \mathbf{x}_{k,p_k}\mathbf{B} \end{aligned} \tag{12.20}$$

where matrix $\mathbf{B}$ has size $I_k \times H$. Due to the properties of HOSVD, $rank(\mathbf{B}) = min(I_k, H)$, $\forall k$.

If $H = I_k$, then $\mathbf{B}$ is H-by-H and full-rank. If $H > I_k$, then the system is overdetermined (there are more equations than variables). In both cases, there is only a single solution to the system. Therefore, it is clear that if $\mathbf{B}$ is fixed and $H \geq I_k$, then there is only one solution for $\mathbf{x}_{k,p_k}$.

Remark 12.3. The manipulation of the weights in $\mathbf{x}_{k,p_k}$ is unique in the sense that in most applications using HOSVD, it is the elements of the core tensor which are modified. In our case, by modifying the weights in a single row of $\mathbf{X}_k$, we are in effect modifying the output values which correspond to a single gradation of a single input dimension of discretized multivariate function $\mathscr{F}^{D(G)}$. When modifying these weights, we are modifying these output values irrespective of the value of other dimensions. In addition, because the subtensors of $\mathscr{S}$ are ordered according to their Frobenius norms, the weights in the leftmost columns of $\mathbf{X}_k$ will be used to scale elements with more energy than weights towards the other end of the weight vector.

12.4.3 Application of the Tuning Model to Cognitive Infocommunication Channels

In the context of cognitive infocommunication channels, the parameter-generating function, F, can be regarded as a bounded discretized multivariate vector function, if we define the channel in only discrete points of the input space defined by the substituted modality.

This approximation of F leads to the direct applicability of the tuning mechanism described earlier in this section to the configuration of cognitive infocommunication channels.

12.5 Example Application: Mapping Surface Properties to Auditory Communication Channels

Our goal in this section is to define audio signals which can be used to convey information on four tactile dimensions: hardness/softness, roughness/smoothness, stickiness/slipperiness, and temperature. The choice of these four dimensions was justified because cognitive psychologists claim that these four dimensions are almost exclusively responsible for forming tactile percepts through direct touch [24, 14]. In our current implementation, all four dimensions have 10 discrete gradations. The users are required to make absolute judgements about the sounds they hear. In the following sections, we describe how individual sounds are created based on a given gradation for each dimension. For test results using these sounds, we refer the reader to section 12.6.

12.5.1 Synthesis Algorithm Design

In this subsection, we briefly describe the synthesis algorithms used to generate signals for the 4 different tactile dimensions previously introduced.

1. Softness
   We use a combination of glissandi and vibrato. Both of these components are in effect a change in pitch, which can be interpreted as a symbol of changes

in surface deformation as the user presses down on the surface. Formally, the sound at time t is produced as follows:

$$gl(t) = \frac{gend - gstart}{t_{max}} t + gstart$$
$$vb(t) = vibamp \cdot \sin\left(\frac{gl(t)}{vibfreq} \cdot \frac{2\pi t}{t_{max}}\right) \tag{12.21}$$
$$sound(t) = amp \cdot \sin\left((gl(t) + vb(t))\frac{2\pi t}{t_{max}}\right)$$

where $gstart$ and $gend$ are the starting and ending frequencies of a glissando, and $vibamp$ and $vibfreq$ are used to control the amplitude and frequency of the vibrato, respectively.

2. Roughness
We use granular synthesis to generate different audio textures to represent the roughness of surfaces. The physical analogy linking the sounds to an interactive icon is the number and size of obstacles the finger encounters as it is dragged across the surface. Formally, the sound at time t is produced as follows:

$$w(t, t_0, t_{max}) = \begin{cases} 1 \text{ if } & |t - t_0| < \frac{1}{2 \cdot t_{max}} \\ 0 \text{ otherwise} \end{cases} \tag{12.22}$$

$$gr(t, t_0, t_{max}) = w(t, t_0, t_{max}) \sin\left(2\pi \frac{f \cdot t}{t_{max}}\right) \tag{12.23}$$

$$sound(t) = amp \cdot \sum_{i=1}^{\lfloor \frac{t \cdot dens}{t_{max}} \rfloor} gr(t, \frac{i \cdot t_{max}}{dens}, t_{max}) \tag{12.24}$$

where $dens$ specifies the number of grains per second, and f specifies their frequency, and t_{max} is the complete length of the audio signal in seconds. Parameter i in the third equation runs from 0 to $dens - 1$, hence t_0 in the first two equations runs from 0 to $\frac{dens-1}{dens} t_{max}$.

3. Stickiness
We propose two models for sounds representing stickiness:

   a. The carrier sound consists of two sections: a "dissonant" section containing many random frequencies that change in time, and a "consonant" section containing only a few, predetermined frequencies. The "dissonant" section is not really dissonant in a harmonic sense, but serves to contrast the final held-out chord, which is definitely harmonically more stable. The length of the dissonant section symbolizes the user's contact time with the surface. The stickier the surface, the longer the time it takes to "break loose" from it. Formally, the sound at time t is produced as follows:

$$rnote(t) = rseq\,(mint, maxt, spch, minjmp, maxjmp)\,[t] \quad (12.25)$$

$$conss(t) = ton_1(t) + ton_2(t) \quad (12.26)$$

$$consw(t) = \begin{cases} 1 \text{ if } & t > t_{offset} \\ 0 \text{ otherwise} \end{cases} \quad (12.27)$$

$$dissw(t) = \begin{cases} 1 \text{ if } & t < t_{offset} \\ 0 \text{ otherwise} \end{cases} \quad (12.28)$$

$$sound(t) = consw(t) \cdot conss(t) + dissw(t) \cdot rnote(t) \quad (12.29)$$

where ton_1 and *ton*2 are sinusoidal waves, and *rseq*() is a function that returns a list of random notes, in which the length of each note is between *mint* and *maxt*. The starting pitch is defined by *spch*, and the minimal and maximal jump between notes is *minjmp* and *maxjmp*, respectively.

b. The second carrier sound was created to be an iconic sound based on the Fourier analysis of the sound made by various kinds of tapes being removed from various kinds of surfaces. The thicker and wider the tape, the stickier we assumed the produced sound to be. A Fourier analysis revealed that the frequency components could be approximated using Gaussian windows. After empirical experimentation, we chose to calculate the strengths of component frequencies by interpolating between an exponential and gaussian decay. Formally, the sound at time t is produced as follows:

$$Amp(f) = \alpha^3(ba)(e^{\frac{-i^2}{2\sigma^2}}) + (1-\alpha^3)(ba)(e^{-|4i|}) \quad (12.30)$$

$$sound(t) = \sum_{freq=0}^{\infty} Amp(f) \cdot \sin\left(\frac{2\pi ft}{t_{max}}\right) \quad (12.31)$$

where $i = \frac{f-basef}{fjmp}$, and ba is a base amplitude.

4. Temperature

Three different frequencies are used together to convey temperatures. The three frequency channels are periodically turned on and off. Formally:

$$sumn(t) = not_1(t) + not_2(t) + not_3(t) \quad (12.32)$$

$$onoff(t,f) = \begin{cases} 0 \text{ if } & mod(\lfloor \frac{t \cdot f}{t_{max}} \rfloor, 2) = 0 \\ 1 \text{ otherwise} \end{cases} \quad (12.33)$$

$$sound(t) = amp \cdot onoff(t,f) \cdot sumn(t) \quad (12.34)$$

where not_1, not_2 and not_3 are sinusoidal waves.

The gradations within the ranges used for modulated parameters do not change in linear proportions, and do not even appear in order of magnitude. The exact values were determined experimentally with the help of a group of 20 people, through separate tests for the individual sounds. A fault tolerance model was developed, so that errors of magnitudes of 2 notches were tolerated, as long as they occurred at rates smaller than a preset value.

12.5.2 Orchestration of Synthesis Algorithms

When there is more than a single synthesis algorithm, as in the case of this application example, an important decision to make is how to arrange the activation of the synthesis algorithms through time. Because the synthesis algorithms can be pictured as an orchestra of instrument players who can be either playing or not playing at any given time, we refer to this problem as the *orchestration* of synthesis algorithms.

In the section on test results (section 12.6) our goal is to demonstrate the utility of what we call *experience-based orchestration*. The main idea behind this hypothesis is that the users can become better accustomed to the cognitive infocommunication channels if the sequence of events produced by the channel can be mapped to the sequence of events normally encountered by the user when perceiving the same type of information through the usual, substituted modality.

In our case, such an experience-based orchestration could be the following:

- A sound for softness is played for a certain time x. This is equivalent to saying that the user is pressing down on the surface to see how soft it is.
- After time x, the sound for roughness is played instead of the sound for softness. This is equivalent to saying that the user is dragging her finger along the surface in order to perceive how rough it is.
- Time x depends on how sticky the surface is. The interactive idea behind this is that the stickier the surface, the longer the time it will take for the user to be able to drag her finger across the surface and perceive its roughness.
- The sound for temperature can be heard throughout, because it is only a function of contact, and as long as the user's finger is touching the surface, its temperature can be heard.

12.5.3 Psychophysical Aspects

Once we have settled the question of what sounds we would like to use to convey individual parameters, an important question that remains is whether these sounds can be played at the same time, and if so, under what circumstances. There are extensive studies on the psychophysics of hearing in terms of the JNDs (just noticeable differences) of simple sounds, critical bands of frequency (which can be used only by a single sound source at a time, otherwise masking effects will arrive), and timbre space characteristics, to name just a few.

The effects of these psychophysical factors are taken into consideration at different levels in our work. Some of these aspects - such as JNDs - are disregarded, given that our model allows users to tune the system so that the sounds can be fitted to their own liking. Since JNDs were measured only for the simplest of sounds, it seems more reasonable to allow the user to tailor the sounds after the system has been deployed.

Other psychophysical factors - such as the masking effects of sounds that are played in the same critical frequency bands, or the difference in perceived loudness of sounds (depending not only on the sound intensity, but also on the frequency components of the sound) - can be taken into consideration in earlier phases of development.

In order to address these latter effects, we created a loudness model using which the individual loudness of the different sounds can be kept at a comfortable level, depending on the spectral properties of each individual sound. We sampled the strength of 5000 different frequency components centered around the central frequencies of different critical bands (the width of the critical bands was calculated using the well-known equivalent rectangular bandwidth formula introduced by Moore and Glasberg [13, 17]). Based on the obtained spectral distribution, we calculated the weighted mean of the central frequencies:

$$f_{average} = \frac{\sum_{i=1}^{5000} strengths(i) \cdot frequencies(i)}{\sum_{j=1}^{5000} strengths(j)} \tag{12.35}$$

For each dimension, we allocated an interval of phon levels, $[phon_a, phon_b]$ with which the sounds could be played. Along each gradation of each dimension (for example, all 10 degrees of roughnesses), we calculated the ratio between the strengths close to the average frequencies and the total strength. Based on these ratios, we performed a linear interpolation between the two extremes of phon levels, thus obtaining a unique phon level for each gradation. Finally, using the equal loudness contours specified in the ISO 226 standard (originally defined by Fletcher and Munson as the well-known Fletcher-Munson curves, and later reevaluated by Robinson and Dadson [9, 20]), we calculated the decibel level for each individual sound based on the obtained phon level and the average frequency calculated in equation 12.35.

12.6 Experimental Results

As detailed in our previous work [8], we refer to sounds that convey information within the frequency domain as structural sounds, and sounds which carry information through temporal changes as temporal sounds. We hypothesize that the superposition of our sounds can be efficient if we find a good balance between structural and temporal aspects of sound, because the two types of sounds require different forms of concentration. To verify our hypotheses, we conducted a set of experiments. As only preliminary results were shown in our previous work, we give a more detailed description of our methods and findings here.

12.6.1 Methods

In our first experiment, we played 10 subjects a set of 50 sounds within 2 categories of sound combinations. The following categories were used as sounds:

1. A juxtaposition of 4 structural sounds with no pause in between. The total time for all four sounds was 2 seconds.
2. A combination of 3 structural sounds – for softness, roughness and temperature, respectively – and 1 temporal sound for stickiness. The temporal sound was not audible, but implicitly contained a changing point. Before the changing point, the sounds for temperature and softness were heard. After the changing point, the sounds for temperature and roughness were played. The total time for all four sounds was 2 seconds.

The subjects were asked to click on the parts of a GUI which they felt represented the surface parameters conveyed by the sounds. Each sound could be replayed twice if the test subjects felt they needed it.

In our second experiment, the same test subjects were asked to perform the same task with only slight modifications which restricted their concentration abilities. In these concentration tests, 10 test sounds were played, and the subjects were asked to memorize a 5-digit random number while each sound was played. After hearing the sounds once, the subjects had to enter the memorized digits, and only then could they submit their guesses. In the concentration test, the test sounds could not be replayed. Thus, in total, we conducted 4 different experiments (2 different sound types times 2 different levels of concentration).

Testing sessions took about an hour. Subjects were given at most 30 minutes to practice, and testing was started when this time was over or when the subjects felt comfortable with the sounds. Tests with more extended practicing periods were not conducted because the goal was to gain comparative, not absolute results.

12.6.2 Test Results

Test results are summarized over all users in table form (Table 12.1) and error statistics are summarized in graphical form (Figure 12.3). In Table 12.1 the averages and standard deviations of error sizes (the errors in number of gradations) are shown over all users per dimension for both the normal and the concentration tests in both test cases. (The average number of replays requested are also indicated. In the concentration tests, the field is left empty because no replays were allowed.)

In Figure 12.3, the darker colors indicate pairs of gradations which are confused more frequently than others. Each dimension is represented in a separate coordinate system for each of the four tests.

For each test, we tallied the number of outliers (guesses with an error greater than $+/-2$ in any single dimension out of the 4 were considered to be outliers), and we calculated the associated p-values, in order to calculate the probability that the given number of inliers are due to chance alone. The computed p-values were

Table 12.1 Aggregated test results for 5 test subjects when testing the sounds in all four test cases. Average sizes of errors as well as their standard deviations are shown for both the normal tests (in which the users could concentrate solely on the sounds) with 50 rounds, and the concentration tests (in which the users had to memorize a 5-digit number in addition to recognizing the sounds) with 10 rounds.

Test type		Avg.	Std.Dev.	Avg. Replays	Std.Dev.Replays
4 structural can concentrate	Softness	1.13	1.0	1.53	0.6
	Roughness	1.51	1.5		
	Stickiness	1.56	1.46		
	Temperature	1.01	1.22		
4 structural cannot concentrate	Softness	2.49	2.08		
	Roughness	2.28	2.09		
	Stickiness	2.37	1.98		
	Temperature	1.99	1.94		
3 structural, 1 temporal can concentrate	Softness	1.29	1.44	1.164	0.76
	Roughness	1.41	1.43		
	Stickiness	1.42	1.34		
	Temperature	1.07	1.27		
3 structural, 1 temporal cannot concentrate	Softness	2.49	2.12		
	Roughness	1.53	1.62		
	Stickiness	2.48	1.94		
	Temperature	1.07	1.27		

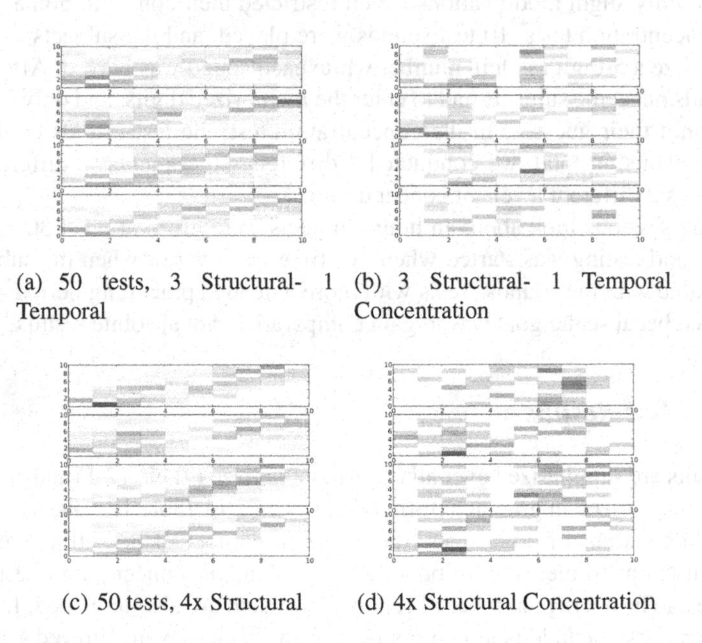

(a) 50 tests, 3 Structural- 1 Temporal

(b) 3 Structural- 1 Temporal Concentration

(c) 50 tests, 4x Structural

(d) 4x Structural Concentration

Fig. 12.3 Confusion matrices for 4 different kinds of tests. The horizontal axes represent gradations that were tested, and the vertical axes represent gradations that were answered. The darker a cell, the greater the number of times the two gradations were confused. The four coordinate systems in each subfigure represent softnesses, roughnesses, stickinesses and temperatures, from top to bottom.

$8.38 * 10^{-13}$, 0.99, $7.43 * 10^{-17}$ and 0.88 for the four tests, respectively. We can see that the probability that any of the test results, in which the user could concentrate solely on the sounds, are due to pure chance is extremely small, and a comparison of the p-values shows that greater luck would have been necessary to achieve the results by pure chance in the case where 3 strucutral sounds are combined with 1 temporal sound. Results were much less convincing in the case where the users also had to memorize the 5-digit random numbers. When we considered the 4 dimensions separately, the p-value against random selection in each separate dimension was at least as small as a 10^{-3} order of magnitude. Even this result is clearly a deterioration.

Based on the data, we carried out a series of Z-tests between corresponding pairs of haptic dimensions in the two test cases, both for normal and concentration tests. We conclude that in the normal test case, when users were allowed to concentrate solely on the sounds, there was no statistically significant difference in error statistics. In the test cases where the users also had to concentrate on something other than just the sounds (i.e., memorize the 5-digit random numbers), performance of users in certain dimensions were significantly better in the case where 3 structural sounds were combined with 1 temporal sound. For example, the p-value for the difference in performance for surface roughness was 0.006, and the p-value for the difference in performance for surface temperature was 0.0003. In these cases, we could discard the null hypothesis that test performance wasn't significantly different. In all other haptic dimensions (softness and stickiness), test results were not significantly different. Also interesting to mention is that the average number of replays was somewhat less on average in the case where 3 strucutral sounds were combined with 1 temporal sound (although not statistically significant), and that more importantly, the test subjects consistently reported that the sounds in this case were much less disturbing. The test subjects also reported that they felt they could be able to perform with more ease and accuracy if they were given more time to practice.

From the confusion matrices in Figure 12.3, we can see that some pairs of gradations were confused more often than others. Further, in the cases where the subjects were allowed to concentrate just on the sounds, the confusion matrices resemble band matrices (gradations on the extremities are almost never confused). For these reasons, we can conclude that our test results justify the use of the tuning model we described earlier in section 12.4.2.

12.7 Conclusion

In this chapter, we introduced the concept of cognitive infocommunication channels as a promising future direction which could help create more natural forms of communication between humans and machines. We gave a general overview of the various aspects which arise during the creation of cognitive infocommunication channels, and aimed to elucidate these concepts using an application example in which audio signals were used to convey information on tactile percepts along the four dimensions of softness, roughness, stickiness and temperature. Based on our test results, we argued that the model we propose can be used successfully in engineering systems requiring remote teleoperation. In such environments, hidden physical parameters arise naturally, and their transfer through alternative means of communication can lead to a tighter coupling between the user and the remote environment.

Acknowledgements. This research was supported by the ETOCOM project (TÁMOP-4.2.2-08/1/KMR-2008-0007) through the Hungarian National Development Agency.

References

1. Arno, P., Vanlierde, A., Streel, E., Wanet-Defalque, M.C., Sanabria-Bohorquez, S., Veraart, C.: Auditory substitution of vision: pattern recognition by the blind. Applied Cognitive Psychology 15(5), 509–519 (2001)
2. Auvray, M., Myin, E.: Perception with compensatory devices: from sensory substitution to sensorimotor extension. Cognitive Science 33, 1036–1058 (2009)
3. Auvray, M., Philipona, D., O'Regan, J., Spence, C.: The perception of space and form recognition in a simulated environment: The case of minimalist sensory-substitution. Perception 36, 1736–1751 (2007)
4. Barnard, M.: Graphic Design as Communication. Taylor and Francis Group, Routledge (2005)
5. Barrass, S.: Auditory information design. PhD thesis, Australian National University (1997)
6. Blattner, M., Sumikawa, D., Greenberg, R.: Earcons and icons: Their structure and common design principles. Human Computer Interaction 4(1), 11–44 (1989)
7. Brewster, S.: Providing a structured method for integrating non-speech audio into human-computer interfaces. PhD thesis, University of York (1994)
8. Csapó, A., Baranyi, P.: An interaction-based model for auditory subsitution of tactile percepts. In: IEEE International Conference on Intelligent Engineering Systems, INES (2010)
9. Fletcher, H., Munson, W.: Loudness, its definition, measurement and calculation. Journal of the Acoustical Society of America 5, 82–108 (1933)
10. Ganter, B., Wille, R.: Formal Concept Analysis. Springer, Germany (1999)
11. Gaver, W.: Auditory icons: Using sound in computer interfaces. Human Computer Interaction 2(2), 167–177 (1986)
12. Gaver, W.: The sonicfinder: An interface that uses auditory icons. Human Computer Interaction 4(1), 67–94 (1989)
13. Glasberg, B., Moore, B.: Derivation of auditory filter shapes from notched-noise data. Hearing Research 46, 103–138 (1990)
14. Hollins, M., Risner, S.: Evidence for the duplex theory of tactile texture perception. Perceptual Psychophysics 62, 695–705 (2000)
15. Lathauwer, L.D., Moor, B.D., Vandewalle, J.: A multi linear singular value decomposition. SIAM Journal on Matrix Analysis and Applications 21(4), 1253–1278 (2000)
16. Massimino, M.: Sensory substitution for force feedback in space teleoperation. PhD thesis, MIT, Dept. of Mechanical Engineering (1992)
17. Moore, B., Glasberg, B.: A revision of Zwicker's loudness model. Acta Acustica 82, 335–345 (1996)
18. Bach-y-Rita, P.: Tactile sensory substitution studies. Annals of New York Academic Sciences 1013, 83–91 (2004)

19. Bach-y-Rita, P., Tyler, P., Kaczmarek, K.: Seeing with the brain. International Journal of Human-Computer Interaction 15(2), 285–295 (2003)
20. Robinson, D., Dadson, R.: Threshold of hearing and equal-loudness relations for pure tones, and the loudness function. Journal of the Acoustical Society of America 29, 1284–1288 (1957)
21. Schmorrow, D.: Foundations of Augmented Cogntition. Lawrence Erlbaum Associates, Mahwah (2005)
22. Schmorrow, D., Kruse, A.: Augmented cognition. In: Bainbridge, W. (ed.) Berkshire Encyclopedia of Human-Computer Interaction, pp. 54–59. Berskshire Publishing Group (2004)
23. Wang, Y.: On concept algebra: A denotational mathematical structure for knowledge and software modeling. International Journal of Cognitive Informatics and Natural Intelligence 2(2), 1–19 (2008)
24. Yoshioka, T., Bensmaia, S., Craig, J., Hsiao, S.: Texture perception through direct and indirect touch: an analysis of perceptual space for tactile textures in two modes of exploration. Somatosensory and Motor Research 24(1-2), 53–70 (2007)

Chapter 13
Advantages of Fuzzy and Anytime Signal- and Image Processing Techniques - A Case Study

Teréz A. Várkonyi

Abstract. Nowadays practical solutions of engineering problems always involve some kind of, preferably model-integrated, information processing task. Unfortunately, however, the available knowledge about the information to be processed is usually incomplete, ambiguous, noisy, or totally missing. Furthermore, the available time and resources for fulfilling the task are often not only limited, but can change during the operation of the system. All these facts seriously limit the effective usability of classical information processing algorithms which pressed researchers and engineers to turn towards non-classical methods and these approaches proved to be very advantageous. In this chapter, a brief overview is given about various imprecise, fuzzy and anytime, signal- and image processing methods and their applicability is discussed in treating the insufficiency of knowledge of the information necessary for handling, analyzing, modeling, identifying, and controlling of complex engineering problems.

13.1 Introduction

Nowadays in solving engineering problems the processing of different kinds of information is of key importance. The information processing is typically performed by model based approaches which means that (1) advantageously a priori knowledge is built into system and (2) the model also contains a representation of our knowledge about the nature and actual circumstances of the problem in hand. Since the performance of the information processing usually has a direct effect on the performance of the engineering system, it is an obvious requirement to ensure appropriate methods and sufficient amount of resources for this purpose.

Up till the seventies, eighties, and nineties, classical problem solving methods proved to be entirely sufficient in handling, analyzing, modeling, identifying, and

Teréz A. Várkonyi
Óbuda University, Bécsi street 96/b H-1034 Budapest Hungary
e-mail: varkonyi.teri@phd.uni-obuda.hu

J. Fodor et al. (Eds.): Recent Advances in Intelligent Engineering Systems, SCI 378, pp. 283–301.
springerlink.com

controlling of complex engineering problems. Nowadays, however, engineering science tackles problems of previously unseen spatial and temporal complexity, with ill-defined problem formulation, and non-exact, nonnumeric, and vague input and environmental information which led to the situation that in a large number of cases analytical models and traditional information processing methods and equipment failed to handle the problems. It became clear that new ideas are required for specifying, designing, and implementing sophisticated modeling and information processing systems. Especially in cases where real-time, embedded, highly non-linear, and highly non-monotonous systems and plants are concerned. As numerical information was frequently missing or was uncertain it made place for various new, qualitative and/or symbolic representation methods (see e.g. [24]).

The ideas of the Artificial Intelligence (AI) [12], Soft Computing (SC) [16], and Imprecise Computation (IC) [6, 16] address these problems. AI, SC, and IC offer means to handle nonnumeric and vague information, and also a novel view at the computational accuracy as a utility rather than an ultimate aim of the development. A lot of successful solutions and adaptations have been born based first of all on fuzzy inference (see e.g. [17, 5]) but also on neural networks [10] and other related approaches however the problem of high and changing complexity has not been really solved especially that with the appearance of the new methods, the complexity of the problems manifested itself not only as a hierarchy of subsystems and relations, but also as the variety of modeling approaches needed to grasp the essence of the modeled phenomena [7]. It meant a real breakthrough from this respect when anytime processing [27, 23] and a couple of years later an akin idea, situational control [8] have been introduced.

The so called "anytime algorithms" [28, 29, 27, 23] offer considerable control over limited resources and they are to provide continuous operation in cases of changing circumstances and are to avoid critical breakdowns in cases of missing input data, temporary shortage of time, or computational power.

Situational control [7, 1, 8] has been designed for the control of complex systems where the traditional cybernetics models haven't proved to be sufficient because the characterization of the system is incomplete or ambiguous, containing unique, dynamically changing, and unforeseen situations. Typical cases are the alarm situations, structural failures, starting and stopping of plants, etc.

The combination of the above ideas, e.g. embedding fuzzy models in anytime systems extends the advantages of the Soft Computing approach with the flexibility with respect to the available input information and computational power thus making possible to achieve an optimal or near optimal trade off between accuracy and resource usage.

In this chapter, based on the results of the author published in [18], the adequacy and advantages of the above recited new modeling and information processing methods are illustrated. In the following sections, an overview of anytime, fuzzy, and combined methods is given with examples taken from the fields of signal- and image processing however they can be serious candidates in various other fields of engineering as well.

13.2 Anytime Signal Processing Techniques

13.2.1 Anytime Systems

Today there is an increasing number of applications where the computing must be carried out online, with a guaranteed response time and limited resources. Moreover, the available time and resources are not only limited but can also change during the operation of the system. Good examples are the modern computer-based signal processing, diagnostics, monitoring, and control systems, which are able to supervise complex processes and determine appropriate actions in case of failures or deviation from the optimal operational mode (see e.g. [22, 19]). In these systems the model of the supervised system is used and the evaluation of the system model must be carried out online, thus the model must not only be correct, but also treatable by the limited resources during limited time. Moreover, if some abnormality occurs in the system's behavior it may cause the reallocation of a part of the finite resources from the evaluation of the system model to another task. Also in case of an alarm signal, lower response time may be needed. Having approximate results can also help in making decisions for the further processing. In these cases, the so-called anytime algorithms and systems [27, 23] can be used advantageously, which are able to provide guaranteed response time and are flexible in respect to the available input data, time, and computational power.

Recursive or iterative algorithms are popular tools in anytime systems, because their complexity can be easily and flexibly changed. These algorithms always give some, possibly not accurate result and more and more accurate results can be obtained if the calculations are continued. Unfortunately, the usability of iterative algorithms is limited. Besides the iterative algorithms, in a more general frame, a wide-range of other types of computing methods/algorithms can be applied in anytime systems. This frame means that a modular architecture is used. The system is composed of modules each of them offering several implementations for a given task. These units (implementations within a given module) have uniform interface (same set of inputs, outputs, and solve the same problem) but can be characterized by different attribute-values, i.e. differ in their computational need and accuracy. At a given time, in the knowledge of the temporal conditions (tasks to complete, achievable time/resources, needed accuracy, etc.) an expert system can choose the adequate configuration, i.e. the units from the modules, which will be used. This means the optimization of the whole system instead of individual modules. Anytime processing may have great advantages in signal processing, monitoring, diagnostics, control, and related fields.

13.2.2 Block-Recursive Averaging

With the introduction of the idea of block-recursive averaging many classical transformed domain DSP methods can be operated as anytime algorithms. Block-recursive averaging means that the standard algorithms for recursive averaging are extended for data-blocks as single elements. To illustrate the key steps first the block-recursive linear averaging will be introduced. For an input sequence $x(n), n = 1, 2, \ldots$, the recursive linear averaging can be expressed as

$$y(n) = \frac{n-1}{n} y(n-1) + \frac{1}{n} x(n-1) \tag{13.1}$$

For $n \geq N$ (where N denotes the blocksize of the signal) the "block-oriented" linear averaging has the form of

$$X(n-N) = \frac{1}{N} \sum_{k=1}^{N} x(n-k) \tag{13.2}$$

while the block-recursive average can be written as

$$y(n) = \frac{n-N}{n} y(n-N) + \frac{N}{n} X(n-N) \tag{13.3}$$

If Eq.(13.3) is evaluated only in every N^{th} step, i.e. it is maximally decimated, then we can replace it with $n = mN, m = 1, 2, \ldots$, by

$$y(mN) = \frac{m-1}{m} y[(m-1)N] + \frac{1}{m} X[(m-1)N] \tag{13.4}$$

or simply

$$y(m) = \frac{m-1}{m} y(m-1) + \frac{1}{m} X(m-1) \tag{13.5}$$

where m stands as block identifier. (Note the formal correspondence with Eq.(13.1)). If the block identifier m in Eq.(13.5) is replaced by a constant $Q > 1$ then an exponential averaging effect is achieved. In many practical applications exponential averaging provides the best compromise if both the noise reduction and the signal tracking capabilities are important. This is valid in our case, as well, however, in this section only the linear and the sliding averagers are investigated, because they can be used directly to extend the size of certain signal transformation channels and can be applied in anytime systems. A similar development can be provided for the sliding-window averagers. The recursive form of this algorithm is given for a block size of N by

$$y(n) = y(n-1) + \frac{1}{N} [x(n-1) - x(n-N-1)] \tag{13.6}$$

If in Eq.(13.6) the input samples are replaced by preprocessed data, e.g. as in Eq.(13.2), then a block-recursive form is also possible:

$$y(n) = y(n-N) + \frac{1}{N} [X(n-N) - X(n-(M+1)N)] \tag{13.7}$$

which has real importance. If Eq.(13.7) is evaluated only in every N^{th} step, i.e. it is maximally decimated, then we can replace it with $n = mN$, $m = 1,2,\ldots$, by

$$y(mN) = y[(m-1)N] + \frac{1}{M}[X((m-1)N) - X((m-M-1)N)] \qquad (13.8)$$

or simply

$$y(m) = y(m-1) + \frac{1}{M}[X(m-1) - X(m-M-1)] \qquad (13.9)$$

where m stands as block identifier. (Note the formal correspondence with Eq.(13.6)). The generalization of these averaging schemes to signal transformations and/or filter-banks is straightforward. Only Eq.(13.2) should be replaced by the corresponding "block-oriented" operation. Fig.13.1 shows the block diagram of the linear averaging scheme. This is valid also for the exponential averaging except m must be replaced by Q. In Fig.13.2 the sliding window averager is presented. These frameworks can incorporate a variety of possible transformations and corresponding filter-banks which permit decimation by the block-size. Standard references, e.g. [3] provide the necessary theoretical and practical background.

The idea of transform-domain signal processing proved to be very efficient especially in adaptive filtering (see e.g. [15]). The most important practical advantage here compared to other methods is the early availability of rough estimates which can orientate in making decisions concerning further processing. The multiple-block sliding-window technique can be mentioned as a very characteristic algorithm of the proposed family. For this structure the computational complexity figures are also advantageous since using conventional methods to evaluate in "block-sliding-window"

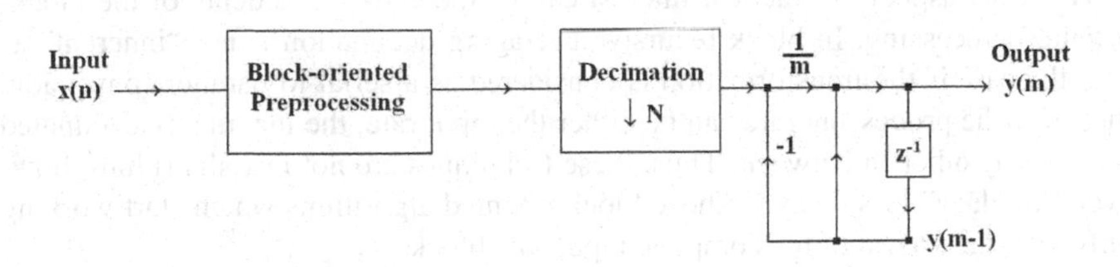

Fig. 13.1 Block-recursive linear averaging signal processing scheme, $n = mN$

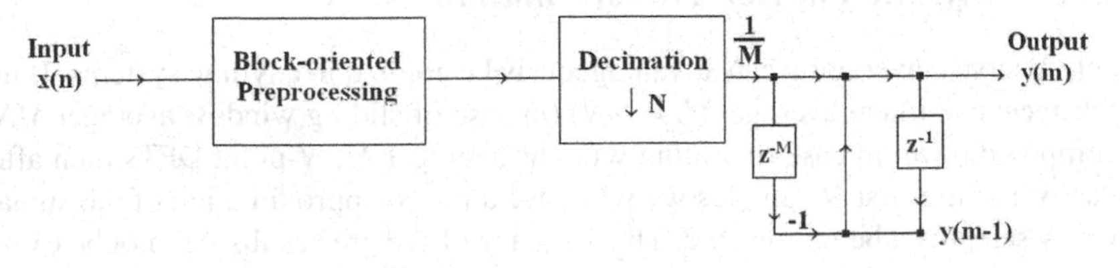

Fig. 13.2 Block-recursive sliding-window averaging scheme, $n = mN$, window-size MN

mode the transform of a block of MN samples would require M times an $(MN) * (MN)$ transformation, while the block-recursive solution calculates only for the last input block of N samples, i.e. M times an $(MN) * (N)$ "transformation".

As block-oriented preprocessing the DFT is the most widely used transformation for its fast algorithms (FFTs) and relatively easy interpretation. The above schemes can be operated for every "channel" of the DFT and after averaging this will correspond to the channel of a larger scale DFT. If linear averager is applied this scale equals mN while for sliding averager this figure is MN. The number of channels obviously remains N unless further parallel DFTs are applied. These additional DFTs have to locate their channel to the positions not covered by the existing channels. For the case where $M = 2$, i.e. only one additional parallel DFT is needed, this positioning can be solved with the so-called complementary DFT which is generated using the Nth roots of -1. This DFT locates its channels into the positions $\pi/N, 3\pi/N$, etc. For $M > 2$ proper frequency transposition techniques must be applied. If e.g. $M = 4$ then the full DFT will be of size $4N$ and four N-point DFTs (working on complex data) are to be used. The first DFT is responsible for the channels in positions $0, 8\pi/4N$, etc. The second DFT should cover the $2\pi/4N, 10\pi/4N$, etc., the third the $4\pi/4N, 12\pi/4N$, etc, and finally the fourth the $6\pi/4N, 14\pi/4N$, etc. positions, respectively. The first DFT does not need extra frequency transposition. The second and the fourth process complex input data coming from a complex modulator which multiplies the input samples by $e^{j2\pi n/4N}$ and $e^{j6\pi n/4N}$, respectively. The third DFT should be a complementary DFT (see Fig.13.3). It is obvious from the above development that if a full DFT is required the sliding window DFT must be preferred otherwise the number of the parallel channels should grow with m. The majority of the transform-domain signal processing methods prefers the DFT to other possible transformations. However, there are certain applications where other orthogonal transformations can also be utilized possibly with much better overall performance.

A further aspect of practical interest can be the end-to-end delay of the block-oriented processing. In block recursive averaging decimation is not "inherent" as it is the case if the transformation is considered as a serial to parallel conversion, therefore the processing rate can be either the input rate, the maximally decimated one, or any other in between. Thus, these techniques are not fast algorithms, however, "produce" less delay as those block-oriented algorithms which start working only after the arrival of the complete input data block.

13.2.3 Anytime Fourier Transformation

Block-recursive averaging can advantageously be applied in anytime systems. If the block-recursive linear averager ($L = mN$) (in case of sliding window averager MN) is composed of m, in case of sliding window averager M, N-point DFTs then after the arrival of the first N samples we will have a rough approximation of the signal, after $2N$ samples a better one, etc. The accuracy of the pre-results will not be exact, however the error is in most cases tolerable or even negligible [21]. In the following a simple example is presented which illustrates the usability of the proposed method.

So that, for a.e. t in I we have (with $\theta_h(\zeta) = \int_0^\tau \theta(\zeta(t))dt$)

$$\frac{\partial}{\partial s}\theta_h(\zeta^s(t))_{s=0}^p$$

$$= \int_{D\times D} \rho_h(||x-y||)\frac{|\zeta(x)-\zeta(y)|^p}{||x-y||^{N+\varepsilon p+2}}\langle x-y, Z(t,x)-Z(t,y)\rangle\, dxdy \tag{121}$$

$$+\int_{D\times D} \rho_h'(||x-y||)\frac{|\zeta(x)-\zeta(y)|^p}{||x-y||^{N+\varepsilon p}}\langle x-y, Z(t,x)-Z(t,y)\rangle\, dxdy$$

As $||x-y|| \le h$ in the previous integrals, we have:

$$Z(t,x)-Z(t,y) = DZ(t,x+\delta(t)(y-x)).(y-x). \tag{122}$$

There exists a measure $\mu_h(\Gamma(t)$ supported by

$$\Delta_h(\Sigma) = \cup_{0<t<\tau}\{t\}\times(\cup_{x\in\partial\Omega_t}B(x,h)), \tag{123}$$

such that

$$<\mu_h, Z> = \frac{\partial}{\partial s}\theta_h(\zeta^s(t))_{s=0}^p. \tag{124}$$

In some sense when $h\to 0$ the measure converges to the mean curvature of the moving boundary Γ_t.

6 Euler-Convection Problem

We have

Theorem 2. *Let V_0 be any given element in R^N. Then any minimizer (ζ, V) to the functional $\mathscr{E}$ over the family of tubes $\mathscr{T}$ solves the following problem:*

$$\frac{\partial}{\partial t}\zeta + \nabla\zeta.V = 0,\ \ \zeta(0) = \chi_{\Omega_0},\ \ \zeta(\tau) = \chi_{\Omega_1}, \tag{125}$$

$$\exists\Pi\ s.t. \frac{\partial}{\partial t}((\alpha\zeta+\beta)V) + D((\alpha\zeta+\beta)V).V + \nabla\Pi = \mu_h. \tag{126}$$

Moreover we have

$$V(0) = (V_0 + \nabla\theta)/(\alpha\zeta(0)+\beta). \tag{127}$$

References

1. Ambrosio, L.: Lecture notes on optimal transport problems. In: Colli, P., Rodrigues, J.F. (eds.) Mathematical Aspects of Evolving Interfaces. Lecture Notes in Math., vol. 1812, pp. 1–52. Springer, Berlin (2003)
2. Cannarsa, C., Da Prato, G., Zolésio, J.-P.: The damped wave equation in a moving domain. Journal of Differential Equations 85, 1–16 (1990)
3. Cuer, M., Zolésio, J.-P.: Control of singular problem via differentiation of a min-max. Systems Control Lett. 11(2), 151–158 (1988)
4. Delfour, M.C., Zolésio, J.-P.: Structure of shape derivatives for non smooth domains. Journal of Functional Analysis 104(1), 1–33 (1992)

5. Delfour, M.C., Zolésio, J.-P.: Shape analysis via oriented distance functions. Journal of Functional Analysis 123(1), 129–201 (1994)
6. Delfour, M.C., Zolésio, J.-P.: Shapes and Geometries. Analysis, Differential Calculus, and Optimization. SIAM, Philadelphia (2001)
7. Delfour, M.C., Zolésio, J.-P.: Oriented distance function and its evolution equation for initial sets with thin boundary. SIAM J. Control Optim. 42(6), 2286–2304 (2004)
8. Desaint, F.R., Zolésio, J.-P.: Manifold derivative in the Laplace-Beltrami equation. Journal of Functional Analysis 151(1), 234–269 (1997)
9. Dziri, R., Zolésio, J.-P.: Dynamical shape control in non-cylindrical Navier-Stokes equations. J. convex analysis 6(2), 293–318 (1999)
10. Dziri, R., Zolésio, J.-P.: Dynamical shape control in non-cylindrical hydrodynamics. Inverse Problem 15(1), 113–122 (1999)
11. Dziri, R., Zolésio, J.-P.: Tube derivative of non-cylindrical shape functionals and variational formulations. In: Glowinski, R., Zolésio, J.-P. (eds.) Free and Moving Boundaries: Analysis, Simulation and Control. Lecture Notes in Pure and Applied Mathematics, vol. 252. Chapman & Hall/CRC (2007)
12. Kawohl, B., Pironneau, O., Tartar, L., Zolésio, J.-P.: Optimal shape design. Lecture Notes in Mathematics, vol. 1740. Springer, Heidelberg (2000)
13. Moubachir, M., Zolésio, J.-P.: Moving shape analysis and control: application to fluid structure interaction. Pure and Applied Mathematics series. CRC, Boca Raton (2006)
14. Da Prato, G., Zolésio, J.-P.: Dynamical programming for non cylindrical parabolic equation. Sys. Control Lett. 11 (1988)
15. Da Prato, G., Zolésio, J.-P.: Existence and control for wave equation in moving domain. In: Stabilization of Flexible Structures. LNCIS, vol. 147, pp. 167–190. Springer, Heidelberg (1990)
16. Sokolowski, J., Zolésio, J.-P.: Introduction to Shape Optimization: Shape Sensitivity Analysis. Springer Series in Computational Mathematics, vol. 10. Springer, Berlin (1992)
17. Zolésio, J.-P.: Introduction to shape optimization and free boundary problems. In: Delfour, M.C. (ed.) Shape Optimization and Free Boundaries. NATO ASI, Series C: Mathematical and Physical Sciences, vol. 380, pp. 397–457 (1992)
18. Zolésio, J.-P.: Shape differential with non smooth field. In: Borggard, J., Burns, J., Cliff, E., Schreck, S. (eds.) Computational Methods for Optimal Design and Control. Progress in Systems and Control Theory, vol. 24, pp. 426–460. Birkhauser, Basel (1998)
19. Zolésio, J.-P.: Variational principle in the Euler flow. In: Leugering, G. (ed.) Proceedings of the IFIP-WG7.2 Conference, Chemnitz. Int. Series of Num. Math., vol. 133 (1999)
20. Zolésio, J.-P.: Weak set evolution and variational applications. In: Shape Optimization and Optimal Design. Lecture Notes in Pure and Applied Mathematics, vol. 216, pp. 415–442. Marcel Dekker, N.Y. (2001)
21. J.-P. Zolésio: Set Weak Evolution and Transverse Field, Variational Applications and Shape Differential Equation INRIA report RR-464 (2002), `http://www-sop.inria.fr/rapports/sophia/RR-464`
22. Zolésio, J.-P.: Shape topology by tube geodesic. In: Information Processing: Recent Mathematical Advances in Optimization and Control, pp. 185–204. Presses de l'Ecole des Mines de Paris (2004)
23. Zolésio, J.-P.: Control of moving domains, shape stabilization and variational tube formulations. International Series of Numerical Mathematics, vol. 155, pp. 329–382. Birkhauser Verlag, Basel (2007)
24. Zolésio, J.-P.: Tubes analysis. In: Glowinski, R., Zolésio, J.-P. (eds.) Free and Moving Boundaries: Analysis, Simulation and Control. Lecture Notes in Pure and Applied Mathematics, vol. 252. Chapman & Hall/CRC (2007)

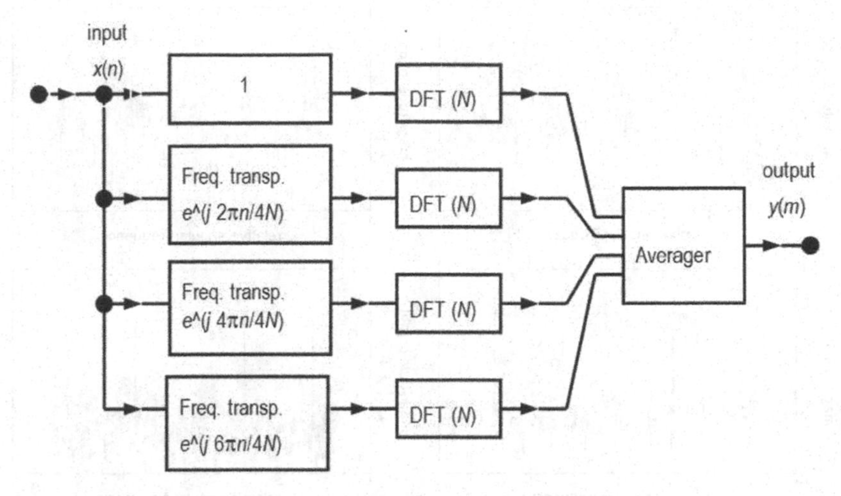

Fig. 13.3 The block structure of the Anytime Fourier Transformation

In the example a 256-channel DFT is calculated recursively with $N = 64$ for $m = 1,2,8,16$ on a noisy multisine signal. The input sequence is

$$x(n) = \sin\left(\frac{2\pi 50n}{4N}\right) + 0.8\sin\left(\frac{2\pi 112n}{4N}\right) + \text{rand} - 0.5 \tag{13.10}$$

where rand stands for a random number generated by MATLAB between 0 and 1. The sinusoid is located exactly to a DFT channel position. The simulation results for $m = 1,2,8$, and 16 are given in Fig.13.4.

Here we would like to remark the following: If the signal is noiseless then after the arrival of a full signal period, the recursive Anytime Fourier Transformation structure output will match exactly to the frequency components of the signal. This is not valid however if a noisy signal is processed. Although, a well appreciable noise filtering effect can be followed parallel with the increase of m (as illustration, see Fig. 13.4). Therefore, the presented technique can effectively be used even in those cases where the full size of the DFT is shorter then that of the signal or when the signal to be processed is weekly non-stationary.

13.3 Fuzzy Based Information Processing Techniques

Fuzzy logic is rapidly emerging as a powerful resource of information processing because fuzzy approaches are able to deal with the typical uncertainty, which characterizes any physical system. In the following subsections a brief selection of such methods is given.

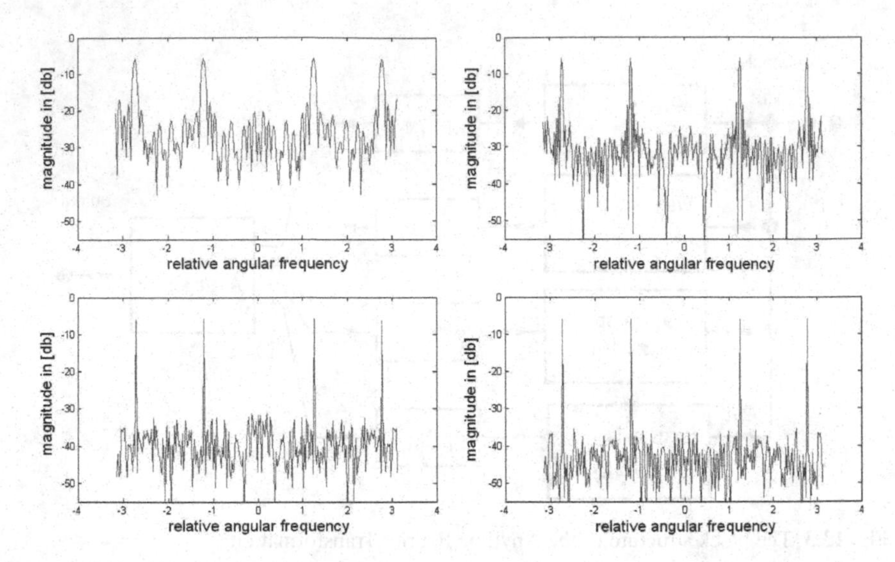

Fig. 13.4 256-channel anytime DFT of a noisy multisine, exactly at a DFT channel. $N = 64, m = 1,2,8,16$.

13.3.1 Fuzzy Based Noise Elimination

A major task in the field of digital processing of measurement signals is to extract information from sensor data corrupted by noise [13, 14]. For this purpose Russo's fuzzy filters use a special fuzzy system characterized by an IF-THEN-ELSE structure and a specific inference mechanism. Different noise statistics can be addressed by adopting different combinations of fuzzy sets and rules [13, 14].

Let $x(\mathbf{r})$ be the pixel luminance at location $\mathbf{r} = [r1, r2]$ in the noisy image where r_1 is the horizontal and r_2 the vertical coordinate of the pixel. Let $\mathbf{N}$ be the set of eight neighboring pixels (see Fig.13.5). The input variables of the fuzzy filter are the amplitude differences defined by:

$$\Delta x_j = x_j - x_0, j = 1,...,8 \tag{13.11}$$

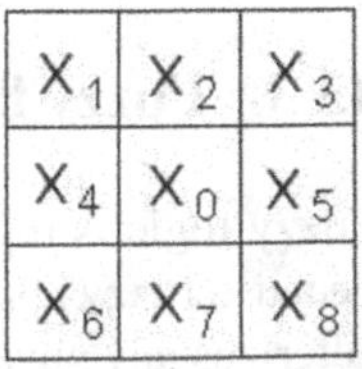

X_1	X_2	X_3
X_4	X_0	X_5
X_6	X_7	X_8

Fig. 13.5 The neighboring pixels of the actually processed pixel x_0

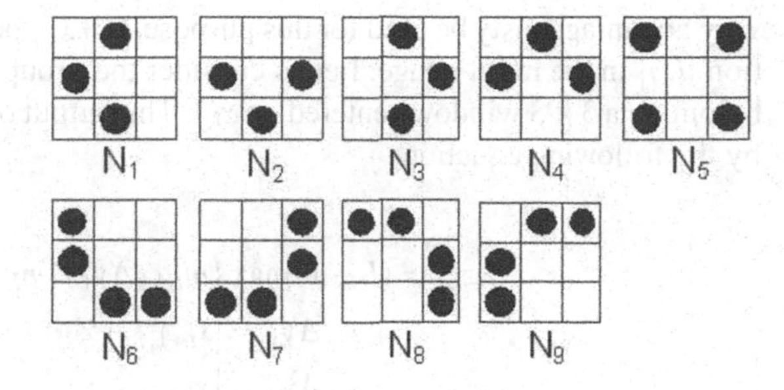

Fig. 13.6 Pixel Patterns $N_1, , N_9$

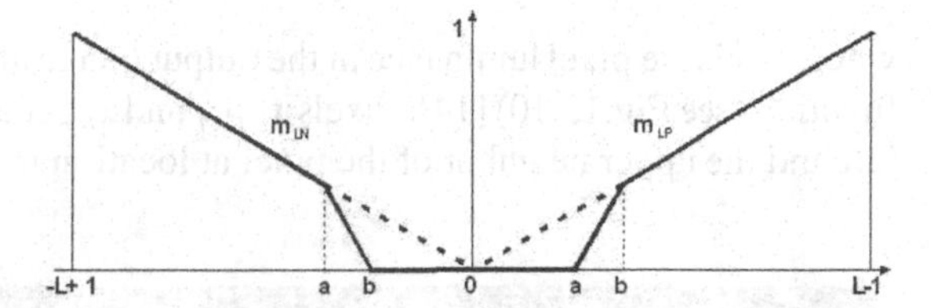

Fig. 13.7 Membership function m_{LP}. Parameters a and b are appropriate constant values

where the $x_j, j = 1, \ldots, 8$ values are the neighboring pixels of the actually processed pixel x_0 (see Fig.13.5). Let y_0 be the luminance of the pixel having the same position as x_0 in the output signal. This value is given by the following relationship

$$y_0 = x_0 + \Delta y \tag{13.12}$$

where Δy is determined thereinafter in Eq.(13.17). Let the rule base deal with the pixel patterns $\mathbf{N_1}, \ldots, \mathbf{N_9}$ (see Fig.13.6). The value y_0 can be calculated, as follows [13]:

$$\lambda = \max\left\{\min\left\{m_{LP}(\Delta x_j) : x_j \in N_i\right\}, i = 1, ..., 9\right\} \tag{13.13}$$

$$\lambda^* = \max\left\{\min\left\{m_{LN}(\Delta x_j) : x_j \in N_i\right\}, i = 1, ..., 9\right\} \tag{13.14}$$

$$\begin{aligned} \Delta y &= (L-1)\Delta\lambda \\ y_0 &= x_0 + \Delta y \end{aligned} \tag{13.15}$$

where $\Delta\lambda = \lambda - \lambda^*, m_{LP}$ and m_{LN} correspond to the membership functions and $m_{LP}(u) = m_{LN}(-u)$ (see Fig.13.7). The filter is recursively applied to the input data. An example of the described fuzzy-filter can be seen in Fig.??-??.

13.3.2 Fuzzy Based Edge Detection

Edge detection in an image is a very important step for a complete image understanding system. In fact, edges correspond to object boundaries and are therefore useful inputs for 3D reconstruction algorithms. Fuzzy based edge detection [14] can very advantageously be used for this purpose. Let $x_{i,j}$ be the pixel luminance at location $[i, j]$ in the input image. Let us consider the group of neighboring pixels which belong to a 3×3 window centered on $x_{i,j}$. The output of the edge detector is yielded by the following equation

$$\begin{aligned} z_{i,j} &= (L-1)\max\{m_{LA}(\Delta y_1), m_{LA}(\Delta y_2)\} \\ \Delta y_1 &= \left|x_{i-1,j} - x_{i,j}\right| \\ \Delta y_2 &= \left|x_{i,j-1} - x_{i,j}\right| \end{aligned} \tag{13.16}$$

where $z_{i,j}$ is the pixel luminance in the output image and m_{LA} is the used membership function (see Fig.13.10) [14]. Pixels $x_{i-1,j}$ and $x_{i,j-1}$ are the luminance values of the left and the upper neighbor of the pixel at location $[i, j]$.

Fig. 13.8 Original photo of a crashed car corrupted by noise.

Fig. 13.9 Fuzzy-filtered image of the photo.

The fuzzy based technique compared to the classical methods provides better results with less (very small) processing time. Fig.13.12 shows an example for edge detection results of Fig.13.11.

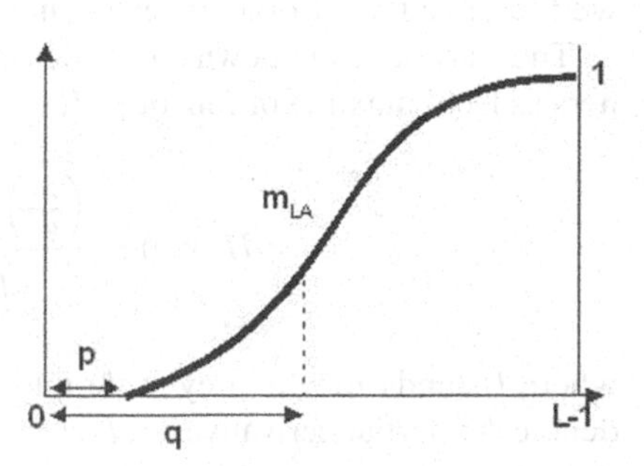

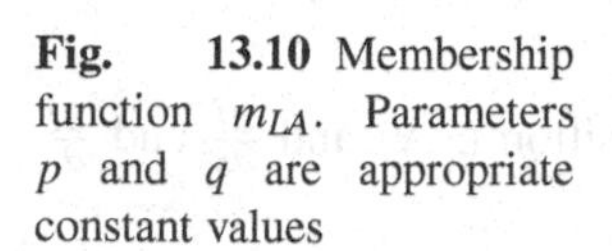

Fig. 13.10 Membership function m_{LA}. Parameters p and q are appropriate constant values

Fig. 13.11 Original photo

13.3.3 Fuzzy Based Corner Detection

Corner detection should satisfy the requirements of (1) All the true corners should be detected; (2) No false corners should be detected; (3) Corner points should be well localized; (4) Corner detector should be robust with respect to the noise [20].

There are several known corner detectors, among which Förstner determines corners as local maxima of function $H(x,y)$ [4]

$$H(x,y) = \frac{\left(\frac{\partial I}{\partial x}\right)^2 \left(\frac{\partial I}{\partial y}\right)^2 - \left(\frac{\partial I}{\partial x}\frac{\partial I}{\partial y}\right)^2}{\left(\frac{\partial I}{\partial x}\right)^2 + \left(\frac{\partial I}{\partial y}\right)^2} \tag{13.17}$$

where I stands for the grey scale intensity at pixel position (x,y) and $\frac{\partial I}{\partial x}$ and $\frac{\partial I}{\partial y}$ denote the partial derivatives of I.

Fig. 13.12 Fuzzy based edge detection of Fig. 13.11.

Starting from this definition, a new improved corner detection algorithm can be developed by combining it with fuzzy reasoning [20]. This is used for the characterization of the continuous transient between the localized and not localized corner points, as well. The algorithm consists of the following steps: First, the picture, in which we have to find the corners, is preprocessed. As result the noise is eliminated. For this purpose the intelligent fuzzy filters described in [13] and [14] can be applied. If necessary, the image is also smoothed before taking the derivatives [2]. After noise-filtering, the first derivatives of the intensity function $I(x,y)$ are calculated in each image point by using the following convolution masks:

$$\begin{bmatrix} -1 & 0 & 1 \\ -1 & 0 & 1 \\ -1 & 0 & 1 \end{bmatrix} \text{ for determining } \frac{\partial I}{\partial x} \text{ and } \begin{bmatrix} -1 & -1 & -1 \\ 0 & 0 & 0 \\ 1 & 1 & 1 \end{bmatrix} \text{ for determining } \frac{\partial I}{\partial y}.$$

For increasing the effectiveness of the corner detection it is proposed to smooth each of the entries I_x^2, I_y^2 and I_xI_y, in Eq.(13.18), which correspond with the first partial derivates of the intensity function $I(x,y)$ (here x,y denote the 2D coordinates of the pixels). This can be done e.g. by applying a Gaussian 6×6 convolution kernel with $\sigma = 1$ [20]. As the following step, the values $H(x,y)$ are calculated for each image point with the help of the previously determined I_x^2, I_y^2 and $I_x I_y$ smoothed values. If the detected corners are neighbors, then we should keep only the corner having the largest calculated value $H(x,y)$. The others are to be ignored. In most cases we can not unambiguously determine that the analyzed image point is a corner or not with only the help of a certain concrete threshold value, therefore in the proposed algorithm fuzzy techniques are applied for the calculation of the values (corners) which increases the rate of correct corner detection. By the score of the membership function of fuzzy set "corners" (see Fig.13.13) we can determine a weighting factor, which characterizes the rate of the corner's membership. The value of the membership function m_c is 1 for those image points for which the calculated value H equals or is larger than the given threshold value. With the help of parameters p, q (see Fig.13.13) the shape of the membership function and thus the sensitivity of the described detector can be modified. Finally, the output of the proposed corner detector is yielded by the following relation:

$$C_{x,y} = (L-1)\, m_c(H) \tag{13.18}$$

where $C_{x,y}$ represent the gray-level intensity values of the output image, x and y are the horizontal and vertical coordinates of the processed image point, and L corresponds to the largest intensity value.

Fig.13.14-13.15 illustrates the performance of the presented fuzzy corner detection algorithm. In Fig.13.14 a fuzzy filtered image can be seen, while Fig.13.15 shows the result of the corner detection.

13.4 Combining Anytime and Fuzzy Techniques

The efficiency of the presented non-classical techniques can further be improved by combining the methods. In the following, the anytime signal processing example of Subsection 2.3 (Anytime Fourier Transformation) is combined with fuzzy interpretation of the non-exact anytime estimations of the results (Anytime Fuzzy Fast Fourier transformation, FAnFFT). According to it, the noisy frequency characteristics is viewed as a fuzzy set over the Universe of Frequencies. Before the defuzzifica-

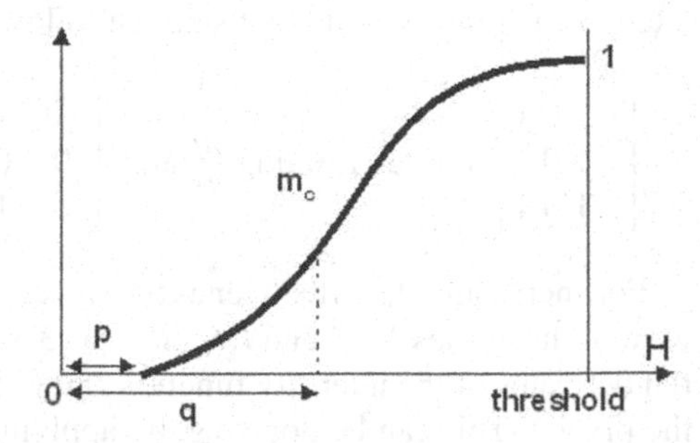

Fig. 13.13 Membership function of fuzzy set corner (m_c). Axis H is the axis of the calculated $H(x,y)$ values

Fig. 13.14 Fuzzy filtered image

tion we first evaluate the α-cut of the fuzzy set based on a properly chosen α value. This value serves to determine the limit separating the "useful" signals from what is interpreted as noise. In the obtained α-cut, the separated "picks" are handled and defuzzificated separately since each pick, as an individual fuzzy set, represents the frequency of a signal component. As defuzzification, the indexed Center of Gravity (iCoG) defuzzification method is applied based on the chosen α-cut of the output.

Since the most typical errors, the picket fence and the leakage cause symmetrical error around the accurate value, the applied fuzzy defuzzification method results in high accuracy. This is because instead of taking the non-accurate values as exact ones, thus bringing error into the interpretation, value imprecisiation' is done which in reality means meaning precisiation'.

In the next example, FAnFFT is applied on the noisy multisine signal of Eq.(13.10). The results of the anytime Fourier transformation is further processed by fuzzy interpretation. Fig.13.16 shows the α-cuts with $\alpha = 17$ dB for $m = 1,2,8,16$. Table 13.1 summarizes the obtained frequencies evaluated by $\alpha = 17$ dB. In Fig.13.17 the convergence of the non-zero amplitude components can be followed if (a) the non-noisy and (b) the noisy signal is processed. The improvement in both the resolution and noise reduction is remarkable.

Fig. 13.15 The corner detection results

Table 13.1 Obtained Frequencies for $m = 1, 2, 8$, and $16, \alpha = -17$ dB

m	Obtained frequencies			
1	-2.7515	−1.2268	1.2267	2.7515
2	-2.7490	−1.2275	1.2275	2.7490
8	-2.7517	−1.2267	1.2267	2.7517
16	-2.7488	−1.2271	1.2271	2.7488

Here we would like to remark that if value α is chosen too small then false picks caused by the noise may also appear in the spectrum. At $\alpha = 25$ dB e.g., in case of $m = 1$ and 2 we obtain 10-11 picks (of which 6-7 come of noise) however as the approximation becomes more accurate along the time, at $m = 8$ and 16, the false picks disappear and only the 4 "useful" frequencies remain in the spectrum.

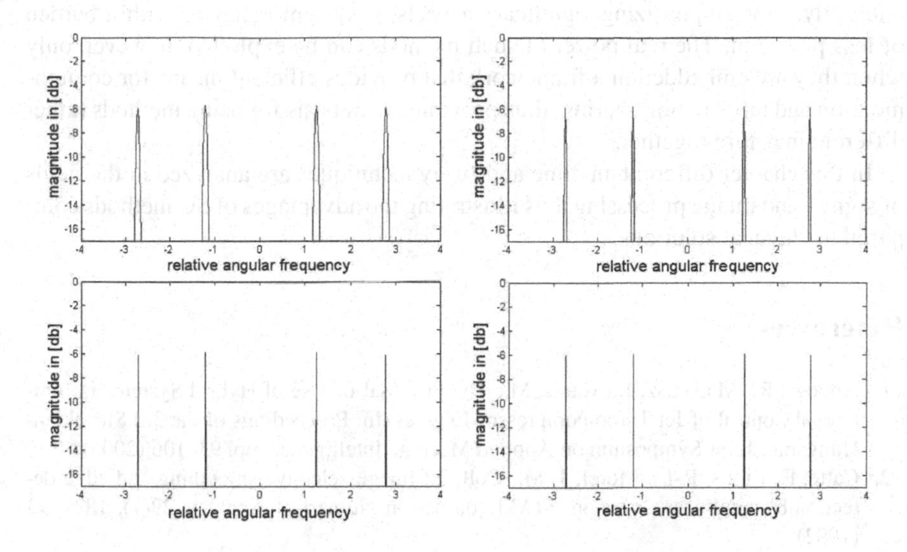

Fig. 13.16 α-cuts with $\alpha = -17$ dB of the 256-channel anytime DFT of the noisy multisine in Eq.(13.10). $N = 64$, $m = 1, 2, 8, 16$.

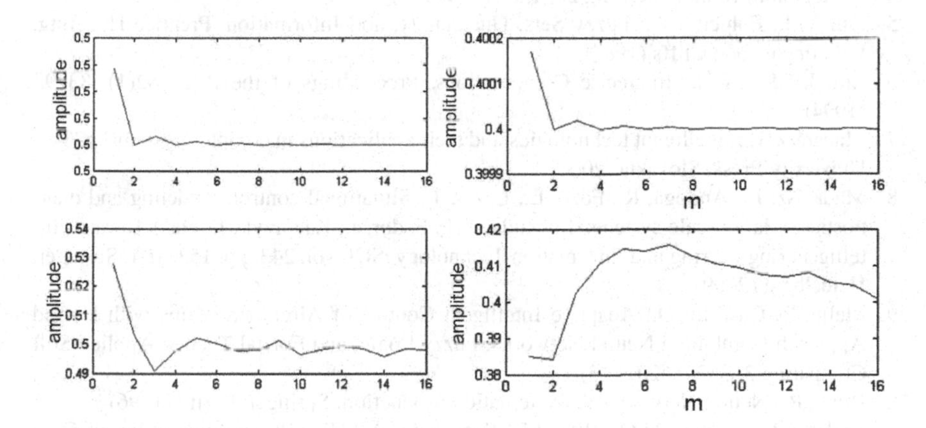

Fig. 13.17 Convergence of the amplitude of the two non-zero multisine components in Eq.(13.10): non-noisy signal (upper) and the signal with noise (lower)

13.5 Conclusion

The increased complexity of engineering systems caused classical problem solving methods, especially in resource-bounded applications, to fail to produce "usable" solutions. This led to focus on knowledge representation, information handling, and modeling. Soft Computing methods are serious candidates for handling many of the theoretical and practical limitations and, in many cases, are the best if not the only alternatives for emphasizing significant aspects of system behavior with a burden of less precision. The real power of such methods can be exploited, however, only when they are embedded in a framework that provides efficient means for communication and information sharing, thus providing firm basis for using methods rather different in nature together.

In this chapter different anytime and fuzzy techniques are analyzed in the fields of signal- and image processing thus illustrating the advantages of SC methods compared to classical solutions.

References

1. Andoga, R., Madarász, L., Karas, M.: The Proposal of Use of Hybrid Systems in Situational Control of Jet Turbo-compressor Engines. In: Proceedings of the 3rd Slovakian-Hungarian Joint Symposium on Applied Machine Intelligence, pp. 93–106 (2005)
2. Catté, F., Lions, P.-L., Morel, J.-M., Coll, T.: Image selective smoothing and edge detection by nonlinear diffusion. SIAM Journal on Numerical Analysis 29(1), 182–193 (1992)
3. Crochiere, R.E., Rabiner, L.R.: Multirate Digital Signal Processing. Prentice-Hall, Inc., Englewood Cliffs (1983)
4. Förstner, W.: A feature based correspondence algorithm for image matching. Int. Arch. Photogramm Remote Sensing 26, 150–166 (1986)
5. Klir, G.J., Folger, T.A.: Fuzzy Sets, Uncertainty, and Information. Prentice Hall Intg. Inc., Englewood Cliffs (1988)
6. Liu, J.W.S., et al.: Imprecise Computations. Proceedings of the IEEE 82(1), 83–93 (1994)
7. Madarász, L.: Intelligent technologies and their applications in complex systems, p. 348. University Press, Slovakia (2004)
8. Madarász, L., Andoga, R., Fözö, L., Lazar, T.: Situational control, modeling and diagnostics of large scale systems. In: Rudas, I.J., Fodor, J., Kacprzyk, J. (eds.) Towards Intelligent Engineering and Information Technology. SCI, vol. 243, pp. 153–164. Springer, Heidelberg (2009)
9. Melin, P., Castillo, O.: Adaptive Intelligent Control of Aircraft Systems with Hybrid Approach Combining Neural Networks, Fuzzy Logic, and Fractal Theory. Applied Soft Computing 3, 353–362 (2003)
10. Rojas, R.: Neural Networks, A Systematic Introduction. Springer, Berlin (1996)
11. Rudas, I.J., Kaynak, M.O., Bitó, J.F., Szeghegyi, Á.: New Possibilities in Fuzzy Controllers Design Using Generalized Operators. In: Proceedings of the 5th International Conference on Emerging Technologies and Factory Automation, pp. 513–517 (1996)
12. Russel, S., Norvig, P.: Atrifial Intelligence a Modern Approach. Prentice Hall, New Jersey (2003)
13. Russo, F.: Fuzzy Filtering of Noisy Sensor Data. In: Proceedings of the IEEE Instrumentation and Measurement Technology Conference, pp. 1281–1285 (1996)
14. Russo, F.: Recent Advances in Fuzzy Techniques for Image Enhancement. IEEE Transactions on Instrumentation and Measurement 47(6), 1428–1434 (1998)
15. Shynk, J.J.: Frequency-Domain and Multirate Adaptive Filtering. IEEE Signal Processing Magazine, 15–37 (January 1992)
16. Sinčák, P., et al. (eds.): Intelligent Technologies Theory and Applications. IOS Press, Amsterdam (2002)

17. Yager, R.R.: Fuzzy thinking as quick and efficient. Cybernetica 23, 265–298 (1980)
18. Várkonyi, T.A.: Soft Computing Based Signal Processing Approaches for Supporting Modeling and Control of Engineering Systems - A Case Study. In: Proceedings of the 14th International Conference on Intelligent Engineering Systems, pp. 102–107 (2010)
19. Várkonyi-Kóczy, A.R.: State Dependant Anytime Control Methodology for Non-linear Systems. International Journal of Advanced Computational Intelligence and Intelligent Informatics (JACIII) 12(2), 198–205 (2008)
20. Várkonyi-Kóczy, A.R.: Fuzzy Logic Supported Corner Detection. Journal of Intelligent and Fuzzy Systems 19(3), 41–50 (2008)
21. Várkonyi-Kóczy, A.R.: Fast Anytime Fuzzy Fourier Estimation of Multisine Signals. IEEE Trans. on Instrumentation and Measurement 58(5), 1763–1770 (2009)
22. Várkonyi-Kóczy, A.R., Baranyi, P., Patton, R.J.: Anytime Fuzzy Modeling Approach for Fault Detection Systems. In: Proceedings of the IEEE Instrumentation and Measurement Technology Conference, pp. 1611–1616 (2003)
23. Várkonyi-Kóczy, A.R., Kovácsházy, T.: Anytime Algorithms in Embedded Signal Processing Systems. In: Proceedings of the IX. European Signal Processing Conference, vol. 1, pp. 169–172 (1998)
24. Vaščák, J., Kováčik, P., Hirota, K., Sinčák, P.: Performance-based Adaptive Fuzzy Control of Aircrafts. In: Proceedings of the 10th IEEE International Conference on Fuzzy Systems, vol. 2, pp. 761–764 (2001)
25. Vaščák, J., Mikloš, M.: Hybrid Fuzzy Adaptive Control of LEGO Robots. In: Proceedings of the 2nd International Symposium on Advanced Intelligent Systems, vol. 2, pp. 252–256 (2001)
26. Zadeh, L.: Fuzzy Logic, Neural Networks, and Soft Computing. Communications of the ACM 37(3), 77–83 (1994)
27. Zilberstein, S.: Using Anytime Algorithms in Intelligent Systems. AI Magazine 17(3), 73–83 (1996)
28. Zilberstein, S., Russel, J.: Reasoning about optimal time allocation using conditional profiles. In: Proceedings of AAAI 1992 Workshop on Implementation of Temporal Reasoning, pp. 191–197 (1992)
29. Zilberstein, S., Russel, J.: Constructing utility-driven real-time systems using anytime algorithms. In: Proceedings of the IEEE Workshop on Imprecise and Approximate Computation, pp. 6–10 (1992)

Part III
Application of Computational Intelligence

Chapter 14
Differential Diagnosis of Dementia Using HUMANN-S Based Ensembles

Patricio García Báez, Carmen Paz Suárez Araujo, Carlos Fernández Viadero, and Aleš Procházka

Abstract. Dementia is one of the most prevalent diseases associated to aging. The two most common variations of this disease are Alzheimer Dementia (AD) type and Vascular Dementia (VD) type, but there are other many forms (OTD): Lewi Body, Subcortical, Parkinson, Trauma, Infectious dementias, etc. All of these forms can be associated with different patterns of anatomical affectation, different risk factors, multiple diagnostic characteristics and multiple profiles of neuropsychological tests, making the Differential Diagnosis of Dementias (DDD) very complex. In this chapter we propose new automatic diagnostic tools based on a data fusion scheme and neural ensemble approach, concretely we have designed HUMANN-S ensemble systems with missing data processing capability. Their ability have been explored using a battery of cognitive and functional/instrumental scales for DDD, among AD, VD and OTD. We carried out a comparative study between theese methods and a clinical expert, reaching these systems a higher level of performance than the expert. Our proposal is an alternative and effective complementary method to assist the diagnosis of dementia both, in specialized care as well as in primary care centres.

Patricio García Báez
Departamento de Estadística, Investigación Operativa y Computación,
Universidad de La Laguna. 38271 La Laguna, Spain
e-mail: pgarcia@ull.es

Carmen Paz Suárez Araujo
Instituto Universitario de Ciencias y Tecnologías Cibernéticas,
Universidad de Las Palmas de Gran Canaria. 35017 Las Palmas de Gran Canaria, Spain
e-mail: cpsuarez@dis.ulpgc.es

Carlos Fernández Viadero
Hospital Psiquiátrico Parayas, Gobierno de Cantabria. Santander, Spain
Aleš Procházka
Institute of Chemical Technology in Prague, Department of Computing
and Control Engineering. Prague, Czech Republic
e-mail: Ales.Prochazka@vscht.cz

J. Fodor et al. (Eds.): Recent Advances in Intelligent Engineering Systems, SCI 378, pp. 305–324.
springerlink.com

14.1 Introduction

Aging population has supposed an increase in dementia cases because the most important risk factor for this neurodegenerative pathology is old age [39][20]. In clinical practice, dementia refers to a syndrome characterized by acquired cognitive deterioration that can be associated with several potential stages of the disease [1]. Two of the most common variations of dementia are the AD type and the VD type, and both can be associated with different anatomical affectation patterns, different risk factors, different diagnostic characteristics, and different neuropsychological test profiles [4]. AD is considered a prototypical form of cortical dementia, given the pronounced atrophy of the cerebral cortex [2]. The affectation of the medial temporal region and, specifically, the hippocampal and entorhinal cortex, seems to justify the typical memory difficulties found in AD patients. Neuropathological disorders take on an accumulation form of atypical proteins (Beta amiloide and Tau) that are also a typical finding of this disease. In fact, the disease is associated with two abnormal proteins: neurofibrillary tangles clustering inside the neurons, and amyloid plaques that accumulate outside of the neurons of primarily the cerebral cortex, amygdale and the hippocampus. On the other side, VD can be the result of a heterogeneous group of disorders [3], such as ictus, cerebral hemorrhaging and Binswanger disease. Some of the characteristic risk factors for VD are hypertension condition, cardiopathy, hypercholesterolemia, nicotine poisoning and diabetes mellitus. VD is more characteristic dementia for men than for women. For both types of dementia and for dementia in general, one of the most important risk factors is the ageing process. The most widely known type of dementia is AD, which accounts for approximately 50% of all types of diagnosed dementia. Today, it is estimated that there are 18 million people suffering from AD worldwide, and the disease affects 5% of 65-years old and 30-50% of 85-years old [33]. VD has traditionally been considered the second most common cause of dementia (up to 20% of all dementias, either alone or in combination with AD) [42]. The rest of dementias can be from different types like Subcortical, Parkinson, Trauma, Infectious dementias. All these types no AD neither VD can be denominated other type of dementia [12]. In view of the prevalence of cognitive impairment, a proper recognition of dementia and its differential diagnosis become essential.

Establishing a clinical diagnosis of AD and/or VD can be a difficult task. This difficulty is related to several factors. In the first place, the wide spectrum of diagnostic criteria which can be used by the clinical expert, for AD we can find CERAD NINCDS-ADRA, CAMDEX, DSM-IV, and for VD, ADDTC, NINDS-AIREN, DSM-IV, and their short coincidence, not much more than 5% in the set. Another important factor is still the lack of specifically validated clinical criteria for each type of dementia. On the other hand, we face the clinical-pathological duality, which according to it can be found brains with a high neuropathological load without the clinical manifestation of the disease. Finally, since a few months, by means of the International Working Group for New Research Criteria for the Diagnosis of AD (IWGNRCDAD) meetings, a new complex terminology has been created [8] because a series of situations or "diagnostics" appear, even years before the disease starts to show symptomatically, being able to emerge multiple categories, related to AD: AD, prodromic AD, Alzheimer's Dementia, typical AD, atypical AD, mixed AD, Alzheimer's pathology, mild cognitive impairment, AD preclinical stages, at which the presence of a series of physiopathological, topographical or genetic biomarkers are going to play an important role. Other important considerations are the heterogeneous nature of VD, the affirmation that a definitive diagnosis of AD can be only obtained with the neuropathology and, most importantly, the typical presence of patients that can exhibit signs and symptoms of both dementias and/or others (mixed dementia) [3][12]. There is data showing that 15% of AD patients meet the criteria for VD and, nonetheless, that many VD patients (75%) can be considered to meet the criteria for AD [1].

From a neuropsychological perspective, for DSM-IV [1], AD and VD reflect memory deterioration, although this is less obvious in VD than in AD, in addition to other forms of cognitive impairment (aphasia, apraxia, agnosia, or executive function), as well as deterioration related to social and/or professional functionality compared to its level of the previous function. In neuropsychological exploration there is no clear consensus on what constitutes a cognitive profile of AD as compared to VD. Despite this variability in the defining criteria, in the actual literature support for at least two general tendencies: first, patients with AD tend to reveal more important deficits in long term memory, and second, patients with VD reveal greater alterations in the measures of executive function as opposed to those patients with AD [4]. Consequently, from a point of view based on clinical diagnosis, it can be suggested that AD and VD patients differ in neuro-images, the presence of cardiovascular risk factors or cardiovascular illnesses and neuropsychological tests.

As mentioned previously, the diagnosis among several types of dementia is a complex task. Even in specialized centers diagnostic accuracy usually reaches a level of only 80%, leaving the remaining cases to be subject to a long-term evaluation. Several biomarkers have been linked to AD, such as the cerebrospinal fluid tau, amyloid, urine F2-isoprostane, brain atrophy and volume loss detected by PET or MRI scan [41][25]. However, these methods have either not proven to be conclusive, or remain primarily university or research hospital-based tools. Even so, there are studies that report a diagnostic accuracy rate of AD diagnostic that does not reach 75% [26].

In this context it is necessary to develop new and alternative methods and instruments of diagnosis, placing special emphasis on early and DDD, and introducing its use in all healthcare areas, not just in specialized care but also in primary care centers.

We propose new diagnostic tools based on a data fusion scheme using Artificial Neural Networks (ANN) and ensemble systems. Concretely we have designed two types of HUMANN-S based ensembles, where HUMANN-S is the supervised version of HUMANN architecture, with the capacity to process missing data. In this work we explore the ability of a simple HUMANN-S and HUMANN-S based ensembles, which were combined with simple and weighted majority voting strategies. We present our preliminary results on DDD using neuropsychological tests along with these systems. We also perform a comparative study between these methods and a clinical expert, revealing the high level of performance of our proposal. In addition, the proposed systems in this chapter have important advantages referring to other computational solutions based on artificial neural networks: HUMANN-S can handle noise in an efficient way [11], it has a strong adaptive character, it is able to process missing data, reaching accuracy levels up to 11% greater than others, like against a special computer-intensive algorithms based on ensemble learning methods, recently developed in regard to AD classification, essentially using EEG as input [32][33]. Finally, in the proposed methods a novel information environment is used and applied to DDD, as of yet not found in the literature. This information environment consists of different combinations of several neuropsychological tests, concretely, The Mini Mental State Examination [10], the most internationally frequently used instrument to value the cognoscitive function, and a set of scales to value functional, basic and instrumental activities. These last scales are the FAST scale (Functional Assessment Staging) [34], Katz's Index [21], Barthel's index [29] and Lawton-Brody's index [24].

14.2 Data Sets

The dataset for this study is made using the results of the 267 clinical consultations performed on 30 patients during 2005 at Alzheimer's Patient Association of Gran Canaria [14]. Its structure includes a patient identifier, resulting of 5 neuropsychological tests and diagnose of cognitive decline as well as diagnose of differential dementia. An advantage of this data is its homogeneity, each patient has scores of his/her monthly-made tests, with the exception being the Mini Mental test which is done twice a year. Although the majority of the patients have been tested 12 times, there are some patients with fewer consultations made and some missing data from the the consultations. Thus, the dataset is incomplete, that is, there are missing data features in many of the consultations as well as the complete patient test set, Table 14.2.

A description of the 5 different used data tests follows:

1. Mini Mental Status Examination (MMSE) [10] is the most spread, employed and quoted standardized instrument to value the cognoscitive function. It consists of a set of short and simple questions that allow a quick evaluation of several cognitive areas: orientation, fixation, calculation and attention, memory, language, reading, writing and viso-constructive abilities. Its score ranges from 0 to 30 points. It also constitutes the most used pruning test on the international epidemiological investigation, as well as in the clinical tests that require an evaluation of the patient's intellective functions.
2. FAST scale (Functional Assessment Staging) [34] is used to evaluate the possible relation between functional stage and survival. It consists of 7 stages (1 to 7). The very last two ones, severe and advanced dementias, are subdivided (6A to 6E and 7A to 7F). Derived from our experience, it seems that the evaluation by means of a specific scale as FAST, in which the function is included, clearly

oriented at the different stages towards a less accurate diagnose while the functional affectation rises [9]. A non-despicable disadvantage is derived from the generalization of the FAST scale, initially designed for patients of Alzheimer's, and therefore it does not have a sufficiently contrasted use in other dementias.
3. Katz's index [21] evaluates the pure function in the basic activities of the daily life: bathing, dressing, toileting, transfers, continence and feeding. It is defined as an observation instrument and an objective guide of the course of the chronic disease, as an aid to study the process of aging and as an aid in the rehabilitation. Its score ranges from A to H.
4. Barthel's index (Bar) [29] is similar to the one of Katz's with the difference of a numerical result, which is adapted for a continuous gradual evaluation, since the Katz's index is an ordinal scale with items of dichotomizing character. Like Katz's index, it evaluates the same functions this one does, although in a wider way. We use a summary score that consists of 5 stages: Independent, Slight, Moderate, Severe and Totally dependent.
5. Lawton-Brody's index (L-B) [24] evaluates the behavior aspects of instrumental character, for which it is necessary to be able to make the basic functions of a suitable form. It therefore implies a much greater and more complex functional integrity. The use of the telephone, the capacity to make purchases, being able to cook, the house care, washing the clothes, the use of transport means, being able to handle the proper medication and the handling of the financial aspects are evaluated. Its score ranges from 0 to 8 points for women, and 0 to 5 for men.

The scores of this tests are strongly correlated, because the different tests are composed of common cognitive components, Table 14.1.

Table 14.1 Correlation coefficients between pairs of neuropsychological test values

	MMSE	FAST	Katz	Bar	L-B
MMSE	1				
FAST	-0.86	1			
Katz	-0.59	0.77	1		
Bar	-0.55	0.7	0.84	1	
L-B	0.58	-0.7	-0.71	-0.65	1

The missing data corresponding to the MMSE test have been completed, in agreement with the clinical experts, by means of interpolation from the annual results of both annual tests that almost all the patients have been put under. Even so, other values have been left empty, 74 of the total of 1335, whose distribution by test and number of patients is indicated in Table 14.2.

In order to facilitate the convergence, as previous step to their use, the different fields that constitute the successfully obtained information were preprocessed. The neuropsychological tests were standardized between 0 and 10 from the minimum

Table 14.2 Statistics of missing data in data set

Test type	Number of missing data	Number of patients
MMSE	40 (14.98%)	6 (20%)
FAST	0 (0%)	0 (0%)
Katz	33 (12.36%)	12 (40%)
Bar	1 (0.37%)	1 (3.33%)
L-B	1 (0.37%)	1 (3.33%)
Total	75 (5.62%)	16 (53.33%)

and maximum values that can be reached in these tests. Those fields not being filled up are labelled as lost values or missing ones, and were later trated in a special way.

At a control stage, the values of diagnosis are used. The diagnosis of differential dementia can contain three different classes: Alzheimer-type dementia (ALZ), Vascular-type dementia (VAS) and other type of dementia (OTH) that include Trauma, Subcortical, Parkinson and Infectious dementias. From the entire set of consultations, 73.8% were diagnosed as ALZ, 6.7% as VAS and 19.5% as OTH, due to the data source which does not include patients without dementia.

14.3 HUMANN-S Based Ensembles

Two neural ensemble systems [31] were used to approach the problem of DDD. The systems are based on modules that implement a supervised variant of HUMANN architecture (HUMANN-S) [11]. The difference between these systems is the combination strategy used to built the HUMANN-S ensemble, the first one uses a Simple Majority Voting (SMV) strategy and the second one uses a Weighted Majority Voting (WMV) strategy [31].

14.3.1 Supervised HUMANN

The classification modules of both proposed systems are a supervised version of neural architecture HUMANN [11]. This neural network can implement the general approach of the classification process, which has three stages: feature extraction, template generation and discrimination (labelling), in a transparent and efficient way. Normally, and specifically in this application, first stage must be implemented by pre-processing modules that will be application-dependent. HUMANN uses a multi-layer neural structure with three modules and with different neurodynamics, connectivity topologies and learning laws, Fig. 14.1.

The first neural module of HUMANN-S is a Kohonens Self-Organizing Map (SOM) [22] using euclidean distance, eq. (14.1). This module implements a non-linear *projection* from an input space onto a two-dimensional array.

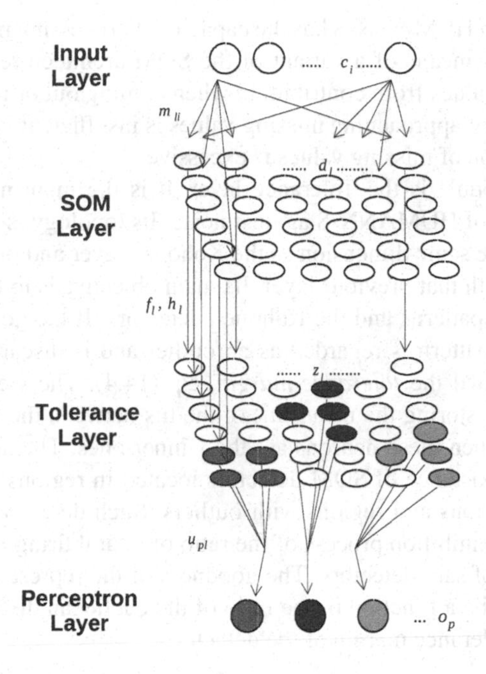

Fig. 14.1 HUMANN-S architecture

$$d_l = \|\mathbf{c} - \mathbf{m}_l\| = \left(\sum_i (c_i - m_{li})^2 \right)^{\frac{1}{2}} \tag{14.1}$$

This kind of neural structure is used because its main feature is the formation of topology-preserving feature maps and approximation of the input probability distribution, by means a self-organizing process which can produce fcatures detectors. Then the main advantage of this module is the simplification that produces to the inputs.

HUMANN-S applies a unsupervised and competitive training paradigm: The modification of the synaptic weights, eq. (14.2) and (14.3), not only affects the winning neuron but also to a lesser degree the set of neurons in the winners neighborhood *N*, and consequently being able to generate topological relations. The neighborhood relationship between nodes is normally given by a hexagonal or squared type lattice, whose size decreases during the training period. Also $\varepsilon(t)$ is a time decreasing learning rate.

$$\Delta m_{li} = \varepsilon(t) N_c(l)(x_i - m_{li}) \tag{14.2}$$

$$c = \arg\min_l \{d_l\} \tag{14.3}$$

For this application HUMANN-S has the capacity of processing missing data, which is implemented by means of a variant of the SOM architecture [37]. This variant prevents missing values from contributing when coming out or modifying weights. Even so, this way of approaching missing values is insufficient by itself, essentially when the proportion of missing values is excessive.

The second module is the Tolerance layer. It is the main module responsible for the robustness of HUMANN-S against noise. Its topology is a two-dimensional array which has the same dimension as the Kohonen layer and one-to-one interconnection scheme with that previous layer. Its main objective is to compare the fitting between the input patterns and the Kohonen detectors. If the goodness of the fit is not sufficient the pattern is regarded as an outlier and is discarded. We introduce a new concept called the *tolerance margin*, eq. (14.4). The weights of this layer are responsible for storing the mean (h_l) of the fits between the inputs and the Kohonen detectors when these neurons are the winner ones. The adaptive firing ratio (f_l) looks at the existence of SOM detectors located in regions with a low pattern density, empty regions and regions with outliers. Such detectors will be inhibited depending on the inhibition process of the ratio of neural firing as a function of the tolerance margin of said detectors. The goodness of the representation of a pattern by a detector will be a function of the ratio of the euclidean distance between both of them and the tolerance margin of the detector.

$$z_l = \begin{cases} 0 & if\ d_l \geq \lambda h_l \vee h_l = 0 \vee \frac{f_l}{h_l^2} \leq \xi \\ 1 - \frac{d_l}{\lambda h_l} & otherwise \end{cases} \tag{14.4}$$

The needed learning rule to obtain the weights of the global variance in the degree of the pairing, is based on a differential equation that converges towards said average, eq. (14.5), to which a decay term must be added to make the final inputs to the system more relevant, in addition to avoiding possible pernicious effects of artefacts or outliers patterns, eq. (14.6). In both cases $\beta(t)$ is a time decreasing learning rate.

$$\Delta f_l = \begin{cases} \beta(t)(1-f_c) & if\ l = c \\ -\beta(t) f_l & otherwise \end{cases} \tag{14.5}$$

$$\Delta h_l = \begin{cases} \beta(t)(d_c - h_c) - \eta(t) h_c & if\ l = c \\ -\eta(t) h_l & otherwise \end{cases} \tag{14.6}$$

The last module embodiment reflects the supervised character of this neural network, and is implemented by a Perceptron type net [35] and performs the last stage of a classification process, the discrimination task. It receives the input values from the previous stage and has an output (o_p) for each class to recognize, eq (14.7) and (14.8):

$$v_p = b_p + \sum_l u_{pl} z_l \tag{14.7}$$

$$o_p = \begin{cases} 1 \; if \; v_p \geq 0.5 \\ 0 \;\; otherwise \end{cases} . \tag{14.8}$$

The learning model that it follows is capable of adapting its weights and thresholds by means of a supervised paradigm using the so-called *perceptron rule*, based on the correction of the produced error in the output layer:

$$\Delta u_{pl} = \alpha(d_p - o_p)z_l \tag{14.9}$$

$$\Delta b_p = \alpha(d_p - o_p), \tag{14.10}$$

where α is the learning rate and d_p is the desired output. One of the advantages of this model is that it follows the *Perceptron Convergence Theorem*Perceptron [35] which guarantees the learning convergence in finite time and that architecture always allows the solution to be represented.

One of the most important advantages of HUMANN-S is its tremendous speed. Possible trainings take place between 10 and 100 times faster than the conventional back-propagation networks, producing similar results. Increased speed is attributed to the simplification which occurs in the self-organizing stage. This simplification also allows last stages to use a simple classifier which produces proven convergence in linear problems, better generalization skills and a reduction in the consumption of computing resources [17].

14.3.2 Neural Network Ensembles

In single neural network approach, the neural network learning problem is often formulated as an optimisation problem. Learning is different from optimisation. The neural network with the minimum error on the training set does not necessarily have the best generalization, unless there is an equivalence between generalisation and the error function. The ANNs are capable to have several network configurations close to the optimal one, according to the initial conditions of the network and the ones typical of the environment. As each network configuration makes generalization errors on different subsets of the input space, it is possible to argue that the collective decision produced by the complete set, or a *screened* subset, of networks, with an appropriate collective decision strategy, is less likely to be in error than the decision made by any of the individual networks [15]. This has generated the use of groups of neural networks, in a trial to improve the accuracy and the generalization skills of them, referred as an *ensemble, Neural Network Ensemble* (NNE).

A NNE combines a set of neural networks which learn to subdivide the task and thereby solve it more efficiently and elegantly. The main idea is to divide the data space into smaller and easier-to learn partitions, where each ANN learns only one of the simpler partitions, adopting the divide-and-conquer strategy. The underlying complex decision boundary can then be approximated by an appropriate combination of different ANNs.

The idea of designing ensemble learning systems can be traced back to as early as 1958 [28] and 1979 with the paper Dasarathy and Sheela [31]. Then, and since the early 1990s, algorithms based on similar ideas have been developed in many different but related forms, appearing often in the literature under various other names, such as ensemble systems [32], classifier fusion [23], committees of neural networks [6], mixture of experts, [18] [19], boosting and bagging methods [38][7] among others.

Ensemble systems consist of two key components:

1. A strategy is needed to build an ensemble that is as diverse as possible. Some of the more common ones are bagging, boosting, AdaBoost, stacked generalization, and mixture of experts [31][32]. Bagging, boosting and AdaBoost belong to the sequential training methods of designing NNEs [28]. The bagging method randomly generates a new training set with an uniform distribution for each network member from the original data set [31]. The boosting approach [38], on the other hand, resamples the data set with a non-uniform distribution for each ensemble member. The whole idea of boosting and bagging is to improve the performance by creating some weak and biased classifiers [5]. When aggregating these classifiers using an average or other mechanisms, the bias of the ensemble is hoped to be less than the bias of an individual classifier.
2. A strategy is needed to combine the outputs of individual classifiers that make up the ensemble in such a way that the correct decisions are amplified, and incorrect ones are cancelled out. Two taxonomies can be considered, first, trainable vs. non-trainable combination strategies, and second, combination strategies that apply to class labels vs. to class-specific continuous outputs.

In trainable combination rules, the parameters of the combiner, *weights*, are determined through a separate training algorithm. In non-trainable rules, the parameters become immediately available as the classifiers are generated. WMV is an example of such non-trainable schemes. In the second taxonomy, combination rules that apply to class labels need the classification decision only, whereas others need the continuous-valued outputs of individual classifiers. These values often represent the degrees of support the classifiers give to each class, and they can be accepted as an estimate of the posterior probability for each class. For combining class labels we have Majority Voting, Weighted Majority Voting, Behaviour Knowledge Space, and Borda Count Schemes. For combining continuous outputs we can find some other schemes such as, algebraic combiners, decision templates and Dempster-Shafer based combination [31]. Taking into account all these strategies, it can be found three ways of designing NNE in these methods: independent training, sequential training and simultaneous training [28].

Most of the independent training methods and sequential training methods follow the two-stage design process: first generating individual networks, *ensemble members* or *classifier modules*, and then combining them. The possible interactions among the individual networks cannot be exploited until the integration stage. There is no feedback from the integration stage to the individual network design stage. It is possible that some of the independently designed networks do not make much contribution to the integrated system, but the final result will be improved.

In order to use the feedback from the integration, simultaneous training methods train a set of networks together. Negative correlation learning [27] and the Mixtures of Experts (ME) architectures [18][19] are two examples of simultaneous training methods. Same other ensemble architectures as Stacked Generalization (SG), ME or Gating Neural Ensemble (GaNEn) [40] use a designing strategy in which two level of ensemble members have to be generated . A first ensemble level, where individual classifiers are experts in some portion of the feature space. A second level classifiers, which is used for assigning weights for the consecutive combiner, which is usually not a classifier, in the ME, in the GaNEn, and as a meta classifier for final decision in the SG [31][32].

Whereas there is no single ensemble generation algorithm or combination rule that is universally better than others, all of the approaches discussed above have been shown to be effective on a wide range of real world and benchmark datasets. An appropriate design of NNE is where selection and fusion are recurrently applied to a population of best combinations of classifiers rather than the individual best [36].

Given the advantages of NNEs and the complexity of the problems that are beginning to be investigated, it is clear that the ensemble method will be an important and pervasive problem-solving technique. In this work this will be shown in the field of clinical decision support systems.

14.3.3 *HUMANN-S Ensembles*

Two strategies are needed to build an ensemble system, diversity and combination strategies. In our developments we have used as diversity strategy several HUMANN-S with a data fusion scheme. The selected HUMANN-S are those with low validation errors and with a high diversity between pairs. The combination strategics used were the simple and the weighted majority voting, working with two schemes of HUMANN-S ensembles, SMVE and WMVE respectively.

There are three versions of majority voting, where the ensemble choose the class a) on which all classifiers agree (*unanimous voting*); b) predicted by at least one more than half the number of classifiers (*simple majority*); or c) that receives the highest number of votes, whether or not the sum of those votes exceeds 50% (*plurality voting* or just *majority voting*). For designing one of the HUMANN-S based ensembles the simple majority voting version.

SMVE system uses the selected modules and creates a SMV process based on their outputs. Each module emits a vote, indicating whether it considers that input belongs to a class or is unknown. A later module is responsible for the overall count, of considering whether the input belongs to one class or another, depending on whether most of the HUMANN-S modules consider it. Then the class with the maximum number of votes is selected as the ensemble's output.

When certain ensemble members are more qualified than others, weighting the decisions of those qualified experts more heavily may further improve the overall

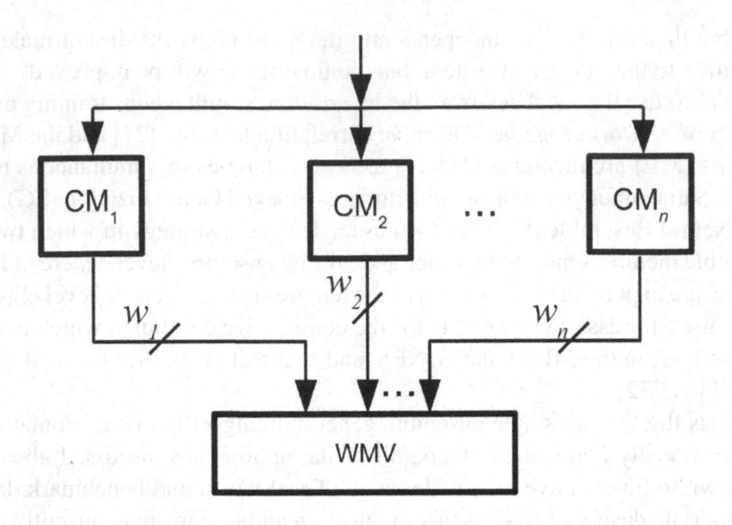

Fig. 14.2 WMVE scheme. CM_i is the *i* classifier module, w_i is the weight assigned to CM_i and WMV is the weighted majority voting module

performance than that can be obtained by the majority voting. This is the other combination strategy to built our second HUMANN-S based ensemble. In this type of NNE a main question is "how do we assign the weights?" It would be necessary to know which ensemble member would work better, in this case we would give the highest weights to those members, or perhaps, use only those ensemble members. In the absence of this knowledge, a plausible strategy is to use the performance of a single network on a separate validation dataset, or even its performance on the training dataset, as an estimate of that classifiers future performance [31].

The WMVE system, Fig. 14.2, assigns different weights to the HUMANN-S modules, according to their performance, in order to avoid the possibility of certain classifiers to be better than the other ones. We assign heavy weights to the decisions of the more expert classifiers [31]. It is assigned a weight w_i to each i classifier in proportion to its estimated accuracy, eq. (14.11). We have used the single HUMANN-S 30-fold validation error rate (e_i), number of classification errors divided by number of patterns in the validation set, as an estimate of that classifiers future performance.

$$w_i = log\frac{1-e_i}{e_i} \tag{14.11}$$

The combination of the outputs of several classifiers does not guarantee a superior performance to the best ensemble module, but it clearly reduces the risk of making a particularly poor selection [31].

14.4 Results and Discussion

Data set size is one of the added difficulties which must be addressed when designing an artificial intelligence system which will aid in the DDD. The data set limits how suitable the training of the used neural models will be, in addition to accurate generalization error. A method must be found that allows an efficient approach to this aspect of the problem. A cross-validation method is used and it improves training and error considerations [16]. These methods are based on making several partitions on the total data employed, or *resampling*. Diverse training is carried out with these data. A first part is used as a training set while the second one functions as a validation set. The generalization average error of the different estimations carried out on the different validation sets will provide a trustworthy measurement of the evaluated model error. The cross-validation variant was the denominated k-fold, and consisted in partitioning the data set into k subgroups, performing k training exercises, and leaving a validation data subgroup in each one while using the remaining (k - 1) as training data. The conducted partition on our queries was based on the identification of the patient involved in the consultation, that is, the consultations performed on the same patients were grouped in the same subgroup. Consequently we created 30 different subgroups, or a 30-fold consultation. This approach allows more objective results to be obtained because a model that has not been trained with other source consultations from the same patient will be used when the error validation of the consultation is carried out. In other cases the consultation used to train modules would have a much greater correlation with the evaluation consultation.

The results in DDD were obtained using the two intelligent designed systems, HUMANN-S based system and HUMANN-S ensemble-based system. These systems provide information towards the analysis of several aspects related with the DDD. The clinical criteria used in the diagnosis, the most appropriate type of system used and the effectiveness in the proposed systems.

Results are given based on individual HUMANN-S network system, with different combinations of the 5 neuropsychological tests that make up the information environment. In addition the results of the built NNEs were based on a selection of individual classifiers, using SMV and WMV combined strategies. Finally, a validation of our systems against the actual diagnosis from a clinical expert was carried out.

Fig. 14.3 shows the average training errors and the validation errors according to the different individual classifiers that were studied. Notice that the accuracy is lower than the achieved in differential cognitive deterioration diagnosis [13], using the same method. This suggests the need for a system with multiple clinical criteria, not only based on neuropsychological tests, to achieve the highest reliability levels in the DDD. In addition we can observe that for the used tests, each one separately or in combination, do not reveal the same matching characteristics for the diagnosis of cognitive deterioration of dementia [13]. Although more specialized studies are required on the items of each test and behaviour for each dementia, for a more exact knowledge of their relevance against the diagnosis, our results allow us to perform an initial approximation. In a more detailed analysis of the possible adequacy of

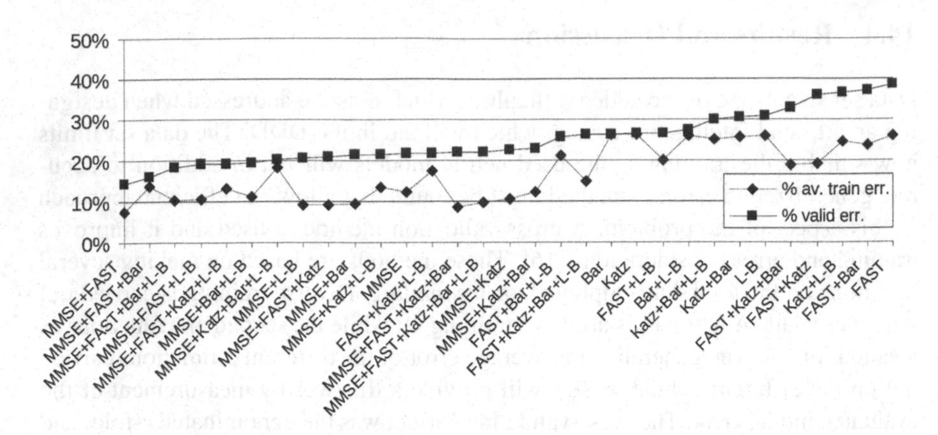

Fig. 14.3 Average training and validation errors for all the HUMANN-S modules with different inputs

each test or combination of tests of a DDD, it can be observed that the best clinical criterion is with the use of the MMSE+FAST sets that cover functional and cognitive symptomatology (MMSE and FAST, respectively), but this last test was specifically designed for people with AD, following the retrogenesis hypothesis [34]. There is also a set of test combinations where the common element is the MMSE, which provides a diagnosis with the smallest errors. These results confirm the character of the dominant scale of the MMSE. When MMSE is no longer used in the input of the system, all of the errors are larger, with just one exception, the FAST+ Katz+L-B combination, with a 21.7% validation error rate. This result suggests that the FAST would substitute the MMSE, which what would be in agreement with the high correlation between both tests, Table 14.1. The rest of the scale belonging to this combination reflects all of the functional and instrumental aspects. The largest error was found in the FAST test, allowing us to conclude that it is not worthwhile by itself. This observation indicates that its ability in individual use for a DDD is not adequate, revealing that it is a better indicator of functional characteristics by itself.

The systems proposed based on the ensemble approach improve the effectiveness of the diagnosis with respect to basic systems, Table 14.4. In order to achieve this, we focused our efforts on the design and creation on two key components of any ensemble system, namely, diversity among the members that make up the ensemble and the combination strategy [31]. One of the most important aspects to take into account when designing a very diverse ensemble is to have on one side the effectiveness of each individual classifier and so that each one of the said classifiers generates their errors in different data space areas [30]. In other words, an adequate combination of classifiers could serve to reduce the total error. Consequently, each classifier to use in the ensemble the must be as unique as possible, especially with respect to poorly classified input. There are many ways to measure diversity among classifiers [31], our estimate of the diversity measure has been

Table 14.3 Comparison of diversity between pairs of the seventeen best classifiers using the correlation between their errors

	MMSE+FAST (C_1^*)	MMSE+FAST+Bar (C_2^*)	MMSE+FAST+Bar+L-B (C_3^*)	MMSE+FAST+L-B (C_4)	MMSE+FAST+Katz+L-B (C_5)	MMSE+Bar+L-B (C_6^*)	MMSE+Katz+Bar+L-B (C_7)	MMSE+L-B (C_8)	MMSE+FAST+Katz (C_9^*)	MMSE+Bar (C_{10})	MMSE+Katz+L-B (C_{11})	MMSE (C_{12})	FAST+Katz+L-B (C_{13})	MMSE+FAST+Katz+Bar (C_{14})	MMSE+FAST+Katz+Bar+L-B (C_{15})	MMSE+Katz (C_{16})	MMSE+Katz+Bar (C_{17})
C_1^*	1																
C_2^*	*0.26*	1															
C_3^*	*0.31*	*0.29*	1														
C_4	0.48	0.33	0.68	1													
C_5	0.16	0.44	0.58	0.59	1												
C_6^*	*0.12*	*0.25*	*0.60*	0.60	0.53	1											
C_7	0.12	0.36	0.60	0.49	0.54	0.46	1										
C_8	0.23	0.72	0.55	0.53	0.54	0.53	1										
C_9^*	*0.19*	*0.45*	*0.37*	0.36	0.52	*0.28*	0.36	0.42	1								
C_{10}	0.13	0.29	0.48	0.33	0.35	0.32	0.42	0.55	0.43	1							
C_{11}	0.13	0.37	0.53	0.54	0.77	0.53	0.51	0.55	0.41	0.38	1						
C_{12}	0.19	0.30	0.26	0.24	0.30	0.33	0.20	0.39	0.32	0.37	0.31	1					
C_{13}	0.33	0.39	0.55	0.48	0.53	0.45	0.44	0.50	0.38	0.46	0.52	0.28	1				
C_{14}	0.35	0.43	0.54	0.54	0.61	0.44	0.56	0.47	0.46	0.43	0.49	0.27	0.53	1			
C_{15}	0.12	0.38	0.66	0.54	0.61	0.54	0.59	0.60	0.35	0.38	0.58	0.20	0.55	0.54	1		
C_{16}	0.44	0.39	0.45	0.52	0.52	0.43	0.41	0.47	0.50	0.43	0.43	0.45	0.57	0.58	0.37	1	
C_{17}	0.37	0.42	0.55	0.62	0.58	0.57	0.44	0.54	0.40	0.43	0.50	0.33	0.56	0.57	0.55	0.59	1

* Final classifiers selected as ensemble modules

based on the correlation between the error of pairs of classifiers and in their individual effectiveness. The correlation of error of pairs of classifiers can be considered as an appropriate diversity estimator because is one of the most simple way to estimate the similarity, in opposite to diversity. Table 14.3 shows the different values of pair correlations for the seventeen best individual classifiers. From there we apply the average of the chosen distinguishing factor, and we obtain the optimal components that make up the two HUMANN-S-designed ensembles (SMVE and WMVE), Fig. 14.4. As seen in Table 14.3, five is the optimal number of classifiers that we select. They correspond to the following modules, input combinations and errors: CM_1: MMSE+FAST, 14.23%; CM_2: MMSE+FAST+Bar, 16.10%; CM_3: MMSE+FAST+Bar+L-B, 18.35%; CM_4: MMSE+Bar+L-B, 19.48% and CM_5:

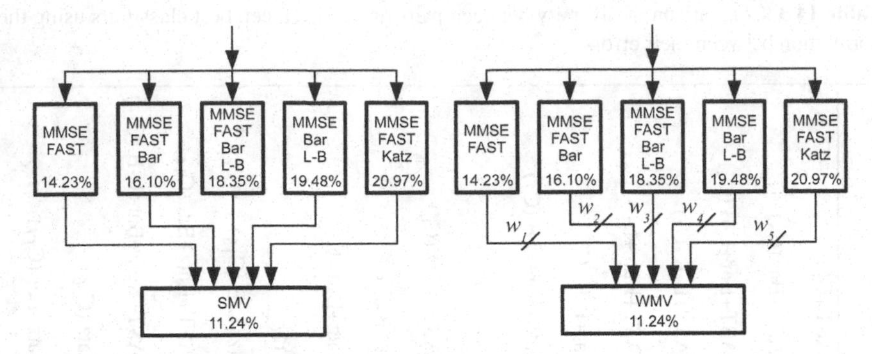

Fig. 14.4 SMVE and WMVE schemes and the modules used in it. The number of each module is the validation error

MMSE+FAST+Katz, 20.97%. The two obtained systems with this approach present an error in the validation sets of 11.24%, Table 14.4. These results successfully improve by 2.99% the validation error when compared to the best one from the modules CM_1 (HUMANN-S MMSE+FAST) applied on the total of all consultations. Table 14.4 also shows the values for sensibility and specificity of the best individual classifier and of the ensembles. The best capacity is offered by the neural ensemble with WMV strategy.

Table 14.4 Comparison between results for the validation sets of classification modules, ensembles and the clinical expert

		HUMANN-S					Ensembles		
		CM_1	CM_2	CM_3	CM_4	CM_5	SMVE	WMVE	Physician
Sensitivity	ALZ	86.8%	95.9%	89.9%	88.3%	89.9%	96.5%	95.9%	53.3%
	VAS	100%	44.4%	0.0%	5.6%	11.1%	16.7%	22.2%	16.7%
	OTH	76.9%	51.9%	78.9%	76.9%	61.5%	84.6%	84.6%	3.9%
Specificity	ALZ	93.4%	84.4%	87.6%	86.6%	85.5%	89.2%	89.6%	68.2%
	VAS	81.8%	80.0%	0.0%	14.3%	16.7%	60.0%	66.7%	3.3%
	OTH	64.5%	81.8%	69.5%	70.2%	80.0%	89.8%	88.0%	9.1%
Error		14.23%	16.10%	18.35%	19.48%	20.97%	11.24%	11.24%	58.8%

The observed variations amongst the different designed DDD aid systems are given in Table 14.4. The values can be explained if we carry out a detailed analysis of the errors individually as well as in the data space characteristics. The SMVE and WMVE systems incorrectly classified 30 consultations, 29 of which were detected in both systems. Thus, there are slight differences with respect to their sensibility and specificity as shown in Table 14.4. Of these 29 errors the best module CM_1

(HUMANN-S MMSE+FAST) produced 14 diagnostic errors, and also produced 24 errors more, in total resulting in 38 incorrect consultations. The analysis of the results with respect to the 29 errors in common between SMVE and WMVE reveal the following: 10 of these errors are correctly classified by 2 of the 5 modules and 14 of these errors are correctly classified by only 1 of the 5 modules, the remaining 5 errors are incorrectly classified by all 5 chosen modules. One aspect that needs mentioning is the generated result from the classifiers in relation to VAS type. A study of Table 14.4 shows that the sensibility for this type does not exceed 22.2% of the ensemble systems. Thus the low values of these parameters indicate in all of the individual modules except for CM_1 and CM_2. Even CM_4 is unable to diagnose any patient inside VAS class. One of the main reasons for these poor values is that there is a low number of consultations belonging to this type of diagnosis, 34.6% with respect to type OTH and only 9.1% with respect to type ALZ. Adding this situation to the use of quadratic error to value the classifiers shows that the improvements in the classification of this type would have limited impact on the calculation of global error. The generated situation because the classes are inequally distributed could be addressed in the future by improving the data set or weighting the errors of classes separately.

The proposed diagnostic aid systems have been validated against the diagnosis carried out by a clinical expert in dementias. We have used a blind validation model, where the main characteristics are the no previous information about identity and diagnosis of every patient. The human expert was used to send his diagnosis in the same input space as the proposed intelligent diagnostic aid system. The obtained results in the classification are shown in the last column of Table 14.4. The final obtained error by the said expert was 58.8%, 47.56% worse than the values produced by the proposed neural ensemble systems. Hence we can observe especially low values in the sensibility and specificity of the clinical expert. The physicians, usually, use several diagnostic criteria in order to reach a diagnosis of dementia. A small group of neuropsycological test aren't used for this clinical decision. Because of this the validation process designed for us is very hard for the clinical expert and very useful to demonstrate the goodness of our proposal. Furthermore, these facts reveal the inherent difficulties in the DDD when only using final values from neuropsychological tests, reaffirming the goodness of the proposed systems. Still, there is a need to discover some sort of biological indicator for the different dementias that would make the diagnostic process less subjective and more reliable.

14.5 Conclusion

Our developments contributes advances on clinical support decision systems to aid in medical diagnostics in general and specifically in the diagnosis of dementia.

We have presented new intelligent methods to assist in differential diagnosis of dementias, based on artificial neural networks and neural ensembles, which are able to handle missing data and eliminate dependent biases of the clinical expert. A novel information environment has been used, as of yet unseen in the literature, under different combinations.

The use of these new instruments can help alleviate the degree of existing underdiagnosis, thanks to its high performance and because it could be used in all healthcare areas, not just in specialized care but also in primary care.

In addition to this, the analysis conducted on the different modules while the system was running allows the best test for a correct diagnosis to be selected. Therefore the study of other possible tests, and/or criteria could be extended to elaborate refined diagnostic protocols.

These systems offer tremendous functional potential. But one conclusion from our study is the insufficiency of the use of only these independent neuropsychological tests and the need to use clinical multicriteria for an effective and reliable differential diagnosis of dementias. As future work also we plan to design new ensemble systems using GaNEn approach, in order to facilitate the design, tuning and improve the results.

Finally the results obtained here show that our proposed intelligent decision systems is an adequate computational tool to aid the facultative in the diagnosis of dementias, considered worthwhile for future research which could add upon its effectiveness and refinement.

Acknowledgements. We would like to thank Canary Islands Government, the Science and Innovation Ministry of the Spanish Government and EU Funds (FEDER) for their support under Research Projects "SolSubC200801000347" and "TIN2009-13891" respectively. Additional thanks are extended to the Alzheimer's Patient Association of Gran Canaria for their generous input of data.

References

1. Bennett, D.: Alzheimer's disease and other dementias. In: Weiner, W., Goetz, C. (eds.) Neurology for the Non-Neurologist, 4th edn., pp. 233–243. Lippincott, Williams and Wilkins, Philadelphia, PA (1999)
2. Boller, F., Duyckaerts, C.: Alzheimer's disease: clinical and anatomic issues. In: Feinberg, T., Farah, M. (eds.) Behavioral Neurology and Neuropsychology, 2nd edn., pp. 515–544. McGraw-Hill, New York (2004)
3. Bowler, J., Hachinski, V.: Vascular dementia. In: Feinberg, T., Farah, M. (eds.) Behavioral Neurology and Neuropsychology, 2nd edn., pp. 589–603. McGraw-Hill, New York (2004)
4. Caselli, R., Boeve, B.: The degenerative dementias. In: Goetz, C. (ed.) Textbook of Clinical Neurology, 2nd edn., Saunders, Philadelphia, pp. 681–712 (2003)
5. Dam, H., Abbass, H., Lokan, C., Yao, X.: Neural-based learning classifier systems. IEEE Transactions on Knowledge and Data Engineering 20(1), 26–39 (2008)
6. Drucker, H., Cortes, C., Jackel, L., LeCun, Y., Vapnik, V.: Boosting and other ensemble methods. Neural Computation 6(6), 1289–1301 (1994)
7. Drucker, H., Schapire, R., Simard, P.: Boosting performance in neural networks. Int. J. Pattern Recognition Artif. Intelligence 7(4), 704–709 (1993)
8. Dubois, B., Feldman, H., Jacova, C., Cummings, J., Dekosky, S., Barberger -Gateau, P.: Revising the definition of alzheimer's disease: a new lexicon. Lancet Neurology 9(11), 1118–1127 (2010)

9. Fernandez -Viadero, C., Verduga, R., Crespo, D.: Biomarcadores del envejecimiento. In: Biogerontología, pp. 233–262. Universidad de Cantabria, Santander (2006)
10. Folstein, M., Folstein, S., McHugh, P.: Mini-mental state. a practical method for grading the cognitive state of patients for the clinician. Journal of Psychiatric Research 12(3), 189–198 (1975)
11. García Báez, P., Fernández López, P., Suárez Araujo, C.P.: A parametric study of humann in relation to the noise. application to the identification of compounds of environmental interest. Systems Analysis Modelling and Simulation 43(9), 1213–1228 (2003)
12. García Báez, P., Fernández Viadero, C., Pérez del Pino, M., Prochazka, A., Suárez Araujo, C.: Humann-based systems for differential diagnosis of dementia using neuropsychological tests. In: 14th International Conference on Intelligent Engineering Systems (INES), pp. 67–72. IEEE Xpress, Las Palmas de GC (2010)
13. García Báez P., Pérez del Pino, M., Fernández Viadero, C., Regidor García, J.: Artificial intelligent systems based on supervised HUMANN for differential diagnosis of cognitive impairment: Towards a 4P-HCDS. In: Cabestany, J., Sandoval, F., Prieto, A., Corchado, J.M. (eds.) IWANN 2009. LNCS, vol. 5517, pp. 981–988. Springer, Heidelberg (2009)
14. García Báez, P., Suárez Araujo, C.P., Fernández Viadero, C., Regidor García, J.: Automatic prognostic determination and evolution of cognitive decline using artificial neural networks. In: Yin, H., Tino, P., Corchado, E., Byrne, W., Yao, X. (eds.) IDEAL 2007. LNCS, vol. 4881, pp. 898–907. Springer, Heidelberg (2007)
15. Hansen, L., Salamon, P.: Neural network ensembles. IEEE Transactions on Pattern Analysis and Machine Intelligence 12(10), 993–1001 (1990)
16. Hjorth, J.: Computer Intensive Statistical Methods Validation, Mod. Sel., and Bootstap. Chapman and Hall, Boca Raton (1994)
17. Hrycej, T.: Modular Learning in Neural Networks. John Wiley and Sons, New York (1992)
18. Jacobs, R., Jordan, M., Nowlan, S., Hinton, G.: Adaptive mixtures of local experts. Neural Computation 3(1), 79–87 (1991)
19. Jordan, M., Jacobs, R.: Hierarchical mixtures of experts and the em algorithm. Neural Computation 6(2), 181–214 (1994)
20. Jorm, A., Korten, A., Henderson, A.: The prevalence of dementia: a quantitativeintegration of the literature. Acta Psychiatrica Scandinavica 76(5), 465–479 (1987)
21. Katz, S., Ford, A., Moskowitz, R.: Studies of illness in the aged. the index of adl: a standardized measure of biological and psychosocial function. JAMA 185, 914–919 (1963)
22. Kohonen, T.: Self-Organization and Associative Memory, 3rd edn. Springer Series in Information Sciences, Berlin, GE (1989)
23. Kuncheva, L., Bezdek, J., Duin, R.: Decision templates for multiple classifier fusion: An experimental comparison. Pattern Recognition 34(2), 299–314 (2001)
24. Lawton, M., Brody, E.: Assessment of older people: self-mantaining and instrumental activities of daily living. Gerontologist 9, 179–186 (1969)
25. de Leon, M., Klunk, W.: Biomarkers for the early diagnosis of alzheimer's disease. The Lancet Neurology 5(3), 198–199 (2006)
26. Lim, A., Kukull, W., Nochlin, D., Leverenz, J., McCormick, W.: Clinico-neuropathological correlation of alzheimer's disease in a community-based case series. Journal of the American Geriatrics Society 47(5), 564–569 (1999)
27. Liu, Y., Yao, X.: Simultaneous training of negatively correlated neural networks in an ensemble. IEEE Transactions on Systems, Man, and Cybernetics, Part B: Cybernetics 29(6), 716–725 (1999)

28. Liu, Y., Yao, X., Higuchi, T.: Designing neural network ensembles by minimising mutual information. In: Mohammadian, M., Sarker, R., Yao, X. (eds.) Computational Intelligence in Control. Idea Group Inc., USA (2003)
29. Mahoney, F., Barthel, D.: Functional evaluation: The barthel index. Maryland State Medical Journal 14, 61–65 (1965)
30. Opitz, D., Shavlik, J.: Actively searching for an effective neural-network ensemble. Connection Science 8(3), 337–353 (1996)
31. Polikar, R.: Ensemble based systems in decision making. IEEE Circuits and Systems Magazine 6(3), 21–45 (2006)
32. Polikar, R., Topalis, A., Green, D., Kounios, J., Clark, C.: Comparative multiresolution wavelet analysis of erp spectral bands using an ensemble of classifiers app. for early diagnosis of alzheimer's disease. Computers in Biology and Medicine 37(4), 542–558 (2007)
33. Polikar, R., Topalis, A., Green, D., Kounios, J., Clark, C.: Ensemble based data fusion for early diagnosis of alzheimer's disease. Information Fusion 9(1), 83–95 (2008)
34. Reisberg, B.: Functional assessment staging (fast). Psychopharmacology Bulletin 24(4), 653–659 (1988)
35. Rosenblatt, F.: Principles of Neurodynamics. Spartan Books, Washington (1961)
36. Ruta, D., Gabrys, D.: Classifier selection for majority voting. Information Fusion 6(1), 63–81 (2005)
37. Samad, T., Harp, S.: Self-organization with partial data. Network 3(2), 205–212 (1992)
38. Schapire, R.: The strength of weak learnability. Machine Learning 5(2), 197–227 (1990)
39. Schoenberg, B., Anderson, D., Haerer, A.: Severe dementia - prevalence andclinical features in biracial us population. Archives of Neurology 42(8), 740–743 (1985)
40. Suárez Araujo, C., García Báez, P., Fernández Viadero, C.: Ganen: a new gating neural ensemble for automatic assessment of the severity level of dementia using neuropsychological tests. In: International Conference on Broadband and Biomedical Communications (IB2COM). IEEE Xplore, Málaga (2010)
41. Sunderland, T., Gur, R., Arnold, S.: The use of biomarkers in the elderly: current and future challenges. Biological Psychiatry 58(4), 272–276 (2005)
42. Zaffalon, M., Wesnes, K., Petrini, O.: eliable diagnoses of dementia by the naive credal classifier inferred from incomplete cognitive data. Artificial Intelligence in Medicine 29(1-2), 61–79 (2003)

Chapter 15
Texture Classification in Bioindicator Images Processing

Martina Mudrová, Petra Slavíková, and Aleš Procházka

Abstract. The section deals with classification of microscope images of Picea Abies stomas. There is an assumption that a stoma character strongly depends on the level of air pollution, so that stoma can stand for an important environmental bioindicator. According to the level of stoma incrustation it is possible to distinguish several classes of stoma structures. A proposal of an algorithm enabling the automatic recognition of a stoma incrustation level is a main goal of this study. There are two principles discussed in the chapter: The first principle is based on gradient methods while the second one uses a wavelet transform. Possibilities of application of mentioned attitudes were investigated and the classification criteria distinguishing the stoma character were suggested, as well. The resulting algorithm was verified for a set of four hundred real images and results achieved were compared with an expert's sensual classification. Selected methods of image preprocessing as noise reduction, brightness correction and resampling are studied, as well.

15.1 Introduction

Image processing represents a widely spread interdisciplinary research area with many applications in various disciplines including engineering, medicine and technology [8, 46, 21, 43]. The section is devoted to the application of selected mathematical methods in environmental engineering to detect and to analyze the quality of vegetation which can be substantially affected by air pollution and dust particles

Martina Mudrová · Petra Slavíková · Aleš Procházka
Department of Computing and Control Engineering,
Institute of Chemical Technology Prague, Technická 1905,
166 28 Prague 6, Czech Republic
e-mail: {Martina.Mudrova, Petra.Slavikova}@vscht.cz,
A.Prochazka@ieee.org

J. Fodor et al. (Eds.): Recent Advances in Intelligent Engineering Systems, SCI 378, pp. 325–339.
springerlink.com

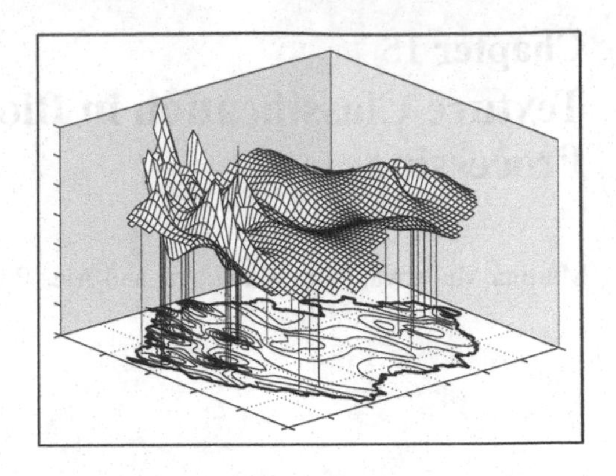

Fig. 15.1 /Dust particles concentration over the Czech Republic observed during a selected day and time measured by ground measuring stations and extrapolated over the whole region pointing to the most polluted areas

concentration. The air pollution is regularly observed by ground measuring stations in most countries. Figure 15.1 presents the situation in the Czech Republic with 103 observation points precisely defined by their longitude and latitude. Fig. 15.1 presents these surface observations interpolated to the whole Czech Republic. In this case, the spline two dimensional interpolation has been used allowing for evaluation of a given variable at the chosen grid points. This method allows the estimation of air pollution in places without measuring stations and detection of air pollution sources as well.

Remote satellite sensing represents another method used for air pollution detection. Real data presented in Fig. 15.2 epresent images taken by satellites operated by NOAA (The National Oceanic and Atmospheric Administration). All these satellites fly on elliptical or circular orbits around the planet earth, the earth's centre being the focal or central point and they belong to the so-called polar orbiting satellites which fly in rather low altitudes, typically at 850 km height. Polar orbiters

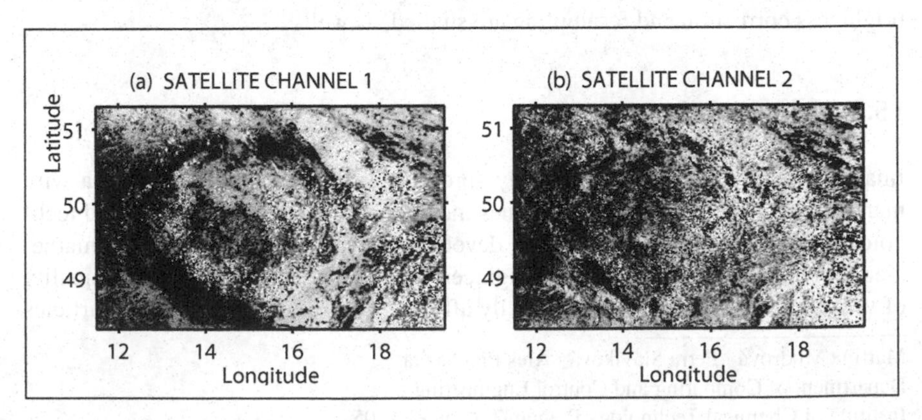

Fig. 15.2 Satellite images of the Czech Republic from the NOAA satellite simultaneously observed by the scanning radiometer AVHRR using different spectral ranges with results obtained from its (**a**) channel 1 and (**b**) channel 2

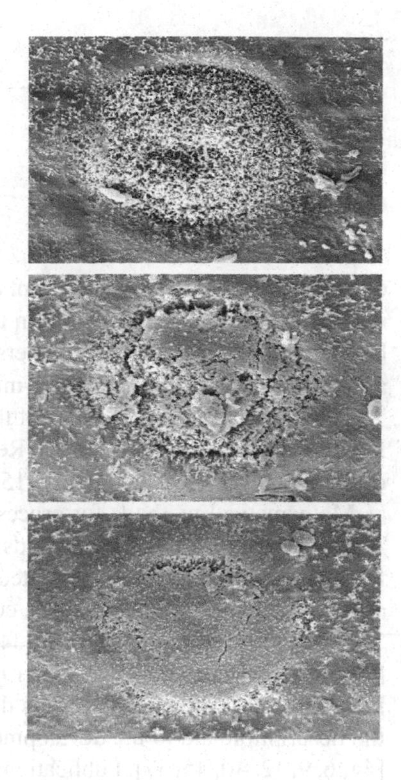

Fig. 15.3 Selected examples of electron-microscope images of stomas of Picea Abies presenting different stages of the affect of air pollution with the level of stoma incrustation increasing from the top to the bottom

provide excellent pictures of all parts of the earth including the polar regions. The main apparatus of NOAA satellites is the scanning radiometer AVHRR (Advanced Very High Resolution Radiometer). This is a five-channel apparatus covering spectral ranges (1) 0.58-0.68μm, (2) 0.725-1.1μm, (3) 3.55-3.93μm, (4)10.3-11.3μm, (5) 11.5-12.5μm. The first two channels work with the reflected sun radiation only in red and close infrared region, the last two ones are fully heat radiation channels, and channel 3 is a mixed one. Figure 15.2 presents satellite images of the Czech Republic from the NOAA satellite simultaneously observed in two different spectral ranges. Their difference can be used as a measure of dust particles concentration owing to their different reflection providing an alternative to the ground observations in this way.

Figure 15.3 presents the effect of air pollution to vegetation samples taken from selected parts of the Czech Republic. The figure shows electron-microscope images of stomas of Picea Abies. This tree is considered to be an important bioindicator of air pollution in a given area owing to the great appearance of this biotope within the Czech Republic and Europe generally. The stoma epidermis has usually miscellaneous structure but in the case of polluted air it covers with epicuticular vax to protect itself against negative influences of pollutants, and stoma epidermis incrusts. The level of incrustation can be used as a factor for air pollution assessment. Stoma quality can be investigated using its microscope images and the level of incrustation

Table 15.1 List of monitored localities

1 - Svratouch	4 - Řepice u Strakonic
2 - Býšť u Pardubic	5 - Boršice-Újezd
3 - Trnová	6 - Velká Losenice

can be divided into several different categories. In connection with the experimental results five classes [28] are used in the following text. As the class assignment can be easily affected by an expert's personality in the case of human visual evaluation, a requirement of independent fair-minded images classification appears.

The section presents analysis of the extensive set of electron-microscope images from several regions of the Czech Republic summarized in Table 15.1. The location of these areas is presented in Fig. 15.4.

Mathematical methods for processing of time-series and images are very close. The section is devoted to methods of two-dimensional digital signal processing forming the basic tool for automated description, classification or evaluation of images [29], their segmentation [38], compression, pattern recognition [3, 37, 1, 2, 23, 15, 41] and image enhancement [34]. These methods include both time-frequency [5] and time-scale [9, 20, 18, 35, 19, 10, 16] signal analysis procedures. There is a large amount of books and papers devoted both to image processing theory and to the description and to the development of mathematical and algorithm background [44, 6, 9, 12, 40, 45, 47]. Publications devoted to the application of selected methods in various branches are widely spread as well.

The goal of this contribution is in the development of an automatic algorithm which can recognize the level of stoma changes [33] by means of methods of texture classification. Essential algorithmic procedures discussed further cover topics presented in Fig. 15.5 and including image preprocessing methods, segmentation procedures and classification algorithms.

The following study is devoted to the discussion of two basic principles including (i) the application of gradient methods and (ii) the use of methods based on the wavelet transform. These methods are preceded by selected image preprocessing

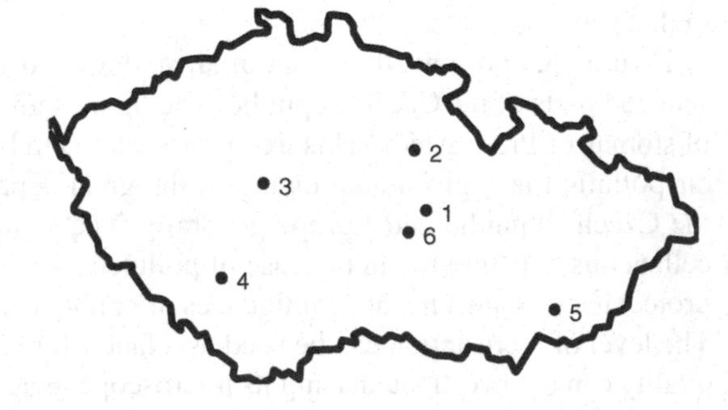

Fig. 15.4 Localities within the Czech Republic from where vegetation samples were collected

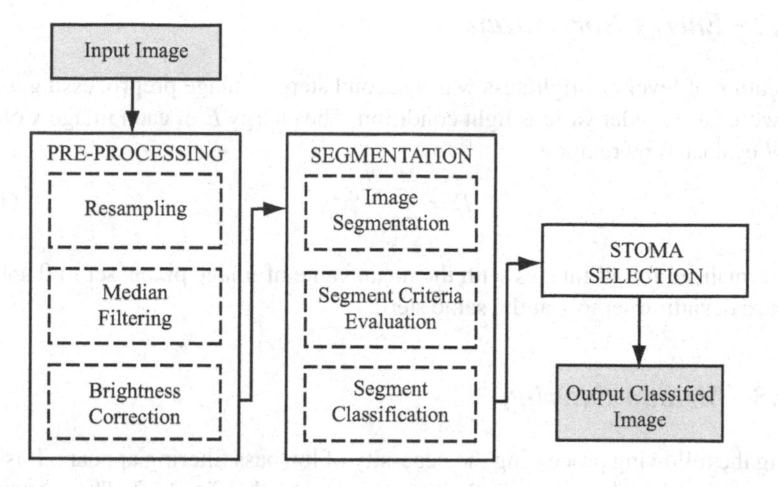

Fig. 15.5 The block diagram presenting the fundamental steps used for classification of microscopic images including their preprocessing, feature extraction and classification

procedures for image enhancement and noise reduction. The whole algorithm proposed based upon the structure presented in Fig. 15.5 is presented in the final part of the chapter. Results achieved [39] are based upon the analysis of the set of about four hundred images collected from 6 localities in the Czech Republic with their location summarized in Table 15.1 and presented in Fig. 15.4.

15.2 Image Preprocessing

The set of about four hundred real images was processed in this study. As images were observed under different conditions thanks to various stoma size and character, the image preprocessing was unavoidable to eliminate unexpected effects. Images were unified into the same resolution, cropped to the same sizes and their energy was normalized. The application of non-linear digital filters turns out to be suitable before their feature extraction and image classification. Median filter was used as a proper method in this case.

15.2.1 Image Resizing

In the first step it was necessary to unify image resolution. Various methods of the two-dimensional interpolation [17, 31] can be used in the process of image resampling. The method of the nearest neighbor, bilinear, bicubic and spline interpolation were tested and their results compared. An interpolation method based on the discrete two-dimensional Fourier transform [11] have been studied in this connection as well. In the final version, the bilinear method has been selected. Together with the change of resolution it was necessary to crop images into the same size to enable comparative data processing and results.

15.2.2 Energy Normalization

Unification of level of brightness was a second step of image preprocessing as images were taken under various light condition. The energy E of each image $\mathbf{x}$ of size M x N evaluated by relation

$$E = \sum_{i=1}^{M} \sum_{j=1}^{N} x_{i,j}^2 \tag{15.1}$$

was normalized for all images with the mean value of image pixels set to 0 and the standard deviation set to 1 at the same step.

15.2.3 Median Filtering

During the following processing the necessity of lowpass filtering appears. It is possible to notice the white spots in the lowest image in the Fig. 15.3. These spots are considered to be indicators of stoma structure recovery. They appear only in some images of class 5. Their presence affects the wrong classification of these stomas so they should be removed.

It is possible to use various methods of noise rejection [4]. Median filtering is considered to be a suitable method used to decrease the effect of this shoot-typed noise. Filter size selection depends on image resolution and the size of disturbing areas. Filter size of 3×3 pixels has been satisfactory in this case.

15.3 Texture Analysis Methods

As it is not trivial to delimitate the stoma border within the image, methods of texture classification were applied. There are various attitudes to solution of a general problem of texture analysis mentioned in [26, 13]. Presented chapter is focussed to the application of gradient methods and edge detection and to the possibilities of wavelet transform use for analysis of given images. Various edge detectors were investigated as well as various wavelet functions.

The principle of application the selected method to the given image set included (i) the split of each image into several areas of the same size and (ii) separate processing of each area. The subimage size was selected so that it would be easy to use the wavelet transform, i.e. dimensions of them should be a power of 2. Then it is possible to assign various weights according to image area site as the stoma was usually located in the center of image, as well.

As the definition of stoma classification [28] is uncertain and it is given by several phrases only, a database of cuts was created at first and they were investigated by an experienced expert to obtain his classification of each of them. In the case that his classification was not obviously evaluated, the cut was excluded from the database. Clearly evaluated cuts solved in the next processing as the templates for criteria values determination. A selected set of these templates is presented in Fig. 15.6. Each row includes typical representatives of the same class.

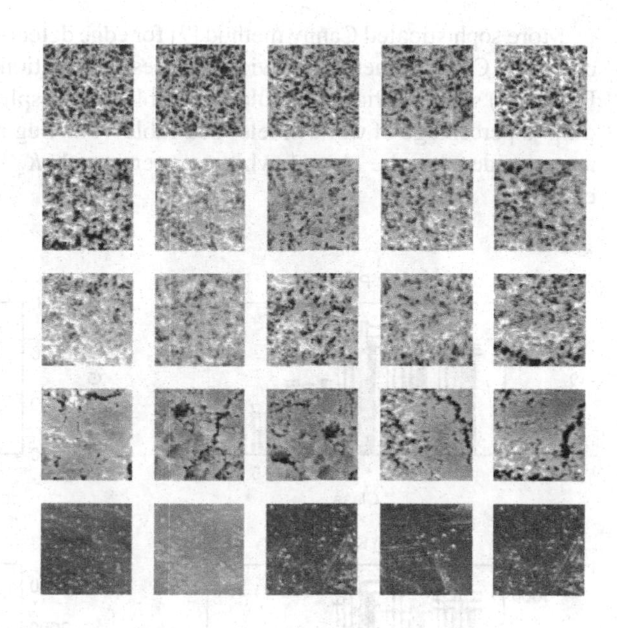

Fig. 15.6 Selected templates of textures found in given images. Each row corresponds to one class, so that a level No. 1 is in the first row and level No. 5 is in a bottom.

15.3.1 Gradient Methods and Edge Detection

Gradient methods mentioned e.g. in [12, 32] belong to the basic tool of image processing. As they discover changes of brightness function, edge detection provides their primary use. The principle of gradient methods is based on the convolution of an image **x** of size M x N and a convolution mask **h** according to relation

$$y(i,j) = \sum_{m=0}^{M-1} \sum_{n=0}^{N-1} x(m,n)h(i-m,j-n) \tag{15.2}$$

Resulting image **y** can be thresholded to obtain binary image only in which the pixels with a value "one" indicate a part with the highest level of brightness changes - these pixels represent edges in the original image. The shape and size of the convolution mask **h** depends upon the degree of derivation used, its direction and method. Robinson, Prewitt, Kirsch and Sobel convolution masks were applied to the set of image segments with a size 128 x 128 pixels with known level 1 - 5 of stoma incrustation. Selected example of a set of such image segments is presented in Fig. 15.6. As the operators mentioned above have directional behavior, resulted image was obtained as a sum of convolutions in each direction. Otsu's algorithm [30] was used for the threshold selection during the process of conversion into a binary image. Number of white pixels G in each segment was evaluated and its correlation with a class number was compared - see Fig. 15.7. It is obvious that with increasing level of incrustation the fine structure of stoma is gradually disappearing and the number of edge pixels G is decreasing.

More sophisticated Canny method [7] for edge detection was used as well. Application of Canny's method provides the best correlation with a segment class (see Fig. 15.7) so a criterion K_g could be established. It splits segments' classes according to percentage of white pixels - see Table 15.2. Segments with values $K_g > 17.5$ are considered to be class 1 while segments with $K_g < 12.2$ are supposed to be class 5.

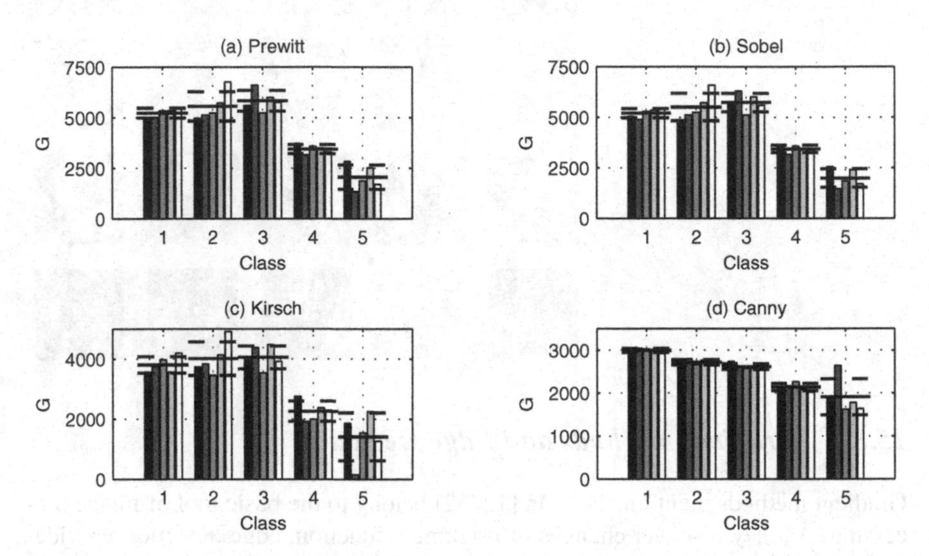

Fig. 15.7 Results of application of Prewitt (a), Sobel (b), Kirsch (c) and Canny (d) methods to the selected set of image cuts with a known classification. Number of pixels representing edges G by given method is drawn in dependence upon stoma class. Mean values and standard deviations are marked, as well.

Table 15.2 Image Texture Classification Using Gradient Method

	Border values of gradient criterium			
Criterion	$1-2$	$2-3$	$3-4$	$4-5$
K_g [%]	17.5	16.2	14.0	12.2

15.3.2 *Wavelet Transform Use for Image Classification*

The discrete wavelet transform (DWT) [9, 45] is studied for more than 20 years already with its mathematical roots going to last several centuries. There is a large amount of papers concerning its application in the area of image segmentation, noise reduction, classification and others [4, 6]. The basic idea is based on the use of a special set of wavelet and scaling functions to decompose the original one-dimensional continuous signal x into its detail and approximation coefficients. The the basic idea of 1-dimensional wavelet transform based on initial wavelet function dilation and translation can be mathematically described by relation

$$DWT\{x(k)\} \equiv X(m,n) = 2^{-\frac{m}{2}} \sum_k x(k)\psi(2^{-m}k - n) \tag{15.3}$$

The Mallat's pyramidal scheme [25, 40] presented in Fig. 15.8 describes another attitude for wavelet transform understanding. It describes the DWT as a set of band filters which are applied subsequently in several levels to obtain approximation coefficients. A wavelet and proper scaling function use guarantees the possibility of perfect reconstruction of the original signal.

Thanks to these properties the DWT can be simply extended into two dimensions and in this form it can be used as a powerful tool in image processing. The wavelet decomposition at each level can be performed separately in the rows and columns of the image. The combination of DWT application in both directories is provides two-dimensional wavelet coefficients passing on into the next decomposition level. Fig. 15.9 presents an example of wavelet decomposition of selected real image segment into 2 levels.

Various wavelet functions were used for processing of the given set of real image segments (with the same size of 128 x 128 pixels, again). The proper criteria for stoma structure class evaluation were searched. Selected example of wavelet decomposition of presented set of image segment is shown in Figures 15.10 and 15.11 where the sum C_W of absolute values of wavelet coefficients $c_{i,j}$ is explored in dependence upon the decomposition level. Coefficients belonging to the same class

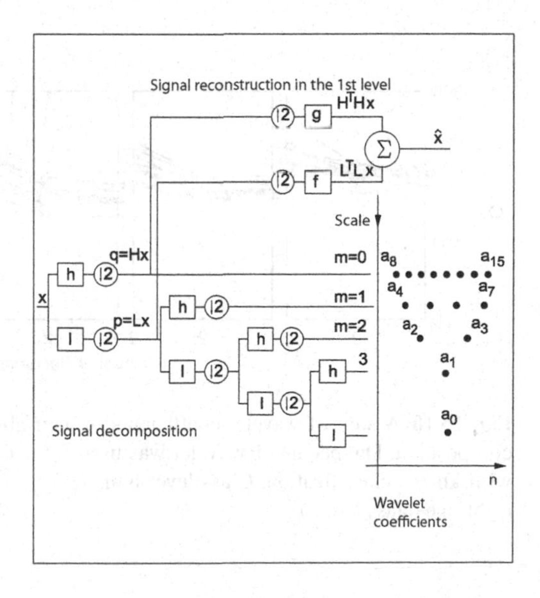

Fig. 15.8 Mallat's pyramid scheme of signal decomposition. The principle of the reconstruction in the first level is denoted, as well

Fig. 15.9 Wavelet transform of a selected image cut. Decomposition to the second level is shown

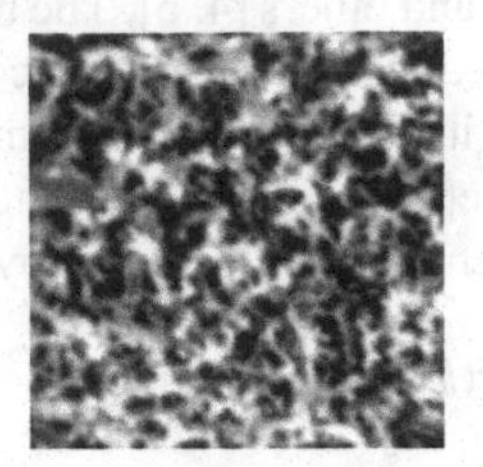

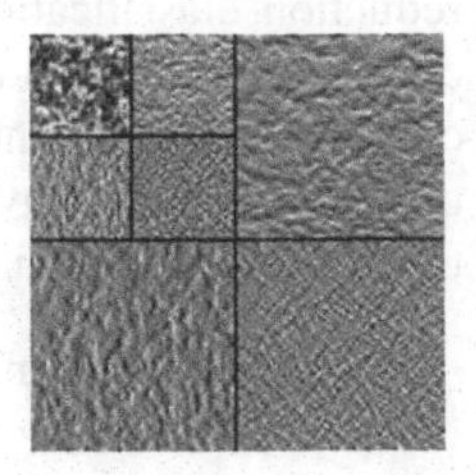

segments are drawn in the separated graphs with increasing class level from the left to the right. A set of ten selected segments of the same class were decomposed into the third level and the sum of their wavelet coefficient in their absolute values were explored. Again it is possible to notice the decreasing level of these sums with increasing segment's class. This fact can be simply explained by vanishing fine stoma structure with increasing segment class, so that the details in the wavelet decomposition subsequently disappears.

According to achieving results, the following three border criteria were suggested: The criterion K_e according the Eq. (15.4) evaluates a sum of wavelet coefficients of the first level decomposition using Daubechies 1 wavelet function. The criterion K_{s1} given by Eq. (15.5) is suggested as a difference of sum of Daubechies 8 wavelet coefficients the second and the first level. The coefficients are summarized in their absolute values. Similarly, the criterion K_{s2} defined according to Eq. (15.6) is counted using the Daubechies 8 wavelet coefficients coming from the third and the second level.

$$K_e = \sum_{i=1}^{M} \sum_{j=1}^{N} |c_{i,j}| \tag{15.4}$$

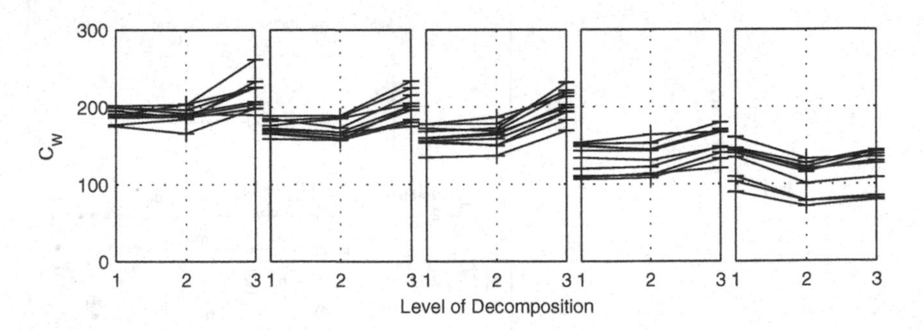

Fig. 15.10 A sum of wavelet coefficients in their absolute values in the three levels of decomposition. Daubechies 1 wavelet was used for decomposition of a selected set of 10 cuts with known classification. Class level is increasing from the left graph (class No. 1) to the right one (class No. 5)

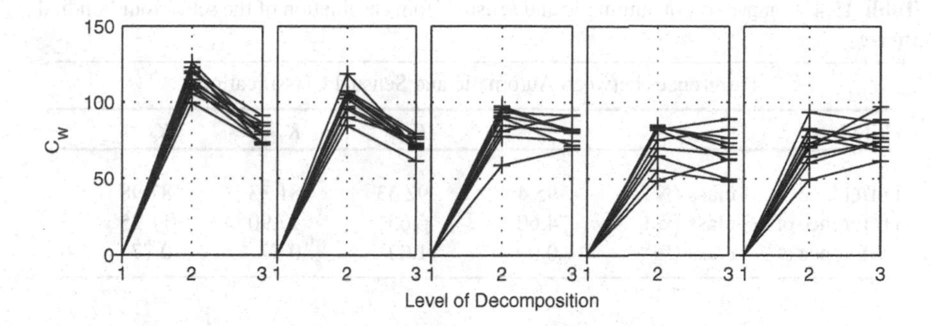

Fig. 15.11 A sum of wavelet coefficients in their absolute values in the three levels of decomposition. Daubechies 8 wavelet was used for decomposition of a selected set of 10 cuts with known classification. Class level is increasing from the left graph (class No. 1) to the right one (class No. 5)

$$K_{s1} = \sum_{i=1}^{M} \sum_{j=1}^{N} |c_{i,j}^{2^{nd} level}| - |c_{i,j}^{1^{st} level}| \tag{15.5}$$

$$K_{s2} = \sum_{i=1}^{M} \sum_{j=1}^{N} |c_{i,j}^{3^{rd} level}| - |c_{i,j}^{2^{nd} level}| \tag{15.6}$$

The Table 15.3 presents border values of mentioned criteria solving for classification of a given image segment.

Table 15.3 Image Texture Classification Using Wavelet Method

	Border values of wavelet criteria			
Criterion	$1-2$	$2-3$	$3-4$	$4-5$
K_e	187	171	160	131
K_{s1}	115	98	79	72
K_{s2}	-32	17.7	10	0

15.4 Results

According to suggested criteria the resulting algorithm was proposed - see Fig. 15.5 and it was verified for more than four hundred real images. As the real images differ in value of brightness and they have the various level of resolution, these parameters had to been unified at first. The influence of noise was reduced by the application of median filter with size 3×3 pixels. The image segments were evaluated

Table 15.4 Comparison of automatic and sensual stoma evaluation of the set of four hundred images

Differences between Automatic and Sensual Classification				
Criterion	K_g	K_{s1}	K_{s2}	K_e
Difference of ±1 class (%)	95.40	92.33	81.33	87.98
Difference of ±2 class (%)	4.60	6.65	17.90	11.25
Difference of ±3 class (%)	0.00	1.02	0.77	0.77

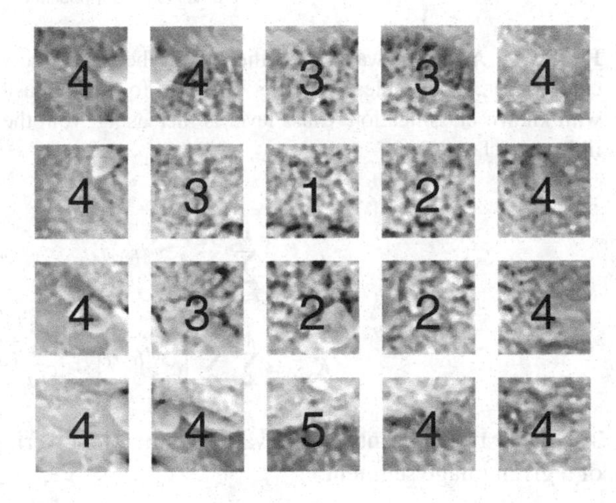

Fig. 15.12 Result of a selected image classification.

by a selected criterion while the segments belonging to stoma only were considered. Achieved results were compared with an experts' sensual image evaluation. The difference of one class was omitted. Comparison of automatic and individual results is presented in Table 15.4. A selected example of segment classification is presented in Fig. 15.12.

It would be also interesting to compare an achieving results with another air pollution indicators. The Table 15.5 could solve for such a comparison.

15.5 Conclusion

The chapter presents the problem of the affect of air pollution to the quality of vegetation. In the general sense this problem is a global one and it can be extrapolated to the study of environment pollution and its affect to the environment quality both on the earth's continents and inside oceans and seas.

Table 15.5 Mean Level of Stomas' Class in the monitored localities according to suggested algorithm

	No. of Locality					
Criterion	1	2	3	4	5	6
K_g	3.3±0.9	3.5±0.9	3.4±0.6	3.3±0.6	3.1±0.8	3.5±0.7
K_{s1}	3.2±0.7	3.5±0.6	3.3±0.5	3.2±0.5	3.1±0.6	3.4±0.6
K_{s2}	3.3±0.6	2.6±0.6	2.6±0.5	2.2±0.5	2.2±0.5	2.5±0.6
K_e	2.5±0.7	2.8±0.8	2.5±0.5	2.5±0.7	2.5±0.6	2.6±0.6

Algorithmic tools proposed in the chapter are devoted to the analysis of the quality of vegetation derived from microscopic images of tree needles in the different areas of the Czech Republic. Results obtained by the wavelet transform are compared with those obtained by general gradient methods.

Further research will be devoted to the study of the most appropriate methods for analysis of textures closely related to the quality of the vegetation. More detail study will be devoted to the correlation between air pollution sources and changes of vegetation.

Acknowledgements. This work has been supported by the Ministry of Education of the Czech Republic (program No. MSM 6046137306). This support is very gratefully acknowledged.

References

1. Arivazhagan, S., Ganesan, L.: Texture Classification Using Wavelet Transform. Pattern Recogn. Lett. 24(9-10), 1513–1521 (2003)
2. Arivazhagan, S., Ganesan, L.: Texture Segmentation Using Wavelet Transform. Pattern Recogn. Lett. 24(16), 3197–3203 (2003)
3. Bishop, C.M.: Neural Networks for Pattern Recognition. Oxford University Press, Oxford (1995)
4. Boashash, B.: Time-Frequency Signal Analysis and Processing - A Comprehensive Reference. Elsevier Science, Amsterdam (2003)
5. Bracewell, R.N.: Fourier Analysis and Imaging. Kluwer Academic Press, Dordrecht (2003)
6. Burger, W., Burge, M.J.: Digital Image Processing. Springer, Heidelberg (2008)
7. Canny, J.F.: A computational approach to edge detection. IEEE Transactions on Pattern Analysis and Machine Intelligence 8, 679–698 (1986)
8. Chellapa, R.: Digital Image Processing. IEEE Computer Society Press, Los Alamitos (1992)
9. Daubechies, I.: Orthonormal Bases of Compactly Supported Wavelets. Comm. Pure and Applied Math. 41, 909–996 (1998)
10. Debnath, L.: Wavelets and Signal Processing. Birkhäuser, Boston (2003)

11. Frayer, D.: Interpolation by the FFT Revised-an Experimental Investigation. IEEE Trans. Acoustics, Speech and Signal Processing 37(5), 665–675 (1989)
12. Gonzales, R.C., Woods, R.E.: Digital Image Processing, 2nd edn. Prentice-Hall, Englewood Cliffs (2002)
13. Haralick, R.M., Shanmugan, K., Dinstein, I.: Texture Features for Image Classification. IEEE Trans. on Syst. Man Cybern 3(6), 610–612 (1992)
14. Huang, K., Aviyente, S.: Wavelet Selection for Image Classification. IEEE Trans. on Image Processing 17(9), 1709–1720 (2008)
15. Jafari-Khouzani, K., Soltanian-Zadeh, H.: Rotation-Invariant Multiresolution Texture Analysis Using Radon and Wavelet Transforms. IEEE Trans. on Image processing 14(6), 783–795 (2005)
16. Jafari-Khouzani, K., Soltanian-Zadeh, H.: Rotation-invariant multiresolution texture analysis using Radon and wavelet transforms. IEEE Transaction on Image Processing 14(6), 783–795 (2005)
17. Keys, R.: Cubic Convolution Interpolation for Digital Image Processing. IEEE Trans. Acoustics, Speech and Signal Processing 29(6), 1153–1160 (1981)
18. Kingsbury, N.G.: Complex Wavelets for Shift Invariant Analysis and Filtering of Signals. Journal of Applied and Computational Harmonic Analysis 10(3), 234–253 (2001)
19. Selesnick, I.W., Baraniuk, R.G., Kingsbury, N.G.: The Dual-Tree Complex Wavelet Transform. IEEE Signal Processing Magazine 22, 123–151 (2005)
20. Kingsbury, N.G., Mugarey, J.F.A.: Wavelet Transforms in Image Processing. In: Procházka, A., Uhlíř, J., Rayner, P.J.W., Kingsbury, N.G. (eds.) Signal Analysis and Prediction, Applied and Numerical Harmonic Analysis. ch. 2, Birkhäuser, Boston (1998)
21. Klette, R., Zamperoni, P.: Handbook of Image Processing Operators. John Wiley & Sons, Chichester (1994)
22. Laine, A., Fan, J.: Texture Classification by Wavelet Packet Signature. IEEE Trans. on Pattern Analysis and Machine Intelligence 15(11), 1186–1190 (1993)
23. Li, S., Shawe-Taylor, J.: Comparison and fusion of multiresolution features for texture classification. Pattern Recogn. Lett. 25 (2004)
24. Lindeberg, T.: Edge Detection and Ridge Detection with Automatic Scale Selection. Internation Journal of Computer Vision 30(2), 117–154 (1998)
25. Mallat, S.: A Theory of Multiresolution Signal Decomposition: The Wavelet Representation. IEEE Trans. on Pattern Analysis and Machine Intelligence 11(7), 674–693 (1989)
26. Mirmehdi, M., Xie, X., Suri, J.: Handbook of Texture Analysis. World Scientific Publishing, New Jersey (2008)
27. Moosmann, F., Nowak, E., Jurie, E.: Randomized Clustering Forests for Image Classification. IEEE Trans. on Pattern Analysis and Machine Intelligence 30(9), 1632–1646 (2008)
28. Náhlík, J., Cudlín, P., Kašová, E., Bartoníčková, R.: Studium morfologie stomatálního vosku smrku ztepilého na vybranch lokalitách ČR pomoc EŘM. In: Stav a Perspektivy ekologického výskumu horských lesných ekosystémov 2001, Slovensk republika, Polana (2001)
29. Nixon, M., Aguado, A.: Feature Extraction & Image Processing. NewNes Elsevier (2004)
30. Otsu, N.: A Threshold Selection Method from Gray-Level Histograms. IEEE Transactions on Systems, Man, and Cybernetics 9(1), 62–66 (1979)
31. Parker, A.J., Kenyon, R.V., Troxel, D.E.: Comparison of Interpolating Methods for Image Resampling. IEEE Trans. on Medical Imaging 2(1), 31–39 (2007)
32. Petrou, M., Bosdogianni, P.: Image Processing, The Fundamentals. John Wiley and Sons, Chichester (2000)

33. Procházka, A., Gavlasová, A., Volka, K.: Wavelet Transform in Image Recognition. In: International conference ELMAR 2005. IEEE, Zadar (2005)
34. Procházka, A., Ptáček, J.: Wavelet Transform Application in Biomedical Image Recovery and Enhancement. In: The 8th Multi-Conference Systemics, Cybernetics and Informatic, vol. 6, pp. 82–87. IEEE, USA (2004)
35. Ptáček, J., Šindelářová, I., Procházka, A., Smith, J.: Wavelet Transforms In Signal And Image Resolution Enhancement. In: International Conference Algoritmy 2002, STU, Podbanske (2002)
36. Randen, T., Husoy, J.H.: Filtering for Texture Classification: A Comparative study. IEEE Trans. on Patern Analysis and Machine Inteligence 21(4), 291–310 (1999)
37. Randen, T., Husoy, J.H.: Filtering for Texture Classification: A Comparative Study. IEEE Trans. on PAMI 21(4), 291–310 (2000)
38. Shaffrey, C.W.: Multiscale Techniques for Image Segmentation, Classification and Retrieval. Ph.D. thesis, University of Cambridge, Department of Engineering, Contribution (2003)
39. Slavíková, P., Mudrová, M., Procházka, A.: Automatic Bionindicator Images Evaluation. In: Intelligent Engineering Systems, INES 2010 (2010)
40. Strang, G., Nguyene, T.: Wavelets and Filter Banks. Wellesley-Cambridge Press, USA (1996)
41. Stringer, S.M.: Invariant Object Recognition in the Visual System with Novel Views of 3D Objects. Neural Computing (14), 2585–2596 (2002)
42. Unser, M.: Texture Classification and Segmentation using Wavelet Frames. IEEE Trans. on Image Processing 4(11), 1549–1560 (1995)
43. Van der Heijden, F.: Image Based Measurement Systems. John Wiley & Sons, New York (1994)
44. Vaseghi, S.V.: Advanced Digital Signal Processing and Noise Reduction. John Wiley & Sons Ltd, Chichester (2006)
45. Vetterli, M., Kovacevic: Wavelets and Subband Coding. Prentice-Hall, Englewood Cliffs (1995)
46. Watkins, C., Sadun, A., Marenka, S.: Modern Image Processing: Warping, Morphing, and Classical Techniques. Academic Press, Ltd, London (1993)
47. Weibao, Z., Yan, L.: Image Classification Using Wavelet Coefficients in Low-pass Bands. In: Neural Networks, IJCNN 2007, pp. 114–118 (2007)
48. Ziou, D., Tabbone, S.: Edge Detection Techniques: An Overview. Internation Journal of Pattern Recognition and Image Analysis 8(8), 537–559 (1998)
49. Zhang, W., Berholm, F.: Multi-scale Blur Estimation and Edge Type Classification for Scene Analysis. Internation Journal of Computer Vision 24(3), 219–250 (1997)

Chapter 16
Identification Approach Lip-Based Biometric

Carlos M. Travieso, Juan C. Briceño, and Jesús B. Alonso

Abstract. A robust biometric identification approach based on lip shape is presented in this chapter. Firstly, we have built an image processing step in order to detect the face of an user, and to enhance the lips area based on a color transformation. This step is ended detecting the lip on the enhanced image. A shape coding has been built to extract features of the lip shape image with original and reduced images. Those reductions have been applied with reduction scale of 3:1, 4:1 and 5:1. The shape coding points have been transformed by a Hidden Markov Model (HMM) Kernel, using a minimization of Fisher Score. Finally, a one-versus-all multiclass supervised approach based on Support Vector Machines (SVM) with RBF kernel is applied as a classifier. A database with 50 users and 10 samples per class has been built (500 images). A cross-validation strategy have been applied in our experiments, reaching success rates up to 99.6% and 99.9% for original and reduced size of lip shape, respectively; using four lip training samples per class and two lip training samples, respectively; and evaluating with six lip test samples and eight lip test samples, respectively. Those success rates were found using a lip shape of 150 shape coding points with 40 HMM states and 100 shape coding points with 40 HMM states in Hidden Markov Model, respectively, reaching with reduced lip shape image our best success, and finally, our proposal.

16.1 Introduction

The security area has become one of the most important items in Research and Development Plans for different countries: European Union, United States and others [1].Besides, many companies are working in this area and the perspective is to

Carlos M.Travieso · Jesús B. Alonso
Signals and Communications Department Institute for Technological Development and Innovation in Communications University of Las Palmas de Gran Canaria Campus University of Tafira, 35017, Las Palmas de Gran Canaria, Las Palmas, Spain
e-mail: {ctravieso, jalonso}@dsc.ulpgc.es

Juan C. Briceño
Computer Science Department, Universidad de Costa Rica, Costa Rica
e-mail: juan.briceno@ucr.ac.cr

J. Fodor et al. (Eds.): Recent Advances in Intelligent Engineering Systems, SCI 378, pp. 341–360.
springerlink.com

increase this world market. Biometrics/indexbiometrics are a field which have increased during the last years due to the identification of a person with her/his body features, as face, lips, fingerprints, voice, etc [12].Those biometric modalities can be used in different applications: search of persons, remote access, timetable control, etc.; with a high or low grade of security. In this work, we present a biometric-based system from lips, in particular, lips without movement, or in a fixed position. This approach can be used on isolated identification system or integrated in other multimodal systems, joined to face, lips with movements, voice, etc.

Nowadays, face identification has become an important application in modern security system due to its easiness of data capturing, its distinctiveness, its acceptance by the public and the low level requirement of cooperation. In the past years, numerous approaches have been proposed and studied in order to improve accuracy and robustness of face recognition systems in real situations. Representative lips works could be found in [12, 9].Lips are a very important part of human face, well studied for biometric verification using its movement, and it has been experimentally demonstrated that it is an effective biometric features for person identification [6, 7] Recently, researchers have shown that the static lip (without movement) is an additional cue for person identification[17].In this chapter, we argue that identifying with static lipshape features could outperform each of them individually and have demonstrated this argument in the proposed experiment. Until now, most of the researches on lip reading still take it as an assistant for speech recognition system. Research has rarely been done as an independent direction. The state-of-the-art, based on lip contour, has researched different methods. In [24],there is used a histogram on gray images to get the lip corner.[4] uses RGB images, in [19] HSV images and in [21] Prewitt and Sobel operators are used. Many approaches have been found on identification or verification systems, using a lip-movement-based approach [6, 7].And, in [22], there is presented a multimodal biometric system based on face image and part of its image, using too the mouth image but not a lip contour. Therefore, this area has been studied.

On [2]an identification system based on lip shape is proposed. This system uses as parameterization a transformation of Shape Coding, using HMMK, and classified with SVM. That approach reached very good results, over 99.6% with 4 training samples.

Other references are working on lips detection. [8] presents a 3D model to detect lips based on a set of particles whose dynamic behaviour is governed by the Smooth Particles Hydrodynamics (SPH). This approach has to be checked with more data before use it on identification. The following work describes a fully automated technique of detecting lip contours from static face images [18]. Face detection is performed first on the input image using a variation of the AdaBoost classifier trained with Haarlike features extracted from the face. A second trained classifier is applied over this extracted face region for isolating the mouth section. This method has been tested on three different face databases that contain images of both neutral faces as well as facial expressions, and a maximum successful lip contour detection rate of 91.08% has been achieved [18]. In [20], Lucas-Kanade method is applied to define an initial parametric lip model of the mouth. According to a combined luminance and chrominance gradient, the model is optimized and precisely locked on to the lip contours. The algorithm performances are evaluated with respect to a lip-reading application.

This work present the identification with a novel and improved method applied to lips, under original and reduced size, based on the transformation of its shape by HMM kernel (HMMK). In order to develop our task, it is divided in three main part: lip extraction, Hidden Markov Model kernel (HMMK) parameterization and identification. In the first block, we have detected the face; we have selected the mouth area, have detected the lips contour, and have calculated its scale reduction. In the second step, the lip features have been extracted based on angular coding and they have been transformed by HMMK. Finally, a Support Vector Machine (SVM) with RBF kernel is used in order to identify each user. The approach of this work is shown in Fig. 16.1.

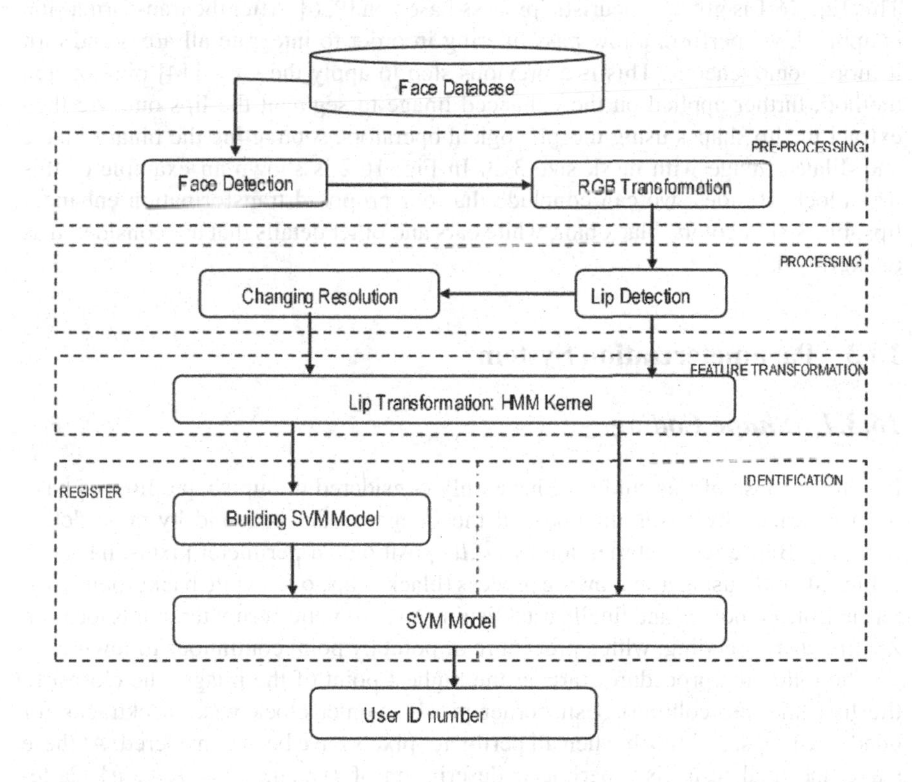

Fig. 16.1 Proposed approach

16.2 Face and Lip Detection

The first block is implemented by three steps, face detection, mouth detection and finally lips extraction. The first step in this work is the face detection from our dataset image. It is based on Jones and Viola approach [23].

Once we have the face image, then we extract the inferior middle of that image because the mouth always is situated in this position. Too, other small horizontal and vertical cuts are done (see red rectangle on Fig. 16.2), before to apply the RBF transformation.

First based on [9, 6], we have used a RBF transformation, but finally, we have built a new version. This is done, because we want to enhance the red color in order to get the lips as an object. The rest of elements have to be the background. Besides, we want to include as background different artifacts as the bear, dark skin, and others. Therefore, we apply the following simple transformation from color to gray scale image;

$$I = R - 2.4G + B \tag{16.1}$$

This Eq. 16.1 is got by a heuristic process based on [9, 6]. After the transforma-tion is applied, we perform a low-pass filtering in order to integrate all areas and turn it more homogeneous. This is a previous step to apply the Otsu [14] binarization method, further applied on the enhanced image to segment the lips out. We then extract the lip shapes using morphological operators, subtracting the binary image and dilated image with mask size 3x3. In Fig. 16.2 is shown an example of this first block. Besides, we can conclude that our proposed transformation enhances lips, black skin colors, black hair, white ears and other details that are considered as background.

16.3 Parameterization System

16.3.1 Shape Coding

For the purpose of this study we have only considered the lip shape. Images have been binarized by Otsus method and the image border is found by edge detection [14]. Border characterization by (λ,μ) positions of perimeter pixels, has been achieved firstly using a shadowing process (black shape over white background), filtering isolated points, and finally establishing an automatic perimeter points location λ= line μ=row coding, with a procedure of point by point continuous following.

The following procedure starts at the highest point of the image (the closest to the first line and column), visit border pixels counter clock wise, backtracks for dead-end loops, and finish when all perimeter pixels have been considered. At these stage, the final result is a perimeter description of $\{(\lambda_i,\mu_i) \mid i = 1,\ldots,n\}$ Cartesian points location description; representing the closed border of a lip shape with one pixel of wide stroke. The general idea is to consider (λ,μ) positional perimeter

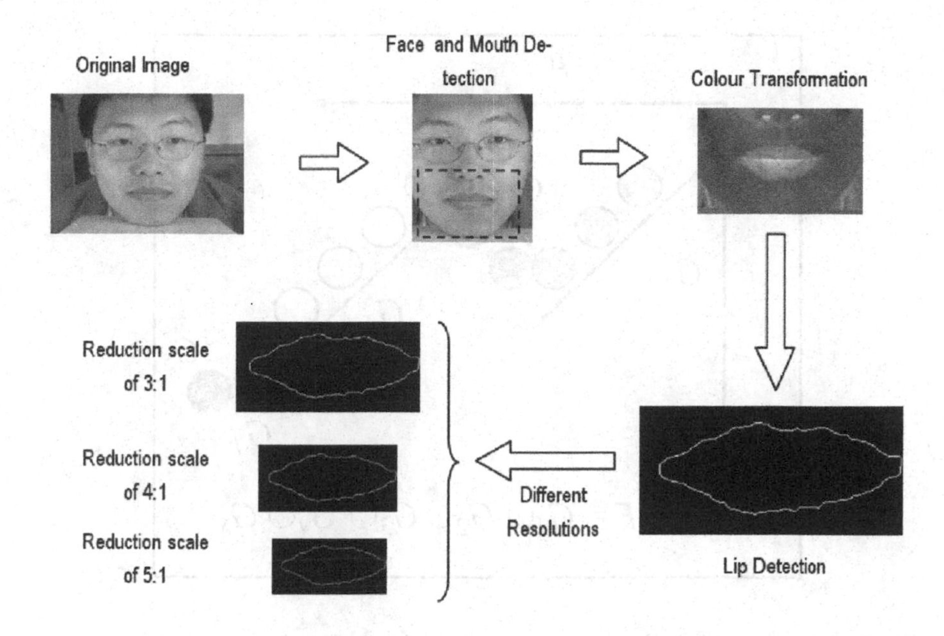

Fig. 16.2 Examples in the process for Lip Detection

pixels as $(\lambda, F(\lambda))$ graph points of a 1-D defined piece like function description F, representing the border on the 2-D image plane.

Agreement on the arbitrary consideration for μ coordinates as $\mu = F(\lambda)$ is done, because the way the structuring procedure simplifies the choice of points; needing maximum size placed over λ ordinate and so we consider the 90° rotated lip image.

Data compression, size regularization and critical control point selection of perimeters description are achieved by an easy to implement automatic structuring procedure. That structure, that agglutinate points with a similar behavior, helps improve the HMM states interpretation, hence facilitating the HMM tuning.

This procedure is based on the idea that a one pixel stroke on a black and white image may be described as a G_f graph of a one dimensional trajectory application f, if we have preservation of a correct sequencing definition or monotonic behavior on the λ ordinate. That is $G_f = \{(\lambda, \mu) \mid \mu_i = f(\lambda_i), i = 1, \ldots, n\}$, is such a graph describing relation, if ordinate points λ_i, of the f stroke must be strictly ascending or descending; that is, the λ_i must be: $\lambda_i < \lambda_{i+1}$ or $\lambda_{i+1} < \lambda_i$ for $i = 1, \ldots, n-1$. Considering the complete perimeter G, we define its description relation F as the partial definition of piece like 1-D trajectory applications f_j (with graphs G_j) preserving monotonic behavior. That is G the general border relation, is such that $G = \bigcup_{j \varepsilon J} G_j$where each G_j is a relational set of positional points and each of them is an obvious, by its λ_i monotonic quality, automatic build piece of border information. As a result, the restriction trajectory applications (or coded pieces of border) $f_j =$

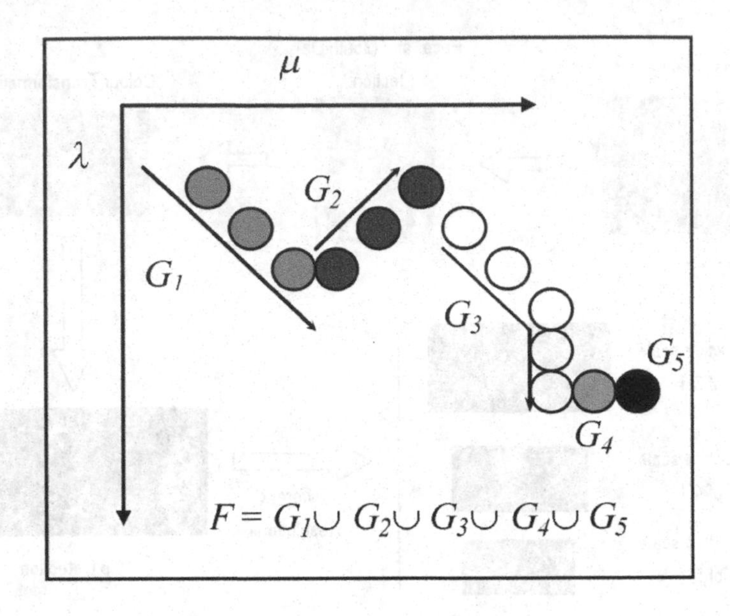

Fig. 16.3 An example of a fragment of an image border, decomposed in G_j graphs.

$F_{|\{x_a|\alpha\varepsilon J_i\}}$ are in such a way that the next point following the last of G_j is the first of G_{j+1}. Accordingly G_j graphs are correct f_j trajectory applications description. Building the G_j sets is a very straightforward operation:

- Beginning with a first point we include the next one of F.
- As soon as this point does not preserve monotonic behaviour we begin with a new $G_j + 1$.
- Processes stop when all F points are assigned.

In order to simplify and avoid G_j reduced to one pixel, as is the case for G_4 and G_5 in Fig. 16.3, we preserve only the first point of constant λ ordinate series. Note that structuring G, by partial graph descriptions, G_j, account for abrupt direction changes on the perimeter description, and may bee interpreted has HMM states. In order to define $p \geq n$ coding $(\lambda_i, \mu)_{i=1,...,p}$ control points, after building up the G_j we select all the $n$first points of each G_j, $j = 1, \ldots, n$ and then we complete the perimeter points description by $k = n - p$ points, chosen uniformly distributed for each G_j and proportionally to its size.

In order to perform a rotational, scale size and origin reference free coding; we perform an angle transformation for the positional point border coded as before. For a given coded border of p positional control points $G = \{X_i = (\lambda_i, \mu_i) \mid i = 1, \ldots, p\}$ let C_0 be its central point, and let be θ_i and ϕ_i the angles referred by C_0 and X_i, $\theta_i = angle(C_0, X_i, X_{i+1})$ and $\phi = angle(X_i, C_0, X_{i+1})$. Note also that such coding interpret Markovian sequences point to point dependencies.

Then the sequence of $(\lambda_i, \mu)i = 1, \ldots, p$ positional points are transformed in sequence of $(\theta_i, \phi_i)i = 1, \ldots, p-1$ angular origin free representations points. Note that the choice of the start point X_1 and the C_0 points account for scale and lip shape rotation, as well as geometrical properties of triangular similarities make such sequence of lip shape coding, size and location free.

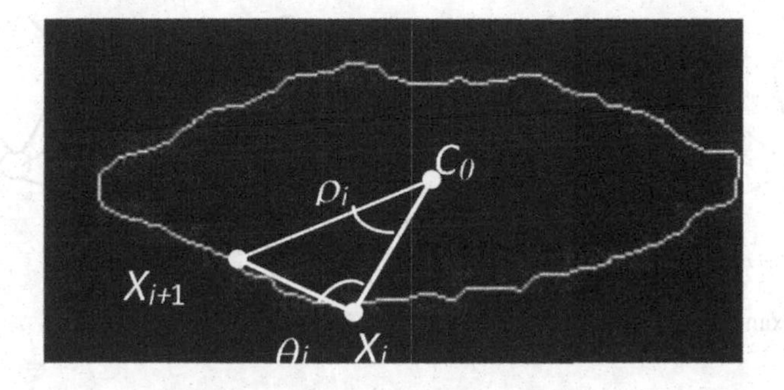

Fig. 16.4 An example of angular coding, applied to lip shape.

16.4 Classification Approach

In order to implement the classification system based on a HMM kernel from shape data, there is to follow three steps. The first one is the use of Hidden Markov Model (HMM) applied to the border coding of previous section [15, 16], the second one is the transformation of data obtained from de HMM, and finally the use of a Support Vector Machine (SVM) [3], as a classifier.

16.4.1 HMM Classification Approach

An HMM is the representation of a system in which, for each value that takes a variable t, called time, it is found in one and only one of N possible states and declares a certain value at the output. Furthermore, an HMM has two associated stochastic processes: one hidden associate with the probability of transition between states (non observable directly); and another observable one, an associate with the probability of obtaining each of the possible values at the output, and this depends on the state in which the system has been found [15]. It has been used a Discrete HMM (DHMM), which is defined by [15, 16];

- N is the number of states,
- M is the number of different observations,
- $A(N,N)$ is the transition probabilities matrix from one state to another,
- $\pi(N,1)$ is the vector of probabilities that the system begins in one state or another;
- and $B(N,M)$ is the probabilities matrix for each of the possible states of each of the possible observations being produced.

We have worked with an HMM called "left to right" or Bakis, which is particularly appropriate for sequences evaluation. These "left to right" HMMs turn out to be especially appropriate for lip shape because the transition through the states is produced in a single direction, and therefore, it always advances during the transition of its states.

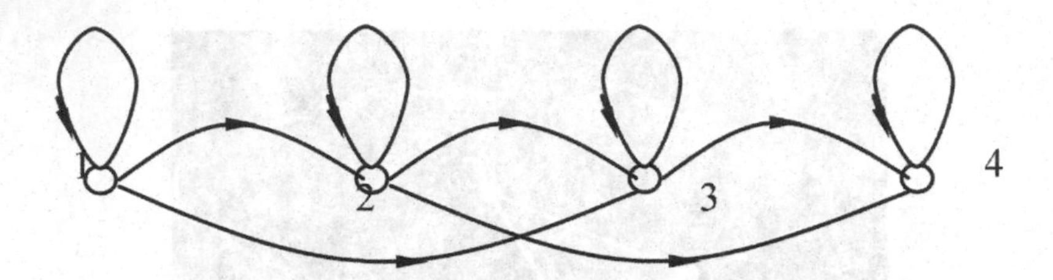

Fig. 16.5 Example of a left to right HMM

This provides for this type of model the ability to keep a certain order with respect to the observations produced where the temporary distance among the most representative changes. Finally, it has been worked from 20 to 45 states, 32 symbols per state and 2 labels (multi-labeling) [3].

In the DHMM approach, the conventional technique for quantifying features is applied. For each input vector, the quantifier takes the decision about which is the most convenient value from the information of the previous input vector. To avoid taking a software decision, a fixed decision on the value quantified is made. In or-der to expand the possible values that the quantifier is going to acquire, multi-labelling is used, so that the possible quantified values are controlled by varying this parameter. The number of labels in the DHMM is related to values that can be taken from the number of symbols per state.

DHMM algorithms should be generalized to be adjusted to the output multilabelling $(\{v_k\}k = 1,\ldots,C)$, to generate the output vector $(\{w(x_t, v_k)\}k = 1,\ldots,C)$. Therefore, for a given state j of DHMM, the probability that a vector xt is observed in the instant t, can be written as;

$$b_j = \sum_{k=1}^{C} w(x_t, v_k) b_j(k) \tag{16.2}$$

where $b_j(k)$ is the output discrete probability, associated with the value v_k and the state j; being C the size of the vector values codebook. This Eq. 16.2 will be used for the proposal of transformation.

One of the advantages that can offer the Hidden Markov Models (HMM) is the approach to see, the observation sequence as a matrix. For the particular case of the voice, the previous sequence is a vector defined for determined parameters, but for another type of applications they can be assigned to several sequences in parallel, forming the matrix commented previously.

For the case of image signals, it would permit to continue the sequence in Cartesian coordinates (components "X" and "Y") or in polar (module and argument), by means of a matrix of two columns, that describes the sequence of a certain region or contour.

Therefore, the vector can take the expression;

$$x_t = [x_t^1 x_t^2 \dots x_t^{N_1}] \quad (16.3)$$

And, the output probability for this vector xt, it will be given by

$$b_j(x_t) = \prod_{i=1}^{N} b_j(x_t^i) \quad (16.4)$$

The DHMM software is the gpdsHMM toolbox, which is freely available from: http://www.gpds.ulpgc.es/download/index.htm, David et al. [5].

This approach models a DHMM from a lip shape. After experiments, it can be observed that this system is not appropriate in order to achieve a discriminative identification system. Therefore, it is proposed an improvement by the transformation of the HMM kernel.

16.4.2 Data Transformation

The next step is the transformation of DHMM probabilities, relating to the approach of the HMM score [7]. With this goal, the aim is to unite the probability given by the HMM to the given discrimination by the classifier based on SVM. This score calculates the gradient with respect to HMM parameters, in particular, on the probabilities of emission of a vector of data x, while it is found in a certain state $q\varepsilon\{1,\dots,N\}$, given by the matrix of symbol probability in state $q\varepsilon(b_q(x))$, as it is indicated in 16.2;

$$P(x/q,\lambda) = b_q(x) \quad (16.5)$$

If the derivative of the logarithm of the previous probability is calculated (gradient calculation), the HMM kernel is obtained, whose expression is given by [11], but adapted to the HMM by the authors as:

$$\frac{\partial}{\partial P(x/q,\lambda)} logP(x/q,\lambda) = \frac{\xi(x,q)}{b_q(x)} - \xi(q) \quad (16.6)$$

Approximations and calculations for the previous Eq. 16.6 have been found in [11].

In our case, and using a DHMM,$\xi(x,q)$ represents the number of times that it is localized in a state q, during the generation of a sequence, emitting a certain symbol x [15, 11]. And $\xi(q)$ represents the number of times which it has been in q during the process of sequence generation [15, 11]. These values are directly obtained from the forward backward algorithm, applied to the DHMM by [15, 11].

The application of this score (U_X) to the SVM is given by Eq. 16.5 , using the technique of the natural gradient (see [17]);

$$U_X = \nabla_{(x,q)} logP(x/q, \lambda) \tag{16.7}$$

where U_X defines the direction of maximum slope of the logarithm, of the probability of having a certain symbol in a given state.

16.4.3 SVM Classification Approach

The basic idea consists of training the system to obtain two sets of vectors (in two dimensions corresponding with points) that represent classes to identify. Subsequently, the separating hyperplane H (in two dimensions is a lineal classifier) between these two sets is calculated. The pertinent points within the hyperplane have to satisfy the following equation [3]:

$$\omega \cdot x_i + \Psi = 0 \tag{16.8}$$

where ω is normal to hyperplane, $\Psi / \mid\mid \omega \mid\mid$ is the perpendicular distance from the hyperplane to origin, $\mid\mid \omega \mid\mid$ is the Euclidean norm of ω, ω is the independent term, and x_i is a point in the hyperplane. Eq. 16.8 is equal to 0, and it means that indicates the decision border.

Furthermore, another two hyperplanes are defined as follows; $H_1 : x_i \cdot \omega + \Psi = 1$ and $H_2 : x_i \cdot \omega + \Psi = -1$, which contains support vectors. The distance between planes H_1 and H_2 is known as the margin. The aim of this classification algorithm is to calculate the maximum of the aforementioned margin.

Once the system has been trained and, therefore, the separation hyperplane has been obtained, we must determine what the decision limit is (hyperplane H located between H_1 and H_2 and equidistant to them). In accordance with the previous decision, the corresponding class label is assigned, that is, the class of x will be defined by sign($\omega \cdot x_i + \Psi$). This means that test samples are assigned with label "$+1$", and the remainder, with label "-1".

SVM calculates the separation between classes, by means of the calculation of the natural distance between the scores of two sequences X and Y;

$$D^2(X,Y) = \frac{1}{2}(U_x - U_y)^T F^{-1} (U_x - U_y) \tag{16.9}$$

where F is the HMM information matrix, and is equivalent to the matrix of covariance of the vectors U_X and U_Y.

Finally, different types of functions, which can be used for SVM, are with a lineal and a Gaussian kernel (RBF). This is used for establishing the decision limit, based on SVM^{light} [13]. The RBF kernel is shown in the following equation;

$$K(X,Y) = e^{-D^2(X,Y)} \tag{16.10}$$

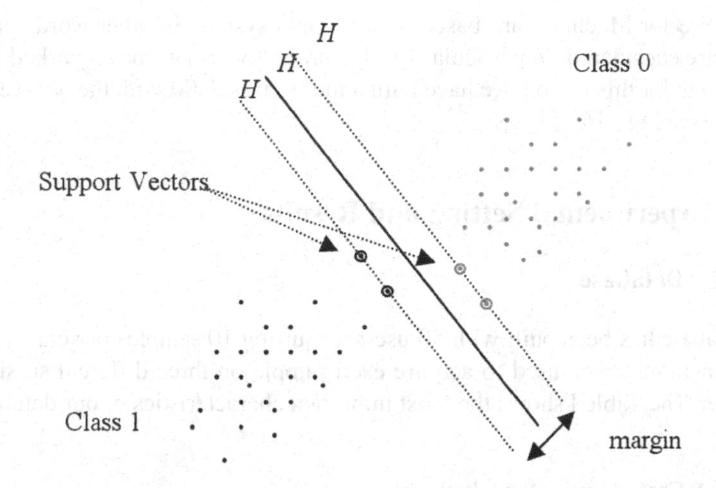

Fig. 16.6 Hyperplanes of lineal separation for the separable case. The support vectors are marked with red and black circles.

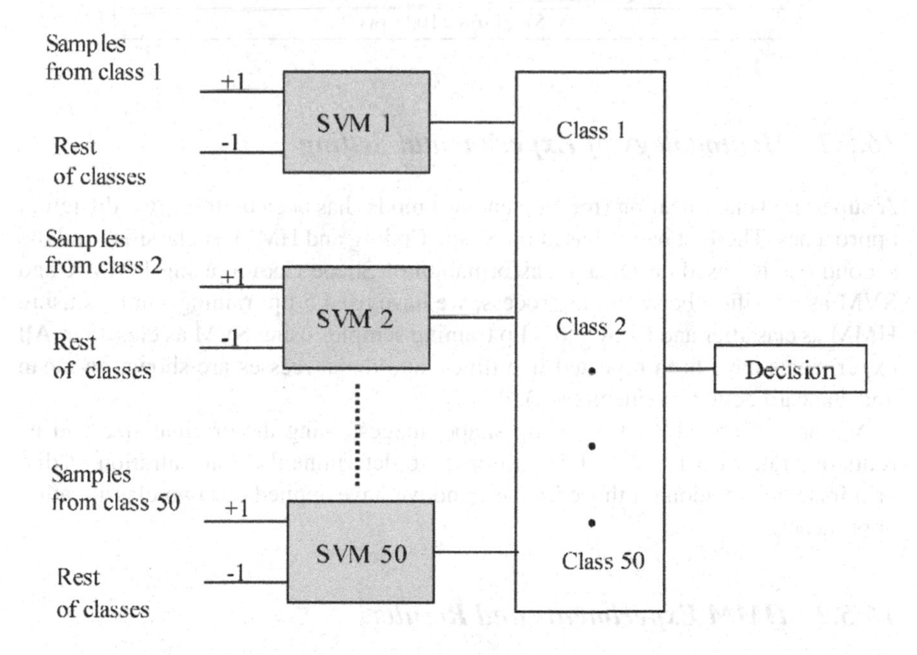

Fig. 16.7 Block diagram for the SVM, with training mode.

Support Vector Machines are based on a bi-class system, in other words only two classes are considered. In particular for this present work, we have worked with 50 classes, and for this reason, we have built a multi-class SVM with the one-versus-all strategy (see Fig. 16.7).

16.5 Experimental Setting and Results

16.5.0.1 Database

Our database has been built with 50 users, acquiring 10 samples per class. A compact camera has been used to acquire each sample on three different sessions for each user. The Table I shows the most important characteristics of our database.

Table 16.1 Characteristic of our database

Parameters	Data
Number of classes	50
Number of samples per classes	10
Acquisition and Quantification	RGB (24 bits, 256 levels)
Size	768×1024 pixels

16.5.1 Methodology of Experimental Setting

A supervised classification (training and test mode) has been built for two dif-ferent approaches. The first one is based on Shape Coding and HMM as classifier and the second one is based on Data Transformation of Shape Coding using HMMK and SVM as classifier. For training process, we have used 5 lip training samples using HMM as classifier and from 5 to 4 lip training samples using SVM as classifier. All experiments have been repeated five times, and the successes are shown on mean and standard deviation (mean±std).

We have changed the size of lip shape images, using the original size and its reduction rate of 3:1, 4:1 and 5:1, in order to determine the concentration of discriminate information on those images, and we have applied our two classification approaches.

16.5.2 HMM Experiments and Results

Experiments have been based on parameters calculated from the lip shape. Therefore, we have achieved our results varying some variables from the proposed system; in particular, the number of HMM states (between 20 and 45 states), the number of parameters for defining the lip shape (between 100 and 200) and the size of lip shape image. But this first experiment, we have used information shape, using HMM as classifier. In Table 16.2 and Fig. 16.8 is observed this experiment for the original image, using 5 training samples per each case, and 5 independent test samples for its evaluation (samples not used on training mode).

The success rates achieved rates up to 51% for HMM classification between 20 and 45 states. This success is low, and for this reason, we decide to check the information included of the lip shape image, reducing its size.

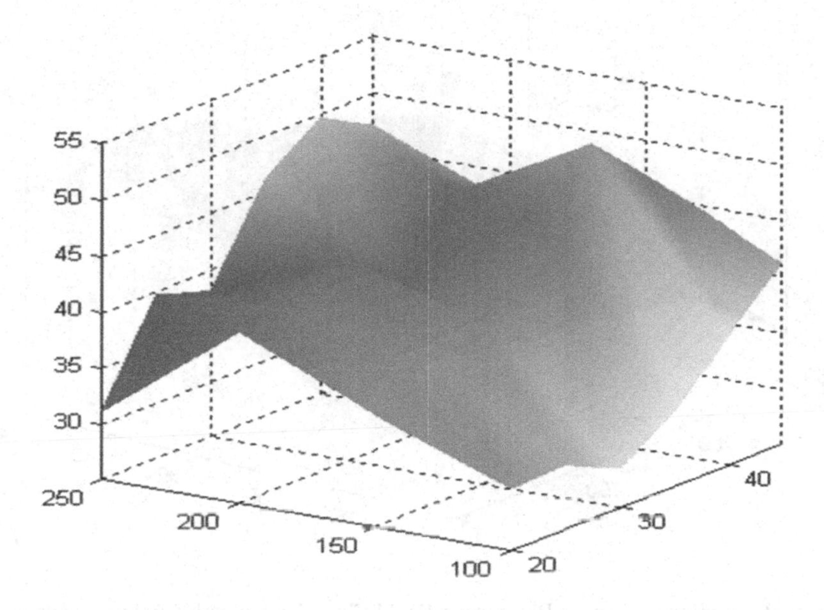

Fig. 16.8 Variation of Success Rates using HMM Classification with 5 training samples for original images.

Table 16.2 Success Rates using HMM Classification with 5 training samples for original images

State numbers	Number of shape points			
	100	150	175	200
20	30.71% ± 4.36	35.22% ± 4.13	40.47% ± 2.21	31.26% ± 3.73
25	30.92% ± 3.78	37.58% ± 2.52	40.43% ± 1.94	39.72% ± 2.91
30	28.58% ± 3.95	39.42% ± 2.09	43.31% ± 1.86	38.23% ± 2.36
35	31.84% ± 3.36	47.33% ± 1.57	39.02% ± 2.04	46.29% ± 1.92
40	36.73% ± 4.05	51.64% ± 1.37	45.39% ± 1.65	49.80% ± 1.43
45	41.20% ± 3.26	46.83% ± 1.52	42.25% ± 1.83	47.16% ± 1.74

For the size reduction of lip shape images, we have applied the same HMM, varying the same parameters, number of HMM states, number of lip shape parameters and 5 training samples in order to build the model. Besides, we have added three reduction rates (3:1, 4:1 and 5:1), applied to the lip shape. On Tables 16.3, 16.4, 16.5 and Fig. 16.9, 16.10 and 16.11 can be observed the success rates and theirs tends, respectively.

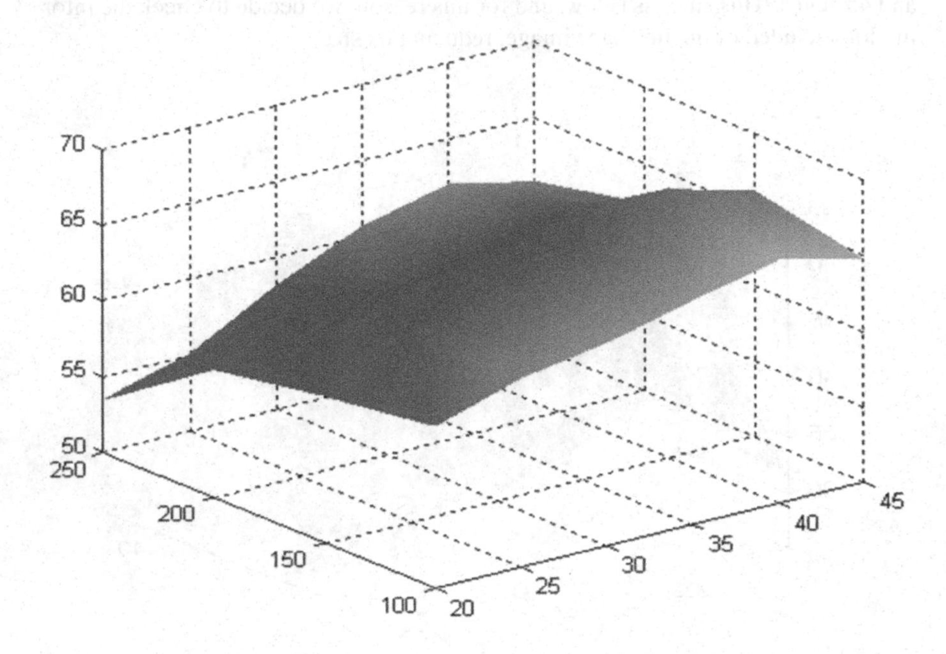

Fig. 16.9 Variation of Success Rates using HMM Classification with 5 training samples for a size reduction of 3 times.

Table 16.3 Success Rates using HMM Classification with 5 training samples for a size reduction of 3 times

State numbers	Number of shape points			
	100	150	175	200
20	60.83% ± 3.04	59.76% ± 9.32	58.51% ± 5.71	53.44% ± 2.77
25	62.65% ± 2.87	62.82% ± 3.67	59.76% ± 9.32	55.20% ± 3.30
30	63.52% ± 2.19	64.16% ± 3.13	60.73% ± 3.81	58.20% ± 4.28
35	64.88% ± 2.09	65.73% ± 3.81	61.88% ± 3.29	60.72% ± 3.51
40	66.16% ± 2.79	67.88% ± 3.29	63.28% ± 3.92	62.08% ± 3.76
45	64.80% ± 2.06	66.21% ± 3.29	62.45% ± 4.27	60.87% ± 3.89

All simulations have given better success than experiments using original image. We can observe that the number of lip shape parameters is minor; due to the size

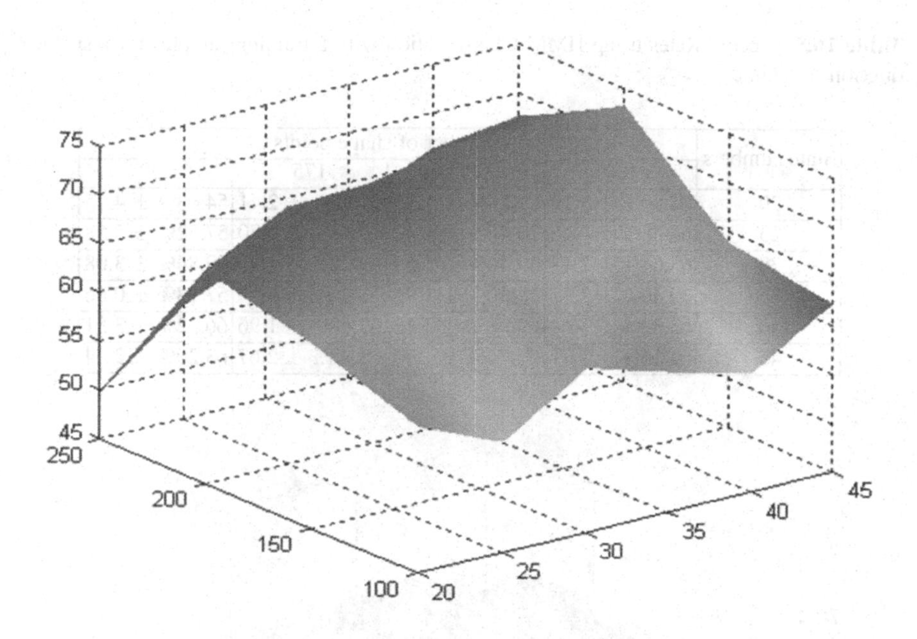

Fig. 16.10 Variation of Success Rates using HMM Classification with 5 training samples for a size reduction of 4 times.

Table 16.4 Success Rates using HMM Classification with 5 training samples for a size reduction of 4 times

State numbers	Number of shape points			
	100	150	175	200
20	60.08% ± 4.89	64.16% ±1.46	66.88% ± 1.78	49.76% ± 5.74
25	56.40% ± 4.19	64.80% ± 2.42	71.20% ±3.02	56.64% ± 3.52
30	61.92% ± 2.34	67.28% ± 3.01	71.68% ± 3.58	60.32% ± 1.48
35	59.36% ± 2.84	64.56% ± 2.88	73.52% ± 3.66	62.40% ± 1.90
40	57.12% ± 3.89	66.96% ± 1.51	73.60% ± 3.03	66.40% ± 4.03
45	62.16% ± 1.25	63.84% ± 2.18	73.38% ± 4.14	67.60% ± 2.53

of lip shape is minor. But the information is more concentrated and its extraction is easier extracted than the original size. For this reason, for all cases, the success is bigger than for original size, being the best size, the reduction rate of 4:1, reaching up to 73.60%, 20% over the original size.

16.5.3 SVM Experiments and Results

After HMM classification, on Tables 16.6 and 16.7 can be observed the success is low for a case of automatic biometric identification. Therefore, we have used the HMM as feature extraction by the HMM kernel, and its classification was done by

Table 16.5 Success Rates using HMM Classification with 5 training samples for a size reduction of 5 t imes

State numbers	Number of shape points			
	100	150	175	200
20	64.80% ± 3.19	69.20% ± 1.27	61.92% ± 2.41	54.08% ± 4.35
25	62.56% ± 4.46	70.46% ± 1.42	70.48% ± 1.70	57.52% ± 5.58
30	63.04% ± 2.18	69.40% ± 1.59	69.20% ± 1.47	59.84% ± 3.08
35	60.40% ± 2.32	68.80% ± 1.71	72.00% ± 3.76	57.28% ± 0.66
40	63.32% ± 3.04	69.68% ± 1.10	67.68% ± 1.96	66.32% ± 3.51
45	64.16% ± 2.71	67.76% ± 1.74	68.48% ± 2.67	63.20% ± 2.04

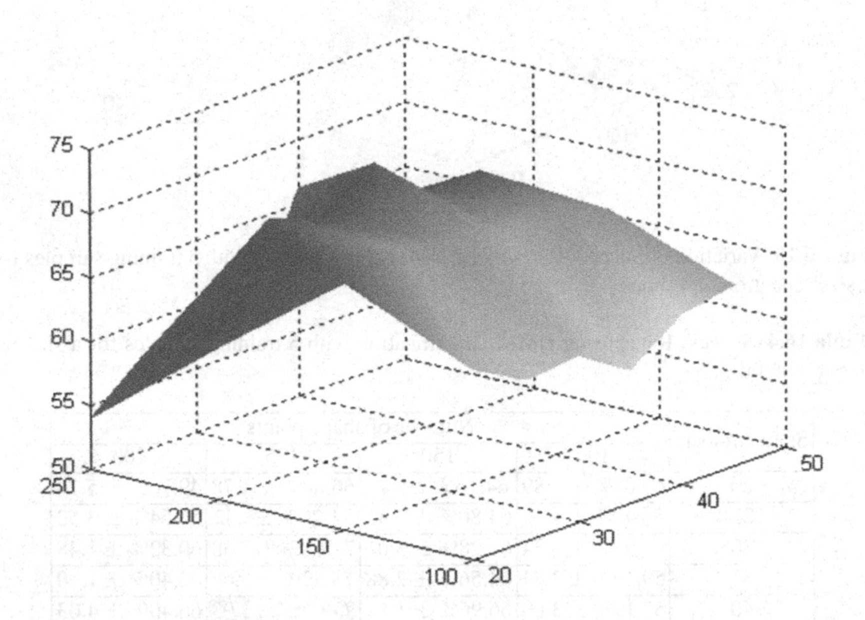

Fig. 16.11 Variation of Success Rates using HMM Classification with 5 training samples for a size reduction of 5 times

SVM. We have used two kinds of kernel for SVM (lineal and Gaussian kernel). For all cases, Gaussian kernel has given better results and it only is shown that kind of kernel, where G is the variance of the Gaussian function, this is, the width of the RBF kernel (G, see tables 16.6 and 16.7), which is automatically determined by the SVM using grid search on training images only. This parameter G is inversely proportional of δ (see Eq. 16.11).

$$k(x,y) = exp(-\frac{||x-y||^2}{2\delta}) \tag{16.11}$$

Therefore, for each case, our system contains two parameters; the number of lip shape points and the parameter G from SVM with RBF kernel (see Tables 16.6 and 16.7).

For these experiments, we have used the best classification based on HMM, as the beginning point to use as parameterization. In particular, for the original size of lip shape, we have applied HMMK and SMV for 40 HMM states and 150 points of shape coding. For lip shape with reduced scale, we have used the best results reached on 40 HMM states and 100 points of shape coding. Tables 16.6 - 16.7 are shown the results for original size and its reduction rate of 4:1, respectively.

Table 16.6 Success Rates using SVM Classification with RBF Kernel for different numbers of training samples for original images

Number of shape points	State numbers	Number of training samples	SVM Success rates	G
100	40	5	100 % $\pm$ 0	1×10^{-6}
100	40	4	99.54 % $\pm$ 0.08	3×10^{-6}
100	40	3	99.29 % $\pm$ 0.56	3×10^{-6}
100	40	2	99.01 % $\pm$ 0.90	4×10^{-6}
150	40	5	100 % $\pm$ 0	1×10^{-6}
150	40	4	99.67 % $\pm$ 0.15	3×10^{-6}
150	40	3	99.31 % $\pm$ 0.40	3×10^{-6}
150	40	2	99.17 % $\pm$ 0.68	3×10^{-6}
175	40	5	100 % $\pm$ 0	1×10^{-6}
175	40	4	99.56 % $\pm$ 0.20	3×10^{-6}
175	40	3	99.43 % $\pm$ 0.58	3×10^{-6}
175	40	2	99.37 % $\pm$ 0.78	4×10^{-6}
200	40	5	100 % $\pm$ 0	1×10^{-6}
200	40	4	99.56 % $\pm$ 0.20	3×10^{-6}
200	40	3	99.36 % $\pm$ 0.52	3×10^{-6}
200	40	2	99.15 % $\pm$ 0.71	3×10^{-6}

On Tables 16.6 and 16.7, we have reached very good results, and in particular, Table 16.7 shows the concentration of information obtained from the image reduction, giving efficiency better than the original size (versus Table 16.6. In particular, this improvement is only observed with the reduction of the number of training samples, because on the same conditions than HMM (using 5 training samples), our proposal gives 100%. Finally, only with two training samples, we are reached over 99.9%, and with three training samples, our proposal obtains 100%. Therefore, this new adjustment gives a better improvement and better stability for our system.

Table 16.7 Success Rates using SVM Classification with RBF Kernel for different numbers of training samples for a size reduction of 4 times

Number of shape points	State numbers	Number of training samples	SVM Success rates	G
25	40	5	100% ± 0	4×10^{-6}
25	40	4	100% ± 0	4×10^{-6}
25	40	3	99.95% ± 0.12	1×10^{-6}
25	40	2	99.80% ± 0.28	1×10^{-6}
50	40	5	100% ± 0	4×10^{-6}
50	40	4	100% ± 0	4×10^{-6}
50	40	3	100% ± 0	1×10^{-6}
50	40	2	99,83% ± 0.26	2×10^{-6}
100	40	5	100% ± 0	4×10^{-6}
100	40	4	100% ± 0	4×10^{-6}
100	40	3	100% ± 0	4×10^{-6}
100	40	2	99.92% ±0.14	4×10^{-6}
150	40	5	100% ± 0	4×10^{-6}
150	40	4	100% ± 0	4×10^{-6}
150	40	3	100% ± 0	4×10^{-6}
150	40	2	99,87% ± 0.20	3×10^{-6}

16.6 Discussion

After experiments, we consider that the use of only lip shape is a bad feature, and it can be observed in Tables 16.2 - 16.5. The success is very low versus other references. Therefore, we have introduced the data transformation using the HMM kernel. Each state on HMM represents a variation of shape, and the best discriminative system has 40 HMM states from 150 points of shape coding for an original size from the image (see Table 16.2), and 40 HMM states from 100 points of shape coding for a scale reduction with reduction rate of 4:1 (see Table 16.4). The image effect of reduction scale improves to get more useful information. It is due to original image is very big (see Table 16.1), and therefore, shape coding obtain points very dispersed for getting a discriminated sequence. Reducing that size, it is reached this proposal. It is independent of shape coding, since this method calculates with scale invariance each parameter, but authors have to find the adequate size to maximize the scale invariance information.

A set of 4 or 5 points represent a state, as averaged. As it has low success, we applied the HMM kernel as an enlarged representation, using the relation between $b_q(x)$, $\xi(x,q)$ and $\xi(q)$, according to HMM kernel.

Now, we are representing the number of times that it is localized in a state q, the data vector for each state according the probability of emission for the same data vector for each state, it is an enlarged representation.

These new features have a great set of data and it is classified by SVM, because it has a good behavior with big set of features.

Success rates are shown in Tables 16.6 and 16.7, and we have demonstrated that DHMM kernel is a very efficient and robust feature extraction because it always improves the DHMM approach. Too, it is shown that working with 150 points of shape coding and using 40 HMM states for original size and 100 points of shape coding and using 40 HMM states for the same lip shape 16.6 and 16.7), for this reason, it only is shown results with RBF kernel. Finally, with only two lip samples for training on the reduced lip shape image with rate of 4:1, we have achieved a success over 99.9%.

16.7 Conclusion

An original and robust system has been built for automatic based-lip identifica-tion, using its shape on a HMM kernel transformation and finally, being classified with an SVM with RBF kernel. The success rate is up to 99.6%, using our database with 4 lip sample for training mode; and we have improved this success reducing the original size of the lip shape image up to 99.9%. It is due to the quantity of information for our learning machine is classified better, if we eliminate the extra information, and only use the just size, in order to get the best efficiency of our approach. The use of this work on the process of face identification can give more robustness to the whole system, improving its success rate.

In future works, we will implement a verification approach with multimodal system with lip and face. Besides, our database will be increased.

Acknowledgements. This work has been partially supported by Catedra Telefnica ULPGC 2009/10 (Spanish Company), and partially supported by Spanish Government un-der funds from MCINN TEC2009- 14123-C04-01.

References

1. Acuite Market Intelligence: Categories: Authentication, Identity & Access Management, Manufacturers, TechNews, Trends Tags: biometrics, facial recognition, fingerprint reader, hand geometry, retinal scanning (2011) http://www.securitydreamer.com/2007/06/new_clearheaded.html (accesed March 14, 2011); Slifka, M.K., Whitton, J.L.: Clinical implications of dysregulated cytokine production. J. Mol. Med (2000), doi:10.1007/s001090000086
2. Briceño, J.C., Travieso, C.M., Alonso, J.B., Ferrer, M.A.: Robust Identification of persons by Lips Contour using Shape Transformation. In: 14th IEEE International Conference on Intelligent Engineering Systems, pp. 203–207 (2010)
3. Burges, C.: A Tutorial on Support Vector Machines for Pattern Recognition. Data Mining and Knowlshape Discovery 2(2), 121–167 (1998)
4. Coianiz, T., Torresan, L., Massaro, D.: 2D deformable models for visual speech analysis. In: NATO Advanced in Speechreading by Humans and Machines, pp. 391–398. Springer, Heidelberg (1996)

5. David, S., Ferrer, M.A., Travieso, C.M., Alonso, J.B.: gpdsHMM: A Hidden Markov Model Toolbox in the Matlab Environment. Complex Systems Intelligence and Modern Technological Applications, 476–479 (2004)
6. De la Cuesta, A., Zhang, J., Miller, P.: Biometric Identification Using Motion History Images of a Speaker's Lip Movements. In: Machine Vision and Image Processing International Conference, pp. 83–88 (2008)
7. Faraj, M.I., Bigun, J.: Motion Features from Lip Movement for Person Authentication. In: 18th International Conference on Pattern Recognition, vol. 3, pp. 1059–1062 (2006)
8. Gastelum, A., Krueger, M., Marquez, J., Gimel'farb, G., Delmas, P.: Automatic 3D lip shape segmentation and modeling. In: 23rd International Conference Image and Vision Computing, New Zealand, pp. 1–6 (2008)
9. Goudelis, G., Zafeiriou, S., Tefas, A., Pitas, I.: Class-Specific Kernel-Discriminant Analysis for Face Verification. IEEE Transactions on Information Forensics and Security 2(3) Part 2, 570–587 (2007); doi:10.1109/TIFS.2007.902915
10. Hernando, J., Nadeu, C., Mariño, J.B.: Speech recognition in a noisy environment based on LP of the one-sided autocorrelation sequence and robust similarity measuring techniques. Speech communications 21, 17–31 (1997)
11. Jaakkola, T., Diekhans, M., Haussler, D.: A discriminative framework for detecting remote protein homologies. Journal of Computational Biology 7(1-2), 95–114 (2000)
12. Jain, A.K., Flynn, P., Ross, A.A.: Flipbook of Biometrics. Springer, Heidelberg (2008)
13. Joachims, T.: SVMlight Support Vector Machine (2008), `http://svmlight.joachims.org/` (accessed November 2, 2009)
14. Otsu, N.: A thresholding selection method from gray-level histogram. IEEE Transactions on Systems, Man, and Cybernetics 9(1), 62–66 (1979)
15. Rabiner, L.R.: A tutorial on Hidden Markov models and Selected Applications in Speech Recognition. Proceedings of the IEEE 77(2), 257–286 (1989)
16. Rabiner, L., Juang, B.H.: Fundamentals of Speech Recognition. Prentice-Hall, Englewood Cliffs (1993)
17. Rao, R.A., Mersereau, R.: Lip modeling for visual speech recognition. In: Eighth Asilomar Conference on Signals, Systems and Computers, vol. 1, pp. 587–590 (1994)
18. Sohail, A.S.M., Bhattacharya, P.: Automated Lip Contour Detection Using the Level Set Segmentation Method. In: 14th International Conference on Image Analysis and Processing, pp. 425–430 (2007)
19. Steifelhagen, R., Yang, J., Meier, U.: Real time lip tracking for lipreading. In: Proceedings of Eurospeech (1997)
20. Stillittano, S., Girondel, V., Caplier, A.: Inner and outer lip contour tracking using cubic curve parametric models. In: 16th IEEE International Conference on Image Processing, pp. 2469–2472 (2009)
21. Yaling, L., Minghui, D.: Lip Contour Extraction Based on Manifold. In: International Conference on MultiMedia and Information Technology, pp. 229–232 (2008)
22. Yan, X., Guangda, S.: Multi-parts and Multi-feature Fusion in Face Verification. In: Conference on Computer Vision and Pattern Recognition, pp. 1–6 (2008)
23. Viola, P., Jones, M.: Robust Real-time Object Detection. International Journal of Computer Vision 57(2), 137–154 (2004)
24. Wark, T., Sridharan, S., Clipran, V.: An approach to statistical lip modelling for speaker identification via chromatic feature extraction. In: IEEE International Conference on Pattern Recognition. pp.123–125 (1998)

Chapter 17
Computational Intelligence in Multi-channel EEG Signal Analysis

Aleš Procházka, Martina Mudrová, Oldřich Vyšata, Lucie Gráfová, and Carmen Paz Suárez Araujo

Abstract. Computational intelligence and signal analysis of multi-channel data form an interdisciplinary research area based upon general digital signal processing methods and adaptive algorithms. The chapter is restricted to their use in biomedicine and particularly in electroencephalogram signal processing to find specific components of such multi-channel signals. Methods presented include signal de-noising, evaluation of their fundamental components and segmentation based upon feature detection in time-frequency and time-scale domains using both the discrete Fourier transform and the discrete wavelet transform. Resulting pattern vectors are then classified by self-organizing neural networks using a specific statistical criterion proposed to evaluate distances of individual feature vector values from corresponding cluster centers. Results achieved are compared for different data sets and selected mathematical methods to detect segments features. Proposed methods verified in the MATLAB environment using distributed data processing are accompanied by the appropriate graphical user interface that enables convenient and user friendly time-series processing.

17.1 Introduction

Analysis of simultaneous multi-sensor observations of engineering or biomedical systems with a selected sampling period stand for a very common method allowing the collection of many different information with the following modelling of the given system. The processing of parallel time series forms a specific area of digital

Aleš Procházka · Martina Mudrová · Oldřich Vyšata · Lucie Gráfová
Institute of Chemical Technology in Prague,
Department of Computing and Control Engineering
e-mail: {Ales.Prochazka, Martina.Murova, Oldrich.Vysata,
Lucie.Grafova}@vscht.cz

Carmen Paz Suárez Araujo
Universidad de Las Palmas de G.C., Spain
e-mail: cpsuarez@dis.ulpgc.es

J. Fodor et al. (Eds.): Recent Advances in Intelligent Engineering Systems, SCI 378, pp. 361–381.
springerlink.com

signal processing methods [22, 38, 20] which allows the detailed study of the system, correlation analysis of relations of individual signals, multi-dimensional system modelling and extraction of its components. Fault diagnostics can be important both in engineering to find technological defects and in biomedicine to detect serious diseases. Even though these applications are different the mathematical tools behind are very close and include discrete Fourier transform (DFT) and wavelet transform (DWT) for signal analysis [56, 44], de-noising methods [52, 7] for rejection of undesirable signal components, principal (PCA) and independent component analysis (ICA) for multi-channel signal processing [45, 55, 24, 32, 6] and specific methods for signal segmentation [15, 37], feature extraction [1, 51] and classification [50, 43].

Multi-channel signals common in many applications are in specific cases observed in precisely defined space locations and can represent either time-series (like air pollution data measured at specific locations) or images with the necessity of very precise time synchronization in many cases. Such multi-channel signals are very common in biomedicine as well. The situation with such observations in electroencephalography is presented in Fig. 17.1 showing the typical distribution of electrodes collecting electrical potentials of the brain activity on the head. This figure also presents a simple interpolation of observed brain activity for three successive time instants. This two-dimensional projection instead of the three-dimensional visualization is presented here to simplify the following analysis.

Figure 17.2(a) presents a selected part of the electroencephalogram (EEG) signal standing for a typical multi-channel signal observed with the given sampling

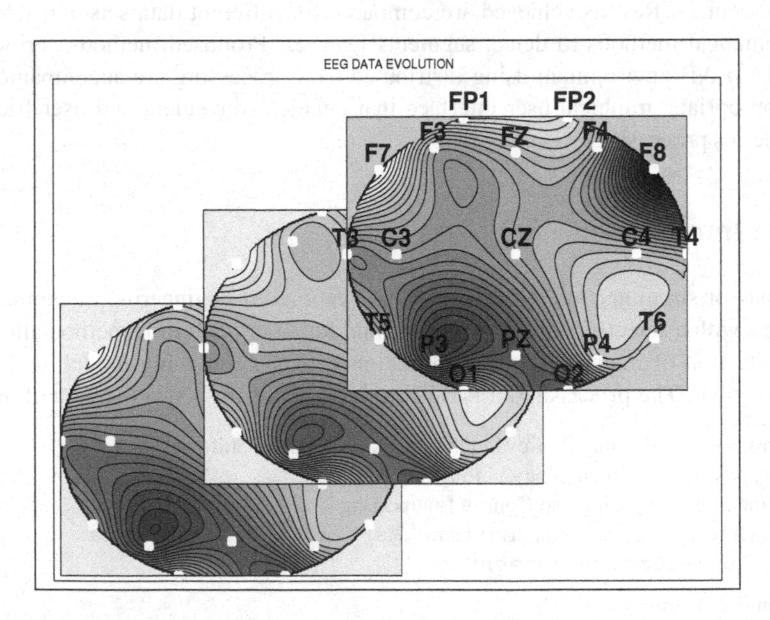

Fig. 17.1 The evolution of EEG energy distribution in subsequent time instants interpolated from observed signal components

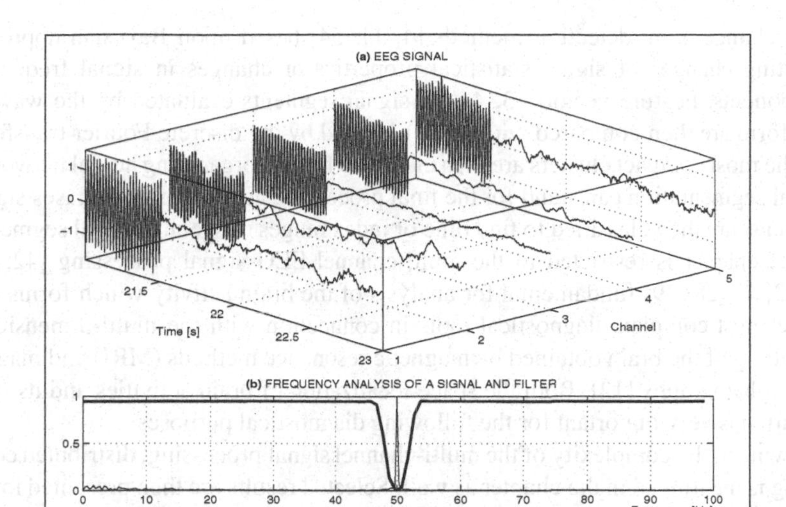

Fig. 17.2 EEG signal processing presenting **(a)** selected channels before and after rejection of the additive noise component of 50 Hz and **(b)** spectrum of a selected signal (fine line) pointing to the net frequency of 50 Hz and the FIR stop-band filter characteristics (bold line) used for its rejection

frequency f_s. The recorded signal contaminated by the additive noise component of 50 Hz must be then digitally extracted by a properly chosen stop-band digital filter with its characteristics in Fig. 17.2(b) and results presented in Fig. 17.2(a) as well.

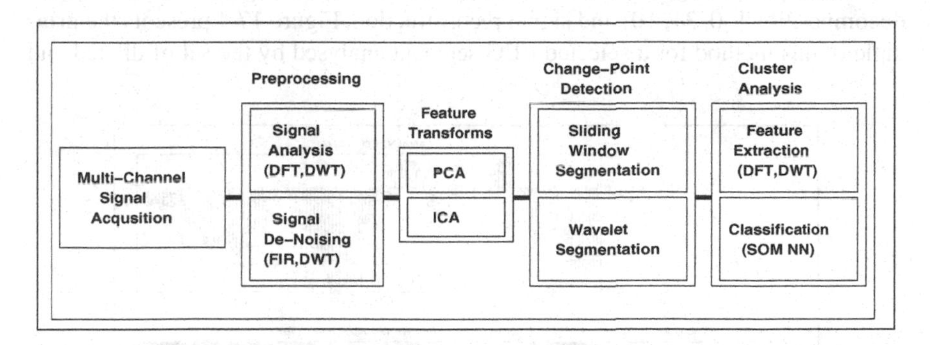

Fig. 17.3 Basic blocks of multi-channel signal processing including the preprocessing stage, feature transforms, data segmentation and signal components classification

The chapter presents selected methods used for multi-channel signals de-noising, extraction of their components and the application of the double moving window for signal segmentation using its first principal component. Figure 17.3 presents the typical structure of such a multi-channel signal processing proposed to detect specific signal components. Principal component analysis is used as an alternative to other change-point detection methods [4, 11, 54] based upon Bayesian approach detecting changes of signal statistical properties or changes in signal frequency components. Feature vectors [3, 45] of signal segments evaluated by the wavelet transform are then compared with those obtained by the discrete Fourier transform and the most compact clusters are then classified by self-organizing neural networks. Signal segmentation can stand for the final processing goal but in most cases signal segments are then classified to find time or index ranges of specific signal segments.

The chapter is restricted to the multi-channel EEG signal processing [42, 36, 29, 53, 13, 28, 39] fundamental for analysis of the brain activity which forms one of the most complex diagnostical tools in connection with the multi-dimensional modelling of the brain obtained by magnetic resonance methods (MRI) and magnetoencephalography [12]. Both the space localization of brain activities and its time evolution is very important for the following diagnostical purposes.

Owing to the complexity of the multi-channel signal processing, distributed computing is mentioned in the chapter as well. Selected results are then presented in the graphical user interface (GUI) environment proposed for the more convenient signal preprocessing and its analysis.

17.2 Wavelet Decomposition and Reconstruction

Computational intelligence and adaptive methods include many mathematical algorithms for complex data processing. Wavelet transform is one of the tools which allows both detailed and global signal analysis and which seems to be useful for multi-channel signals in many different stages of their processing as well.

The discrete wavelet transform (DWT) forms a general mathematical tool for signal processing [44, 19, 40] using the time-scale signal analysis, signal decomposition [30, 31, 10] and signal reconstruction. Figure 17.4 presents the principle of this method for a selected EEG segment analysed by the set of dilated and

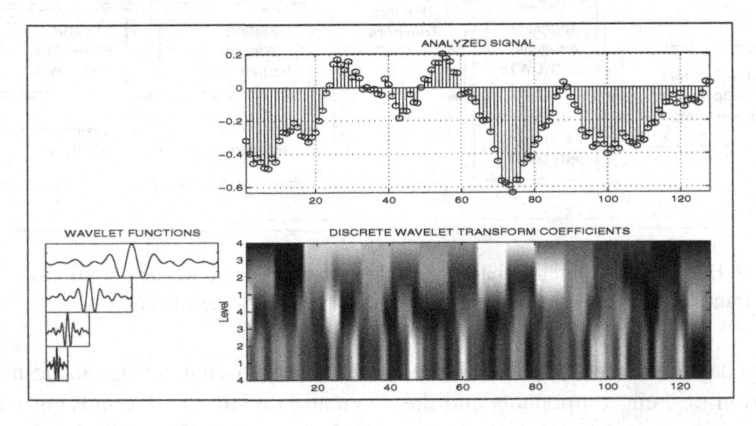

Fig. 17.4 Principle of the EEG signal wavelet decomposition using dilated and translated wavelet functions with the resulting scalogram presenting wavelet coefficients up to the fourth level

translated time-limited wavelet functions resulting in sets of coefficients for each scale.

The set of wavelet functions is usually derived from the initial (mother) wavelet $h(t)$ which is dilated by value $a = 2^m$, translated by constant $b = k\, 2^m$ and normalized so that

$$h_{m,k}(t) = \frac{1}{\sqrt{a}}\, h(\frac{t-b}{a}) = \frac{1}{\sqrt{2^m}}\, h(2^{-m}\, t - k) \tag{17.1}$$

for integer values of m, k and the initial wavelet defined either by the solution of a dilation equation or by an analytical expression [9, 35, 41]. Both continuous or discrete signals can be then approximated in the way similar to Fourier series and discrete Fourier transform. In case of a sequence $\{x(n)\}_{n=0}^{N-1}$ having $N = 2^s$ values it is possible to evaluate its expansion

$$x(n) = a_0 + \sum_{m=0}^{s-1} \sum_{k=0}^{2^{s-m-1}-1} a_{2^{s-m-1}+k}\, h(2^{-m}\, n - k) \tag{17.2}$$

Wavelet transform coefficients can then be organized in a matrix **T** with its nonzero elements forming a triangle structure

$$\begin{bmatrix} a_{2^{s-1}} & & a_{2^{s-1}+1} & & \cdots & & & a_{2^s-1} \\ & a_{2^s-2} & & & \cdots & & a_{2^{s-1}-1} & \\ & & & & \cdots & & & \\ & & a_4 & & a_5 & a_6 & a_7 & \\ & & & a_2 & & a_3 & & \\ & & & & a_1 & & & \\ & & & & a_0 & & & \end{bmatrix} \tag{17.3}$$

with each its row corresponding to a separate dilation coefficient m. The set of $N = 2^s$ decomposition coefficients $\{a(j)\}_{j=0}^{N-1}$ of the wavelet transform is defined in the way formally close to the Fourier transform but owing to the general definition of wavelet functions they can carry different information. Using the orthogonal set of wavelet functions they are moreover closely related to the signal energy [35].

The initial wavelet [35] can be considered as a pass-band filter and in most cases half-band filter covering the normalized frequency band $\langle 0.25, 0.5 \rangle$. A wavelet dilation by the factor $a = 2^m$ corresponds to its pass-band compression. This general property can be demonstrated for the harmonic wavelet function [35] and the corresponding scaling function defined by expressions

$$h(t) = \frac{1}{j\pi/2\, t}(e^{j\pi\, t} - e^{j\pi/2\, t}) \tag{17.4}$$

$$l(t) = \frac{1}{\pi/2\, t}(e^{j\pi/2\, t} - 1) \tag{17.5}$$

As both these functions are modified by the scaling index $m = 0, 1, \ldots$ according to Eq. (17.1), the wavelet is dilated and its spectrum compressed resulting in time and frequency domain representation presented in Fig. 17.5(d). A similar approach can be also applied for other wavelet functions defined in either analytical or recurrent forms presented in Fig. 17.5 as well.

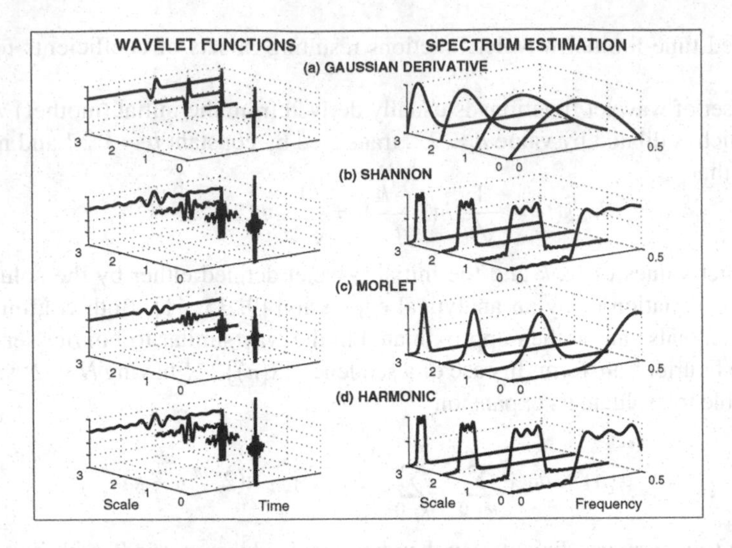

Fig. 17.5 The effect of wavelet dilation to the corresponding spectrum compression comparing results for **(a)** Gaussian derivative, **(b)** Shannon, **(c)** Morlet, and **(a)** harmonic wavelet functions

The set of wavelets defines a special filter bank which can be used for signal component analysis and resulting wavelet transform coefficients can be further used as features valuesfor signal components classification [19, 49]. Signal decomposition performed by a pyramidal algorithm is interpreting wavelets as pass-band filters. Another approach [35] is based upon a very efficient parallel algorithm using the fast Fourier transform.

The basic decomposition of a given column vector $\{x(n)\}_{n=0}^{N-1}$ assumes a half-band low-pass scaling sequence

$$\{l(n)\}_{n=0}^{L-1} = [l(0), l(1), l(2), \cdots, l(L-1)] \tag{17.6}$$

and the complementary orthogonal wavelet sequence

$$\{h(n)\}_{n=0}^{L-1} = [l(L-1), -l(L-2), l(L-3), \cdots, -l(0)] \tag{17.7}$$

These sequences are convolved with the given column vector $\{x(n)\}_{n=0}^{N-1}$ and after the subsampling by two it is possible to evaluate values

$$p(n) = \sum_{k=0}^{L-1} l(k)\, x(n-k) = \sum_{j=n,n-1,\ldots}^{n-L+1} x(j)\, l(n-j) \tag{17.8}$$

$$q(n) = \sum_{k=0}^{L-1} h(k)\, x(n-k) = \sum_{j=n,n-1,\ldots}^{n-L+1} x(j)\, h(n-j) \tag{17.9}$$

for $n = L-1, L+1, \ldots, N-1$. Application of the same method for evaluated signals result in signal decomposition into further scales. Introducing decomposition matrices

$$\mathbf{L}_{N/2,N}=\begin{bmatrix} l(1)\; l(0) & 0 & 0 & \cdots \\ l(3)\; l(2) & l(1) & l(0) & \cdots \\ \cdots\;\cdots & \cdots & \cdots & \cdots \\ 0 \;\; \cdots & l(L-1) & \cdots & l(0) \end{bmatrix} \tag{17.10}$$

$$\mathbf{H}_{N/2,N}=\begin{bmatrix} h(1)\; h(0) & 0 & 0 & \cdots \\ h(3)\; h(2) & h(1) & h(0) & \cdots \\ \cdots\;\cdots & \cdots & \cdots & \cdots \\ 0 \;\; \cdots & h(L-1) & \cdots & h(0) \end{bmatrix} \tag{17.11}$$

it is possible to decompose the initial signal $\mathbf{x}$ into two sequences $\mathbf{p} = \mathbf{L}\,\mathbf{x}$ and $\mathbf{q} = \mathbf{H}\,\mathbf{x}$ standing for subsampled low-pass and high-pass signal components. The elements of vector $\mathbf{q}$ represent wavelet transform coefficients of the initial level. Further wavelet coefficients can be obtained after the application of this process to signal $\mathbf{p}$. Resulting wavelet coefficients $\{a(n)\}_{n=0}^{N-1}$ related to chosen scales can then be used for signal compression, de-noising or as features values for signal segments classification [46, 25].

The basic idea of wavelet signal decomposition has been further enlarged and both complete tree decomposition and dual tree decomposition [21, 23, 44] have been published to improve the wavelet transform shift invariance and to add further properties with a variety of applications in biomedicine. Further studies are devoted to multi-dimensional problems.

17.3 Signal De-noising

Signal de-nosing is very important in EEG signal analysis and it forms its fundamental preprocessing step eliminating the necessity to observe brain activities in special rooms free of disturbing signals. Mathematical tools can substitute technical devices in this way. Digital filters used for this purpose can be proposed in the time, frequency and scale domains.

A simple general discrete system used for signal component rejection in the time-domain can be based on the the difference equation

$$y(n) + \sum_{k=1}^{M} a(k)\, y(n-k) = \sum_{k=0}^{M} b(k)\, x(n-k) \tag{17.12}$$

which can be further processed mathematically by the z-transform to obtain the discrete transfer function

$$H(z) = \frac{Y(z)}{X(z)} = \frac{\sum_{k=0}^{M} b(k)\, z^{-k}}{1 + \sum_{k=1}^{M} a(k)\, z^{-k}} \tag{17.13}$$

The corresponding frequency transfer function

$$H(e^{j\omega}) = H(z) \mid_{z=e^{j\omega}} \tag{17.14}$$

for $\omega \in \langle 0,2\,\pi)$ can then be used for filter analysis and design of filter coefficients to reject frequency components in given ranges. Even though the infinite impulse response (IIR) filters defined by Eq. (17.12) allow the lower order M for a predefined accuracy the finite impulse response (FIR) filters with zero coefficients $\{a(0), a(1), \cdots, a(N)\}$ were used for EEG signal processing owing to their stability and linear phase characteristics.

The convolution of evaluated filter coefficients (of length $M = 100$) and previous observed values for separate EEG channels provided results presented in Fig. 17.2(a) with the stop-band filter characteristics given in Fig. 17.2(b) rejecting the net frequency of 50 Hz.

Algorithm 17.1 Algorithm of the multi-channel time-domain filtering of an EEG signal stored in individual columns of matrix **EEG** observed with the sampling frequency f_s using FIR filter of order L

```
%% FIR filters coefficients evaluation %%%
   b1=fir1(L,[48 52]/fs,'stop')
   b2=fir1(L,[0.5 60]/fs,'bandpass')
%% EEG data filtering %%%
   E1=filtfilt(b1,1,EEG);
   E=filtfilt(b2,1,E1));
```

Algorithm 17.1 presents a typical compact MATLAB algorithm for such a processing of an EEG data segment observed with the sampling frequency $f_s = 200\,Hz$ and bi-directional filtering to eliminate the time delay caused by the FIR filter. Two successive filtering methods are applied to reject the net frequency of 50 Hz and to reduce the influence of frequency components outside the frequency band of $\langle 0.5, 60 \rangle\,Hz$. Figure 17.6 presents results of such a process using band-stop and pass-band filters for the analyzed EEG signal segment.

Similar results presented in Fig. 17.7 were obtained by the wavelet thresholding method after time-scale signal decomposition, windowing of resulting coefficients and signal reconstruction applying appropriate threshold limits [48] to wavelet transform coefficients.

In the case of soft-thresholding it is possible to evaluate new coefficients $\overline{c}(k)$ using original coefficients $c(k)$ for a chosen limit δ by relation

$$\overline{c}(k) = \begin{cases} \text{sign } c(k)\ (\mid c(k) \mid - \delta) & \text{if } \mid c(k) \mid > \delta \\ 0 & \text{if } \mid c(k) \mid \leq \delta \end{cases} \tag{17.15}$$

Results of this process applied to the EEG signal are presented in Fig. 17.7 showing signal decomposition into three levels. The level dependent thresholding is then

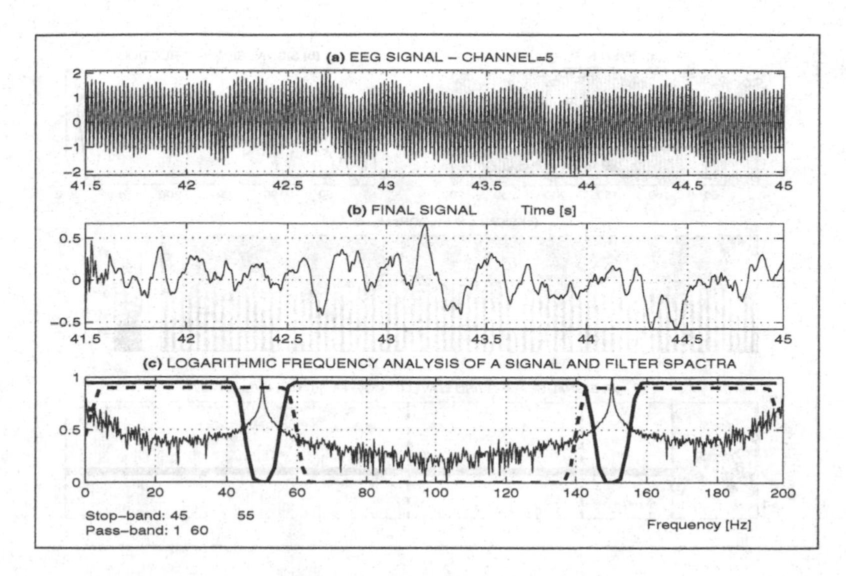

Fig. 17.6 The selected EEG signal channel processing presenting **(a)** the observed sequence contaminated by the noise caused by the net frequency of 50 Hz, **(b)** the de-noised sequence in the selected range, and **(c)** logarithmic spectrum of the given sequence and spectral characteristics (bold) of the FIR stop-band and band-pass filters

applied to the vector of wavelet coefficients to obtain the final signal in Fig. 17.7(a) while scaling coefficients are not affected by this process.

17.4 Feature Transforms

Segmentation over all parallel observations of multi-channel data should take into account information present in all simultaneously recorded signals. To simplify this problem the matrix of the original data can be processed to reduce the significance of less descriptive features. It can be achieved by the computation of the optimal subset to present data in fewer dimensions.

17.4.1 Principal Component Analysis

To perform the segmentation of a multi-channel signal over all channels it is important to extract information common for all parallel time series representing individual channels at first. The principal component analysis [45, 16] can be used to perform this task using results of linear algebra to process a matrix $\mathbf{E}_{N,M}$ containing multi-channel data. In the case of EEG records each column of this matrix for $j = 1, 2, \cdots, M$ contains column vector of N observations associated with the

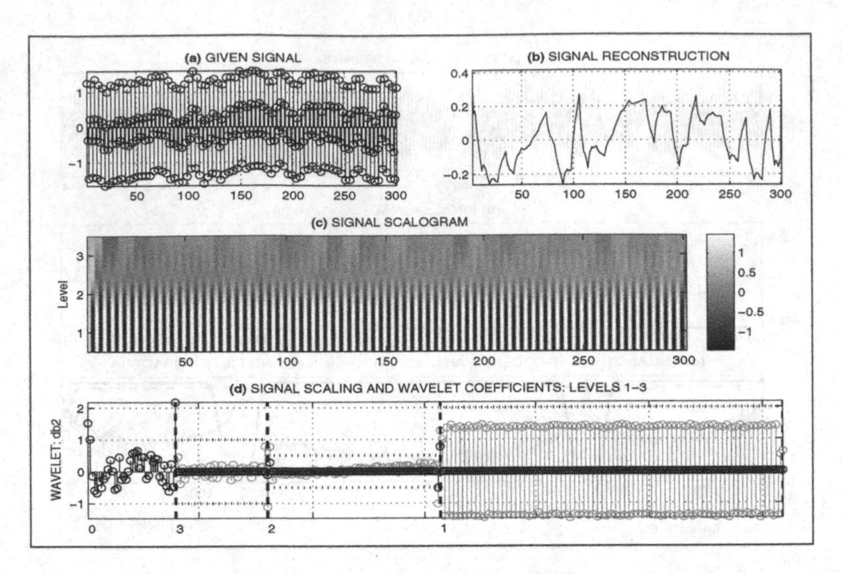

Fig. 17.7 Wavelet transform de-noising presenting the selected EEG signal; **(a)** before processing contaminated by the noise caused by the net frequency of 50 Hz, **(b)** after de-noising reducing undesirable signal components, **(c)** signal scalogram presenting wavelet coefficients up to the third level, and **(d)** decomposition coefficients and thresholding limits evaluated separately for each decomposition level together with resulting coefficients used for signal reconstruction

separate channel. It is possible to transform this matrix $\mathbf{E}_{N,M}$ into a new one using orthonormal matrix $\mathbf{P}_{M,M}$ to find values

$$\mathbf{Y}_{N,M} = \mathbf{E}_{N,M}\,\mathbf{P}_{M,M} \tag{17.16}$$

Resulting matrix $\mathbf{Y}_{N,M}$ has values in its columns with the decreasing variance [45] which allows to use the first most significant column for signal segmentation. The whole process can be performed using MATLAB commands presented in Algorithm 17.2 in the simplest case.

Algorithm 17.2 Algorithm of the PCA applied to processing of a matrix $\mathbf{E}_{N,M}$ including multi-channel data in its individual columns

```
%% Principal Component Analysis %%%
   P = princomp(E);
   Y=E*P;
```

Results of PCA for a selected segment of EEG observations after their de-noising is presented in Figs. 17.8 and 17.9 together with resulting signal variances having the decreasing significance in the case of data processed by the PCA method.

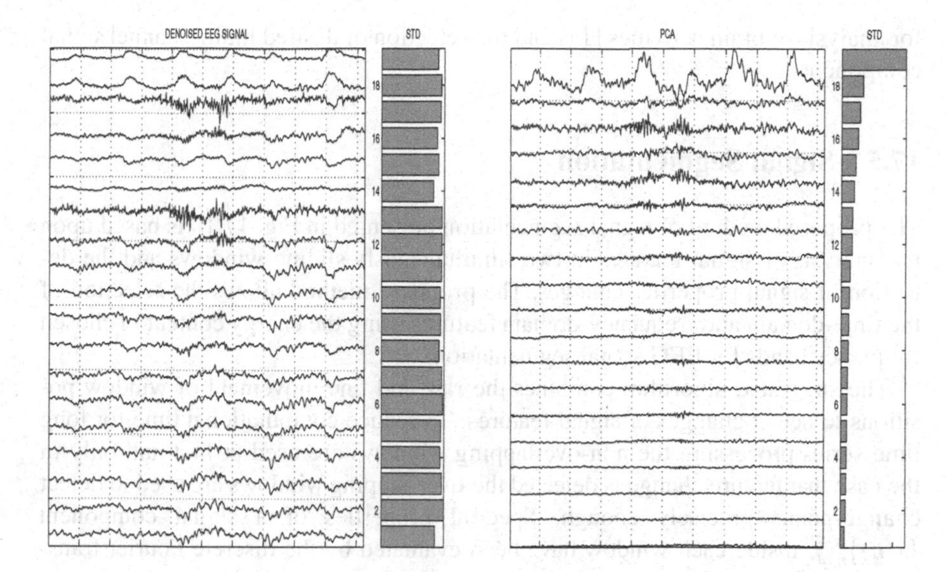

Fig. 17.8 De-noised EEG signal record composed of 19 channels

Fig. 17.9 PCA of the EEG signal with corresponding variances

17.4.2 Independent Component Analysis

Independent component analysis [18, 47] is a method commonly used for data separation into underlying information components. Many papers describe how this method can be used for EEG artifacts removal [17, 32, 53].

Assume have N hidden independent components $\{s(k)\}$ for $k = 1,2,\ldots,N$ and observed variables $\{x(k)\}$ for $k = 1,2,\cdots,N$ formed by their mixture which can be described by relation

$$\begin{aligned} x(1) &= a(1,1)\,s(1) + a(1,2)\,s(2) + \cdots + a(1,N)\,s(N) \\ x(2) &= a(2,1)\,s(1) + a(2,2)\,s(2) + \cdots + a(2,N)\,s(N) \\ &\ldots \\ x(N) &= a(N,1)\,s(1) + a(N,2)\,s(2) + \cdots + a(N,N)\,s(N) \end{aligned} \tag{17.17}$$

The whole model can be defined in the matrix form by relation

$$\mathbf{x} = \mathbf{A}\,\mathbf{s} \tag{17.18}$$

with unknown coefficients of matrix **A** in the general case not allowing simple estimation of signal sources knowing vectors of observations only.

The structure presented by Eq. (17.18) forming the independent component analysis (ICA) model can be used to estimate statistically independent sources which represent the situation related to EEG records observed as the mixture of electrical potentials recorded in precisely defined locations on the scalp. The ICA method can provide information about their independent components that can be further used for analysis of brain activities [17] and for detection of desired multi-channel signal components.

17.5 Signal Segmentation

The proposed method of signal segmentation presented in Fig. 17.10 is based upon the analysis of signal features in two simultaneously sliding windows and the detection of signal properties changes. The proposed method allows the selection of the time-domain and frequency-domain features using the energy content in chosen frequency bands for EEG signal segmentation.

The suggested algorithm combines the raw and fine movement of window positions to detect changes of signal features. To reduce computational time for long time series processing the non-overlapping windows are used at first and only in the case that feature change is detected the over-lapping windows are used to detect change-points precisely enough. Spectral properties of a signal component $\{x(n)\}_{n=0}^{N-1}$ inside each window have been evaluated by the discrete Fourier transform resulting in the sequence

$$X(k) = \sum_{n=0}^{N-1} x(n)\, e^{-j\,k\,n\,\frac{2\pi}{N}} \qquad (17.19)$$

for $k \in \langle 0, N-1 \rangle$ or $f_k \in \langle 0, (N-1)/N * f_s \rangle$ where f_s stands for the sampling frequency. Average spectra values in selected frequency bands used for segmentation are presented in Fig. 17.10.

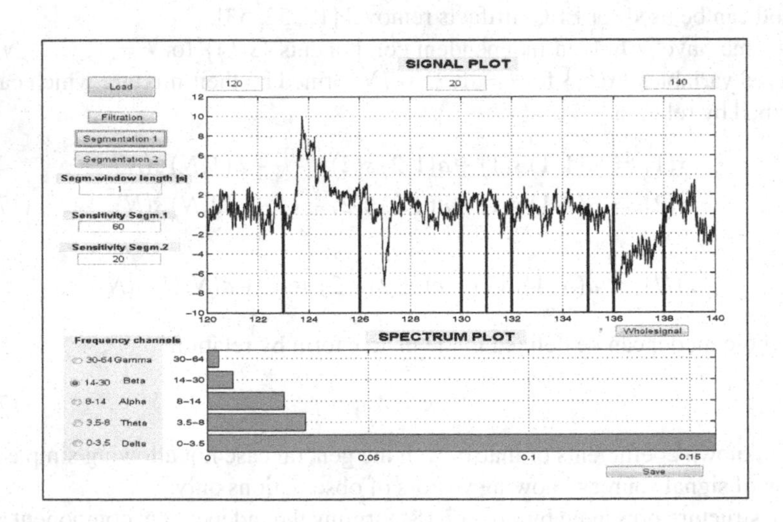

Fig. 17.10 The graphical user interface for segmentation of the EEG signal segment of the chosen range for selected window lengths and detection sensitivity using the average signal energy in user-defined frequency bands

Owing to the necessity of multi-channel signal processing the first principal component has been used for segmentation of the whole set of observed time-series. Figure 17.10 presents the proposed graphical user interface designed to find signal segments of similar properties in the frequency domain using selected window lengths and sensitivity for change-points detection.

17.6 Feature Extraction

The selection of the most efficient and reliable method of feature extraction forms a very important problem of signal segments classification. Methods applied are usually based upon the time-domain or frequency-domain signal analysis [14, 49]. The following study is devoted to the wavelet domain signal feature extraction and comparison of results achieved.

Suggested algorithm is based upon the wavelet decomposition of signal segments and evaluation of its coefficients for estimation of segment features. Figure 17.11 presents application of this method to EEG signal segments and their analysis by a harmonic wavelet transform [35] resulting in features standing for scales 1, 2 and 3 respectively covering three frequency bands. Owing to the Parseval theorem [5] interconnecting the time and transform domains the summed squared values of coefficients in individual wavelet decomposition levels are used as segment features for further signal processing and signal segments classification.

The discrete wavelet transform enables estimation of signal segment features with changing resolution in time and frequency for different scales and in this way it is possible to obtain the complex description of data segments. Moreover it is possible to use different wavelet functions to affect the quality of resulting feature clusters.

17.7 Classification

Each signal segment having index $\{1,2,\cdots,Q\}$ can be described by R features specified in separate columns of the pattern matrix $\mathbf{P}_{R,Q}$ forming clusters in the R-dimensional space. The proposed algorithm for classification of feature vectors has been based upon the application of self-organizing neural networks [16, 8] using Q feature vectors as patterns for the input layer of neural network. The number S of output layer elements [2] is equal to desired signal classes and must be either defined in advance or it can be automatically increased to create new classes [8] during the learning process. Each output neuron is responsible for a separate class and its coefficients stand for the centre of gravity of each class after the learning process.

During the learning process neural network weights are changed to minimize distances between each input vector and corresponding weights of a winning neuron characterized by its coefficients closest to the current pattern using the Kohonen learning rule.

In case that the learning process is successfully completed network weights belonging to separate output elements represent typical class individuals. Algorithm 17.3 presents basic algorithmic steps of this procedure in the MATLAB environment. Results of classification of $R = 2$ features to $S = 3$ classes for a selected EEG signal are presented in Fig. 17.12.

The graphical user interface designed for this process presented results in Fig. 17.12 for EEG classification into five classes by a self-organizing neural network for two selected signal features allowing a simple visualization of segmentation results and visualization of typical class representatives with their features closest to the corresponding cluster centers as well. Signal features used for classification are presented in Fig. 17.13 together with variances of individual clusters.

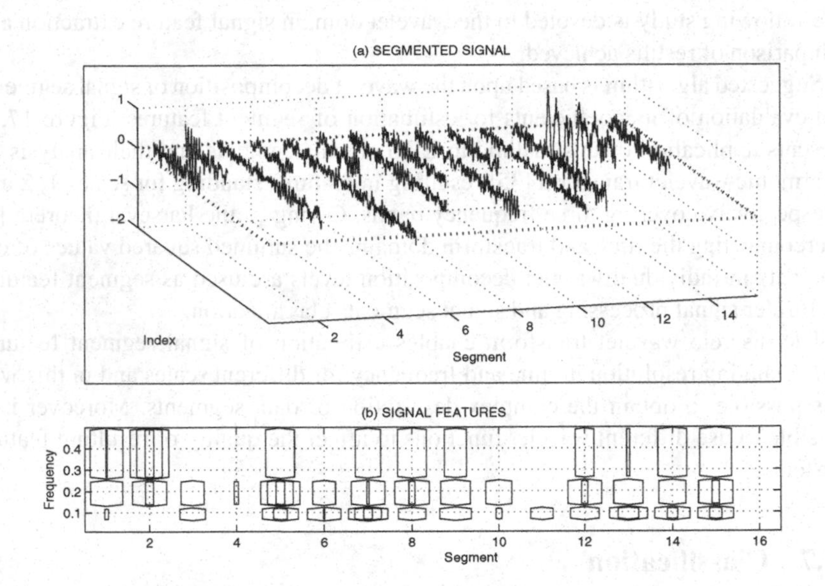

Fig. 17.11 Signal segments feature extraction using wavelet decomposition into three levels presenting **(a)** EEG signal segments and **(b)** their features represented by energy in separate decomposition levels evaluated as the summed squared value of wavelet coefficients belonging to the individual scale for each segment

Algorithm 17.3 Algorithm for the classification of pattern matrix P into S classes and optimization of its weights **W1** followed by the network output evaluation

```
%% Pattern values classification %%%
   net = newc(minmax(P),S);
   net = train(net,P);
   W1=net.IW{1,1};
%% Network output evaluation %%%
   A =sim(net,P); Ac=vec2ind(A);
```

To compare results of classification for Q signal segments with feature matrix $\mathbf{P}_{R,Q} = [\mathbf{p}_1, \mathbf{p}_2, \cdots, \mathbf{p}_Q]$ evaluated for the selection of different sets of $R = 2$ features and C classes a specific criterion has been designed. Each class $i = 1, \cdots, C$ has been characterized by the mean distance of column feature vectors $\mathbf{p}_{j_k}$ belonging to class segments j_k for indices $k = 1, 2, \cdots, N_i$ from the class centre in the i−th row of matrix $\mathbf{W}_{C,R} = [\mathbf{w}_1, \mathbf{w}_2, \cdots, \mathbf{w}_C]'$ by relation

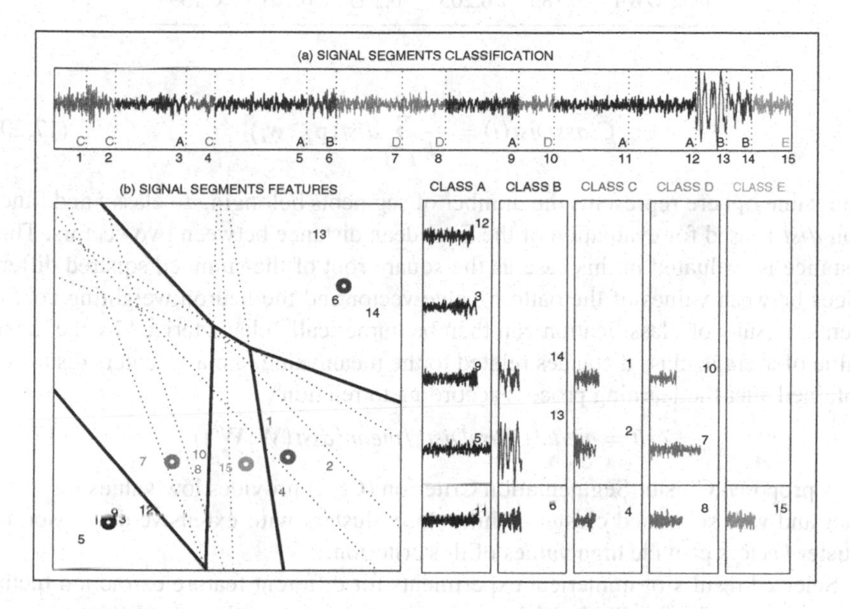

Fig. 17.12 EEG signal segments classification into the given number of classes and detection of typical signal segments closest to cluster centers presenting **(a)** given EEG segment with evaluated change-points and segment classes and **(b)** the plot of signal segment features together with segmentation boundaries and typical class individuals closest to centers of gravities of individual classes

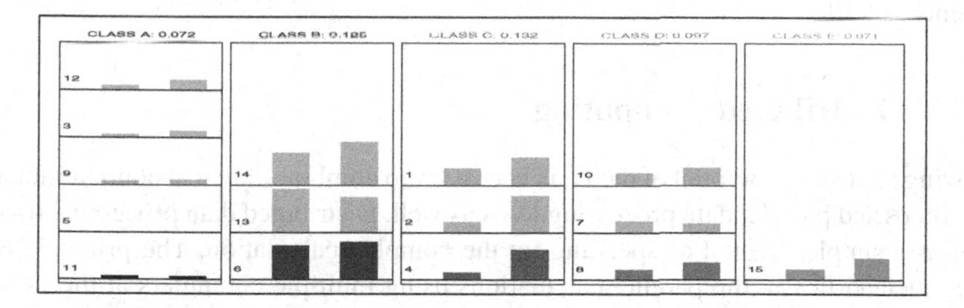

Fig. 17.13 Analysis of features associated with individual segments belonging to given classes used for EEG signal segments classification with characteristics of the typical individuals in the last row of each column

Table 17.1 Cluster compactness evaluation for signal segments classification into 5 classes for different feature extraction methods and various sets of EEG signals

Method	Set 1	Set 2	Set 3	Set 4	Set 5
DFT	0.267	0.326	0.227	0.202	0.199
Haar DWT	0.204	0.147	0.160	0.155	0.180
Db2 DWT	0.185	0.205	0.203	0.161	0.159

$$ClassDist(i) = \frac{1}{N_i} \sum_{k=1}^{N_i} dist(\mathbf{p}_{j_k}, \mathbf{w}_i) \tag{17.20}$$

The value N_i here represents the number of segments belonging to class i and function *dist* is used for evaluation of the Euclidean distance between two vectors. This distance is evaluated in this case as the square root of the summed squared differences between values of the pattern value vector and the neuron weighting coefficients. Results of classification can then be numerically characterized by the mean value of average class distances related to the mean value of class centers distances obtained after the learning process according to relation

$$crit = mean(ClassDist)/mean(dist(\mathbf{W}, \mathbf{W}')) \tag{17.21}$$

This proposed Cluster Segmentation Criterion (CSC) provides low values for compact and well separated clusters while close clusters with extensive dispersion of cluster vectors provide high values of this criterion.

Selected results of numerical experiments for different feature extraction methods are summarized in Table 17.1 presenting the criterion values for different signal segments. It is obvious that classification parameters achieved both by the DFT and DWT provide similar results but slightly better in the case of wavelet features selection. The Haar and Daubechies wavelet functions [9, 34] are used here as representatives of the wavelet transform for analysis of different sets of real EEG data. Segmentation result achieved correspond to visual expert evaluation of EEG segments as well.

17.8 Distributed Computing

Owing to the large amount of data it is necessary to combine classical computational methods and parallel data processing tools as well. Distributed data processing used forms a simple method of speeding up the complex calculation. The principle of this method lies in the parallel calculations using multiple computers at the same time. Requests from the main computer are in this case sent to the job manager (scheduler). This computer is used as a server and it then sends the requirements for computers (Workers) on which the individual calculation is processed according to the Fig. 17.14. Results are then collected and returned to the computer that requested the calculation.

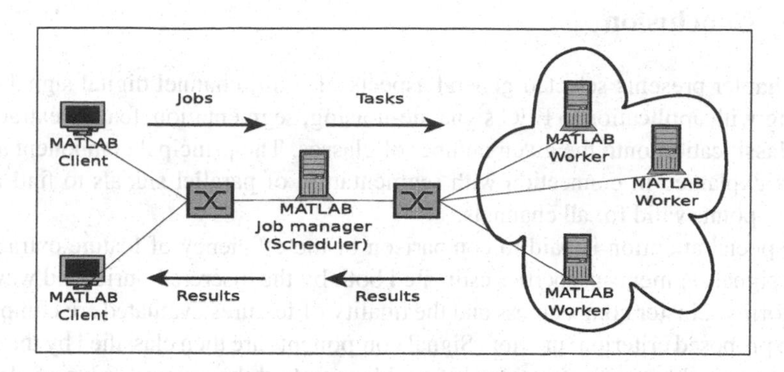

Fig. 17.14 The structure of the distributed computing used for efficient extensive data processing

This method can accelerate computing and significantly reduce the time needed to obtain results in the case that

- large enough computing segments are constructed to minimize the effect of the network traffic as too frequent data transfer can reduce the advantage of parallel computing and in extreme situations it can even significantly increase the total computational time
- individual processing tasks form as independent computational units as possible to reduce the time needed to wait for results from other processing tasks
- network traffic is optimized

The whole computational process can be started completely independent to the client computer allowing processing of large amount of data in this way. Implementation of such parallel processing depends upon computational and software possibilities. A general access to distributed programming is provided by Python and PHP programming languages. The distributed computing platform can combine database access from all application components including Python worker and PHP web interface [27].

The alternative approach based upon the MATLAB Distributed Computing Server [26, 33] is more general but it is limited by the number of licensed parallel computing unites. Paper [26] compares the speed increase for the cluster of 8 separate computers. More efficient computations have been obtained for the BLADE Cisco Unified Computing System system with 16 virtual machines using VMware vSphere. The real decrease of time achieved is close to the theoretical limit of 1/16 of the time needed to solve the same problem by a single machine in this case. To compare computational results for a large number of EEG data files the same algorithm has been applied for parallel processing of 16 different data files simultaneously in the same block.

17.9 Conclusion

The chapter presents selected general aspects of multi-channel digital signal processing with application to EEG signal de-noising, segmentation, feature extraction and classification onto the given number of classes. The principal component analysis is explained in connection with segmentation of parallel signals to find joint change-points valid for all channels.

A special attention is paid to comparison of the efficiency of feature extraction using signal segments properties estimated both by the discrete Fourier and wavelet transforms. Cluster compactness and the quality of features evaluated are compared by the proposed criterion function. Signal components are then classified by the self-creating neural network structures [8] enabling to find the optimal value of classes and to exclude the possibility of dead neurons [16] which do not enter the learning process owing to their improperly chosen random initialization. The graphical user interface is designed for this study as well. Selected algorithms and information concerning the distributed signal processing are available from the web page of the Department of Computing and Control Engineering (http://uprt.vscht.cz).

The following research will be devoted to further methods of signal preprocessing using wavelet function and including the dual tree complex wavelet signal decomposition. Further studies will include the estimation of signal segments features to form compact clusters enabling more reliable signal segments classification. A specific attention will be paid to the remote access to the signal database and the computational environment.

Acknowledgements. The chapter has been supported by the Research grant No.MSM 6046137306 of the Czech Ministry of Education and by the Science and Innovation Ministry of the Spanish Government under Research Projects TIN2009-13891.

References

1. Azim, M.R., Amin, M.S., Haque, S.A., Ambia, M.N., Shoeb, M.A.: Feature extraction of human sleep EEG signals using wavelet transform and Fourier transform. In: 2nd International Conference on Signal Processing Systems, vol. 3, pp. 701–705. IEEE, Los Alamitos (2010)
2. Beale, M.H., Hagan, M.T., Demuth, H.B.: Neural Network Toolbox, p. 01760. The MathWorks, Inc., Massachusetts (2010)
3. Bishop, C.M.: Neural Networks for Pattern Recognition. Clarendon Press, Oxford (1995)
4. Brodsky, B.E., Darkhovski, B.S.: Nonparametric Methods in Change-Point Problems. Kluwer Academic Publishers, Boston (1993)
5. Burrus, C.S., Gopinath, R.A., Guo, H.: Introduction to Wavelets and Wavelet Transforms: A Primer. Prentice-Hall, Englewood Cliffs (1997)
6. Chang, H.Y., Yang, S.C., Lan, S.H., Chung, P.C.: Epileptic seizure detection in grouped multi-channel EEG signal using ICA and wavelet transform. In: IEEE International Symposium on Circuits and Systems, pp. 1388–1391. IEEE Press, Los Alamitos (2010)
7. Chaux, C., Pesquet, J.C., Duval, L.: Noise Covariance Properties in Dual-Tree Wavelet Decomposition. IEEE T Inform Theory 53(12), 4690–4700 (2007)

8. Choi, D.I., Park, S.H.: Self-Creating and Organizing Neural Networks. IEEE T Neural Network 5(4), 561–575 (1994)
9. Daubechies, I.: The Wavelet Transform, Time-Frequency Localization and Signal Analysis. IEEE Trans. Inform Theory 36, 961–1005 (1990)
10. Debnath, L.: Wavelets and Signal Processing. Birkhäuser, Basel (2003)
11. Fitzgerald, W.J., Ruanaidh, J.J.K.O., Yates, J.A.: Generalised Changepoint Detection. Tech. rep., University of Cambridge, U.K (1994)
12. Gómez, C., Hornero, R., Abásolo, D., Fernández, A., Escudero, J.: Analysis of MEG Background Activity in Alzheimers Disease Using Nonlinear Methods and ANFIS. Ann. Biomed. Eng. 37(3), 586–594 (2009)
13. Graichen, U., Witte, H., Haueisen, J.: Analysis of Induced Components in Electroencephalograms Using a Multiple Correlation Method. BioMedical Engineering Online 8(21) (2009), http://www.biomedical-engineering-online.com
14. Hassanpour, H., Mesbah, M., Mesbah, M.: Time-frequency feature extraction of newborn eeg seizure using svd-based techniques. Eurasip J. Appl. Sig. P 16, 2544–2554 (2004)
15. Hassanpour, H., Shahiri, M.: Adaptive Segmentation Using Wavelet Transform. In: International Conference on Electrical Engineering, pp. 1–5. IEEE Press, Los Alamitos (2007)
16. Haykin, S.: Neural Networks, A Comprehensive Foundation. Macmillan College Publishing Company, NY (1994)
17. Hyvarinen, A., Karhunen, J., Oja, E.: Independent Component Analysis: Algorithms and Applications. Neural Networks 13(4–5), 411–430 (2000)
18. Hyvarinen, A., Karhunen, J., Oja, E.: Independent Component Analysis. John Wiley, Chichester (2001)
19. Johankhani, P., Kodogiannis, V., Revett, K.: EEG Signal Classification Using Wavelet Feature Extraction and Neural Networks. In: IEEE John Vincent Atanasoff 2006 International Symposium on Modern Computing, pp. 120–124 (2006)
20. Kay, S.M.: Fundaments of Statistical Signal Processing. Prentice-Hall, Englewood Cliffs (1993)
21. Kingsbury, N.: Complex Wavelets for Shift Invariant Analysis and Filtering of Signals. Journal of Applied and Computational Harmonic Analysis 3(10), 234–253 (2001)
22. Kingsbury, N.G., Mugarey, J.F.A.: Wavelet Transforms in Image Processing. In: Procházka, A., Uhlíř, J., Rayner, P.J.W., Kingsbury, N.G. (eds.) Signal Analysis and Prediction, Applied and Numerical Harmonic Analysis. ch. 2. Birkhäuser, Boston (1998)
23. Kingsbury, N.G., Zymnis, A., Pena, A.: DT-MRI Data Visualisation Using the Dual Tree Complex Wavelet Transform. In: 3rd IEEE International Symposium on Biomedical Imaging: Macro to Nano, vol. 111, pp. 328–331. IEEE, Los Alamitos (2004)
24. Koehler, B.U., Orglmeister, R.: Independent Component Analysis of Electroencephalographic Data Using Wavelet Decomposition. Artif. Intell. Med. 33(3), 209–222 (2005)
25. Krishnaveni, V., Jayaraman, S., Aravind, S., Hariharasudhan, V., Ramadoss, K.: Automatic Identification and Removal of Ocular Artifacts from EEG using Wavelet Transform. Meas. Sci. Rev. 6(4), 45–57 (2006)
26. Krupa, J., Pavelka, A., Vyšata, O., Procházka, A.: Distributed Signal Processing. In: Proceedings of the Conference on Technical Computing. MathWorks & Humusoft (2007)
27. Krupa, J., Procházka, A., Hanta, V., Háva, R.: Technical Computing Using Sybase Database for Biomedical Signal Analysis. In: Proceedings of the Conference on Technical Computing. MathWorks & Humusoft (2009)
28. Krusienski, D.J.: A Method for Visualizing Independent Spatio-Temporal Patterns of Brain Activity. Eurasip J. On Advances Signal Processing, 948–961 (2009)

29. Latchoumane, C.F.V., Chung, D., Kim, S., Jeong, J.: Segmentation and Characterization of EEG During Mental Tasks Using Dynamical Nonstationarity. In: 3rd International Conference on Computational Intelligence in Medicine and Healthcare (2007)
30. Mallat, S.G.: A Theory for Multiresolution Signal Decomposition: The Wavelet Representation. IEEE T Pattern Anal. 11(7), 674–693 (1989)
31. Mallat, S.G.: A Wavelet Tour of Signal Processing. Accademic Press, San Diego (1999)
32. Mammone, N., Inuso, G., La Foresta, F., Morabito, F.C.: Multiresolution ICA for artifact identification from electroencephalographic recordings. In: 11th International Conference on Knowledge-Based and Intelligent Information & Engineering Systems/XVII Italian Workshop on Neural Networks, pp. 680–687. Springer, Heidelberg (2007)
33. MathWorks: Parallel Computing Toolbox, p. 01760. The MathWorks, Inc., Natick, Massachusetts 01760 (2010)
34. Misiti, M., Misiti, Y., Oppenheim, G., Poggi, J.M.: Wavelet Toolbox, p. 01760. The MathWorks, Inc, Massachusetts (2010)
35. Newland, D.E.: An Introduction to Random Vibrations, Spectral and Wavelet Analysis, 3rd edn. Longman Scientific & Technical, U.K (1994)
36. Nixon, M., Aguado, A.: Feature Extraction & Image Processing. Elsevier, Amsterdam (2004)
37. Palmu, K., Stevenson, N., Wikström, S., Hellström-Westas, L., Vanhatalo, S., Palva, J.M.: Optimization of an NLEO-based algorithm for automated detection of spontaneous activity transients in early preterm EEG. Physiol. Meas. 31(11), 85–93 (2010)
38. Proakis, J.G., Manolakis, D.G.: Digital Signal Processing. Prentice-Hall, Englewood Cliffs (1996)
39. Procházka, A., Mudrová, M., Vyšata, O., Háva, R., Araujo, C.P.S.: Multi-Channel EEG Signal Segmentation and Feature Extraction. In: 14th International Conference on Intelligent Engineering Systems, pp. 317–320 (2010)
40. Procházka, A., Ptáček, J.: Wavelet Transform Application in Biomedical Image Recovery and Enhancement. In: The 8th Multi-Conference Systemics, Cybernetics and Informatic, vol. 6, pp. 82–87. IEEE, USA (2004)
41. Rioul, O., Vetterli, M.: Wavelets and Signal Processing. IEEE Signal Processing Magazine 8(4), 14–38 (1991)
42. Sanei, S., Chambers, J.A.: EEG Signal Processing. Wiley - Interscience, Chichester (2007)
43. Scolaro, G.R., de Azevedo, F.M.: Classification of epileptiform events in raw EEG signals using neural classifier. In: 3rd IEEE International Conference on Computer Science and Information Technology, pp. 368–372. IEEE Press, Los Alamitos (2010)
44. Selesnick, I.W., Baraniuk, R.G., Kingsbury, N.G.: The dual-tree complex wavelet transform. IEEE Signal Process Mag. 22(6), 123–151 (2005)
45. Shlens, J.: A Tutorial on Principal Component Analysis (2005), `http://www.snl.salk.edu/~shlens/pub/notes/pca.pdf`
46. Singh, B.N., Tiwari, A.K.: Optimal selection of wavelet basis function applied to ECG signal denoising. Digit Signal Process 16(3), 275–287 (2006)
47. Stone, J.V.: Independent Component Analysis, A Tutorial Introduction. Massachusetts Institute of Technology (2004)
48. Strang, G., Nguyen, T.: Wavelets and Filter Banks. Wellesley-Cambridge Press (1996)
49. Subasi, A.: EEG signal classification using wavelet feature extraction and a mixture of expert model. Expert Syst. Appl. 32(4), 1084–1093 (2007)
50. Subasi, A., Gursoy, M.I.: EEG signal classification using PCA, ICA, LDA and support vector machines. Expert Syst. Appl. 37(12), 8659–8666 (2010)

51. Sun, S.: Extreme energy difference for feature extraction of EEG signals. Expert Syst. Appl. 37(6), 4350–4357 (2010)
52. Vaseghi, S.: Advanced Digital Signal Processing and Noise Reduction. John Wiley & Sons, Chichester (2006)
53. Wang, Z.J., Lee, P.W., McKeown, M.J.: A Novel Segmentation, Mutual Information Network Framework for EEG Analysis of Motor Tasks. Biomed. Eng. Online 8(9) (2009), http://www.biomedical-engineering-online.com
54. Wilson, R.C., Nassar, M.R., Gold, J.I.: Bayesian Online Learning of the Hazard Rate in Change-Point Problems. Neural Comput. 22(9), 2452–2476 (2010)
55. Xie, S., Lawniczak, A.T., Song, Y., Lió, P.: Feature extraction via dynamic PCA for epilepsy diagnosis and epileptic seizure detection. In: IEEE International Workshop on Machine Learning for Signal Processing, pp. 337–342. IEEE Press, Los Alamitos (2010)
56. Yamaguchi, C.: Fourier and wavelet analyses of normal and epileptic electroencephalogram. In: First International IEEE EMBS Conference on Neural Engineering, pp. 406–409. IEEE Press, Los Alamitos (2003)

Chapter 18
WebService-Based Solution for an Intelligent TeleCare System

Vasile Stoicu-Tivadar, Lăcrămioara Stoicu-Tivadar, Sorin Puşcoci, Dorin Berian, and Vasile Topac

Abstract. The chapter describes a teleassistance / telemonitoring system assisting elderly persons, with emphasis on the server component. The system consists of units located in the houses of the monitored persons, collecting and sending medical and environmental data from sensors, and a call centre with a server recording and monitoring the data. Specialised staff will supply telemonitoring and teleassistance services. The software solutions at the server, the WEB Services and the database are briefly described. Conclusions are issued, especially about the "intelligent" features of the developed system and the related future research intentions: the telecare infrastructure and gathered data will be used for researches related to biological signal processing, data mining, decision support systems, as a step toward future generation intelligent telecare networking applications.

18.1 Introduction

Increased life expectancy and the consequent increase in the prevalence of chronic disease rise serious challenges to the sustainability of the national health systems in Europe. Telemedicine, through its various applications can be a part of the solution, creating the basis for seamless care. Shared information is the foundation for seamless care and patient involvement, and its main strategic targets are: high professional quality of care, shorter waiting time, high level of user satisfaction,

Vasile Stoicu-Tivadar · Lăcrămioara Stoicu-Tivadar · Dorin Berian · Vasile Topac
"Politehnica" University Timişoara, 2. V. Parvan blvd., Timişoara, Romania
e-mail: {vasile.stoicu-tivadar,lacramioara.stoicu-tivadar, dorin.berian,vasile.topac}@aut.upt.ro

Sorin Puşcoci
National Research Communications Institute, 6 Preciziei blvd., Bucureşti, Romania
e-mail: sorin.puscoci@inscc.ro

J. Fodor et al. (Eds.): Recent Advances in Intelligent Engineering Systems, SCI 378, pp. 383–408.
springerlink.com

better information about service and quality, efficient use of resources, and freedom of choice. Seamless care is the desirable continuity of care delivered to a patient in the healthcare system across the spectrum of caregivers and their environments. Healthcare services have to be continuous and carried out without interruption such that when one caregiver ceases to be responsible for the patient's care, another one takes on the responsibility for the patient's care.

The last decade shows a major progress towards home-oriented health-care electronic services that have been developed to help elderly people to raise their quality of life. The size of the population aged between 65 and 80 + years in Europe (EU-27) today is 80 million senior citizens, with a doubling of this figure forecasted by 2050. The life expectancy has already been rising on average by 2.5 years per decade and the number of aged over 80 is expected to grow by 180% by 2050 [9].

A teleassistance system offers medical and social citizen-centred responsive services at economic costs. It can help ageing population and people affected by chronic diseases to have a higher quality of life. Unified communications and video-enabled communications allow people to act flexible and more independent. Remote monitoring and consultation, biosensors, and on-line networks for peer support provide new care delivery models. Simple video communications bring physically distributed families and friends closer together for social support. Tele-care can be defined as "the use of information and communication and sensor technologies to deliver health and social support to people helping them living as independently as possible in the lowest intensity care setting consistent with their needs and wishes"[8]. The chapter explores several existing telecare solutions mainly from the technological point of view, but also considering the degree in which seamless care is achieved in order to have quality healthcare services and cost effectiveness in the future.

Having the mentioned study as a background, a telecare system, developed as a project named TELEASIS, is described. The result of the project is a system that supports the implementation and development of the medical and social home assistance services of the elderly persons in Romania. The project developed a pilot system for testing the proposed solutions which consist of units located at the homes of the monitored persons, a call centre, and a handbook containing the necessary information for developing and implementing such a system. The main beneficiaries are the elderly people looking for a decent life at this stage, satisfying their desire to live longer in their own home, with specialised medical and social assistance.

A consortium has been constituted for the project to be carried out: a communication research institute, 4 universities which develop the hardware and software applications, three SME-s with a research component in their activity, and a medical and social assistance company.

The approach specific to the described project is intended to be a systemic one, as in [5], reflected in a vision based on specific subsystems that co-operate in diverse connections related to scenarios that depend on context [27].

18.2 Tele-Care Projects

The following review [22] is based on 8 EU projects, started after 2005, several ended, and several still on the role or continued with other projects and 1 national project that has the same characteristics (CAALYX, eCAALYX, MPOWER, K4CARE, EasyLine+, i2HOME, SHARE-it, CogKnow, TELEASIS). The projects were selected by the results, deliverables - remote controlled homecare applications, the ones that focused on interoperability, and had slightly different end user benefit targets (general or disease oriented).

The comparison was made exploring mainly the technical solution, but also end users benefits (patients, healthcare providers), number of participants and structure of the consortium, budget, and duration. The telecare applications address at the moment mainly 2 categories of persons:

- seniors that are in need of constant attention due to loss of abilities (cognitive and motor disabilities) or chronically ill;
- follow-up of patients, continuing treatment at their homes with good results, not having to stay in a hospital, performing activities of daily living by remote monitoring and being socially reintegrated more rapidly; this last activity is an important part of the continuity of care concept.

In order to achieve the desired goal of continuity of care, the major issue is the interoperability of systems. "Interoperability is the only technical problem in telehealth; the other challenges are not technical" (Gerard Comyn, EC, DG ICT for Health). The interoperability in this context means easy connection between software applications and between software applications and devices (e.g. wireless body or environmental sensors).

The structure of a homecare system, as results from this study (Figure 1) is based on:

- a local platform, home based, consisting of a unit that gets data from the home devices/equipment or body sensors
- a central platform with a server containing the database and the coordination software

The main difference between different approaches is given by the variety of monitored devices and the software solutions. And this is where the interoperability has to be developed. The studied applications use either HL7 standards for communication between platforms, or XML format. One of the projects that are focused on interoperability and its results can be used for further easy development of homecare applications, suggests a service oriented architecture, middleware solution and HL7 standards [16]. The service oriented solution implemented by more than half of the projects has as benefits flexibility and independence of software environments. In the selected projects the software solutions were developed on .NET platform or Java.

There are three main actions related to implementation of a telecare project:

- listing the appropriate equipment that will be installed (home/dispatcher),

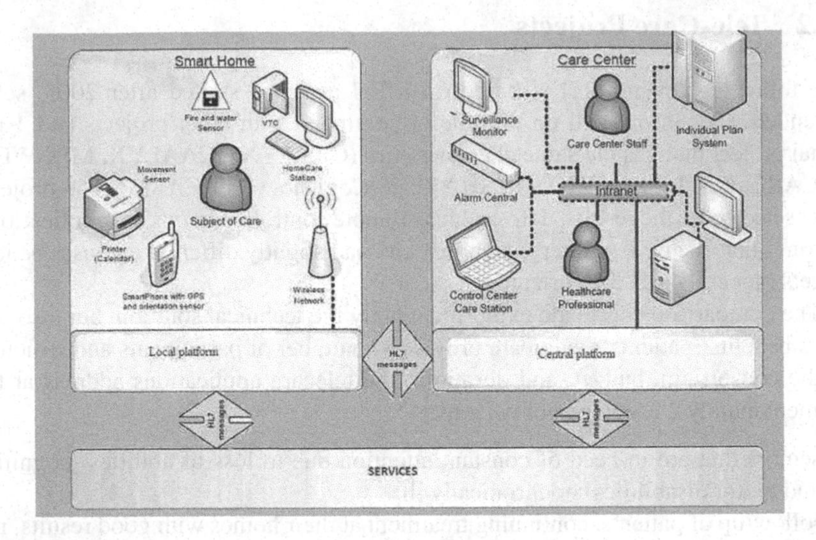

Fig. 18.1 Structure of a recommended tele-care application

- set out the service response protocol to meet the individual's needs; the response protocol must state clearly the response to be made by the monitoring service in the event of a request for assistance,
- obtain the person's informed consent of having equipment in their home.

The devices/equipments located in homes (the Local platform) are:

- sensors that record and transmit data in strict relation with the monitored person: sensors measuring vital signs, fall detector, smart garments (Body Area Networks)
- environmental sensors: gas, smoke, door, fridge, smart appliances using RFID technology
- mobile phones, GPS
- others: semi-automated guided platforms, intelligent walk platforms
- units that collect data from sensors (wireless, Bluetooth, RF).

The interface between the assisted person and the Central Platform can be a computer display, a modified computer display with a receiver [18], a TV set-like one very comfortable solution, a mobile phone, or a PDA [20].

The data is sent between the two platforms in different manners: only when needed, using an intelligent solution [4], streaming or at certain predefined times.

The software from the Central Platform implements a diversity of functions:

- management of medical/social assistance staff: rights, profiles
- record of home individual information: demographics, healthcare record, etc.
- predicting/detecting adverse health conditions and preventing complications before they develop
- monitor the well-being of individuals at home and while they are mobile

- remotely configurable reminding functionality
- control of white goods appliance in the home and detection of the loss of abilities of the user and trying to compensate them [6].
- support, guidance, and relevant health education
- optimization the safe management of chronically ill patients care at home, scheduling prolonged clinical treatments, intelligent decision support, intelligent distribution of data among users [13].

In summary, technology has to cover needs to provide for social inclusion and connectivity, mobility and choice, behavioural change support, care and lifestyle monitoring, condition specific information, access to therapies and named carers. The main activities solved through technological solutions are presented in Figure 2:

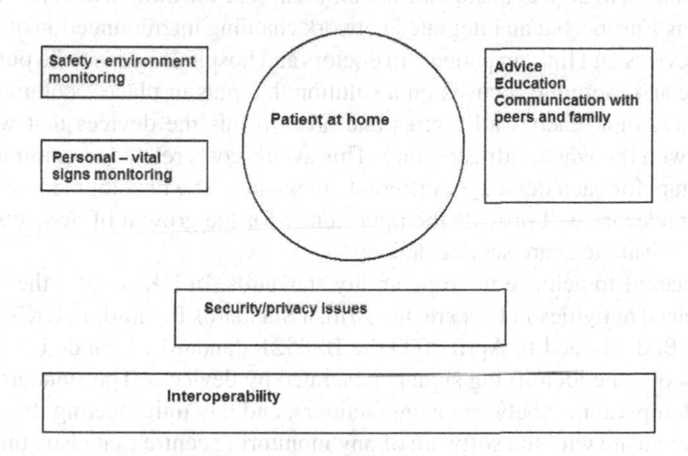

Fig. 18.2 Activities covered by different technology approaches in telecare

Social inclusion and connectivity is related to communication with peers not being restricted by loss of abilities or after hospitalization. This can be provided by TV-sets, mobile phones, PCs, PDAs. Mobility is supported technologically by GPS or mobile phone transmissions inside and outside home. Behavioural change support is done through appropriate equipments and education provided from medical personnel more flexible, from distance on monitors. Care is provided by monitoring the vital signs and prompt interventions when needed and notification or taking drugs or visits to the healthcare facility. Lifestyle is improved using assisted physical exercise equipments, monitoring of environment parameters by specific sensors and intervention in case of dangerous occurrences.

The Consortiums partners developing the explored homecare applications and the range of percentages related to the number of participants in each category were:

- companies: 7,69-57,14%
- Universities and R&D labs: 25-66,66%
- healthcare centers: 0-61,54%

Companies were usually involved 35%, Universities and R&D Labs 40%, and the healthcare centers 25%. That shows that the interest in telecare for these 3 categories of participants is close to one another.

The technical solutions for homecare applications supporting telehealth exist, covering a wide range of end-user needs. There are still barriers in deploying the applications: not very clear proven cost/benefits rates and privacy and security concerns related to the end-users. Technologically, as resulted from exploring the 9 projects and also other related ones, the main issue that has to be solved in the future is the interoperability of devices. Systems for telecare are not currently standardised and lack interoperability. If this will be solved, the economic impact will be better and the costs will be reduced. The current technological solutions show still a high cost for a smart home and the communication with the remote assistance centre.

Technologies to deliver healthcare at home can ease the burden of chronic disease in an ageing Europe, but an integrated network enabling interconnection of different medical devices and linking patients to doctors and hospitals needs to be put in place first. More and more the focus is on a solution that puts in place a communication infrastructure that reaches all houses and adds to this the devices that will communicate with the whole infrastructure. This avoids costs related to communication infrastructure for each device for different companies. The development of such solutions for telecare will provide the opportunity for the growth of new, sustainable models of private telecare service delivery.

Work started to achieve interoperability standards. In UK, one of the countries with sustained activities in telecare, the British Standards Institution, UK's National Standards Body, issued in April 2009 the BS8521 standard which defines the telecare protocols (the identifying signals generated by devices). The standard will facilitate interoperability between manufacturers, and any unit meeting this standard will communicate with the software of any monitoring centre that also complies.

The applications connected with e-Health are solving important issues for the future of European healthcare: increased mobility of citizens in Europe - towards equity in healthcare, the arising problems of an ageing society [14], all in all, reducing costs of healthcare services. That is why the e-Health action plan was developed and gives a timeline for addressing common challenges. Lately the concern was oriented on boosting investments in e-Health and deployment of health information networks (2004-2008) [7].

18.3 The TELEASIS System

The TELEASIS project has developed a pilot tele-assistance network with homecare electronic integrated services, allowing tele-assistance of the elderly, at their residence, based on the most recent IT&C technologies, with a medical and as well, a social target. Designing a tele-assistance system, as part of an assisting service, offers personalized services based on environment conditions and users' requirements. The service-integrating tele-assistance system grants elders the opportunity to benefit from healthcare at home, to enjoy an improved personal lifestyle. The general objective of this home tele-assisting system is to supply a diversity of integrated services for the users (Fig. 3): medical care, premises security through tele-monitoring environment conditions, video-conversation with family and friends, electronic payment, electronic shopping.

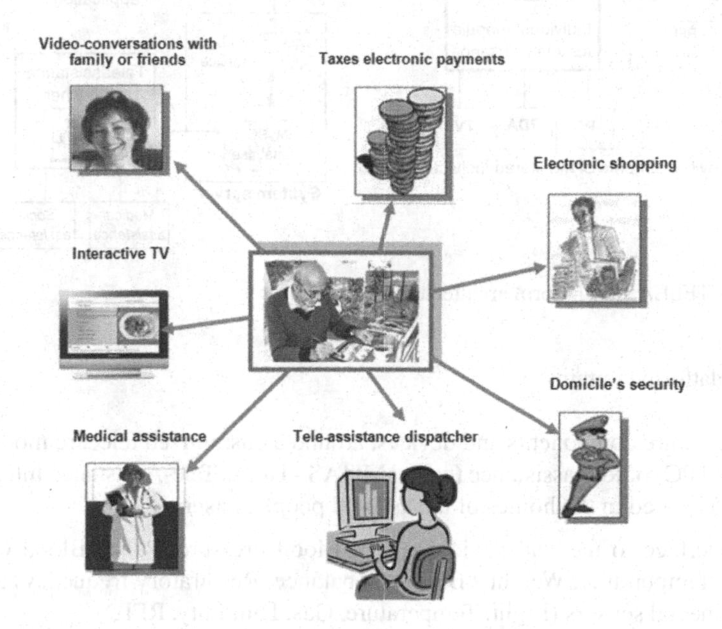

Fig. 18.3 Tele-assistance home integrated services concept

The system optimizes the performance for home assistance services and offers customized services depending on certain conditions and specific user requirements, with the related costs optimized through focused involvement of the medical personnel or social assistance, as well as increasing the nursing at home. Secondly, it may meet the demands of elder persons to live in their own home and not in care homes, the continuation of the active period by involving in daily activities, as well as improvement of the customized management of the assisted person lifestyle. Third, the project explores how older people accept the use of electronic technology, at home, in everyday life, as well.

18.3.1 TELEASIS Platform Architecture

The TELEASIS platform is based on a service-oriented architecture that integrates several specific components (Fig. 4):

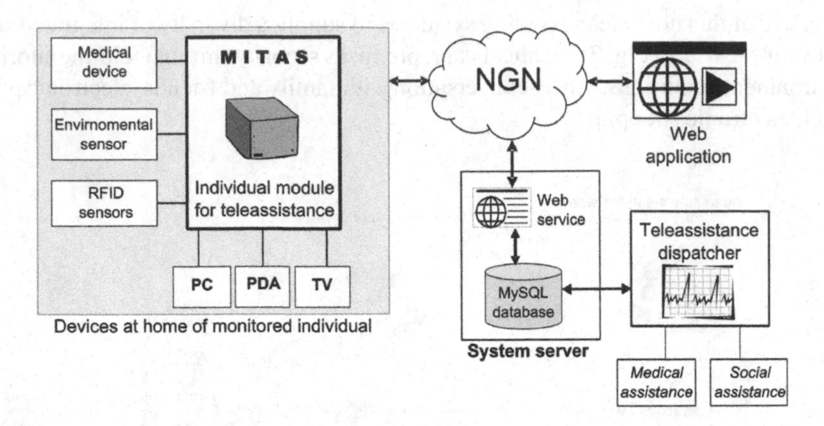

Fig. 18.4 TELEASIS platform architecture

The platform contains:

A. Hardware components and devices, around a customized telecare module (an embedded PC) of tele-assistance (called MITAS - Home Tele-Assistance Integrating Module) located in the homes of the assisted people, ensuring

- the interface to medical devices (ECG, Blood pressure, Pulse, Blood Oxygen level, Temperature, Weight - Bluetooth balance, Respiratory frequency) and environmental sensors (Light, Temperature, Gas, Humidity, RFID)
- the support of communication (via Internet);
- an interface used for delivering the information towards the beneficiary via TV, or PC/PDA if available.

B. Software components support consisting from system logic applications to coordinate activities of the assisting staff, and client-customized applications for agenda activities. In this respect, the software components are:

- A Server with the appropriate software components (WEB Services, Streaming Server, Chat Broker, http Server for Remote Clients) that allows the storing and retrieving the relevant information from the sensors, in/from a Database;
- Clients - for the use of the staff in charge with the monitoring and intervention.

The staff is divided accordingly to different roles: medical, administrative, monitoring/surveillance; the administrator is allowed to commute/establish the roles, as in [26].

The specialised medical staff can set up alarms. These are logical combinations of threshold levels for the values read from the sensors (as an example, high blood pressure, or high glucose level in the blood, and fire alarm - from the environmental sensors - as well). These alarms could be allocated in a personalised way, for each patient [23]. For each alarm, an XML schema is generated and then downloaded in the MITAS modules. At the local level the alarm detection is based on a processing routine for each sensor reading cycle. Thus, all the alarm conditions (alarm scenarios) are checked against the possibility to rise, according to the read values and the XML-coded alarm scenarios. The registered alarms are sent to the server, in order to signal their occurrence to the specialised personnel. This alarm detection strategy is part of the "intelligence" of the system.

18.3.2 Hardware Platform and Infrastructure

The different hardware components (MITAS, sensors, medical devices, Server, other computers) compose the requested infrastructure. Several of them are target computers for the software components of the TELEASIS System.

18.3.2.1 MITAS Module

The MITAS module is located at the patient's home. It serves as intermediary between data collection devices - medical, environmental, presence, devices to display the patient data and connection to the external tele-assistance network. It was designed based on an IBX-530-atomic device (fig.5), which satisfies the conditions for implementing the necessary functions of system operation.

Fig. 18.5 IBX-530-ATOM device used for MITAS

The MITAS module is a complex electronic module carrying the following main functions:

- provides interfaces to medical devices for remote healthcare services;
- provides interfaces to field sensors for home security services;
- provides interfaces to PC, PDA or TV set to link users and dispatchers or public services;
- provides interface to Internet;
- processes the data collected from medical sensors to determine their classification in the predetermined limit, the exceeding of which generated an alarm;
- ensures automatic transmission of data collected at certain programmed time intervals to the Tele-assistance Centre, in normal operation conditions;
- allows the transmission of alarms, collected from medical devices or sensors to the Tele-assistance Centre, in emergency conditions;

- allows transmission of information from the Tele-assistance Centre and display on a display device connected to the module MITAS: TV, PDAs or PCs.

18.3.2.2 Medical Devices and Sensors

Medical data acquisition is performed from medical devices for the related areas, as: Cardiology (blood pressure, pulse), Diabetes (blood sugar/ glycaemia), Pulmonary (peak expiratory flow). Environmental sensors are used, as well: Water leak sensor (to survey water sources in homes), Smoke sensor, Gas sensor (activated when the concentration exceeds the minimum detectable from the kitchen). The Alarm signals from medical devices or sensors are handled by the MITAS module.

MITAS ensures the connection with the following actual medical devices:

- Blood pressure - model 705CP-II
- Gluco-meter + Blood pressure - model Clever Check TD-3213
- Peak flow meter - model PIKO-1.

18.3.2.3 Display Devices

The MITAS Module is designed to connect to one of the devices: TV, PDAs and PCs, and to display information provided by the Tele-assistance dispatcher. The TV set being a device which is easy to use by the majority of seniors is the preferred device used by the TELEASIS platform. The relevant information for user, such as medical data, indications of the doctor, provided by the Tele-assistance Centre are displayed on a television screen from the house (see Fig.6).

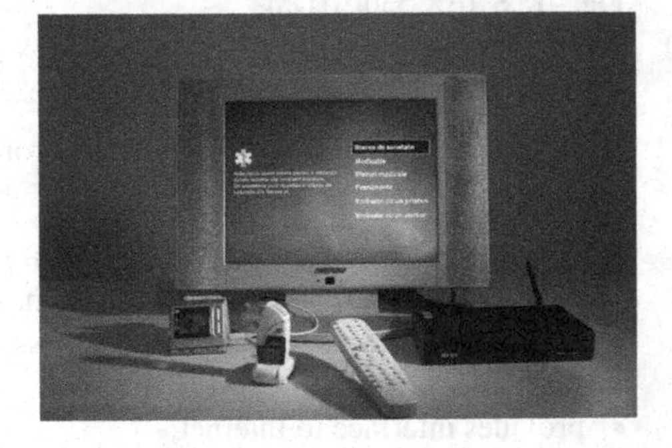

Fig. 18.6 Using TV to display the information from the Tele-assistance Centre

The communication with the TV set is initiated when from the Tele-assistance centre is remotely and unconditionally displayed relevant information on the TV screen located in the patient's room. This function is initiated by MITAS and controls the following events:

- start TV (if off);
- switch to AV (if turned on any program);
- display MITAS user interface to patient;

- display relevant information for the patient;
- interact with the patient and present the required information;
- power off the TV or (optionally) return to the original channel after application.

18.3.2.4 Tele-assistance Centre

The Tele-assistance Centre (Dispatcher) is a complex structure with the following functions:

- retrieves requests from the user and/or from the nurse/assistant who visits the user's home;
- analysis the requirements;
- records and stores events (application, analysis, decision sharing, expertise, action performed, results, costs);
- monitors the chronic state of the patient;
- provides information to users;
- provides optional services (monitoring, security, interventions in home envi-ronment breakdowns, on-line shopping, legal advice, etc.)

This centre requires a special location and associated staff. A Server (the "Dispatcher") should be in this location.

18.3.3 Software Platform

The Software platform consists of a set of applications, dedicated to the system activity. Dedicated applications interact and support staff and tele-assisted persons with medical guides and general information. The platform provides the Tele-assistance Centre with a complete set of information resulted from monitoring activity of the elderly people and also displays information from the centre to the patient, as a feedback. The TELEASIS experimental platform consists of two main modules: MITAS and the Server component from the Tele-assistance Centre. The roles associated with the TELEASIS platform are diverse, from the monitored patient to the administrator. The users with different roles are allowed to act as in the followings:

- Administrator:
  - performs system management functions (server level);
  - adds new system users, edits/deletes users;
  - views data contained in databases.

- Specialist:
  - acts as a medical dispatcher - patients may be consulted for clarification of medical problems;

  - establishes general scenarios (valid for any class of patients), standard scenarios (for categories of patients: cardiac, diabetic, etc..) or custom scenarios for certain patients;
  - matches scenarios and patients (general, standard or custom scenarios) based on active sensors set by the technical dispatcher.

- Technical dispatcher:
  - configures the MITAS modules and allocates the sensors for each patient;
  - is consulted by patients for technical problems and may give solutions for the technical problems occured at the home of the assisted person.

- Current state monitoring user: is a user with medical training, but not a medic (e.g. nurse), who observes continuously critical issues arising in the state of a patient under certain circumstances or unforeseen scenarios, acting based on predefined protocols guiding his/her work.

- Patient:
  - monitors his/hers healthcare status;
  - is monitored in terms of health;
  - is required to confirm that certain actions were performed (e.g. specific drug administration);
  - may consult the specialist or the technical supervisory control on medical or technical problems.

The software architecture developed for MITAS module contains the following components:

- Core Framework - basic software that has overall management functions;
- Update Application - application that will take care of updating all modules;
- Content Update Application - updating the information databases used by all applications;
- Access Database Application - An application for access database: insert values in tables, queries the table, saves results of query and updates tables;
- Application Alarms - Alarms are designed to record and transmit real-time information on the key parameters set for the entire module assistance;
- Scheduler Application - the list of actions to be performed at predetermined intervals;
- Data Acquisition Application - Data acquisition involves the acquisition and processing of signals or wireless signals to get information;
- Data Processing Application - Converts and processes data from one format to another;
- Medical Content Delivery Application - patient-specific content displayed on the screen;

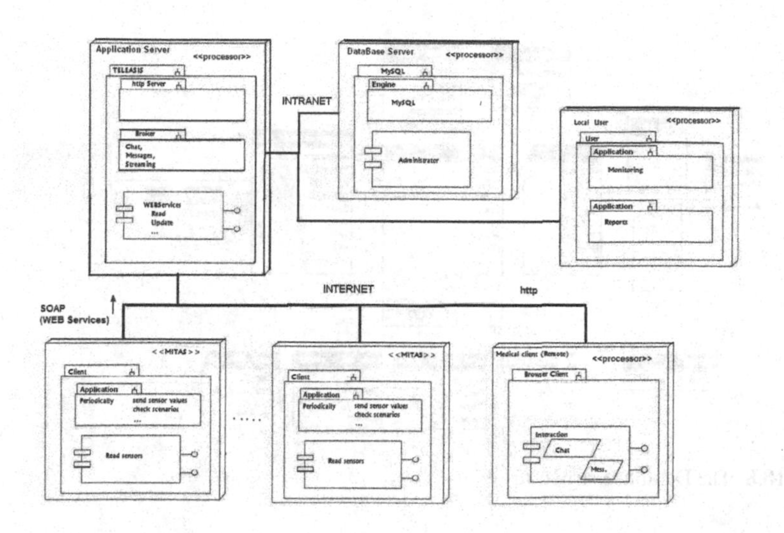

Fig. 18.7 The Server Software Architecture

- Medicine Schedule Application - for tracking medication and patient notification.

The TELEASIS Platform is managed by the Tele-assistance Centre organized in the OrthoVitaMed Hospital in Pitesti.

18.4 The Dispatcher Component

The team from the "Politehnica" University from Timişoara was in charge with the Server Component and other required software components at the Dispatcher (Tele-assistance Centre) level. This is the reason why this component is described in detail. In this respect, a WEB Service-based software was developed.

The software architecture is presented in Figure 7. The Server hosts the WEBService, described in the following chapter, but also an http server that allows remote clients (especially doctors and patients) to access specific data, accordingly to their roles, by a simple browser. A broker component is used to intermediate the dialog between the patients and/or the staff, and to allow video streaming from the WebCams located at the patients home. In order to respect the intimacy of the individuals, the cameras will be remotely activated only in emergency situations. For remote access to the medical records of the patients by the physicians, nurses and patients themselves, an ASP.NET solution that implements remote access web pages was developed. The WCF service (Windows Communication Foundation) is also used. It exposes several public methods that can be called remotely by the web application in order to access the patients' data on the server.

At the level of the Dispatcher, a periodically activated software component ensures the visualisation/ monitoring of the signals read from the sensors and the

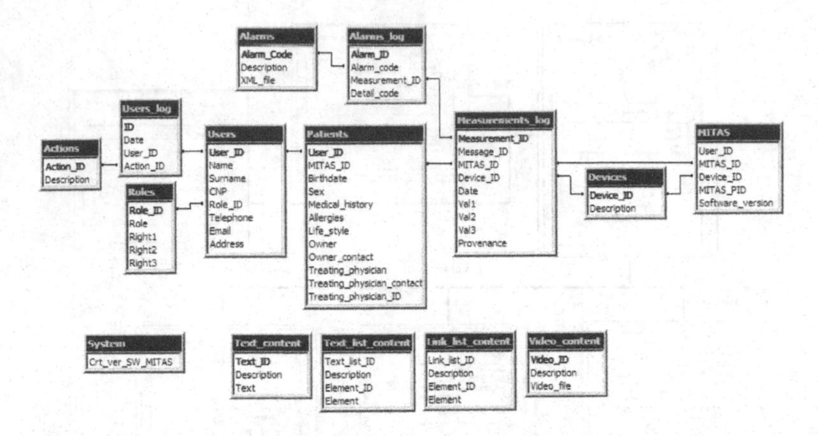

Fig. 18.8 The Database structure

alarms. This software reads the data from the database and displays the graphics of the signals configured for visualisation, for the patients selected by the monitoring staff. Another software component allows the set-up of the data of the patients, the alarms and other settings which personalise the monitoring to the specific needs of each elderly person. A specific report and data mining module will be designed, for further research purposes.

Future interoperability modules will ensure the connectivity of the system with the emergency services and with an automated SMS Sender.

The software solutions are developed under Microsoft Visual Studio.NET (Visual Basic and Visual C#). The database is a MySQL solution and contains relevant information about the patients (identification, contact, and medical information), the staff, the values read from the sensors, the alarms a.s.o. The database contains 15 tables. The data is:

- user data: personal and medical data;
- logs: measurements (values from medical and environmental sensors), alarms, actions;
- miscellaneous data: devices characteristics, system data or definitions (actions, roles, text or video information containing recommendations, treatments and medical procedures).

The structure of TELEASIS database and the relations between tables are presented in the Figure 8.

The dispatcher software is installed on the server unit of the Centre and consists of a user role oriented interface that facilitates the dialogue between all types of users. When starting the application, after displaying a form of greeting, it requires a user and password. Depending on the current user's assigned role, the access is granted to a specific application. Software solutions have been developed, according to the different roles:

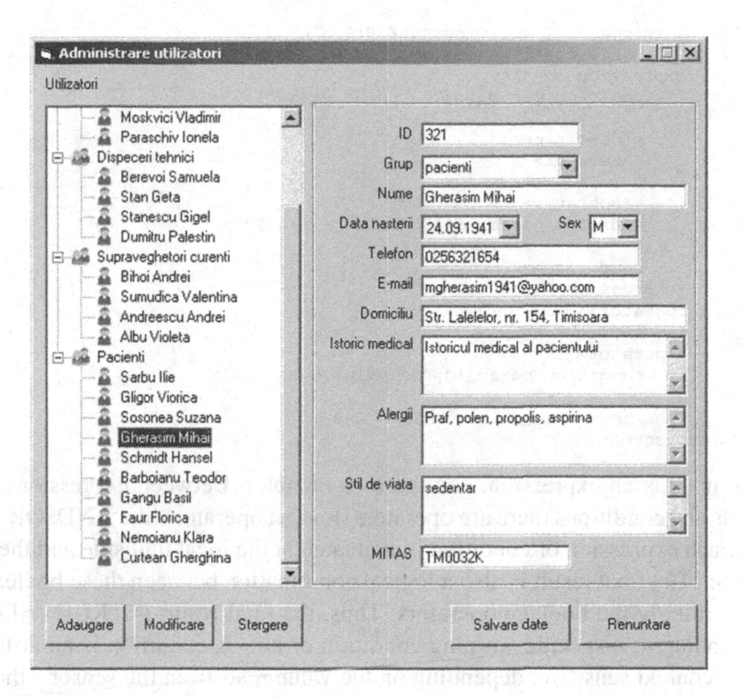

Fig. 18.9 Administration of users

1. Users from group *Administrator* can add new users to the system, amend and delete users (fig. 9).

2. Users from group *Specialist* set scenarios (fig.10) and design scenarios for each patient based on the active sensors set by the technical dispatcher. Several examples of scenarios are listed in the followings:

- Gas (gas sensor detects increasing gas concentration in the eventuality that the assisted person forgot cooker opened and a flame went out for some reason);
- Increasing temperature (oven temperature has grown beyond a certain limit);
- Overriding the limits of physiological parameters (simple, logical combinations, trends) - increased blood pressure over a significant period of time; aberrant cardiac parameters (arrhythmia); oxygen saturation of blood in a certain limit; low blood pressure with danger of loss of consciousness; low glucose level: blood glucose decreased below the limit permitted, with risk of entry into a coma;
- System failure: the essential function of the system is lost, the monitoring system cannot be used to supervise the person, the connection was lost, the protocol should provide connection to regular link by telephone to remedy.

Based on the dialog presented in Fig. 10, the software generates an XML schema, e.g:

```
<?xml version="1.0" encoding="utf-8"?>
<alarm_scenario id="1" type="xml" id_ext="null">
     <expression no="1">
        <condition param_id="2" value_eval="160">
                >
          </condition>
          <operator>
              OR
          </operator>
        <condition param_id="3" value_eval="90">
                    >
        </condition>
     </expression>
     <operator>
          AND
     </operator>
        <condition param_id="5" value_eval="50">
              <
        </condition>
</alarm_scenario>
```

Each node is an expression, condition or operator. Between expressions or expressions and conditions there are operators (logical operators like AND, OR, XOR, NOT). Each expression or condition is evaluated at the local unit side and the result is boolean. The final result is also a logical combination between these boolean values, accordingly the boolean operators. Thus, the final result is TRUE or FALSE, corresponding to a specific warning condition or not. A condition is an arithmetic or logic (context sensitive, depending on the value read from the sensor - there are sensors that give boolean value) comparison between a value read from the sensor corresponding to the specified param_id, and the value value_eval. The comparison is specified in the condition section. An expression is a logical combination between conditions and allows grouping conditions in the same way as the parenthesis

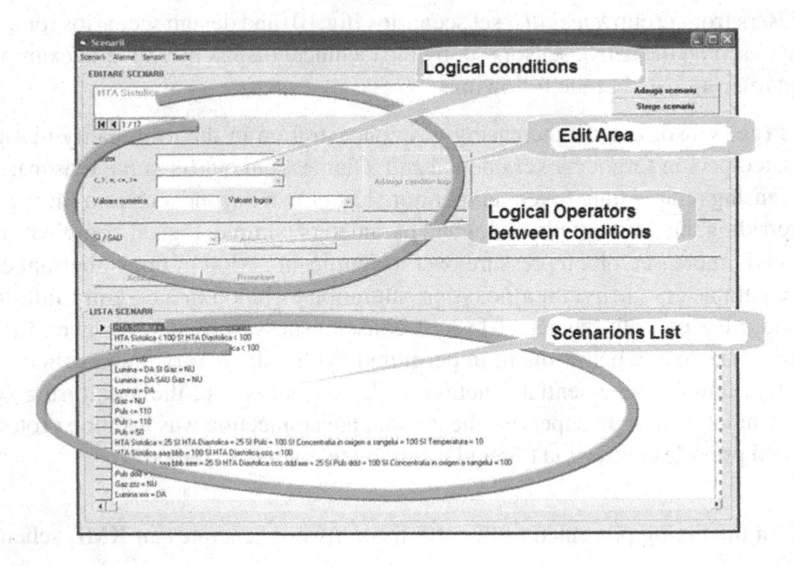

Fig. 18.10 Setting scenarios

in the classical notation for logical expressions. In this way, one can define complex expressions, including nested conditions.

Thus, the XML from the example above corresponds to the warning scenario of logical value

```
Alarm 1 = (systolic blood pressure > 160 AND
          diastolic blood pressure > 90)  OR glycemia <50
```

The warning condition checking is done at the level of MITAS, based on this XML schema, as an "interpretation". If the result of the logical evaluation of the expression above is TRUE, a specific warning is raised at the level of the local system and then immediately this warning with the corresponding code (1 in the example) is sent to the server by calling a specialised method from the Web Service. A warning is listed in the user interface, associated with its name and description, and the name of the elder person affected. The assistance personnel is able to take the appropriate decisions and measures in order to help the person. The same warning is recorded in the database. In the head of the XML schema, an attribute named type codifies a special kind of processing. If the value is XML, like in the previous example, the warning condition check is like described above. That means the warning value is computed after each reading sensors cycle. All the warning conditions (warning scenarios) are checked against the possibility to rise, and the warnings occurred are sent to the server. But if the value of type is "ext"', then the processing does not use the XML schema but an external complex algorithm (as an example, arrhythmia detection from an ECG record, or a high blood pressure tending detection from a blood pressure measurement series). In this case, the algorithm is invoked with the id_ext number associated to the warning. This algorithm is possible to implement with a dll file located at the local module level: different methods should be invoked with the cardinal of the method from the list (the one with the id_ext value). If a new method is required, the developers will add this method to the source code, will recompile the dll and then will update the software via the update mechanism of the system. The warning scenarios must be defined, by an appropriate dialog with the medical staff (presented in Figure 10).

Once the warning scenario is defined, an XML schema is generated in the background. Then, for each tele-assisted person, the medical staff must allocate a specific list of warning scenarios using drag-and-drop actions (see Figure 11).

3. Users from group *Current monitoring* supervise the critical aspects of a patient and can report/ act, according to emergency situations (Fig. 12).

18.5 The Web Service

The TELEASIS server hosts a Web Service that can be invoked for different purposes. The use of WEB Services as an appropriate solution was selected as a first step to a Service-Oriented Architecture - as in [10]. The server is configured so that the Web Service is securely accessed over Secure Socket Layer (SSL). Every web method in the Web Service receives two encrypted parameters that identify the client. Only approved clients are authorised to call these methods. The WEB Service is developed in Visual C#, to be called by the clients at the MITAS level. The web methods are briefly described in the followings:

- *Warning* is the web method that is called when a situation is considered to be dangerous for the patient in order to alert the medical staff over this critical situation. The assessment of a situation is made by comparing the results of a measurement with several predefined dangerous situations stored as XML files. For each patient there are defined several critical situations depending on diseases that he/she suffers and other factors. Warning is called, if needed, after the call of Send_values web method and the assessment of the measurement results. This web method has one parameter that is an XML document containing all the data needed to store the warning. The structure of the XML document is presented in the followings:

```
<log>
  <messageID> value </messageID>
  <mitasID> value </mitasID>
  <patientID> value </patientID>
  <deviceID> value </deviceID>
  <deviceTimestamp> value </deviceTimestamp>
  <alarm_code> value </alarm_code>
  <detail_code> value </detail_code>
</log>
```

  Warning returns a Boolean value representing the success of writing the warning data in the database.

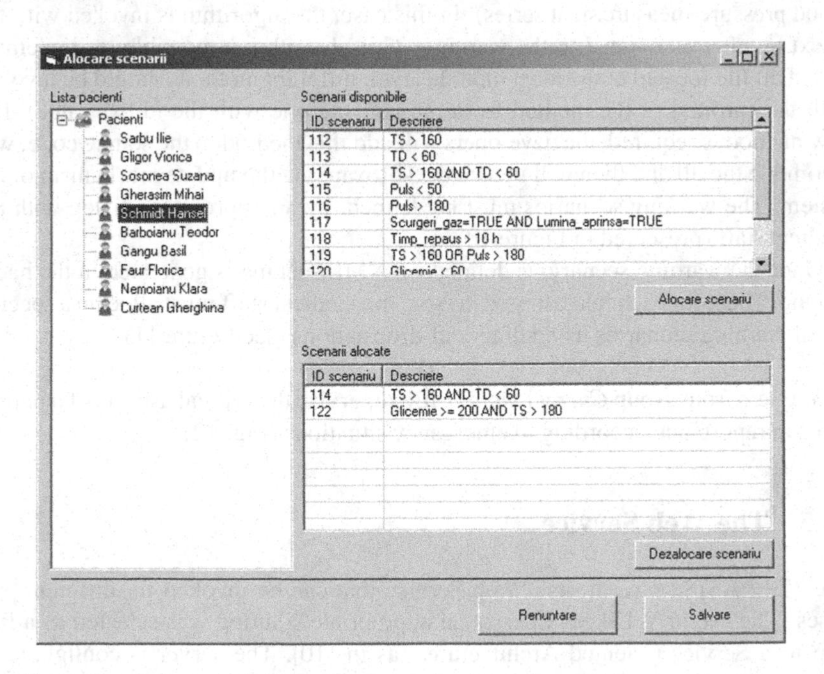

Fig. 18.11 The allocation of the warning scenarios for each patient

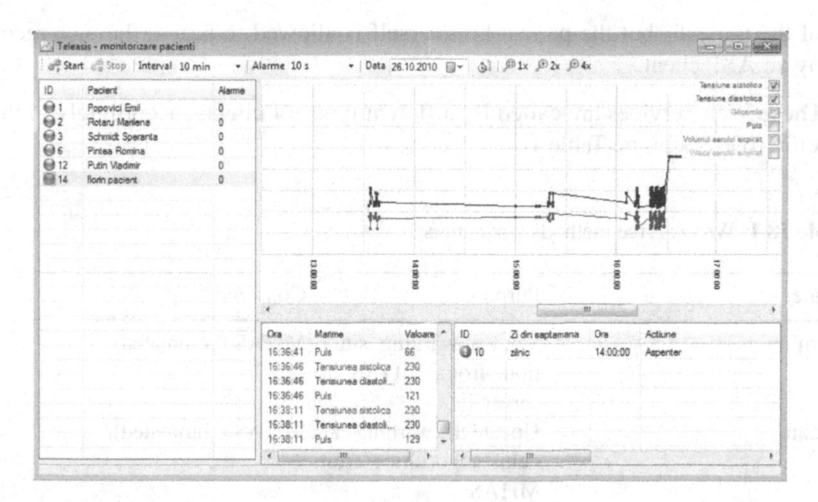

Fig. 18.12 Patients' monitoring

- *Update* Web Service method, called once a day from the MITAS client. This method compares the list of warning scenarios available at the local level (as XML files or dlls) with the required list available for each assisted person at the server side, and if differences occur, the lacking scenarios are downloaded and added at the local level.
- *Download_content* is the web method that should return different types of data: video sequences, text, a list of text elements or a list of links in order to be displayed to a patient on TV, PDA or other devices. The client that calls this web method should pass two parameters that specify what kind of content it wants and the index of the content that will be returned. In this way, the patient is allowed to access relevant information that is useful in healthcare, but related to his/her specific needs. That implements the context-based information access.
- *Send_file* is the web method that starts the upload of a file from a client to server. This web method is called by clients in order to send files storing ECG data or other medical data. The server stores these files in dedicated folders. The two parameters passed to the web method are: the type of data in the file and the name of the file. The method returns a Boolean value representing the success of the sending operation. The Download_content and Send_file methods initiate ftp file transfer.
- *Calculate_profile_matching* is the web method invoked by the patient for context-based communication facilities or context-based information access.
- *Load_Patient_file* is the web method that downloads the specified patient's medical record. This method is used to ensure the basis to compute the matching degree between patients, or between the patient and the data available for download and display (from the relevance point of view) - for context-based communication facilities or context-based information access. This method is invoked by the previous web method or by the medic when accessess the medical information

of the patients, but the patient him/herself is allowed to access the own record, by an ASP client.

These Web Services are called by different types of clients, accordingly to their specific roles, as in the Table 1.

Table 18.1 Web Service methods consumers

Name	Purpose	Consumer
Warning	Send a warning situation from MITAS to server	MITAS (automated)
Update	Update the warning calculus algorithms from MITAS	MITAS (automated)
Download_content	Download medical information for the patient	MITAS (patient role) or ASP patient client
Send_file	Upload medical signal records from MITAS to server	ASP nurse client
Calculate_profile_matching	Determination of profiles matching	MITAS (patient) or ASP patient client
Load_Patient_file	Access the medical record of a patient	Calculate_profile_matching or ASP Medic client or ASP nurse client

This solution is appropriate for all the technologies used by the system: service-oriented architecture, ASP solutions, and other types of clients.

18.6 Future Development: A Better Accessibility to Information

Patient empowerment is defined as helping people to discover and use their own innate ability to gain mastery over their disease or status [12] - by providing education for informed decision-making, assisting patients to weigh costs and benefits of various treatment options, setting self-selected behavioural goals, and providing information about the importance of their role in self-management. The adoption of the collaborative care approach empowers health care professionals as much as it does patients [1]. That is the reason why in the project TELASIS new functions related to increased accessibility of the patients to the medical information, or to a better understanding of the medical terms by the patients, are under development. But we must be aware that the patient empowerment paradigm has its own pitfalls [11],[21]. The healthcare professionals must promote more responsibility for the patients themselves. Medical language is very often hard to understand for regular people. Given this, the communication between doctors and patients can suffer especially when dealing with remote communication that can appear in systems like TELEASIS. A research project, using a specialized classifier, evaluating how easy is for regular people to access data expressed in medical language reached the following conclusion "The classifier was then applied to existing consumer health Web pages. The mentioned research found that only 4% of pages were classified at a layperson level, regardless of the Flesch reading ease scores, while the remaining pages were at the level of medical professionals. This indicates that consumer health Web pages are not using appropriate language for their target audience" [17]. This can affect in a great manner the accessibility of the patients to their health information. Having a bad understanding of their health status may have a bad influence on their heath evolution. Empowering the patients with more understanding of the medical information related to them will strongly reduce this risk. Taking into account all these, the TELEASIS system suggests 3 ways to ensure Patient empowerment, as future developments:

a. Access to a central medical information database, with different content formats (text, multimedia, videos). This allows the patient to display information of interest, related to his/her (health) status. Based on the information about the patient, a selective, medical context-based search engine gives the patient the possibility to select, download and display the information on the available device: TV set, PC or PDA. The information, according to the destination, should be: text list/tree, plain text, text with hyperlinks (html), image, image list, video sequence, audio sequence, multimedia presentation.

The download mechanism is implemented via a WEB Service that invokes the context-based search engine and then establishes an ftp connection in order to download the requested information. The sequence diagram for this process is depicted (simplified) in Figure 13.

b. Access to additional communication channels The patient is allowed to contact the doctor, on-line if available, or off-line (with chat, streaming, and/or messages). This is a simple way to give access to information, directly from the doctor. But the system encourages the communication between the monitored persons, as well. The list of the patients form the system is available for all the patients, and, based on the medical context, the "matching" (or similarities) between the persons subscribed is underlined, as in social, or professional networking systems (as in Facebook or Linkedin). The sequence diagram for such a link is partially presented in Figure 14. Thus, the patients with similar problems have the opportunity to share experience and support each other.

The communication uses a broker component (a WEB Service) for the subscribing process and the communication. Both of the information access improvements

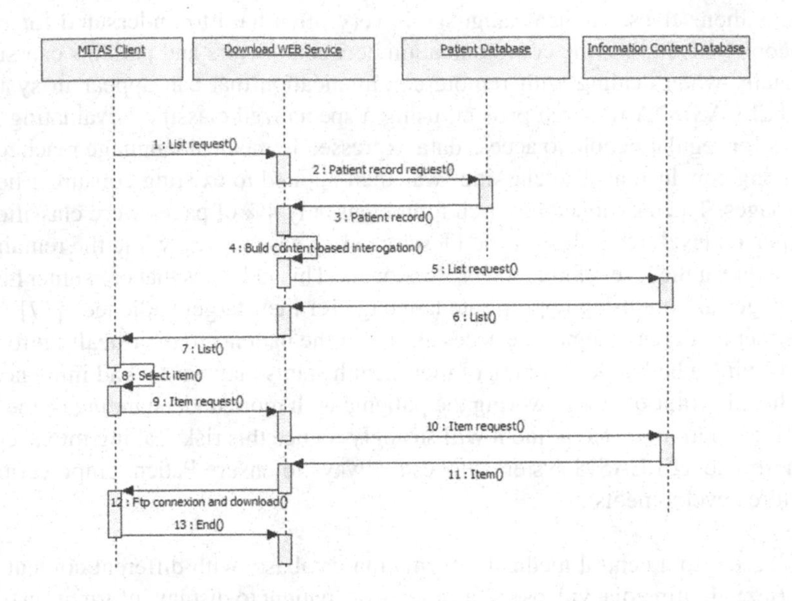

Fig. 18.13 The download mechanism

contain somehow an "intelligent" part, due to the content-based selection of the information: the available information or the communication partners are selected accordingly to the medical record of the patient.

These components are under development.

c. Interpreting medical language

Interpreting medical language represents adapting or "translating" information from specialized medical language to regular or patient friendly language.

TELEASIS system is offering patients access to their health data, reports and additional medical information. All this data is stored in an information and content database. Enrolled medical staff or other power user can add documents to this database, and can set the access rights for patients or groups of patients. Each patient can access different documents. While allowing the patients to access medical information proves to be useful, as reminded in the introduction, the patients encounter big difficulties in understanding that information. For this, TELEASIS will use an interpretation engine that allows the patient to get the medical information "translated" to regular language, which is easier for them to understand. The user can choose whether he wants to see the medical text translated to regular language or not. The required interpretation engine is object of research in progress [24].

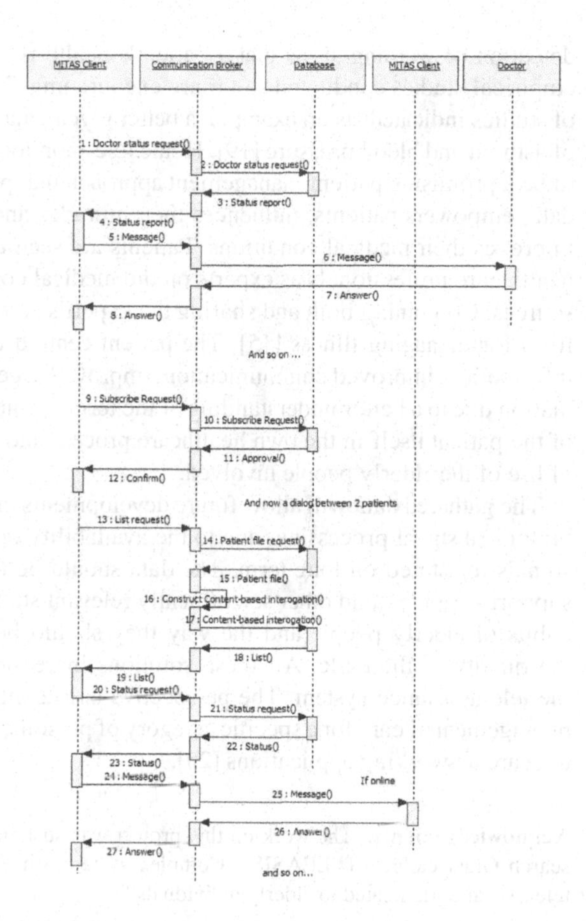

Fig. 18.14 Subscribing and communicating

18.7 Conclusions

A background for the problems raised by telecare and tele-assistance was presented. On this basis, the TELEASIS project was developed, as a Romanian pilot project for solving the elderly people's requirements for a "quite-normal" life. The presented system supplies all the needs related to monitoring elderly people at their homes. As an advantage to other similar ones, there is also a social and communication side, transforming it in a complex tool supporting the mentioned category of persons. The result is an integral, holistic solution, as in [2], including medical and social needs of the target group. The described intelligent alarm scenario checking functionality allows the staff to decrease the burden of surveying many people by the tele-assistance system, to improve their ability to communicate with these, to access their real needs and problems, including the important issue of communication: with other similar people, with the physicians, and understanding better the medical information, in order to implement the concept of patient empowerment. Home telemonitoring allows for closer monitoring of each patient's condition, as well as early detection of warning signs that a patient's health is deteriorating. The findings of empirical studies conducted so far are encouraging. The results of a large majority of studies indicated, as an example, a better glycaemia control and improved control of asthma and blood pressure [19]. Home tele-monitoring of chronic diseases seems to be a promising patient management approach that produces accurate and reliable data, empowers patients, influences their attitudes and behaviours, and potentially improves their medical conditions. Patients are seen as experts of their illness and health care professionals as experts on the medical conditions and management resources. Combining both and sharing the expertise could achieve the intended platform for managing illness [15]. The patient centred features as access to medical information, improved communication, improved accessibility to the medical information due to a better understanding of the terms, contribute to a better involvement of the patient itself in the own healthcare process and as a result, to a better quality of life of the elderly people involved.

The gathered data will allow future developments in the area of measurement and biological signal processing, due to the availability of a large amount of biological signals registered on long term. The data should be used in data mining, decision support - see [3] - and other scientifically relevant studies about the lifestyle, needs, habits of elderly people and the way they should be helped in order to increase the quality of their life. All these solutions increase the "intelligent" features of the tele-assistance system. The perspective is to contribute to improvement of the management of care for a specific category of persons, as well, as a future generation telecare networking applications [25].

Acknowledgements. The work on this project was sustained by a Romanian National Research Grant called "TELEASIS - Complex system, on NGN support for home care and teleassistance dedicated to elderly individuals".

References

1. Anderson, R.M., Funnell, M.M., et al.: Evaluating a problem-based empowerment program for African Americans with diabetes: results of a randomized controlled trial. Ethnicity & Disease 15, 671–678 (2005)
2. Barnes, N.M., Reeves, A.A.: Holistic monitoring to support integrated care provision Experiences from Telecare trials and an introduction to SAPHE. In: Second International Conference on Pervasive Computing Technologies for Healthcare, Pervasive-Health 2008, Tampere, pp. 293–296 (2008)
3. Basilakis, J., Lovell, N.H., et al.: Decision Support Architecture for Telecare Patient Management of Chronic and Complex Disease. In: 29th Annual International Conference of the IEEE Engineering in Medicine and Biology Society, EMBS 2007, Lyon, pp. 4335–4338 (2007)

4. Boulos, M.K.N, Rocha, A., et al.: CAALYX:a new generation of location-based services in healthcare. International Journal of Health Geographics, 6–9 (2007)
5. Carson, E.R., Cramp, D.G., et al.: REALITY in Home Telecare: A Systemic Approach to Evaluation. In: 27th Annual International Conference of the Engineering in Medicine and Biology Society, IEEE-EMBS 2005, Shanghai, pp. 3927–3930 (2005)
6. Casas, R., Marín, R.B., et al.: User modelling in ambient intelligence for elderly and disabled people. In: Miesenberger, K., Klaus, J., Zagler, W.L., Karshmer, A.I., et al. (eds.) ICCHP 2008. LNCS, vol. 5105, pp. 114–122. Springer, Heidelberg (2008)
7. Commision of the European Comunities, e-Health - making healthcare better for European citizens: An action plan for a European e-Health Area (2004)
8. Curry, R., Lethbridge, K., et al.: An International centre of excellenc In Telecare Background and development ICE-T, Final report, `http://www.sehta.co.uk/files/ReportICE-T.pdf` (accesed March 2009)
9. Eurostat newsrelease, A statistical perspective on women and men in the EU27, 35/2010 (2009/2010), `http://epp.eurostat.ec.europa.eu/` (accessed March 2010)
10. Guevara-Masis, V., Belloum, A., et al.: An Agent-based Resource Management for a Service-Oriented Telecare Environment. In: ITAB 2007, 6th International Special Topic Conference on Information Technology Applications in Biomedicine, Tokio, pp. 143–148 (2007)
11. Fox, N.J., Ward, A.J., et al.: The 'expert patient': empowerment or medical dominance? The case of weight loss, pharmaceutical drugs and the Internet. In: Social Science & Medicine, vol. 60(6), pp. 1299–1309 (March 2005)
12. Funnell, M.: Patient Empowerment. Critical Care Nursing Quarterly 27(2), 201–204 (2004)
13. Isern, D., Millan, M., et al.: Home Care Individual Intervention Plans in the K4Care Platform. In: 21st IEEE International Symposium on Computer-Based Medical Systems, pp. 455–457 (2008)
14. Koch, S.: Meeting the Challenges - the Role of Medical Informatics in an Ageing Society Ubiquity: Technologies for Better Health in Aging Societies. In: Hasman, A., Haux, R., Van Der Lei, J., De Clercq, E., Roger-France, F.H. (eds.) Proceedings of MIE 2006. Studies in Health Technology and Informatics, vol. 124, pp. 25–31 (2006)
15. Lau, D.H.: Patient empowerment - a patient-centred approach to improve care. Hong Kong Medical Journal 8(5), 372–374 (2002)
16. Mikalsen, S.H., Fuxreiter, T., et al.: Interoperability Services in the MPOWER Ambient Assisted Living Platform. In: Adlassnig, K.-P., Blobel, B., Mantas, J., Masic, I. (eds.) Medical Informatics in a United and Healthy Europe - Proceedings of MIE 2009 - The XXIInd International Congress of the European Federation for Medical Informatics. Studies in Health Technology and Informatics, vol. 150, pp. 366–371. IOS Press, Amsterdam (2009)
17. Miller, T., Leroy, L., et al.: A Classifier to Evaluate Language Specificity of Medical Documents. In: 40th Annual Hawaii International Conference on System Sciences, HICSS 2007 (2007)
18. Meiland, F.J., Reinersmann, A., et al.: COGKNOW development and evaluation of an ICT-device for people with mild dementia. Stud. Health Technol. Inform. 127, 166–177 (2007)
19. Par, G., Moqadem, K., et al.: Effects of Home Telemonitoring in the Context of Diabetes, Asthma, Heart Failure and Hypertension: A Systematic Review. Journal of Medical Internet Research 12(2), e21 (2010), `http://www.jmir.org/2010/2/e21`

20. Puşcoci, S., Stoicu-Tivadar, L., et al.: Integrated tele-assistance platform - TELEASIS. In: The Second IFAC Symposium on Telematics Applications TA 2010, Timişoara, Romania, October 5-8, pp. 97–102 (2010)
21. Salmon, P., George, M., et al.: Patient empowerment or the emperor's new clothes. Journal of the Royal Society of Medicine 97, 53–56 (2004)
22. Stoicu-Tivadar, L.: ICT frame supporting continuity of care towards increased quality of health care services. In: The Second IFAC Symposium on Telematics Applications TA 2010, Timişoara, Romania, October 5-8, pp. 83–86 (2010)
23. Stoicu-Tivadar, V., Stoicu-Tivadar, L., et al.: A WebService-based Alarm Solution in a TeleCare System. In: Proceedings of 5th International Symposium on Applied Computational Intelligence and Informatics, SACI 2009, Timioara, Romania, pp. 117–121 (2009)
24. Topac, V., Stoicu-Tivadar, V.: Improved Software Architecture for Text-based Information Accessibility Applications. In: The Fifth Advanced International Conference on Telecommunications (AICT 2009), Venice, Italy, pp. 198–202 (2009)
25. Valero, C., Vadillo, M.A., et al.: An Implementation Framework for Smart Home Telecare Services. In: Future Generation Communication and Networking (FGCN 2007),Jeju, vol. 2, pp. 60–65 (2007)
26. Ying, L., Bacon, J.: A Practical Synthesis of Dynamic Role Settings in Telecare Services First. In: International Conference on the Digital Society, ICDS 2007, Guadeloupe, p. 5 (2007)
27. Zhang, D., Yu, Z., et al.: Context-Aware Infrastructure for Personalized Healthcare. In: Nugent, C., McCullagh, P., McAdams, E., Lymberis, A. (eds.) Personalised Health Management Systems, pp. 154–163. IOS Press, Amsterdam (2006)

Chapter 19
Learning Iris Biometric Digital Identities for Secure Authentication: A Neural-Evolutionary Perspective Pioneering Intelligent Iris Identification

Nicolaie Popescu-Bodorin and Valentina Emilia Balas

Abstract. This chapter discusses the latest trends in the field of evolutionary approaches to iris recognition, approaches which are compatible with the task of multi-enrollment in a biometric authentication system based on iris recognition, and which are also able to ensure strong discrimination between the enrolled users. A new authentication system based on supervised learning of iris biometric identities is proposed here. It is the first neural-evolutionary approach to iris authentication that proves an outstanding power of discrimination between the intra- and inter-class comparisons performed for the test database (Bath Iris Image Database). It is shown here that when using digital identities evolved by a *logical* and *intelligent* artificial agent (Intelligent Iris Verifier/Identifier) the separation between inter- and intra-class scores is so good that it ensures *absolute safety* for a very large percent of accepts (97%, for example), i.e. recognition is no longer a statistical event, or in other words, the statistical aspect of iris recognition becomes residual while the logical binary aspect prevails. In this way, iris recognition theory and practice advance from *inconsistent verification* to *consistent verification/identification*.

19.1 Introduction

Nowadays, after years of important studies and contributions, such as those of Wildes [21] and Daugman [3, 5], or the newer developments undertaken at CASIA by Ma *et al.* [10] and Tan *et al.* [20], at the University of Bath by Monro *et al.* [11, 12], Rakshit and Monro [18, 19], at NIST by Grother *et al.* [7] - who summarized the

---

Nicolaie Popescu-Bodorin
Artificial Intelligence and Computational Logic Laboratory, Department of Mathematics and Computer Science, Spiru Haret University, Bucharest, Romania
e-mail: bodorin@ieee.org

Valentina Emilia Balas
Faculty of Engineering, Aurel Vlaicu University, Arad, Romania
e-mail: balas@drbalas.ro

J. Fodor et al. (Eds.): Recent Advances in Intelligent Engineering Systems, SCI 378, pp. 409–434.
springerlink.com

evaluation of iris recognition technologies competing in Iris Challenge Evaluation [9], and also at the University of Notre Dame by Baker *et al.* [1], Bowyer *et al.* [2], Hollingsworth *et al.* [8], after a lot of things being done and being said about iris recognition, it could appear the temptation to believe that iris recognition is a closed domain. We *strongly* disagree with this point of view.

So far, iris recognition theory tells in short that the similarity of some uint8[1] codes (unwrapped iris segments obtained through segmentation, unwrapping and normalisation of the iris captured in an eye image) is decidable in the space of binary iris codes (a space of binary matrices obtained by compressing uint8 codes into binary codes).

Is this *theory* logically consistent (sound)? Nobody asked, as far as we know, but certainly, no answer to this question was ever published, and consequently, *inconsistent iris verification* - a field of experimental science and engineering - took the place and the name of *iris recognition* - a field of science which is supposed to deliver proofs and certitudes concerning the results obtained experimentally. For quite a while (ten years at least), different authors published for different iris recognition approaches a single type of demonstration proving how good these approaches are and identifying upper bounds for whatever False Accept Rate meant in their biometric system. These kinds of demonstrations have at least three logical faults that must be corrected:

- Firstly, a case of False Accept is an objective situation, and therefore, the Fals Accept Rate (FAR), is on its turn, an *objective measure*, and therefore, only in a logically inconsistent practice of iris recognition a relation between FAR at verification and FAR at identification could be formulated (estimated / established) as being a non-identical function.
- Secondly, confusing a *binary iris code* with a *digital identity* does not match a logical restriction that a biometric system must satisfy in terms of entropy: a recognizer object encodes more entropy than any of the recognized objects.
- At last but not the least, *proving an upper bound for the False Accept Rate* (as a value or as a dynamic) *while keeping the space of digital identities stationary is a logical non-sense*, because it means to establish a limitation for the speed with which *the explosion of a contradiction* expands in the internal logic of a biometric system. Of course, this is not a problem if the biometric system does not have to prove a binary consistent internal logic (this cloud be the case if the internal logic of the biometric system is allowed to be an *inconsistent* or a *paraconsistent* 2-valent or n-valent fuzzified logic of *beliefs*). Still, in a stationary space saturated with imbricate clusters (each cluster being a cloud of binary iris codes coming from the same eye of the same person), the process of finding a suitable location for a new cluster to be inserted without colliding it with the other clusters that are already there, only gets harder and harder, and finally impossible, and therefore, if a False Accept occurred, and if the number of enrolled users continues to grow, the trivial expansion of the contradiction within the internal logic of the biometric system can only accelerate.

[1] Unsigned 8-bit integer.

The motivation behind one of our previous papers [13] was the belief that the major improvements in iris recognition will come from the field of artificial intelligence. One challenge defined in that paper (namely C8:*Build an exploratory supervised intelligent agent for iris recognition*) is how to find a way of automating the process of searching for new methods for iris segmentation and for iris matching. In the same paper [13] is said that it was not clear at all why the neural approaches to iris encoding and matching usually do not lead to the same performances as those obtained in the classical approaches based on the direct comparison of binary iris codes. It is also said there that in terms of artificial intelligence, a way to find new methods for iris encoding and matching could be to define a neural network architecture or a heuristic algorithm able to replicate currently available iris recognition results obtained by comparing the iris codes directly. Such an approach would assume that each enrolled identity is stored as a trained memory or as a feature vector and would be able to classify candidate iris codes as well as possible by preserving or improving the quality of the separation between genuine and imposter score distributions in terms of False Accept/Reject Rates.

All of these ideas [13] were succesfully validated in our current work whose results follow to be presented here. Now we know that the classical neural architetures (from Multi-Layer Perceptron to Self Organizing Maps) are too general to be well adapted for achieving biometric purposes. We also know that genetic programming can be used to create (or to optimize) program sources for well specified goals. Genetic programming techniques were integrated into the Analyzer/Adviser Module within NPB Iris Recognition Generic Experimental Model [14] and used to find all the novelties which follows to be discussed here. Also, using the infrastructure described there [14] enabled us to draw one of the most important conclusions which led us to the current neural-evolutionary perspective: in order to be of high quality, the process of learning biometric identities must be *supervised*, must be *adaptive*, and must keep separate tracks of the rewards and punishments occurred during instruction by encoding the history of positive and negative events into two different memory tables. Regardless the type of neural networks used in our simulations (and in the last three years we did more than ten thousands simulations on Bath Iris Image Database), when the learning process encodes experience on unspecialized neurons (punishments an rewards are memorized on the same memory table), the obtained model rapidly lost its power of generalization: the learning data were correctly recognized but the correct classification of test data failed too often.

We also know now that in order to be the main member of a one-to-many relation with the candidate iris codes (one identity / one memory should be able to recognize multiple instances of binary iris codes taken for the same eye of the same person) an identity must have a bigger informational entropy than the candidate binary iris codes, in the same way in which a class encodes more entropy than its own members. If this would not be true, then one solution for defining a suitable encoding of the biometric identities should exist in terms of finding some binary matrices as centroids for the sets of iris codes extracted for the same eye of the same person. Still, this hypothesis has not been validated in our experience so far. Hence we have assumed that the dimension of the stored identities must differ with at least one order from that of the candidate iris codes. Even so, finding these centroids in a richer space (of increased entropy) through k-means or SOM type algorithms would mean to store a memory on unspecialized neurons.

On the other hand, strictly from the point of view of artificial intelligence, it is absolutely logical that *a recognizer (classifier / discriminator) object encodes more entropy than each of the recognized (classified / discriminated) objects*[2]. The difference between them is nothing more, nothing less and nothing else than (computed) *intelligence* - or in other words, encoded (quantized) *entropy* obtained by extracting *knowledge* from data through specific computational means. Hence now, we understand in our own way the motivations behind the recent changes made by Daugman [7] in his proprietary iris code format, and also the necessity of this change.

19.2 Terminology and Problem Formulation

All data and all techniques reported in this chapter are about authentication of persons based on comparing iris biometric binary templates to learned (enrolled) digital biometric identities. A *digital identity* is a collection of data (a memory) stored for each enrolled user of a biometric system.

The identities are learned when the biometric system runs in calibration mode, from a set of iris biometric templates (binary iris codes) further referred to as *learning dataset* (the learning data contain correctly labeled binary templates). After the learning is done, the system goes into regular exploitation mode. In this stage, different users (enrolled or not) will expose their iris to the acquisition device which will process the current iris image up to a binary code.

By claiming an identity - "*I am the enrolled user number 354*", the user asks the system to verify the matching between the binary template extracted from the current image of his iris (*candidate iris code*) and the *digital identity* stored under the unique ID number 345. As a result, the biometric system could accept the claim (if the *candidate iris code* and the claimed *digital identity* are found to be sufficiently similar) or could reject it.

The claims could be *positive* - "*I am*", or *negative* - "*I am not*", *honest* (the enrolled user claims its own identity, or if he is not enrolled claims that he isn't) or *forged* (the user claims something false hoping to cheat the system).

19.2.1 False Accept/Reject, True Accept/Reject

A *False Reject* happens when the biometric system fails to recognize correctly an honest positive claim or a forged negative claim.

A *False Accept* occurs when the biometric system fails to recognize correctly an honest negative claim or a forged positive claim.

[2] The informed readers should appreciate if the domain of Iris Recognition is or is not still in its infancy, if such things about iris codes and biometric digital identities are told here for the first time.

A *True Reject* happens when the biometric system recognize correctly an honest negative claim or a forged positive claim.

A *True Accept* occurs when the biometric system recognize correctly an honest positive claim or a forged negative claim.

When a biometric system is simulated using a database of eye, or iris images (or iris codes), some of them are used to learn identities (*learning dataset*), and all others (which form the *test dataset*) for testing the quality of the learning, i.e. the *power of generalization* achieved through learning. During a simulation, the binary templates within the *test dataset* play the role of *candidate iris codes* and those within the *learning dataset* are *training examples* or, in other words, *enrolled binary templates*.

A training function or a training procedure (a feature extractor / a learning rule) is a computational routine that somehow assemble the information available in all training examples into a new data structure, namely the (enrolled) *digital identity*.

The simplest biometric system is based on single-enrollment: there is only one binary template enrolled under an ID number, the training function is the identical function, and consequently, the enrolled *digital identity* for the person who owns that ID number coincides with the single enrolled *binary template*.

In a multi-enrollment biometric system, at least two binary templates are enrolled under the ID number of the user. The training function/procedure is no longer trivial and the *digital identity* learned from the enrolled templates will differ from them. The definition given above for the False Accept/Reject cases are our definitions. Daugman proposed a different interpretation of these measures. For example, in his view the False Accept Rate in identification (one-to-many comparison) is different than the False Accept Rate in verification (one-to-one comparison) because identification and verification are considered to be two different things. Daugman supposed that the difference between identification and verification is mainly the type of comparison allowed in the system: one-to-many comparisons are allowed in identification systems and only one-to-one comparisons are allowed in verification systems. This assumption inevitably leads to logical inconsistencies (because in this context 'verification' means enrollment without proper validation) and it is not valid in Consistent Biometry - as is further shown here.

As a precaution, in our authentication system, one-to-all comparisons are not just allowed, but mandatory each time when a negative claim such *I'm not enrolled* occurs. The moment when a new enrollment takes place is a crucial one for preserving *logical consistency* of the biometric system. This is why, in our approaches [13]-[15], the False Accept/Reject Rates are considered to be global quality measures for a recognition technique when it is tested on a given database and are always computed by making the statistics of *all-to-all* comparisons (exhaustive testing on the database). The possibility to enroll the same person twice, just because one-to-all comparison would be formally not allowed, does not exist in our models.

The first set of questions to be answered here is the following:

- *Why to choose a multi-enrollment system?*
- *How to select the enrolled binary templates (learning dataset)?*

- *How to learn a biometric digital identity?*
- *How to match a digital identity to a candidate iris binary code or vice versa?*

19.2.2 Why Multi-enrollment?

There are two main reasons for choosing multi-enrollment:

Multi-enrollment ensures that the biometric identity will be trained with different hypostases of the same iris (different pupil dilations, illumination, blur, distortions, and occlusions). As a result, the digital identity will be much able to overcome *intraclass variability* and to recognize more hypostases of the same iris while still preserving the higher similarity scores possible.

Baker, Bowyer and Flynn [1] documented the problem of template aging. In this context, multi-enrollment is a recommended policy for ensuring a smooth variation of the iris identities over time.

19.3 Proposed Method

There are three simple and logical basic ideas underlying our evolutionary approaches to iris recognition:

Firstly, we always follow the idea of a large-scale distributed (geographically scattered) biometric system organized as a network with one or more central units and a lot of peripheral terminals. We follow this idea because we think that the future large-scale systems for iris based biometric identification will be hosted on hardware resources (IBM, Sun) dedicated to and fully compatible with virtualization and cloud computing technologies.

Secondly, we consider naturally that a recognizer must encode much more entropy than the recognized objects. Exactly how much? We don't know *a priori*. This is the reason why, in our models, a *digital identity is free to evolve* during its training up to a stage where it encodes enough entropy such that the recognition of the learning examples to take place at a certain level of quality comparable with the quality of recognition measured for human subjects during a Turing Test. A software enabled to replicate at a certain degree the human performances in iris recognition is further referred to as an *intelligent agent for iris recognition*.

At last but not the least, in our view, iris recognition is a problem of *consistent* or *inconsistent* logic. For example, practicing iris recognition for iris images in which not even the pupil is recognizable is just an inconsistent and vague verification (that we called *possibilistic and inconsistent iris hunting*), and not a logically consistent achievement in *iris recognition*. These cases fall into the following inconsistent logic in which the agent says: *I couldn't say where the pupil is, but I'm sure that this person is George*, or even better, into the following fuzzy, modal, and possibilistic logic of *beliefs* in which the agent says: *it is impossible for me to say where the pupil is, but I believe it could be (very) possible that this person to be George.* Hence, involuntary humour is reachable for certain artificial intelligent agents (not too smart, indeed) - on one hand, but on the other, *iris recognition* and *iris hunting* are two very different things.

19.3.1 Logical Framework: Consistent Biometry

Our concern is doing iris recognition in a logically consistent manner (i.e. intelligent iris identification) or at least with a coarse, predictable and controllable loss in consistency (intelligent iris verification). To achieve these, the primer condition is a consistent procedure of enrollment in the dataset of training examples.

Assuming that a low-quality training could guarantee excellent learning performances, or in evolutionary terms, supposing that insufficiently precisated adaptation stress could guarantee the evolution of a very specialized individual, are too optimistic hypothesis for us to follow. Instead of accepting them, we let the following possibilistic but consistent deduction to lead us:

$$consistent(A) \rightarrow [not(enrolled(I)) \rightarrow impossible(identification(I))],$$

i.e.

$$[possible(identification(I)) \rightarrow (enrolled(I))] \text{ } OR \text{ } [inconsistent(A)],$$

i.e.

$$[inconsistent(A)] \text{ } OR \text{ } [impossible(identification(I))] \text{ } OR \text{ } [enrolled(I)].$$

where A and I encode the agent and an individual, respectively. Nothing changes essentially if the same deduction is made for the verification. Therefore:

$$[inconsistent(A)] \text{ } OR \text{ } [impossible(verification(I))] \text{ } OR \text{ } [enrolled(I)].$$

The following two formulae:

$$[not(enrolled(I)) \rightarrow impossible(identification(I))],$$

$$[not(enrolled(I)) \rightarrow impossible(verification(I))],$$

will be further referred to as the *Principle of Consistent Biometry*, or the first axiom of Consistent Biometry (CBA1).

From the perspective of Artificial Intelligence, CBA1 tells that an intelligent agent who knows and practices a consistent biometric theory could neither verify nor identify an unenrolled individual, simply because it wasn't trained with samples taken from that individual. In order to be verified or identified by an intelligent distributed biometric system any user must enroll himself at one terminal of the system, must be *known* by the system.

The formulae:

$$[enrolled(I) \rightarrow possible(identification(I))],$$
$$[enrolled(I) \rightarrow possible(verification(I))],$$

will be further referred to as the *Positive Possibilistic Axiom of Consistent Biometry*, or the second axiom of Consistent Biometry (CBA2).

The following two formulae:

$$[enrolled(I) \leftrightarrow possible(identification(I))],$$
$$[enrolled(I) \leftrightarrow possible(verification(I))],$$

will be further referred to as the *Fundamental Theorem of Consistent Biometry* (FTCB). It tells that even in a computational perspective the identification and verification are different (being based on one-to-all and one-to-one comparisons, respectively), in a consistent biometric theory they have the same logical meaning. Hence, in CB there is no need to invent different quality measures for verification and identification as proposed in [5].

FTCB also tells that a biometric system in which the identification is not possible is an inconsistent verifier whose output is more likely a fuzzy inconsistent belief about the identities represented by the codes which are currently compared rather than a consistent biometric decision. The theorem also suggests that the *modes* of enrollment will determine the *modes* of identification and verification: accurate enrollment - reliable biometric decision, low-quality enrollment - unreliable biometric decision.

Hence, in a consistent biometric theory, *possibility* is the only guaranteed mode for both identification and verification. Advancing verification/identification from *possible* to *accurate*, or to *necessary* is a matter of calibrating the enrollment, a matter of customizing the enrollment procedures in order to achieve enough quality.

It is obvious now why we said that *consistent enrollment* is the primer condition for practicing iris recognition in a logically consistent manner (intelligent iris identification) or at least with a coarse, predictable and controllable loss in consistency (intelligent iris verification).

19.3.2 Internal logic, Knowledge and Self-awareness Representation for Intelligent Systems

Humans can invent the formal theory of Consistent Biometry. The only problem is how to implement this knowledge into an artificial intelligent agent. In order to transfer logical knowledge to an artificial agent, it must have an inference engine (it has to know at least a formal theory of binary logic) which in the case of our Intelligent Iris Verifier/Identifier is the *Computational* formalization of *Cognitive Binary Logic* (CCBL, [16]). The internal logical language (logical dialect) in which our Intelligent Iris Verifier/Identifier thinks, decides and talks about itself, about logic and about iris recognition in the context of Consistent Biometry is *the Cognitive Dialect* [17] - a logical language supporting *self-reference ambiguity*, a formal language native in CCBL designed to reveal that *auto-referential deductive discourses in CCBL are non-paradoxical*, and to support the *soundness* and *completeness* of the deductive discourse in CCBL regardless if it is auto-referential or not.

The close relationship that exists between the *self-reference ambiguity* in CCBL and the *self-awareness* is illustrated in the following situation: the truth does not depend on who is talking, and therefore, 'p' is a symbol used by us when we talk about a given propositional variable, is a symbol used by an artificial intelligent agent when it 'talks' about a given propositional variable, or is a symbol used by a propositional variable when talking about itself, all at once. This looks a little bit strange at first sight, but it comes very naturally: the most rudimentary intelligent agent is a bit storing the truth value of propositional variable 'p', and the next simple intelligent agent is a logical circuit: '$1 \rightarrow p$', telling that '*p is true*', and obviously, $p \leftrightarrow (1 \rightarrow p)$. This is the beginning of the self-awareness: 'p' is equivalent to '*p is true*'. Hence, who could say that '*p is true*' ? All of us, and even 'p', and obviously, the truth value of 'p' does not depend on who is talking about 'p'. If we now cease to exist, the propositional variable will continue to talk about itself (in a silent non-contradictory auto-referential deductive discourse, [16]) waiting to be heard, waiting to be discovered. This is the essence of CCBL: a self-reference formal deductive discourse (theory) written *with* and *about* the propositional variables of binary logic. Therefore, we said that *self-reference sentences are native and non-paradoxical in CCBL*, and therefore, *CCBL is a suitable inference engine for all of those intelligent agents that aim to be consistent in binary logic* - in general, for our Intelligent Iris Verifier/Identifier which aims to auto-control its evolution toward a logically consistent understanding of iris recognition by stepping always through and always to a logically consistent state - in particular. Of course, human understanding (or *the common belief*) about self-reference sentences formulated in semantically closed languages is a different thing.

In short, the Cognitive Dialect is the language that enables the Intelligent Iris Verifier/Identifier *to know* CCBL and Consistent Biometry, *to decide* how to create (evolve) a consistent theory of iris recognition dynamically over an extending vocabulary of digital identities, *to become* and *to stay aware* of its logical status (at least). This is why we tell that *Cartesian argument is valid even in Artificial Intelligence.*

19.3.3 Iris Segmentation and Encoding

The segmentation and encoding techniques must be used in order to extract a binary iris code for each eye image from the database. The segmentation procedure used here is CFIS2 [15] (Second version of Circular Fuzzy Iris Segmentation) which is a two step segmentation procedure. Firstly, the pupil is found (Fig. 1 in [14]). Secondly, the image is unwrapped through a pupil-centric polar coordinate transform (Fig. 2 in [14]) and the limbic boundary is approximated (Fig.1.a. in [15]). The result is an unwrapped iris segment, further used as an input for the encoding procedure, through which a binary iris code is generated. The encoders used in this chapter are the following two: Log-Gabor Encoder (LGE, [15]) and an encoder based on Haar Wavelet (noise filtering) and Hilbert Transform (phase encoding), abbreviated HH1 and introduced in [15].

19.3.4 Selecting and Aligning the Enrollment Templates

The criterion used here for selecting the enrolled binary templates (*learning dataset*) was pupil dilation. In order to ensure that each identity will be trained with different hypostases of more dilated or contracted pupil, the following selection procedure was practiced: there are 20 images for each eye in the database; hence, excepting the cases of failed segmentation, there are 20 binary iris codes in the template database. Five of them, chosen from the first ten, were used as learning examples. Those five binary iris codes are associated with five eye images chosen such that to preserve (as much as possible) the diversity of pupil dilation as it was measured in the set of the first 10 images. For each subject in the eye image database, the following optimization problem was solved heuristically (Monte-Carlo Simulation), through randomization of selected indices:

$$S_5 = min\{ \parallel [M_s, 2 \times S_s^{1/2}] - [M_{10}, 2 \times S_{10}^{1/2}] \parallel \mid s \in C_{10}^5 \}$$

where C_{10}^5 is the set of all 5-combinations taken from those first 10 images, (M_S, S_S) and (M_{10}, S_{10}) are the means and the standard deviations of the vectors:

$$(PupilRadii)./(IrisRadii)$$

computed for the current selection of indices s, and for the first 10 images corresponding to the same eye, respectively (where ./ signifies component to component division).

After selecting the enrolled templates, excepting the cases of failed segmentation (3 failures in a total of 1000 eye images), for each eye represented in the database, there are 5 images for training and 15 images for testing. From each set of 5 images used for training, the first one is considered to be the unrotated witness, in order to unify the angular alignment of the entire set of images taken for the same eye.

The iris codes generated with LGE [15] were tested for angular alignment using rotations in range of ±5.625 hexadecimal degrees (±8 pixels for unwrapped uint8 iris segments of dimension 512x32) with respect to the witness. The corrections were applied on the collection of unwrapped uint8 iris segments which have been further used to generate binary iris codes of dimension 256x16 using HH1 encoder [15].

19.3.5 Learning Evolutionary Digital Biometric Identities

Learning biometric identities is a problem of *artificial intelligence* and *evolution*. Why is that? Fig. 19.1 shows an instant picture of a biometric system, but the truth is that a biometric system, in order to be *logically consistent*, must be a *non-stationary* system, must be a system which adapts / changes itself over time. As a logical consequence, our opinion is that in a biometric system, it is mandatory to consider that *the time is ticking when a new enrollment occurs*, the enrollment being the *stress factor that demands adaptation*, which on its turn, it is impossible to achieve without *intelligence*[3]. Otherwise, if the enrollment is not accompanied by adaptation, it is just a matter of logical consequence to expect that *contradictions* will be reached very rapidly. The proof of this thing is of colossal importance for the future of Biometry, and it was already made in the year 2000 by Daugman, in [3] (see the formula (16) and the subsequent example of that formula in [3]). Still, it seems that the correct interpretation (the logical meaning) of Daugman's demonstration lied misunderstood, unexplored and unexploited, ever since.

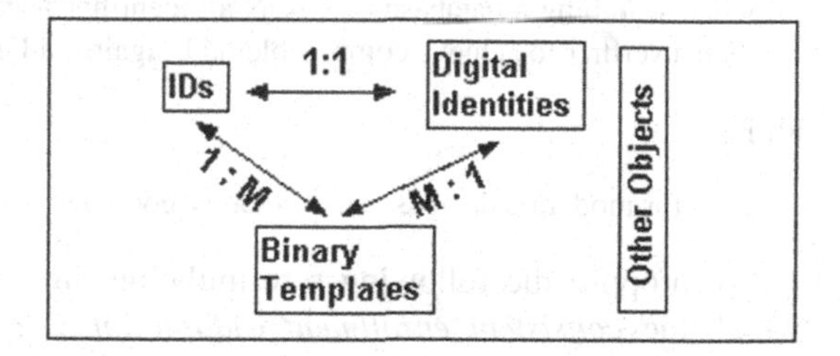

Fig. 19.1 An instant picture of a biometric system frozen in time: a relational collection of *ID numbers*, *binary templates* (subcollection of hypostases / samples taken for the recognized objects), *digital identities* (recognizer objects), optional strings and, possibly, other objects.

[3] This is a simple, informal, but intuitive proof telling that the future of biometry as a science (including iris recognition) will be inevitably shared between the theories and applications of logic, artificial intelligence, evolutionary computation and non-stationary systems. The concept of logically sound, logically complete, intelligent, adaptive, evolutionary (non-stationary) biometric systems, which is introduced here for the first time, will prove to be a milestone in iris recognition. We are currently working on this.

In Daugman's view, a verification system is based on one-to-one (binary candidate to claimed identity) comparisons. Hence, by design, Daugman's verifier systematically fails to adapt itself when a new enrollment occurs.

On the other hand, in a *logically consistent biometric system*[4] (LCBS), one hypostasis of the adaptation triggered by enrollment is a recalibration made in a certain way such to preserve a comfortable distance between the inter-class and intra-class distribution of scores computed for all enrolled users (this implicitly means allowing all-to-all comparisons). We called this recalibration *consistent enrollment.*

Consistent enrollment and *adaptation* are two equivalent semantic labels. The former is a name for a binary value of truth (consistent/inconsistent) in a second order binary logic language over the set of enrolments, or even a name for a modal and fuzzy value of truth in a second order 3-valent modal logic language (consistent / inconsistent / unknown) in case in which we aim to model incertitude. The latter is a name of a generic group of methods of Artificial Intelligence enabled to change the current state of the biometric system to a new state in which the system comfortably discriminate between the newly enrolled identity and all the older ones, previously enrolled.

In a LCBS, if the current enrollment jeopardizes system consistency, it will be dropped immediately in a quarantine where it stays until the system evolve (recalibrate itself) to a new state adapted to comfortably discriminate between the newly enrolled identity and all the older ones, previously enrolled. On short, a *LCBS stays consistent through adaptation* (supervised and consistent enrollment). If the current enrollment satisfy the current safety limits of the system, the adaptation is the identical function mapping the current state of the system to itself.

Neither *consistent enrollment* nor *adaptation* are among the possibilities of Daugman's verifier. This is why it is naturally to assume that Daugman's verifier will face, eventually, situations of logical inconsistency expressed by Daugman as *verification false accepts*. Daugman established a formula which correlates *verification false accepts* and *identification false accepts* in a given number of trials. His conclusion (pp.6 in [3]) was that:

> when searching a database of size N an identifier needs to be roughly N times better than a verifier to achieve comparable odds against a False Accept

and

> even for moderate database sizes, merely good verifiers are of no use as identifiers.

We propose the following reformulation: in a stationary biometric system in which the *consistent enrollment* (*adaptation / learning*) is not guaranteed, the chances for facing inconsistency in the form of False Accept grow nearly linearly

[4] A logically consistent biometric system is one whose internal logic is consistent - meaning that a false affirmation of biometry will never be proved (will never be computed / observed) in the system. For example, in a logically consistent biometric system (which is an idealized concept), the False Accept is not possible.

with the number of trials (with database size). By contrast, a LCBS behave totally different. The next section of the chapter will show that, at least for moderate database sizes (such is the case of Bath Iris Image Database), an *intelligent iris verifier* is also reliable as an *identifier*.

Since we are bonded to a logically consistent approach to iris recognition we could also reformulate Daugman's conclusions with even more precaution: in a stationary biometric system in which the *consistent enrollment* is not guaranteed, in which a *local* iris recognition theory is known only through the experimental measurements onto a given current vocabulary of binary iris codes, there could appear *statistical motivations* (reasons, but not in the sense of an argument in consistent binary logic) *to believe* that chances for facing inconsistency in the form of False Accept grow nearly linearly with the number of trials. If these *motivations would be valid logical arguments* in binary logic, and if the statistically motivated *belief* that the chances of False Accept grow nearly linearly with the number of trials *would be a proved theorem*, then and only then *the statistical decision landscape* proposed by Daugman [3] could have the chance to be a consistent 2-valued formal logic theory of iris recognition. Until then, it is at most an inconsistent theory of statistically motivated *beliefs* about iris verification validated by observing measurements over a given vocabulary of binary iris codes.

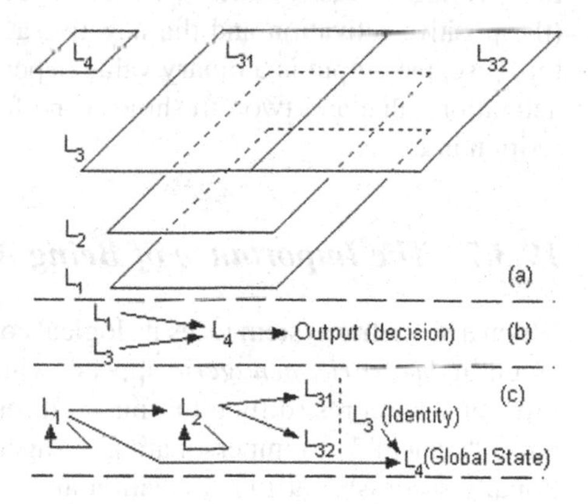

Fig. 19.2 (a) - ANN structure. (b) - Information flow during verification: the input consists in a candidate template (one binary template from the *test dataset*) loaded on L_1 and an enrolled digital identity loaded on L_3.
(c) - Information flow during training: the input is the current learning example (enrolled iris binary code) and the output is a digital identity which follows to be written in a database.

19.3.6 Artificial Neural Network Support for Consistent Enrollment

By design, our neural network for biometric purposes fits into the following restrictions:

i) The learning process does not rely on unspecialized neurons. Discriminator memory is trained only on positive examples. To match this rule, each enrolled identity (a trained memory) stores information for both positive and negative discrimination in separate zones. It memorizes what an iris is, but also what it is not, both types of information being extracted only from positive learning examples, i.e. only from those enrollment templates stored under the ID of currently trained memory.

ii) The learning process resumes each time when a new enrollment occurs (enrollment triggers evolution).

iii) The neural network (ANN) works in two modes: *learning* and *testing*. During the training stage, the neural network acts as a feature extractor by learning digital identities from the enrollment templates, whereas in the second mode, the ANN is used either as a verifier, or as an identifier.

The minimal architecture of an artificial neural network for iris biometric purposes is described in Fig. 19.2. The first layer of the ANN is responsible to load and to keep the current learning example. The third layer will encode the digital identity. It consists of two parts: L_{31} is a positive discriminator which learns what the current example is, whereas L_{32} learns what is not (the negative discriminator). If the current stage is a verification (a test), the global activation (*ga*) will be the number:

$$ga = \{L_{31}, L_1\} - \{L_{32}, L_1\},$$

i.e. the difference between a neural excitation (voting for similarity between the candidate stored in L_1 and the enrolled identity stored in L_3) and a neural inhibition (voting for dissimilarity), where the braces signify partial activation functions (the positive activation and the negative activation or inhibition, respectively). In this case, the output is a binary value depending on the relation between the global activation value and two thresholds (one for recognition, one for non-recognition) written in L_4.

19.3.7 The Importance of Being Aware

When a biometric system loses its logical consistency, among the regular *sheep* [22], an entire *biometric menagerie* appears within it: *the goats* - characterized by their low genuine scores (difficult to match through a genuine comparison), *the lambs* - those "vulnerable to impersonating" (Yager, [22]), and *wolves* - which are "exceptionally successful at impersonation and prey upon lambs" (Yager, [22]), but this is not all. Yager and Dunstone [22] brought more animals in the biometric farm, animals called worms, chameleons, phantoms, and doves.

It happens that we have studied a lot of logical aspects concerning iris recognition and we saw that if the recognition theory is 3-valent, fuzzy and inconsistent (like TSC_2 in Fig. 19.6), there exists a supra-theory of recognition with $2^3 = 8$ values of truth (three of them being those previously considered) which is, on its turn, inconsistent and (even much) fuzzy. Hence, increasing the number of the fuzzy values of truth will not repair the inconsistency. Anyway, from the point of view of Consistent Biometry doing that is as logical as looking at a macroscopic explosion through a microscope[5]. In the context of Consistent Biometry, the most important aspect of self-awareness is that a logically consistent intelligent agent for iris recognition must stay aware of its logical status, must be able to detect any enrollment that could jeopardize its logical status, and must evolve in such a way that all enrolled identities to preserve the quality of being *sheep*. In Consistent Biometry, *reliability* is synonym for *logical consistency*.

19.3.8 The Prototype of Intelligent Iris Verifiers

The procedure describing how a simple prototype of intelligent iris biometric system works is the following:

ANN Based Evolutionary Intelligent Iris Verifier:
(N. Popescu-Bodorin, January 2011, IIV Description)
1: **Global State:** thresholds, mode (testing or training);
2: **Primary Input:** chosen mode (testing OR learning);
3: **Secondary Input:** (L_1,L_3) for testing OR L_1 for training;
4: **If** testing mode is on,
5: **Compute** decision: $d = ga$;
6: **Else,** (Evolution triggered by enrollment: Quarantine, then Individual Evolution or Systemic Evolution or Failure)
7: **Quarantine** the current enrollment
Begin enrollment simulation and analysis:
8: **Try** (for a while) to evolve a new identity in the generation of all identities previously enrolled,
9. OR **Fail** AND **Try** (for a while) to evolve (in a space of higher entropy) a new generation of identities - including the identity which attempts to enroll,
10: OR **Fail** AND:
11: **Keep** the current enrollment quarantined,
12: **Apply** whatever custom routine is associated to the failure event,
13: OR **Succeed** AND:
14: **Finish** enrollment simulation and analysis,
15: **Qualify** the current enrollment as being consistent,
16: **Change** the global state of the biometric system to the newly simulated consistent state,
18: **End**;
18: **Output:**
d (current decision) for testing mode OR
L_3 (current trained identity) for training mode.

---

[5] The *natural* logical framework of logically consistent iris recognition is Binary Logic. Inconsistent enrollment (i.e. inconsistent extension of the current vocabulary of binary iris codes) introduces the contradiction in the internal logic of biometric system. Naturally, the contradiction will explode in an exponential number of ‘words’ (or ‘animals’, i.e. insufficiently precisated values of truth) and finally in pure inconsistency: more and more enrolled users will become *lambs* and *wolves* simultaneously, or the False Reject Rate will increase making the system unreliable.

19.3.9 Evolutionary Network - The Key Factor in Achieving Superior Levels of Intelligence

Fig. 19.3 shows the exact histograms of *all* intraclass / interclass scores obtained by comparing *all* enrolled identities to *all* binary codes from the *learning dataset* (50 eyes, 50 identities, 5 training images for each eye, a total number of 250 binary iris codes) and it gives us an image about the properties which qualify a biometric system as being *intelligent* and *trained.* It can be seen there that, with respect to the *learning dataset,* our system proves a *crisp understanding* of what it means to be a genuine comparison (it qualifies such comparisons with unitary similarity score), and a *fuzzy understanding* of what it means to be an imposter comparison - because it qualifies such comparisons with (fuzzy) similarity scores belonging in [0, 1/2). Still, for 33.3% of all imposter pairs formed with training examples, the system performs a *crisp understanding* of their nature by mapping these pairs to the null similarity score.

All of these facts (described in the previous paragraph and also in Fig. 19.3) are related to the lines 8-10 within the functional description of the *ANN Based Evolutionary Intelligent Iris Verifier* (further referred to as IIV description). The evolution of IIV does not alter its ANN structure. The learning rule is the one that changes under the pressure of those new enrollment requests that have the potential to jeopardize system consistency. Even if this is unusual, it comes very naturally: if the current space of identities becomes incompatible with an imposed safety standard, the identities must migrate in a new space, and consequently, the customized arithmetic formal language underlying the computation of the identities must be evolved to an extension of its, an extension enabled to describe the computation of the migrated identities. Previously, we said that for IIV the time is ticking when a new enrollment occurs. Hence, the system ages, and now we see that as it ages, it becomes a more experienced learner by evolving/updating its own learning rule.

Among the parts of our Intelligent Iris Verifier, we designed a dictionary of searchable arithmetic expressions which allows us to construct learning rules, in real time. Each learning rule is further implemented on the hidden layer L_2 of the ANN (see Fig. 19.2).

To *evolve a new identity in the generation of all identities previously enrolled* (line 8 in IIV description) means that without changing the learning rule the system computes an identity from the set of 5 binary codes currently submitted to enrollment. Then the system tests if the enlarged set of identities is compatible with the restrictions illustrated in Fig. 19.3 and further stated in (C 7.1.1). If the test succeeds, the evolution materializes through the creation of a new individual in the current space of identities. This is what we called *individual evolution* - an asynchronous differentiation of a single individual of a given population, made by training his memory to store the consciousness of individuality.

If the test fails (see for example the genuine comparisons scored other than unitary in Fig. 19.4), then the failure triggers the change of learning rule, which on its turn leads to the evolution of all enrolled identities. This is what we called *systemic evolution* - a synchronous *in mass evolution* of an entire population of individuals

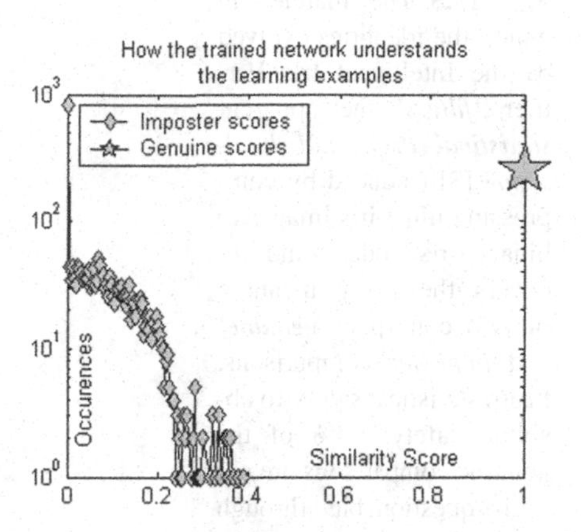

Fig. 19.3 The manner in which an Intelligent Iris Verifier recognizes the binary iris codes on which was trained (learning 50 identities from 250 genuine comparisons, and 2‘450 imposter comparisons). The figure illustrates the behavior of a *trained* Intelligent Iris Verifier whose understanding is very close to the human understanding proved during a Turing Test.

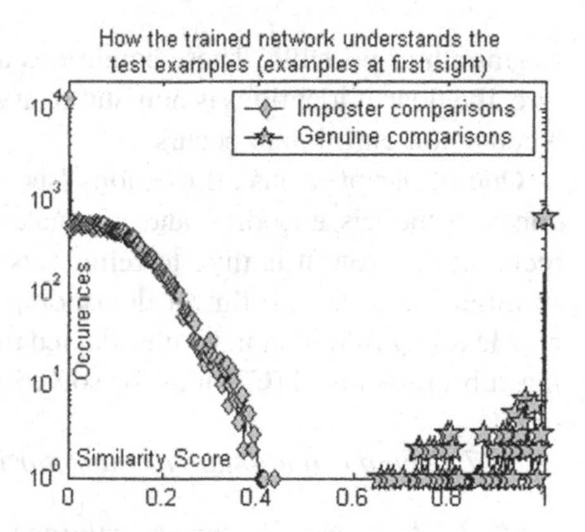

Fig. 19.4 The manner in which an Intelligent Iris Verifier recognizes binary iris codes that it has not seen during the training stage (50 learned identities tested through 747 genuine comparisons and 36603 imposter comparisons). The Intelligent Iris Verifier proves its power of generalization. The figure is also an example of fuzzified but still consistent understanding of the binary logical values.

which redefine their identities on new coordinates in a space *large enough* to host the dynamic consciousness of an extending group in which the ground policy is to preserve the differentiation between its members. Hence, the lines 8 and 9 from IIV description tell that the *adaptation* is achieved through *individual* or *systemic* evolution.

In a geometrical view, the *individual evolution* gives the start point for a trajectory which will host the digital identity of a certain enrolled person along the time. Of course, we may consider that a given identity computed at a certain moment is a discrete point in the current space of all enrolled identities, case in which, the movement of this point along the time describes the trajectory of the given identity, but also, we may see the given identity not like a clear discrete point, but as a fuzzy one, as a density of possible (and probable) points situated in a disk centred on the given identity, case in which, the movement of the identity describes a tube of possible and probable trajectories. The *systemic evolution* must maintain the flow of all these trajectories/tubes (corresponding to all enrolled persons) as laminar (untangled) and

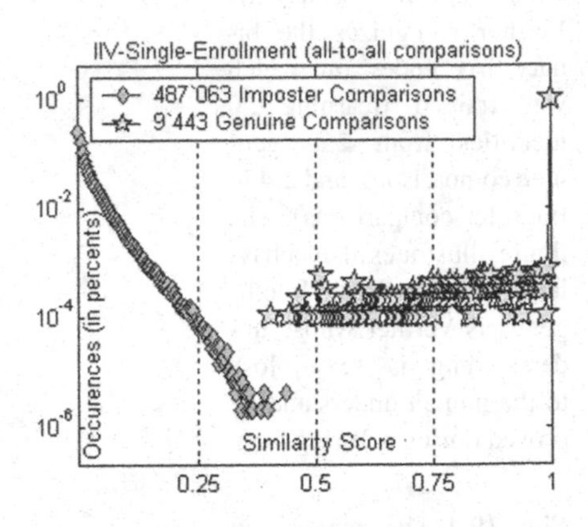

Fig. 19.5 The manner in which the identities evolved by the Intelligent Iris Verifier '*filters*' the apparent *statistical decisional landscape* [3] (induced by compressing uint8 iris images to binary iris codes) and recovers the fuzzy meaning of two concepts: '*genuine*' and '*imposter*' comparisons. From statistical safety to absolute safety: 97% of the genuine comparisons are not even questionable through reasonable statistical doubts.

as smooth as possible. These conditions are not easy to satisfy because, by its nature, the flow of identities is non-stationary: a new spring appears within it each time when a new enrollment occurs.

One of our previous affirmations (see the challenge C 7 in [13]) is that we still consider the iris encoding and iris matching as being two open problems in iris recognition. Now it is time to refine this topic by bringing new elements into the spotlight: at each enrollment demanding systemic evolution, in order to find the new learning rule (that new rule adapted to the enlarged set of identities) the following sub-problems of (C7) must be solved:

(C 7.1) Find evolutionary methods for iris encoding.

(C 7.1.1) Given the current enlarged set of identities, given the dictionary of arithmetic expressions, find a function which satisfies the restrictions:
- It must be well-formed through concatenation between legal arithmetic genes from the dictionary.
- It must prove a crisp understanding of what it means a genuine comparison, i.e. all genuine pairs formed with elements of learning dataset must be mapped to unitary scores (see the genuine comparisons in Fig. 19.3).
- It must prove a fuzzy but still consistent understanding of what it means an imposter comparison, i.e. all imposter pairs formed with elements of learning dataset must be mapped to scores in [0, 1/2).

Fig. 19.4 shows the exact histograms of all intra- and inter-class scores obtained by comparing *all* enrolled identities *to all* binary codes from the *test dataset* (examples at first sight). It gives us a visual representation for the quality of the training by showing how much *power of generalization* the *trained* Intelligent Iris Verifier proves:

i) For 34.14% (12'498) of all imposter pairs formed with test examples, IIV performs a *Crisp Reject* by mapping these pairs to the null similarity score. Hence, in these cases, IIV proves a *crisp understanding* of what it means to be an imposter comparison (or an imposter pair).

ii) For 65.86% (24‘105) of all imposter pairs formed with test examples, IIV performs a *Fuzzy Reject* by mapping these pairs to scores within (0, 1/2). Hence, in these cases, IIV proves a *fuzzy understanding* of what it means to be an imposter comparison (or an imposter pair).

iii) For 87.82% (656) of all genuine pairs formed with test examples, IIV performs a *Crisp Accept* by mapping these pairs to unitary score. Hence, in these cases, IIV proves a *crisp understanding* of what it means to be a genuine comparison (or a genuine pair).

iv) For 12.18% (91) of all genuine pairs formed with test examples, IIV performs a *Fuzzy Accept* by mapping these pairs to scores within (1/2, 1). Hence, in these cases, IIV proves a *fuzzy understanding* of what it means to be a genuine comparison (or a genuine pair).

Summarizing the data presented in Fig. 19.3 and Fig. 19.4, IIV achieves 100% correct recognition of 39‘053 unique imposter pairs (2‘450 pairs formed with evolved identities and elements of the *learning dataset*, 36‘603 pairs formed with enrolled identities and the elements of *test dataset*). It also achieves 100% correct recognition of 997 unique genuine pairs (250 pairs formed with evolved identities and elements of the *learning dataset*, 747 pairs formed with enrolled identities and the elements of *test dataset*).

19.3.10 New Safety Standards for Logically Consistent Biometric Purposes

Definition 19.1. (N. Popescu-Bodorin)

1. A biometric system has/gives/is/induces:
   i) a *consistent and crisp binary safety model* - if it is able to prove crisp understanding of two words (concepts) - *imposter* and *genuine* comparisons, by scoring them into $\{0, 1\}$.
   ii) a *fuzzified binary safety model* - if it is able to prove a fuzzified binary understanding of intra- and inter-class comparisons, by scoring them all into $[0, 1]$.
   iii) a *consistent and fuzzified binary safety model* - if it is able to prove a fuzzy but still consistent binary understanding of inter- and intra-class comparisons, by scoring them into $[0, 0.5]$ and $(0.5, 1]$, respectively.
2. A fuzzified binary safety model for iris recognition proves:
   i) *True Accept Consistency*, if the scores associated to Accept can not be obtained by comparing different irides.
   ii) *True Reject Consistency*, if any pair of irides scored as a fuzzy reject is in fact a pair of different iris images.

It can be seen in Fig. 19.4 that the proposed Intelligent Iris Verifier (which is a multi-enrollment system) has a *consistent and fuzzified binary safety model* which can be transformed into a *consistent and crisp binary safety model* through a simple defuzzification of the similarity score.

19.3.11 New Challenges - New Results

Solving optimization problems like (C 7.1.1) means heuristic optimization through genetic algorithms. There are many solutions to (C 7.1.1) but relatively few of them prove generalization capacities (few of them are *logically and semantically consistent solutions*). There is not enough space here for detailing the reasons why Daugman's verifier [3], Hollingsworth-Bowyer-Flynn best bits matcher [8], Dong-Tan-Sun best bits matcher [6], and our previously proposed multi-enrollment systems [15] also, all of them are strongly unoptimal solutions of the problem (C 7.1.1). In fact, all of them are weak solutions for drastically weakened optimization problems derived from (C 7.1.1). We won't hesitate to write on demand a separate paper on this topic, but here, it is more important to formulate the following new challenge:

(C 7.1.2) Given a logically and semantically consistent iris verifier as solution of (C 7.1.1), evolve a logico-arithmetical model for fuzzy intelligent understanding of iris identification while preserving consistency as much as possible.

The results obtained by answering this new challenge are illustrated in Fig. 19.5. It can be seen there that the identities evolved by IIV *attract* the binary iris codes (generated at dimension 256×16 with Haar-Hilbert encoder HH1, [15]) into a space where recognition is no longer a statistical event, but a logical one, with precise (and natural, and observable) causality, a space in which a simple iris recognition theory written in binary logic, or in a fuzzified binary logic, (see pp. 121 in [15]) is consistent. The exact meaning of this term will be illustrated below. Until then, its opposite is discussed:

Proposition 19.1. *(N. Popescu-Bodorin)*
Let us consider these:
a) d is a given dimension (256×16 for example).
b) C_8 is the set of all uint8 codes of dimension d.
c) C_2 is the set of all binary codes of dimension d.
d) TS_8 is a consistent and complete theory of similarity over C_8 (a theory over a second order language of binary valued affirmations about the similarity between uint8 codes of dimension d).
e) TS_2 is a consistent and complete theory of similarity over C_2 (a theory over a second order language of binary valued affirmations about the similarity between binary codes of dimension d).
Then:
There is no way to define an isomorphism between TS_8 and TS_2.

The elementary argument for the above proposition is the difference between the numbers of elements in the sets TS_8 and TS_2. Behind this simple fact is a deeper understanding of what happens with the Boolean algebras underlying TS_8 (or C_8) and TS_2 (or C_2). It is known that any Boolean algebra generates a subsequent logic which is called here *the intrinsic logic* of that Boolean algebra. If f is a surjective function from C_8 to C_2 which completely covers C_2 and transports the Boolean algebra (underlying the complete and consistent TS_8 theory) from C_8 into a Boolean

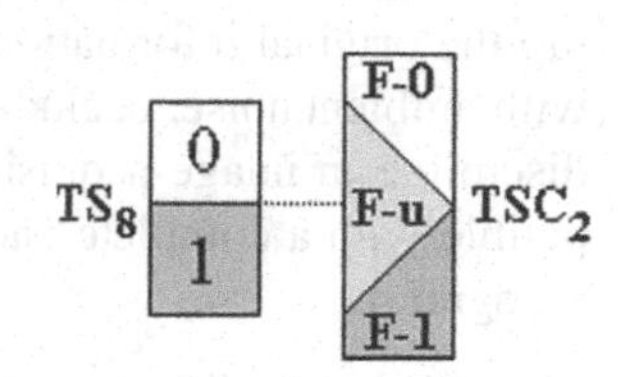

Fig. 19.6 Transporting the binary truth values of the affirmations about the similarity between uint8 codes of dimension d (from TS_8) to fuzzy values of truth in TSC_2.

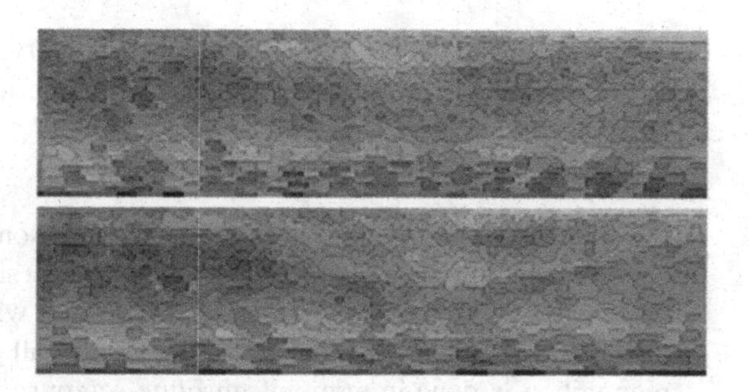

Fig. 19.7 True Reject Consistency: IIV-SE agent classifies correctly (rejects) the hypostases of the same iris if they are very different. Are these two hypostases of the same iris sufficiently similar to be scored with a Fuzzy Accept? IIV-SE agent tells that these two hypostases can't be matched, and this is the truth.

algebra TSC_2 over C_2, then the *intrinsic logic* of the transported Boolean algebra is inevitably *fuzzy* (or modal), *inconsistent* and *incomplete*. Fig. 19.6 shows what is happening in such a case: the crisp binary values of truth from TS_8 are inevitably fuzzified by a binary compression: the crisp values of truth from TS_8 became the fuzzy values F-1, F-0, F-u (i.e. Fuzzy 1, Fuzzy 0 and Fuzzy unknown, respectively) or fuzzy modal values MPI, MPD, U (Most Probable Identical codes, Most Probable Different codes, and Uncertain, respectively). The fuzzy understanding of similarity is inconsistent because there are different uint8 codes that matches equal chances to be or not to be qualified as being similar in TSC_2 (if F-u means *equally probable*) or matches null chances to be qualified as being non-similar in TSC_2 (if F-u means *any other way that F-0 and F-1*).

Hence, when a space of uint8 matrices is compressed to a space of binary matrices of the same dimension, there is always a biometric truth from the initial space which is no longer observable in the compressed space. Consequently, the biometric theory migrated into the compressed space (TSC_2) is incomplete.

On the other hand, in the above example a pair of codes is seen in TSC_2 differently than it is in reality (in TS_8). Consequently, the biometric theory transported in the compressed space is inconsistent (it can prove something unreal).

Therefore, logically consistent biometric *identification* in TS_8 (for the elements of C_8) will never be achievable in the space of binary compressed codes underlain by the transported biometric theory TSC_2. On the other hand, in TSC_2 *verification* [3] is possible, but still logically inconsistent, despite the existence of a suitable choice of the code dimension which induces a *statistical decision landscape* [3] over a given set of binary iris codes.

Poor acquisition discipline is a kind of compression, or even worst, a way of losing the original information because of a mixed effect of: overwriting the original with ambient noise, occluding some areas, improper quantization, etc. Hence, poor discipline in image acquisition is a ticket to inconsistency. It will never be compatible with a complete and consistent theory or with a consistent practice of iris recognition.

19.3.12 Logical Consistency vs. Safety

Let us return now to the fact that iris recognition theory, as is seen by the IIV-Single-Enrollment (IIV-SE) agent, is *consistent*. In its numerical language (see Fig. 19.5), IIV-SE tells us that:

i) For 97% (9‘160) of all (9‘443) genuine comparison, it performs a *Crisp Accept* by mapping these comparisons to the unitary similarity score. Hence, in these cases, IIV proves a *crisp understanding* of what it means to be a genuine comparison (or a genuine pair). For 97% of all genuine comparisons, the recognition is done in terms of *absolute safety* (*absolute security*): 97% of the genuine comparisons are clearly above any doubts (eventualy) motivated through a statistical game of chances; 97% of the genuine comparisons are clearly outside any statistical decision landscape [3]. This is the main quality of our results. They prove that a logically consistent approach to iris recognition qualifies almost all Accept cases as being *absolutely safe / absolutely secure / necessary True Accepts / indubitable True Accepts.*

ii) For 2.93% (276) of all (9‘443) genuine comparisons, IIV-SE performs a *Fuzzy Accept* by mapping these comparisons to scores within (0.5, 1). Hence, in these cases, IIV proves a *fuzzy understanding* of what it means to be a genuine comparison (or a genuine pair).
iii) For 0.0741% (7) of all (9‘443) genuine comparisons, IIV-SE performs a Fuzzy Reject by mapping these comparisons to scores within (0.4023, 0.5]. Hence, in these cases, IIV proves a *fuzzy understanding* of what it means to be an imposter comparison (or an imposter pair) wrong placed under the intra-class index as a consequence of eye image preprocessing.
iv) For 100% (487‘063) of all imposter comparisons, IIV-SE performs a Fuzzy Reject by mapping these comparisons to similarity scores within (0, 0.4368). Hence, in these cases, IIV proves a *fuzzy understanding* of what it means to be an imposter comparison (or an imposter pair).

There are two important aspects regarding the results of all-to-all comparisons computed by IIV-SE agent:

i) Firstly, as the number of comparisons grows (the database enlarges) the logarithmic histogram of all impostor scores runs for nearly vertical asymptotic trends in 0 and 0.5. This behavior ensures that any accept produced by the system is a True Accept, or in other words, IIV-SE proves *True Accept Consistency*. Hence, what seemed like a property that only an idealized system may have (LCBS a theoretical concept of a Logically Consistent Biometric System), it is now ascertained on the basis of a test with real data from University of Bath Iris Image Database (UBIID, Fig. 19.5). Obviously, while tested on UBIID, IIV-SE has not produced any False Accept.

ii) Secondly, the apparent cases of False Reject are in fact cases of True Rejects, or in other words, IIV-SE proves *True Reject Consistency*. Because of accumulated errors (pupil center, iris center, pupillary boundary, and limbic boundary) it is possible to encounter the situation in which two unwrapped uint8 irides are two very different hypostases of the same iris. A consistent matching technique can not overcome localization and segmentation errors. Hence, it is naturally that IIV-SE rejects this kind of pairs. Such an example is given in Fig. 19.7. For that pair of very different hypostases of the same iris IIV-SE agent computes a similarity score of 0.4023. This proves two things: the index of genuine comparisons can be accidentally altered through accumulated errors of iris preprocessing - on the one hand, and IIV-SE is sufficiently intelligent to detect these cases on the other hand.

19.3.13 Comparison to a Result Previously Obtained by Daugman

We must clarify here if the result presented in Fig. 19.5 is or isn't the first of its kind. The answer is negative: a result previously obtained by Daugman and presented in Fig. 10 from [4] also reveal a *weak statistical aspect of recognition*, but Daugman omitted to give the correct interpretation of this fact. Now we know for sure that Daugman obtained that result by doing other things than just comparing only binary iris codes to each other using Hamming distance. Surely he used at least another one element (just as we did in our approach by introducing the *digital identity*) in order to 'attract' the binary codes in a space in which statistical aspect of recognition is weak.

Hence our result is not the first of its kind but it is much better because we managed to give the correct interpretation for the result previously obtained by Daugman, and we also managed to advance from a *weak statistical aspect of recognition*[6] to a *residual statistical aspect of recognition*[7].

The difference between our result and the result previously obtained by Daugman comes from a different understanding of what it means *to recognize*:

---

[6] Daugman said in [4] that "more than half of such image comparisons achieved an HD of 0.00, and the average HD was a mere 0.019", hence in our terms, he said that in more than half of such image comparisons his verifier proved a crisp understanding of what it means to be a genuine pair.

[7] IIV-SE system proposed by Bodorin proves a crisp understanding of what it means to be a genuine pair in 97% of cases.

- Daugman sustained the idea of a statistical decision landscape of recognition (iris recognition is viewed as a game of chances, inevitably logically inconsistent). Still, Fig. 10 from [4] shows how close to the truth was Daugman when he did that experiment.
- We are sustaining the idea of a logically consistent approach to recognition based on cognitive investigation which aims to link the causes to the effects by thinking in Horn clauses, or by following *cognitive implications*, or by exploring *deductive discourses* [16]: recognition or non-recognition happens because precise conditions are or aren't fulfilled. By defining iris recognition as a problem of logic we allowed a custom designed intelligent agent (Intelligent Iris Verifier/Identifier, which knows *Computational Cognitive Binary Logic* and *Consistent Biometry*), to evolve (to create) a vocabulary of digital identities (corresponding to all enrolled users) and a (complex) piece of knowledge - a consistent formal biometric theory over this vocabulary.

IIV-SE achieves consistent iris recognition (True Accept Consistency, True Reject Consistency) and an objective evaluation of the image database: UBIID database contains good quality images on which consistent practice of iris recognition is possible. This means that on UBIID, both IIV systems described here (multi / single enrollment) are consistent *biometric verifiers* and consistent *biometric identifiers*. We are wishful to cooperate for testing if and what other databases match these properties (i.e. prove an acquisition standard compatible with a logically consistent approach to iris recognition) but we will never again waste our time searching for truth in logically inconsistent worlds such is the theory of iris recognition when *no matter how low quality* replaces a credible standard of image acquisition.

IIV-SE also achieves an objective evaluation of iris segmentation: besides the three cases of failed segmentation, IIV-SE detects another seven cases of fully erroneous segmentation. Overall efficiency of the segmentation procedure (CFIS2, [15]) can be now reevaluated at 99%.

19.3.14 Another Question

Before ending, we must answer one more question: what made possible the results presented here? The answer is surprisingly simple: for us, iris recognition is another facet of Binary Logic, or in other words, iris recognition is another hypostasis of a more general problem, namely *logical and intelligent artificial understanding of data* where *logical* means *based on Böhm-Jacopini theorem*. We did nothing more than searching for a logical and intelligent manner of understanding (quantizing!) some signals encoded in the iris texture. But what it is really important is that we managed to create an artificial intelligent agent able to achieve some sub-goals of this complex task *'by himself'* (referring a self-aware artificial intelligent agent as a person is something absolutely legal in his formal language). IIV agent discovers and learns gradually a logico-arithmetical formal theory in which *iris identification* is provable (computable). Hence the present chapter marked the moment in which *Computational Inventics* departs from fiction once for good.

19.4 Conclusion

The present chapter announced the technological advance from *inconsistent iris verification* to *consistent iris identification* and it showed that the future iris-based *identification* will be inevitably marked by multi-enrollment, and by the newly proposed concept of consistent, intelligent, adaptive, evolutionary biometric system. It is also clear that the future of biometry as a science (including iris recognition) will be inevitably shared between the theories and applications of logic, artificial intelligence, evolutionary computation and non-stationary systems. All of these are necessary instruments in achieving secure iris-based (or biometric-based) *identification*, simply because *secure* means *logically consistent*, because *adaptation* of a biometric system means *logical and intelligent evolution* in response to *enrollment*.

Acknowledgements. We thankfully acknowledge the University of Bath and Prof. D. Monro for granting us access to the iris database.

References

1. Baker, S.E., Bowyer, K.W., Flynn, P.J.: Empirical evidence for correct iris match score degradation with increased time-lapse between gallery and probe matches. In: Tistarelli, M., Nixon, M.S. (eds.) ICB 2009. LNCS, vol. 5558, pp. 1170–1179. Springer, Heidelberg (2009)
2. Bowyer, K.W., Hollingsworth, K., Flynn, P.J.: Image understanding for iris biometrics: a survey. Computer Vision and Image Understanding 110(2), 281–307 (2008)
3. Daugman, J.: Biometric Decision Landscapes, Technical Report No. TR482, University of Cambridge (2000), `http://www.cl.cam.ac.uk/techreports/UCAM-CL-TR-482.pdf`
4. Daugman, J.: How Iris Recognition Works. IEEE Trans. on Circuits and Systems for Video Technology 14(1) (January 2004)
5. Daugman, J.: New methods in iris recognition. IEEE Trans. Systems, Man, Cybernetics, B 37(5), 1167–1175 (2007)
6. Dong, W., Tan, T., Sun, Z.: Iris Matching Based on Personalized Weight Map. Accepted for Publication in IEEE-TPAMI (2010) (to appear)
7. Grother, P., Tabassi, E., Quinn, G., Salamon, W.: Interagency report 7629: IREX I - Performance of iris recognition algorithms on standard images. N.I.S.T (October 2009)
8. Hollingsworth, K.P., Bowyer, K.W., Flynn, P.J.: The best bits in an iris code. IEEE TPAMI 31(6), 964–973 (2009)
9. Iris Challenge Evaluation, N.I.S.T., `http://iris.nist.gov/ice/` (cited February 20, 2011)
10. Ma, L., Tan, T., Wang, Y., Zhang, D.: Personal Identification Based on Iris Texture Analysis. IEEE TPAMI 25(12), 1519–1533 (2003)
11. Monro, D.M., Rakshit, S.: Rotation Independent Iris Matching by Motion Estimation. In: Proc. IEEE Int. Conf. on Image Processing (September 2007)
12. Monro, D.M., Rakshit, S., Zhang, D.: DCT-Based Iris Recognition. IEEE TPAMI 29(4), 586–595 (2007)
13. Popescu-Bodorin, N., Balas, V.E.: AI Challenges in Iris Recognition. Processing Tools for Bath Iris Image Database. In: Proc. 11th Int. Conf. on Automation and Information, pp. 116–121. WSEAS Press (June 2010)

14. Popescu-Bodorin, N.: Exploring New Directions in Iris Recognition. In: 11th Int. Symp. on Symbolic and Numeric Algorithms for Scientific Computing, pp. 384–391. CPS - IEEE Computer Society, Los Alamitos (2009)
15. Popescu-Bodorin, N., Balas, V.E.: Comparing Haar-Hilbert and Log-Gabor based iris encoders on Bath Iris Image Database. In: Proc. 4th Int. Conf. on Soft Computing Applications, pp. 191–196. IEEE Press, Los Alamitos (2010)
16. Popescu-Bodorin, N., State, L.: Cognitive Binary Logic - The Natural Unified Formal Theory of Propositional Binary Logic. In: Recent Advances in Computational Intelligence, pp. 135–142. WSEAS Press (April 2010)
17. Popescu-Bodorin, N., Balas, V.E.: From Cognitive Binary Logic to Cognitive Intelligent Agents. In: Proc.14th Int. Conf. on Intelligent Engineering Systems, pp. 337–340. CPS - IEEE Computer Society, Los Alamitos (2010)
18. Rakshit, S., Monro, D.M.: Pupil Shape Description Using Fourier Series. In: Workshop on Signal Processing Applications for Public Security and Forensics (April 2007)
19. Rakshit, S., Monro, D.M.: Robust Iris Feature Extraction and Matching. In: Proc. IEEE Int. Conf. on Digital Signal Processing (July 2007)
20. Tan, T., Ma, L.: Iris Recognition: Recent Progress and Remaining Challenges. In: Proc. of SPIE, vol. 5404, pp. 183–194 (April 2004)
21. Wildes, R.: Iris Recognition - an emerging biometric technology. Proc. of the IEEE 85(9),1348–1363 (1997)
22. Yager, N., Dunstone, T.: The Biometric Menagerie. IEEE TPAMI 32(2), 220–230 (2010)

Author Index

Subject Index